T0191609

Communications in Computer and Information Science 1058

Commenced Publication in 2007
Founding and Former Series Editors:
Phoebe Chen, Alfredo Cuzzocrea, Xiaoyong Du, Orhun Kara, Ting Liu,
Krishna M. Sivalingam, Dominik Ślęzak, Takashi Washio, and Xiaokang Yang

Editorial Board Members

More information about this series at http://www.springer.com/series/7899

Xiaohui Cheng · Weipeng Jing ·
Xianhua Song · Zeguang Lu (Eds.)

Data Science

5th International Conference
of Pioneering Computer Scientists,
Engineers and Educators, ICPCSEE 2019
Guilin, China, September 20–23, 2019
Proceedings, Part I

 Springer

Editors
Xiaohui Cheng
Guilin University of Technology
Guilin, China

Weipeng Jing
Northeast Forestry University
Harbin, China

Xianhua Song
Harbin University of Science
and Technology
Harbin, China

Zeguang Lu
National Academy of Guo Ding
Institute of Data Science
Harbin, China

ISSN 1865-0929   ISSN 1865-0937 (electronic)
Communications in Computer and Information Science
ISBN 978-981-15-0117-3   ISBN 978-981-15-0118-0 (eBook)
https://doi.org/10.1007/978-981-15-0118-0

This Springer imprint is published by the registered company Springer Nature Singapore Pte Ltd.
The registered company address is: 152 Beach Road, #21-01/04 Gateway East, Singapore 189721, Singapore

Preface

As the program chairs of the 5th International Conference of Pioneer Computer Scientists, Engineers and Educators 2019 (ICPCSEE 2019, originally ICYCSEE), it is our great pleasure to welcome you to the proceedings of the conference, which was held in Guilin, China, September 20–23, 2019, hosted by Guilin University of Technology, Guilin University of Electronic Technology, and National Academy of Guo Ding Institute of Data Science. The goal of this conference was to provide a forum for computer scientists, engineers, and educators.

The call for papers of this year's conference attracted 395 paper submissions. After the hard work of the Program Committee, 104 papers were accepted to appear in the conference proceedings, with an acceptance rate of 26.4%. The major topic of this conference was data science. The accepted papers cover a wide range of areas related to Basic Theory and Techniques for Data Science including Mathematical Issues in Data Science, Computational Theory for Data Science, Big Data Management and Applications, Data Quality and Data Preparation, Evaluation and Measurement in Data Science, Data Visualization, Big Data Mining and Knowledge Management, Infrastructure for Data Science, Machine Learning for Data Science, Data Security and Privacy, Applications of Data Science, Case Study of Data Science, Multimedia Data Management and Analysis, Data-Driven Scientific Research, Data-Driven Bioinformatics, Data-Driven Healthcare, Data-Driven Management, Data-Driven eGovernment, Data-Driven Smart City/Planet, Data Marketing and Economics, Social Media and Recommendation Systems, Data-Driven Security, Data-Driven Business Model Innovation, and Social and/or Organizational Impacts of Data Science.

We would like to thank all the Program Committee members, 203 coming from 93 institutes, for their hard work in completing the review tasks. Their collective efforts made it possible to attain quality reviews for all the submissions within a few weeks. Their diverse expertise in each individual research area helped us to create an exciting program for the conference. Their comments and advice helped the authors to improve the quality of their papers and gain deeper insights.

Great thanks should also go to the authors and participants for their tremendous support in making the conference a success.

We thank Dr. Lanlan Chang and Jane Li from Springer, whose professional assistance was invaluable in the production of the proceedings.

Besides the technical program, this year ICPCSEE offered different experiences to the participants. We hope you enjoy the conference proceedings.

June 2019

Xiaohui Cheng
Rui Mao

Organization

The 5th International Conference of Pioneering Computer Scientists, Engineers and Educators (ICPCSEE, originally ICYCSEE) 2019 (http://2019.icpcsee.org) was held in Guilin, China, during September 20–23 2019, hosted by Guilin university of Technology, Guilin University of Electronic Technology, and National Academy of Guo Ding Institute of Data Science.

General Chair

Mei Wang	Guilin University of Technology, China

Program Chairs

Xiaohui Cheng	Guilin University of Technology, China
Rui Mao	Shenzhen University, China

Program Co-chairs

XianXian Li	Guangxi Normal University, China
Liang Chang	Guilin University of Electronic Technology, China
Wei Li	Central Queensland University, Australia
Goi Bok Min	Universiti Tunku Abdul Rahman (UTAR), Malaysia
Xiaohua Ke	Guangdong University of Foreign Studies, China

Organization Chairs

Xiaolian Xie	Guilin University of Technology, China
Yong Ding	Guilin University of Electronic Technology, China
Zhenjun Tang	Guangxi Normal University, China
Donghong Qin	GuangXi University for Nationalities, China
Zeguang Lu	National Academy of Guo Ding Institute of Data Science, China

Organization Co-chairs

Chao Jing	Guilin University of Technology, China
Qin Yang	Guilin University of Technology, China
Yu Wang	Guilin University of Technology, China
Hanying Liu	Guilin University of Technology, China
Shouxue Chen	Guilin University of Technology, China
Jili Chen	Guilin University of Technology, China

Wenpeng Chen	Guilin University of Technology, China
Pinle Qin	North University of China, China
Jianhou Gan	Yunnan Normal University, China
Mingrui Chen	Hainan University, China

Publication Chairs

Hongzhi Wang	Harbin Institute of Technology, China
Guanglu Sun	Harbin University of Science and Technology, China

Publication Co-chairs

Weipeng Jing	Northeast Forestry University, China
Xianhua Song	Harbin University of Science and Technology, China
Xie Wei	Harbin University of Science and Technology, China
Guoyong Cai	Guilin University of Electronic Technology, China
Minggang Dong	Guilin University of Technology, China
Canlong Zhang	Guangxi Normal University, China

Forum One Chairs

Fudong Liu	Information Engineering University, China
Feng Wang	RoarPanda Network Technology Co., Ltd., China

Forum Two Chairs

Pinle Qin	Zhongbei University, China
Haiwei Pan	Harbin Engineering University, China

Forum Three Chairs

Jian Wang	RADI, CAS, China
Weipeng Jing	Northeast Forestry University, China

Forum Four Chairs

Liehuang Zhu	Beijing Institute of Technology, China
Yong Ding	Guilin University of Electronic Science and Technology, China

Forum Five Chairs

Junyu Lin	Information Institute of Chinese Academy of Sciences, China
Haofen Wang	Shanghai Leyan Technologies Co. Ltd., China

Forum Six Chair

Qiguang Miao   Xidian University, China

Forum Seven Chair

Canlong Zhang   Guangxi Normal University, China

Education Chairs

Xiaomei Tao   Guilin University of Technology, China
Hui Li   Guangxi University of Technology, China

Industrial Chairs

Li'e Wang   Guangxi Normal University, China
Zheng Shan   Information Engineering University, China

Demo Chairs

Li Ma   Guilin Medical University, China
Xia Liu   Sanya Aviation and Tourism College, China

Panel Chairs

Yun Deng   Guilin University of Technology, China
Rifeng Wang   Guangxi University of Technology, China
Peng Liu   Guangxi Normal University, China

Post Chair

Panfeng Zhang   Guilin University of Technology, China

Expo Chairs

Chaoquan Chen   Guilin University of Technology, China
Jingli Wu   Guangxi Normal University, China
Jingwei Zhang   Guilin University of Electronic Technology, China

Registration/Financial Chair

Chunyan Hu   National Academy of Guo Ding Institute
                            of Data Science, China

ICPCSEE Steering Committee

Jiajun Bu	Zhejiang University, China
Wanxiang Che	Harbin Institute of Technology, China
Jian Chen	ParaTera, China
Wenguang Chen	Tsinghua University, China
Xuebin Chen	North China University of Science and Technology, China
Xiaoju Dong	Shanghai Jiao Tong University, China
Qilong Han	Harbin Engineering University, China
Yiliang Han	Engineering University of CAPF, China
Yinhe Han	Institute of Computing Technology, Chinese Academy of Sciences, China
Hai Jin	Huazhong University of Science and Technology, China
Weipeng Jing	Northeast Forestry University, China
Wei Li	Central Queensland University, Australia
Min Li	Central South University, China
Junyu Lin	Institute of Information Engineering, Chinese Academy of Sciences, China
Yunhao Liu	Michigan State University, China
Zeguang Lu	National Academy of Guo Ding Institute of Data Sciences, China
Rui Mao	Shenzhen University, China
Qi Guang Miao	Xidian University, China
Haiwei Pan	Harbin Engineering University, China
Pinle Qin	North University of China, China
Zhaowen Qiu	Northeast Forestry University, China
Zheng Shan	The PLA Information Engineering University, China
Guanglu Sun	Harbin University of Science and Technology, China
Jie Tang	Tsinghua University, China
Tian Feng	Institute of Software Chinese Academy of Sciences, China
Tao Wang	Peking University, China
Hongzhi Wang	Harbin Institute of Technology, China
Xiaohui Wei	Jilin University, China
lifang Wen	Beijing Huazhang Graphics and Information Co., Ltd., China
Liang Xiao	Nanjing University of Science and Technology, China
Yu Yao	Northeastern University, China
Xiaoru Yuan	Peking University, China
Yingtao Zhang	Harbin Institute of Technology, China
Yunquan Zhang	Institute of Computing Technology, Chinese Academy of Sciences, China
Baokang Zhao	National University of Defense Technology, China

| Min Zhu | Sichuan University, China |
| Liehuang Zhu | Beijing Institute of Technology, China |

ICPCSEE 2019 Program Committee Members

Program Committee Area Chairs

Wanxiang Che	Harbin Institute of Technology, China
Cheng Feng	Northeast Forestry University, China
Min Li	Central South University, China
Fudong Liu	State Key Laboratory of Mathematical Engineering Advanced Computing, China
Zeguang Lu	National Academy of Guo Ding Institute of Data Science, China
Rui Mao	Shenzhen University, China
Qiguang Miao	Xidian University, China
Haiwei Pan	Harbin Engineering University, China
Qin Pinle	North University of China, China
Zheng Shan	State Key Laboratory of Mathematical Engineering Advanced Computing, China
Guanglu Sun	Harbin University of Science and Technology, China
Hongzhi Wang	Harbin Institute of Technology, China
Yuzhuo Wang	Harbin Institute of Technology, China
Xiaolan Xie	Guilin University of Technology, China
Yingtao Zhang	Harbin Institute of Technology, China

Program Committee Members

Chunyu Ai	University of South Carolina Upstate, USA
Zhipeng Cai	Georgia State University, USA
Richard Chbeir	LIUPPA Laboratory, France
Zhuang Chen	Guilin University of Electronic Technology, China
Vincenzo Deufemia	University of Salerno, Italy
Minggang Dong	Guilin University of Technology, China
Longxu Dou	Harbin Institute of Technology, China
Pufeng Du	Tianjin University, China
Zherui Fan	Xidian University, China
Yongkang Fu	Xidian University, China
Shuolin Gao	Harbin Institute of Technology, China
Daohui Ge	Xidian University, China
Yingkai Guo	National University of Singapore, Singapore
Meng Han	Georgia State University, USA
Meng Han	Kennesaw State University, USA
Qinglai He	Arizona State University, USA
Tieke He	Nanjing University, China
Zhixue He	North China Institute of Aerospace Engineering, China

Contents – Part I

Network

Security

Machine Learning

Contents – Part II

Application

Data Mining

Benign Strategy for Recommended Location Service Based on Trajectory Data

Jing Yang[1], Peng Wang[1,2(✉)], and Jianpei Zhang[1]

[1] College of Computer Science and Technology, Harbin Engineering University,
Harbin, Heilongjiang, China
wpeng68@yahoo.com
[2] College of Information Engineering, Suihua University,
Suihua, Heilongjiang, China

Abstract. A new collaborative filtered recommendation strategy oriented to trajectory data is proposed for communication bottlenecks and vulnerability in centralized system structure location services. In the strategy based on distributed system architecture, individual user information profiles were established using daily trajectory information and neighboring user groups were established using density measure. Then the trajectory similarity and profile similarity were calculated to recommend appropriate location services using collaborative filtering recommendation method. The strategy was verified on real position data set. The proposed strategy provides higher quality location services to ensure the privacy of user position information.

Keywords: Location services · Collaborative filtering recommendation · Trajectory similarity

1 Introduction

Various location-based services (LBS) applications are widely used with the rapid development of location technology and smart devices. According to the user position coordinates, satisfactory recommendations are provided such as location query, search nearby shops, hospitals and attractions [1]. The existing location service architecture is a typical centralized system server model and hotspot recommendations are adopted according to some priority rules. This model has some advantages including simple, direct and fast. However, from the perspective of the recommended location service results, the recommended location service results are too normal. In other words, the users with same location information will receive the same service results which ignoring the user differences. So it is difficult to meet the high individual needs. Moreover, the server in centralized system is vulnerable from malicious attacks with high communication load. It is difficult to ensure the privacy and security of the user. Even the user location information is leaked without the user's perception result in link attack and serious privacy leak. For example, multinational overseas military bases were exposed by running trajectory of sports app. These shortcomings have gradually emerged and attracted the attention of people. Therefore, high quality location service and user privacy security become urgent needs in LBS applications.

© Springer Nature Singapore Pte Ltd. 2019
X. Cheng et al. (Eds.): ICPCSEE 2019, CCIS 1058, pp. 3–19, 2019.
https://doi.org/10.1007/978-981-15-0118-0_1

In order to solve the problem of location data privacy leakage in the location service system, location data privacy protection methods were study and some effective results were obtained [2]. The K-anonymity privacy protection method was proposed by Gruteser and Grunwald [3] firstly to apply in location service. Then more stringent differential privacy protection methods have also been gradually applied in the field of location services [4]. For centralized location server defects, some privacy protection methods were carried out. For example, Chow et al. summarized the location privacy protection scheme for the system structure [5]. To some extent, the spatial anonymity method based on peer to peer structure proposed by Chow et al. could overcome the disadvantages of centralized location server [6]. This method proposed the use of neighboring node location information in a distributed system to implement K-anonymity privacy protection, but privacy among adjacent nodes was ignored. And Shokri et al. proposed a neighboring user service sharing mechanism to reduce the chance of the user position being exposed to the server. A buffer was set to store the past service and used to provide location services for neighboring users, but it faced the cold start and initial LBS service request leakage [7]. In our previous work, model of neighboring nodes recommendation service in distributed systems was established [8], but it had low availability of filtered neighboring users location information and poor quality of location service results.

For the LBS improvement, the user trajectory information were used to construct trajectory information profiles with user preferences based on distributed architecture and density-based clustering algorithm was used to establish a neighbor group in our work. The collaborative filtering recommendation algorithm containing user trajectory similarity and profile similarity was designed to provide service. And homomorphic encryption was used to transfer position data. The privacy protection of user position data sets was fully considered results in that the user position information is effectively protected in our proposed method. In addition, the proposed method made full use of the characteristics of the distributed system. The computing tasks of the recommended service were distributed to the neighbor nodes, which effectively solved the problem of overloading critical nodes in existing methods.

2   Related Work

It is general that the linking attacks occur due to leakage of user location information and the centralized anonymous location service architecture faces serious privacy threats. Aim to ensure high quality services and protect the accurate position for privacy and security, K-anonymous method and differential privacy protection methods are analyzed and applied in researches [9]. Among these methods, the K-anonymous method is the most widely used by k-top or k-mean clustering method [10, 11]. And its main idea is replacing accurate user position with a data set that have at least K positions and the user position. The density-based clustering method was proposed firstly in literature [12]. In the context of our work, the user's position density has important implications. In brief, when the user has a high density of positions within a certain geographical area, it could be explained to a certain extent that the user has more activities in the area. And the service results could be provided precisely in the recommended location. Thus, a neighboring selection method based on density measure was designed to establish neighboring groups for users.

The trajectory data is an ordered collection that combines position information and time information in chronological order. And there are also some research results on privacy protection methods in the field of trajectory data publishing [13–18]. The differential privacy protection method was applied in trajectory data [13]. Trajectory privacy protection based on spatiotemporal relevancy was proposed in trajectory publishing [14]. And personalized trajectory privacy protection method based on a fog structure was proposed in literature [15]. It should be noted that although the trajectory data have been exploited in trajectory data publishing, the introduction of trajectory data to the location service has few been reported in the literature. Compared with the position data, the trajectory data could better describe the user's active area and highlight the sequential characteristics among position points. Therefore, the application of trajectory data provides the possibility of high quality results for location service recommendations.

In addition, there are some location privacy protection methods based on LBS system structure [19–21]. The location data was cached on the mobile client based on distributed systems in literature [19]. When the user requests the LBS, the local cache can be queried so that no location information is exposed to the server. In literature [20], a combination of caching and user collaboration was used to protect location privacy. And each user had a cache to store their recent requests. Its purpose is to avoid users sending requests directly to the server as much as possible. The literature [21] used P2P structure to protect location privacy and reduced the possibility of location privacy leakage by the mobile device capabilities and collaborative recommendation methods. The above research results mainly focused on how to reduce the risk of user privacy leakage in distributed systems. And K-anonymity, caching mechanisms, and other privacy protection methods were used, but the location service recommendation methods and quality issues were ignored.

Collaborative filtering recommendation algorithm was widely used in e-commerce and other fields. The similarity of consumer behavior was used to achieve personalized product recommendation. And the more detailed the information of the user caused the more accurate the recommended results. But the application of collaborative filtering recommendation algorithm is not extensive. Although the leakage of data information privacy still existed in the process of using collaborative filtering recommendation algorithm, some privacy protection methods for collaborative filtering recommendation algorithms were designed including random interference method, grouping anonymous method and so on [22–25]. In literature [8], a simple collaborative recommendation method was used to recommend location services based on a distributed system, but it could not get quality service location results. Therefore, a collaborative filtering recommendation location service method for location service scenarios was designed. Based on the distributed system structure, the profile of user trajectory information was established. And the location information of neighboring users was used to provide location service for requesters. Although the peer-to-peer system structure was adopted, but the social attributes between adjacent users was fully considered in the proposed strategy compared with literature [8]. In our proposed strategy, the density clustering method was used to construct the user trajectory information profile, the collaborative filtering recommendation method was used to recommend the location service, and the homomorphic encryption method was used to encrypt the location data in transmission process to protect privacy security. Thus, the proposed strategy could provide higher quality location services and stringent privacy guarantees.

3 Collaborative Filtering Recommendation Location Service Method

3.1 The Basic Description

A trajectory-oriented collaborative filtering recommendation location service method based on distributed systems was proposed in this paper. The recommendation process of this method includes (a) each user generated their own trajectory profile based on density measure, (b) the own current trajectory information and service request were sent to neighboring users and then neighboring users who meet the density measure requirements recommended the LBS service using collaborative filtering algorithm, (c) finally, the requester used the secondary filtering on the LBS service data set recommended by the neighboring user to obtain its own location service data. Besides, if no satisfactory service result was obtained, the K-anonymous data set was constructed using the virtual trajectory information of the neighboring user and the service request was sent to the location server.

In our strategy, the user's social attributes were considered and the density-based clustering algorithm was used to select neighboring users. In the clustering algorithm, the higher location density in the area represents the more frequent visit of the user to the area results in the higher the recommended location service quality. The collaborative filtering recommendation algorithm adopted in this paper is different from the traditional collaborative filtering algorithm. The traditional collaborative filtering algorithm is a popular commodity recommendation technology. It uses the machine learning technology to recommend products to users by analyzing the historical data of other users in the group [20]. And the trajectory profile data of the neighboring user was used to recommend location service in the proposed collaborative filtering recommendation algorithm. All in all, this strategy designed a collaborative filtering recommendation algorithm based on distributed system structure with the consideration of user trajectory similarity and user profile similarity to recommend LBS service. It overcomes the communication bottleneck and vulnerable defect of the centralized anonymous system structure and improves the quality of location service results.

3.2 Profile Construction

A new trajectory information profile construction method was designed in our work. The trajectory information of each user was used to construct trajectory information profiles including their own trajectory information, subject tags, and score. In the profile construction, the density of the relative position profile was considered preferentially. At the same time, the score of the position point was generated according to the residence time and the access frequency. And finally the trajectory information profile was constructed to represent the user's activity area. The user trajectory information profile model was described as Eq. 1.

$$
\begin{cases}
L = \left\{ (t_1, x_1, y_1, l_1, s_1), (t_2, x_2, y_2, l_2, s_2) \ldots \ldots (t_j, x_j, y_j, l_j, s_j) \right\} \\
T = \{ L_1, L_2, L_3 \ldots \ldots L_k \} \\
D = \sum_{m=1}^{k} \sum_{n=1}^{j} (1 + s_{mn}) / A_{area}
\end{cases}
\tag{1}
$$

Where T is user location profile, the profile T contains k trajectory information and each trajectory in the set consists of several position points. Every point is represented as (t, x, y, l, s), where (x, y) is the positional coordinate, t is the timestamp of the position, l is the subject label of the position and s is the score of the position. And D is the absolute density of the user's location profile to represent the ratio of the number of position points and the score s to the area A_{area} of position information profile. The value of D is proportional to the user's familiarity with the area. And the score function was designed as shown in Eq. 2.

$$
\begin{cases}
\gamma = \alpha z + \beta c & (\alpha > 0, \beta > 0, \alpha + \beta = 1) \\
S(\gamma) = \frac{\gamma}{a + b\gamma} & (a > 0, b > 0)
\end{cases}
\tag{2}
$$

Where z is the user dwell time at the (x, y) position and c is the user's visit number at the $L(x, y)$ position, α and β are the proportion coefficient of the dwell time and the visit number in the score result and $S(\gamma)$ is the user's score in the $L(x, y)$ position. The a and b are constants greater than 0, then the derivative of $S(\gamma)$, $S'(\gamma) = \frac{a}{(a + b\gamma)^2} > 0$, so the function $S(\gamma) = \frac{\gamma}{a + b\gamma}$ is monotonically increasing in $[0, +\infty)$ and the range is $[0, \frac{1}{b})$. The b is used to control the upper limit of the score. In addition, the value of a is used to control the score growth rate because of $S'(0) = \frac{1}{a}$. And the a was set as 0.8, b was set as 0.2 in our work, so the location score was in the range of [0, 5).

The original user trajectory information profile could be constructed using the user trajectory information and score processing on position points. In order to improve the usability of the user trajectory information profile and ensure user privacy security, the data of the user trajectory information profile was further preprocessed in our proposed strategy. Firstly, in the user's trajectory information, the positions with higher individual privacy requirements were processed to fuzzification, such as homes, work units, etc.; secondly, the divergence points with lower scores were deleted, because these divergence points would significantly reduce the absolute density D of the trajectory information profile results in affecting the clustering of the trajectory information profile. In the proposed strategy, the density clustering algorithm [12] was used to identify and delete the divergent point and determine the centroid position of the user trajectory information profile. In the profile, the local density parameter ρ_i and the neighboring distance parameter δ_i of each position point were calculated, where calculation formula of the local density parameter ρ_i was shown in Eq. 3.

$$\rho_i = \sum_j \chi(d_{ij} - d_c) \tag{3}$$

And $\chi(x) = 1$, $x < 0$ and $\chi(x) = 0$, $x > 0$, d_c is set as distance threshold. The local density parameter ρ_i represents the distance from node i is less than the number of d_c nodes. The calculation formula of the neighboring distance parameter δ_i was shown in Eq. 4.

$$\delta_i = \min_{j:\rho_j > \rho_i}(d_{ij}) \tag{4}$$

For the node with the largest local density, the calculation formula of the neighboring distance parameter δ_i was shown in Eq. 5.

$$\delta_i = \max_j(d_{ij}) \tag{5}$$

Those positions with relatively large local densities ρ_i and large δ_i are considered to be the centroid positions of the trajectory information profile. The points with the small local density ρ_i and the large δ_i are abnormal points.

3.3   Collaborative Filtering Recommendation Location Service

In the designed application scenario, the user A was supposed to need LBS service at location l, the user sent request information to the surrounding broadcast. The request information included the user's current real-time trajectory information, its own trajectory location profile and request content. The trajectory profile similarity and profile similarity of neighboring user that meet the density measure requirements were calculated, then the score was predicted based on similarity and the location service with higher score was recommended.

When the user requests the LBS service, it needs to send its own trajectory information and trajectory profile information to the neighboring nodes. However, if these location data are directly broadcast to the neighboring nodes, the leakage risk of certain location information is inevitable. Therefore, the Paillier encryption [26] was used to encrypt the location information in the strategy.

When the user A requests LBS service, a pair of public and private keys (pk, sk) were generated using the encryption scheme (E, D, K) by running the key generation algorithm, then their own trajectory information L' and T' were encrypted, and finally the service request information $R = (pk, E\ (pk, L'), E\ (pk, T'), Q\ (l))$ were sent to neighboring nodes. The final service request information could be formalized as: $R = (pk, E\ (pk, L), E\ (pk, T), Q\ (l))$, where pk is the public key generated using the key generation algorithm for user A, $E\ (pk, L)$ is encrypted user trajectory profile identification and $Q\ (l)$ is the requested content of the topic tag l.

When the neighboring user B accept the request service information $R = (pk, E(pk, L'), E(pk, T'), Q(l))$, pk was used to perform the encryption operation $E(pk, Tb)$ on its own profile T_b to obtain the encrypted trajectory information profile T_b'. The trajectory information profile of user A was encrypted using operation $E\ (pk, T')$ to be T_a'. The core position node of the requester A is l_j. The neighboring user B first calculated the relative density μ of its own profile to the position. And if the density μ is greater than the preset density measure threshold, the neighboring user B sent the recommendation result to user A according to the position profile or the user position. And if it does not, the service request information was not responded.

The relative density formula was shown in Eq. 6.

$$\mu = \sum_i \chi(d_{ij} - d_c) \quad i = \{1, 2, 3 \ldots \ldots k\} \tag{6}$$

Where $\chi(x) = 1$, $x < 0$ and $\chi(x) = 0$, $x > 0$, and d_c is location distance measure, d_{ij} is the distance from point l_i to position, the relative density μ represents the distance of position l_j is less than the number of d_c positions in location profile. The high the relative density μ indicates that the profile of the neighboring user B is dense from the centroid position l_j, which results in the higher quality of the location service. The relative density μ is used as a reference indicator to select location service recommendation result.

Trajectory Information Similarity Measure. When the neighboring user B receives the current trajectory information L_a of user A, the user B inquires the sub-trajectory L_b that is similar to the L' trajectory in his own trajectory information profile, and then recommends the eligible position in L_b to the user A. Obviously, the higher the similarity between the trajectory L_b and the trajectory L_a causes high possibility of service results to satisfy the requester. Trajectory similarity measure has some related research results, but most of these results are based on distance measure between positions with taking into account the trajectory shape, movement characteristics and other factors. And the trajectory similarity measure was mainly used in the trajectory data publishing [16], could not be directly applied in location service application. Thus, a similarity measure based on sub-trajectory matching mechanism was designed in this work. In the similarity measure, the same sub-trajectory of the requester A's current trajectory L_a and the neighboring B's trajectory L_b profile was determined firstly. Then according to the more points with the same time-sequence position in the sub-trajectory, higher the quality of location services would be predict.

Mathematical description of trajectory similarity measure based on sub-trajectory matching mechanism was shown in Eq. 7.

$$sim = \frac{N_c}{N_a} \tag{7}$$

The rating score of recommended location service was shown in Eq. 8.

$$S'_i = S_i * sim \tag{8}$$

Where N_a is the number of position nodes in the user A's sub-trajectory, and N_c is the number of position nodes with the same timing in the sub-trajectory, and the sim is the similarity between N_a and N_c. Obviously, when sim is 1, the two sub-trajectorys are matched completely and the pre-score s_i' of the recommended location service l_i is also exactly the same as the score s_i of the recommended location service l_i.

Position Profile Similarity Measure. Pre-score of recommended locations was calculated using trajectory similarity with only considering the effect of location timing on location service results in the trajectory. The profile of the user's trajectory information to some extent represents the user's social attributes such as activity range and interest preferences. If the time sequence of the location points is ignored, the similar the positions and scores in the trajectory information profiles of the two users illustrate the similar social attributes of the two users, so the predictability of the recommended location service is high. In brief, the users with the same preference are more likely to choose the same location service. From this point of view, the Euclidean similarity measure was adopted in calculating the similarity of profiles. Supposed there were k common position points in the A, B user trajectory information profiles, where the score of i position points were a_i and b_i, respectively. The Euclidean distance of two profiles was shown in Eq. 9.

$$d(x, y) = \sqrt{\sum_{i=1}^{k} (a_i - b_i)^2} \tag{9}$$

The similarity of the two profiles was shown in Eq. 10.

$$sim(x, y) = \frac{1}{1 + d(x, y)} \tag{10}$$

Finally, user B returned the profile similarity, the recommended location service and corresponding pre-score to requester A. Therefore, the user A received a location service candidate set including multiple user recommendations. Supposed there were n location service results in the set and any position point l_i with m recommended users, so the set of attributes (s_j, sim_j) for l_i, where $j = 1,2,3......$m, s_j is the pre-score for l_i, sim_j is the similarity of user B_j and requester A. The score prediction of the l_i position for requester A was shown in Eq. 11.

$$s_i = \sum_{j=1}^{m} s_j sim_j \bigg/ \sum_{j=1}^{m} sim_j \quad (i = 1, 2, 3......n) \tag{11}$$

The service with the highest predicted score in candidate set was selected as the final location recommendation result.

4   Experiment and Result Analysis

In the experiment, the CRAWDAD data set was investigated to analyze our strategy. The CRAWDAD data set contains trajectory data of 536 taxis in San Francisco within a month [27, 28]. And it is widely used in location data analysis and research. The experimental content is mainly divided into two parts. Firstly, the characteristics of profile construction algorithm, LBS service request and response, and collaborative filtering recommendation location service algorithm in the proposed strategy were verified and analyzed. The evaluated indicators mainly included the position information profile characteristics, the number of responding users, the number of service results and the probability of successful recommendation services. In addition, the proposed strategy (TCRLS) was compared with the strategy in [8] (DCRLS). Secondly, the system architecture, algorithm efficiency and communication costs were analyzed and compared with the algorithms proposed in [7] (MobiCrowd) and DCRLS.

4.1   Profile Construction Algorithm Analysis and Verification

The trajectory information of each taxi object was processed for the generation of the trajectory information profile, and the resident point in the trajectory information was used as the location service pre-selection set. Calculated the average moving speed v among neighboring points, and then used v to measure the dwell time of the object in the corresponding area. Obviously when $v = 0$, it means that the object is stationary during the time period and the position point is the resident point. The data of the trajectory information profile were preprocessed, and the local density parameter and the distance parameter of each position were calculated. At the same time, these two parameters were used to determine the position of the centroid and the outliers were deleted. The resident point of the trajectory information profile was finally determined and scored. In the experiment, the profile generation algorithm was used to process the

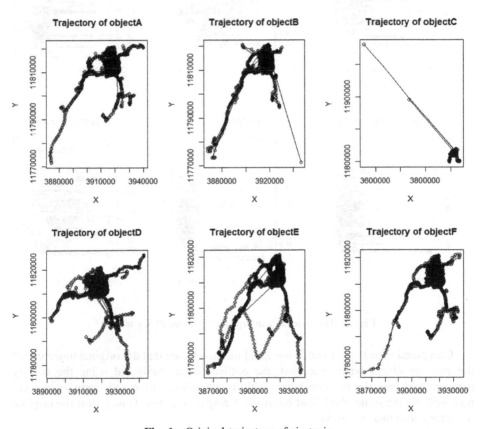

Fig. 1. Original trajectory of six taxis.

position information of 536 taxis, and corresponding position information profiles were generated. The data statistics and analysis were performed and the parameters $\alpha = 0.5$, $\beta = 0.5$, $a = 0.8$, $b = 0.2$. $d_c = 100$. Firstly, the six taxis were randomly selected from 536 taxi objects for instance analysis, and all the object data were given finally.

The six taxi objects were randomly selected as $A \sim F$. Their original trajectory information was shown in Fig. 1.

The proposed position profile generation algorithm was used to process the original trajectory information of the moving object $A \sim F$. The main algorithm contents include statistics and scoring according to the time and visit frequency of the resident points, calculating the local density and neighboring distance value and determination the centroid point and delete the outliers. The corresponding trajectory information profiles were shown in Fig. 2.

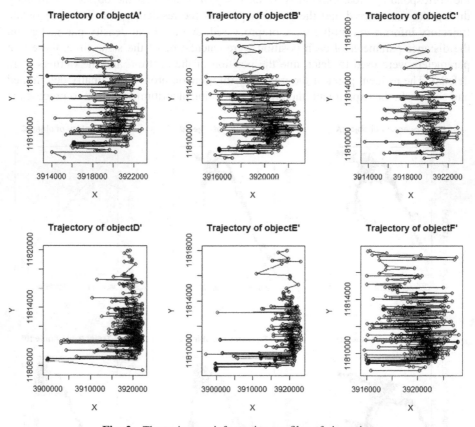

Fig. 2. The trajectory information profiles of six taxis.

Compared with Figs. 1 and 2, we could intuitively see that the original trajectory of the moving object were processed, the outliers were eliminated using the density measure and only the position points with long residence time and many visits were reserved. So the same object had better clustering and higher density than the original trajectory information profile.

In order to analyze the usability of the proposed profile generation algorithm, the data of these six moving objects were further statistically calculated, including the number of position in the original trajectory and the trajectory profile, ratio of position number and absolute density values. These data were shown in Table 1.

Table 1. The data original trajectory and the trajectory profile.

Name	Number of positions in original trajectory	Number of positions in trajectory profile	Ratio of position number	Absolute density
Object A	20543	431	2.1%	0.28
Object B	20159	673	3.34%	0.41
Object C	11616	249	2.14%	0.17
Object D	22694	688	3.03%	0.15
Object E	25611	876	3.42%	0.16
Object F	26165	539	2.06%	0.37

It could be seen from the data in Table 1 that about 2%–3% of the original position points were used to describe the trajectory characteristics of moving objects in the trajectory information profile generation algorithm proposed in this paper and used to provide location services. Combined the Table 1, trajectory profile and location service candidate set, it could be demonstrated that the score function and the density measure could effectively preprocess the original trajectory data and generate higher quality locations service candidate set to improve data operation and storage efficiency. The profile data in the trajectory profile of the same moving object were compared with the related data in position profile of the DCRLS. And the results were shown in Fig. 3.

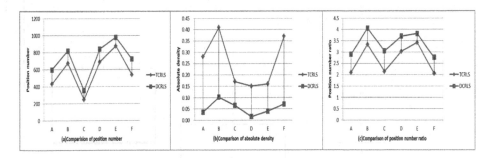

Fig. 3. Comparison of position profile data

From the Fig. 3, we can see that less position information were used to construct the user trajectory information profile in TCRLS with compared with DCRLS, and the higher absolute density showed that TCRLS has better clustering and higher data quality. Therefore, the trajectory information profile generation algorithm proposed in this paper is feasible for the user to recommend LBS service in the designed experiment, which avoids the risk of privacy leakage in the centralized location server.

4.2   Recommendation Algorithm Analysis and Verification

The algorithm TCRLS would be used to perform location service requests and recommendations. In this experimental simulation, the parameters were set the same as in the previous section. First, 100 user positions were randomly selected in the data set as LBS service requesters, at the same time the requester's trajectory information were generated including trajectorys of several known position points. The request contents were set as "Where to go next hour?". It is particularly emphasized that related parameters and service topic could be flexibly set according to different circumstances in practical applications.

The LBS service request and response process was as follows: When the user had the LBS service request, the distance cutoff parameter d_c was set according to the density of regional active and the request was sent, the mobile objects with less than 100 m away from the user in 5 min were filtered; then the density measure of the filtered object was determined according to the parameter d_c and the location service was recommended using the collaborative filtering method. When the parameters d_c were set as 100, 200, and 300 m for three groups of experiments, the number of responding users in each group under 100 random service requests were shown in Fig. 4.

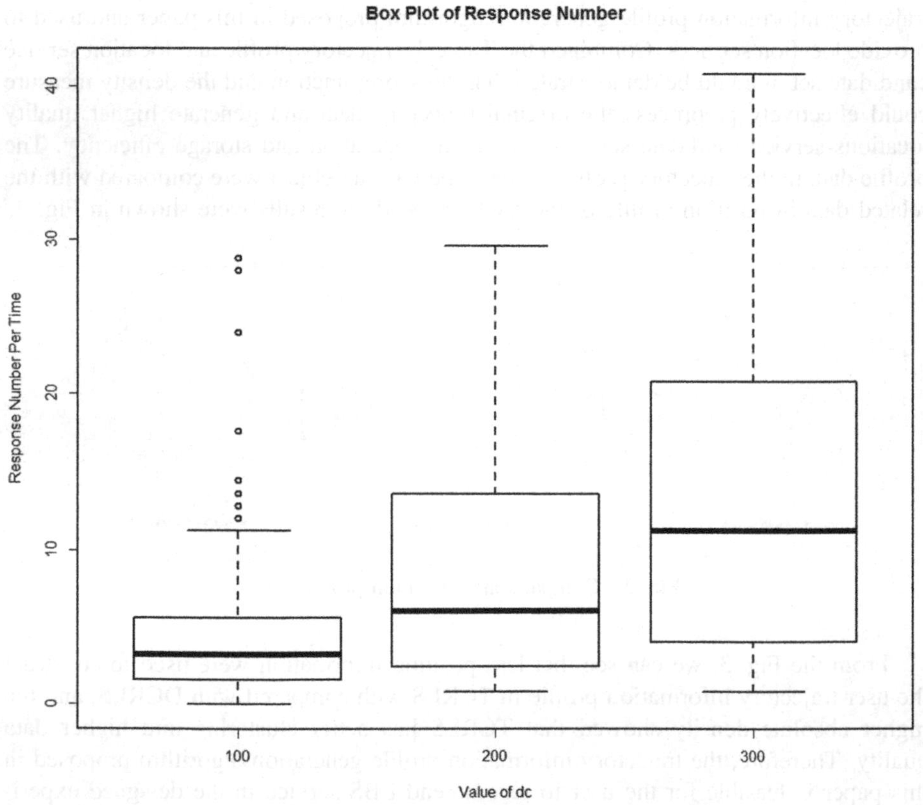

Box Plot of Response Number

Fig. 4. The box plot of response number.

At each service request, the responding users meet density measure requirements in trajectory profile to find similar sub-trajectory and calculate the trajectory similarity and the profile similarity. The location information, pre-score, and profile similarity that meet the requirements were returned to the service requester, and the service requester get a location service candidate set containing the above information. At last, the location services were scored using profile similarity and the location service with highest score was chose.

When the d_c values were 100, 200, and 300 m respectively, the specific statistics for the response number were shown in Table 2.

Table 2. The number of service results.

d_c value	Minimum	Maximum	Average	Failure rate
100	0	1281	135.29	5%
200	5	1090	272.45	0%
300	11	1849	416.4	0%

It is obvious that the number of responding users and the recommendation results increase with the d_c value increases. In the data set, there were only one non-responsive user and five times of no suitable recommended location service result in 100 service requests when the d_c value was 100 m. And there was no failure situation when the d_c value was 200 or 300 m. It shows that it may is not responding user or location service result when the d_c value is small, so the d_c value should be reasonably set according to the actual situation.

4.3   Comparison Among TCRLS, DCRLS and MobiCrowd

The algorithms TCRLS, DCRLS and MobiCrowd all use the distributed system architecture and neighboring users to recommend location services, but these three algorithms are different in system architecture, third-party server dependence, location privacy level, execution efficiency, and communication. Therefore, it is necessary to compare the above points of three algorithms on the CRAWDAD dataset. The system architectures were shown in Table 3. In the initial stage of MobiCrowd algorithm, it needs to access the third-party centralized location server. After accumulating a certain number of location service results, the cache location information was used to provide the location service and the location server still need to be accessed. The TCRLS and DCRLS algorithms used their own location information to recommend location services and only accessed the third-party centralized location server when the recommendation failed. And this situation could be avoided by setting the d_c value properly. The system structures of the two algorithms are basically the same, the difference lies in the location service recommendation method and service quality.

When the three algorithms randomly requested 100 location services under the same conditions, the number of visiting location servers were compared and the comparison was shown in Fig. 5. The parameters were set the same as before and the dc were 100, 200, and 300 m.

Table 3. Architecture comparison.

Item	TCRLS	DCRLS	MobiCrowd
Architecture tiers	2 tiers	2 tiers	3 tiers
Dependence on trusted third party	Low	Low	Medium
Privacy protect among peers	Good	Good	Weak
Location service quality	Good	Medium	Low

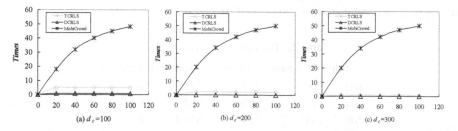

(a) $d_c=100$ (b) $d_c=200$ (c) $d_c=300$

Fig. 5. The comparison for the number of access servers among TCRLS, DCRLS and MobiCrowd.

It could clearly be seen from the Fig. 5 that there are significant differences in the number of access servers among the three algorithms. It needs to access the server in the initial stage of MobiCrowd algorithm, the dependence on the location server was gradually reduced after accumulating service information. In the DCRLS algorithm, it was need to access the location server due to less responding users when the dc was 100 m. In other cases, there was almost no need to access the location server and the dependence on the location server was low. Compared with DCRLS, the recommended location service mechanism was more complicated in the TCRLS proposed in this paper. And the absolute density, trajectory similarity and profile similarity need to be calculated. Therefore, the probability of access server under the same conditions was higher than that of DCRLS.

In the same experimental environment, the average communication costs, time spent on the server and the client of three algorithms were shown in Fig. 6. The communication costs were measured by the number of TCP/IP packets.

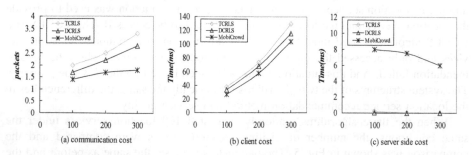

(a) communication cost   (b) client cost   (c) server side cost

Fig. 6. Performance comparison among TCRLS, DCRLS and MobiCrowd with dc = 100, 200, 300

From Fig. 6(a) and (b), it could be seen that the communication load and the time spent on the client of the TCRLS algorithm were higher than the DCRLS and Mobi-Crowd algorithm. This phenomenon could be attributed to the data operation and transmission of TCRLS algorithm mainly occurring among moving objects. While the DCRLS algorithm has relatively less data calculations and the MobiCrowd algorithm data adopts a caching mechanism, so the communication load and the time cost on client of the two algorithms are slightly lower. Comparison for the time spent on the server of three algorithms was shown in Fig. 6(c). With the location service information in the cache gradually increases of MobiCrowd algorithm, the location service information would be shared among the objects. The time spent on the server of MobiCrowd algorithm was gradually reduced and still higher than TCRLS and DCRLS. From the analysis results, the data communication load of the TCRLS algorithm mainly occurs at the client side. However, if the dc value is properly set, the time spent on the server can be ignored. It is consistent with the original intention of our designed algorithm. The designed algorithm provides better location service quality based on distributed systems, balances traffic load, reduces access to location servers and avoids the risk of privacy leaks in existing centralized location service models.

5 Conclusion

In this study, aim to solve the low quality of location service and risk of privacy leakage in existing location service systems, a collaborative filtering recommendation location service strategy oriented to trajectory data was proposed. In this strategy, the trajectory information profile was constructed based on density measure, a collaborative filtering recommendation location service method based on trajectory similarity and profile similarity was designed. And an experimental analysis was performed on the algorithm in the real data set. The analysis shows that the proposed strategy could provide higher quality location service results, do not rely on third-party centralized location servers and overcome the bottlenecks and the shortcomings of attacks in existing centralized system architectures to ensure privacy security of users. In the future work, we will pay more attention to the privacy and security of location information in the strategy and plan to provide a higher data privacy security assurance without increasing the data computing load.

Acknowledgments. We acknowledge the support of the National Natural Science Foundation of China under grant nos. 61672179, 61370083, 61402126; the Natural Science Foundation Heilongjiang Province of China under grant nos. F2015030; the Youths Science Foundation of Heilongjiang Province of China under grant no. QC2016083; the Heilongjiang Postdoctoral Science Foundation no. LBH-Z14071.

References

1. Singh, M.P., Yu, B., Venkatraman, M.: Community-based service location. Commun. ACM **44**, 49–54 (2001)
2. Grutester, M., Grunwald, D.: Anonymous usage of location-based services through spatial and temporal cloaking. In: Proceedings of the 1st International Conference on Mobile Systems, Applications and Services (MobiSys 2003), San Francisco, CA, USA, 5–8 May 2003, pp. 31–42 (2003)
3. Sweeney, L.: k-anonymity: a model for protecting privacy. Int. J. Uncertainty Fuzziness Knowl. Based Syst. **10**, 557–570 (2002)
4. To, H., Ghinita, G., Fan, L., Shahabi, C.: Differentially private location protection for worker datasets in spatial crowdsourcing. IEEE Trans. Mob. Comput. **16**, 934–949 (2017)
5. Chow, C.Y., Mokbel, M.F.: Privacy in location-based services: a system architecture perspective. SIGSPATIAL Spec. **1**, 23–27 (2009)
6. Chow, C.Y., Mokbel, M.F., Liu, X.: Spatial cloaking for anonymous location-based services in mobile peer-to-peer environments. Geoinformatica **15**, 351–380 (2011)
7. Shokri, R., Theodorakopoulos, G., Papadimitratos, P., Kazemi, E., Hubaux, J.-P.: Hiding in mobile crowd: location privacy through collaboration. IEEE Trans. Dependable Secure Comput. **11**, 266–279 (2014)
8. Wang, P., Yang, J., Zhang, J.P.: Protection of location privacy based on distributed collaborative recommendations. PLoS One **11**, e0163053 (2016)
9. Zhang, W., Cui, X., Li, D., Yuan, D., Wang, M.: The location privacy protection research in location-based service. In: Proceedings of the IEEE International Conference on Geoinformatics, Beijing, China, 18–20 June 2010
10. Bayardo, R.J., Agrawal, R.: Data privacy through optimal k-anonymization. In: Proceedings of the IEEE 21st International Conference on Data Engineering (ICDE 2005), Tokoyo, Japan, 5–8 April 2005, pp. 217–228 (2005)
11. Kou, G., Peng, Y., Wang, G.: Evaluation of clustering algorithms for financial risk analysis using MCDM methods. Inf. Sci. **275**, 1–12 (2014)
12. Rodriguez, A., Laio, A.: Clustering by fast search and find of density peaks. Science **344**, 1492–1496 (2014)
13. Cao, Y., Yoshikawa, M.: Differentially private real-time data publishing over infinite trajectory streams. IEICE Trans. Inf. Syst. **E99-D**(1), 68–73 (2016)
14. Seidl, D.E., Jankowski, P., Tsou, M.H.: Privacy and spatial pattern preservation in masked GPS trajectory data. Int. J. Geogr. Inf. Sci. **30**(4), 1–16 (2016)
15. Wang, T., Zeng, J., Bhuiyan, M.Z.A., et al.: Trajectory privacy preservation based on a fog structure in cloud location services. IEEE Access **5**, 7692–7701 (2017)
16. Huo, Z., Meng, X., Hu, H., Huang, Y.: You Can Walk Alone: trajectory privacy-preserving through significant stays protection. In: Lee, S., Peng, Z., Zhou, X., Moon, Y.-S., Unland, R., Yoo, J. (eds.) DASFAA 2012. LNCS, vol. 7238, pp. 351–366. Springer, Heidelberg (2012). https://doi.org/10.1007/978-3-642-29038-1_26
17. Hwang, R.H., Hsueh, Y.L., Chung, H.W.: A novel time-obfuscated algorithm for trajectory privacy protection. IEEE Trans. Serv. Comput. **7**(2), 126–139 (2014)
18. Gao, S., Ma, J., Shi, W., et al.: LTPPM: a location and trajectory privacy protection mechanism in participatory sensing. Wirel. Commun. Mobile Comput. **15**(1), 155–169 (2015)

19. Amini, S., Janne, L., Hong, J., Lin, J., Norman, S., Toch, E.: Cache: caching location-enhanced content to improve user privacy. In: Proceedings of the 9th International Conference on Mobile Systems, Applications, and Services (MobiSys 2011), Bethesda, MD, USA, 28 June–1 July 2011, pp. 197–210 (2011)
20. Shokri, R., Papadimitratos, P., Theodorakopoulos, G., Hubaux, J.P.: Collaborative location privacy. In: Proceedings of the IEEE 8th International Conference on Mobile Adhoc and Sensor Systems, Valencia, Spain, 17–22 October 2011
21. Chow, C., Mokbel, M.F., Liu, X.: A peer-to-peer spatial cloaking algorithm for anonymous location-based services. In: Proceedings of the ACM Symposium on Advances in Geographic Information Systems (ACM GIS 2006), Arlington, VA, USA, 10–11 November 2006, pp. 171–178 (2006)
22. Boutet, A., Frey, D., Guerraoui, R., Jégou, A., Kermarrec, A.M.: Privacy-preserving distributed collaborative filtering. Computing **98**, 827–846 (2016)
23. Chen, K., Liu, L.: Privacy-preserving multiparty collaborative mining with geometric data perturbation. IEEE Trans. Parallel Distrib. Syst. **20**, 1764–1776 (2009)
24. Zhu, T., Ren, Y., Zhou, W., Rong, J., Xiong, P.: An effective privacy preserving algorithm for neighborhood-based collaborative filtering. Future Gener. Comput. Syst. **36**, 142–155 (2014)
25. Huang, Z., Zeng, D., Chen, H.: A comparison of collaborative-filtering recommendation algorithms for e-commerce. IEEE Intell. Syst. **22**, 68–78 (2007)
26. Paillier, P.: Public-key cryptosystems based on composite degree residuosity classes. In: Stern, J. (ed.) EUROCRYPT 1999. LNCS, vol. 1592, pp. 223–238. Springer, Heidelberg (1999). https://doi.org/10.1007/3-540-48910-X_16
27. Piorkowski, M., Sarafijanovic-Djukic, N., Grossglauser, M.: A parsimonious model of mobile partitioned networks with clustering. In: Proceedings of the First International Communication Systems and Networks and Workshops (COMSNETS 2009), Bangalore, India, 5–10 January 2009
28. Domingo-Ferrer, J., Trujillo-Rasua, R.: Microaggregation- and permutation-based anonymization of movement data. Inf. Sci. **208**, 55–80 (2012)

Interest-Forgetting Markov Model
for Next-Basket Recommendation

Jinghua Zhu$^{(\boxtimes)}$, Xinxing Ma, Chenbo Yue, and Chao Wang

School of Computer Science and Technology, Heilongjiang University,
Harbin 150080, Heilongjiang, China
zhujinghua@hlju.edu.cn

Abstract. Recommendation systems provide users with ranked items based on individual's preferences. Two types of preferences are commonly used to generate ranking lists: long-term preferences which are relatively stable and short-term preferences which are constantly changeable. But short-term preferences have an important real-time impact on individual's current preferences. In order to predict personalized sequential patterns, the long-term user preferences and the short-term variations in preference need to be jointly considered for both personalization and sequential transitions. In this paper, a IFNR model is proposed to leverage long-term and short-term preferences for Next-Basket recommendation. In IFNR, similarity was used to represent long-term preferences. Personalized Markov model was exploited to mine short-term preferences based on individual's behavior sequences. Personalized Markov transition matrix is generally very sparse, and thus it integrated Interest-Forgetting attribute, social trust relation and item similarity into personalized Markov model. Experimental results are on two real data sets, and show that this approach can improve the quality of recommendations compared with the existed methods.

Keywords: Markov · Social trust · Next-Basket · Recommendation ·
Interest-Forgetting

1 Introduction

Recommendation system has become a basic function of the online application nowadays, which can recommend products or services that meet user's preferences. Recommendation system can discover the internal relationship between users and items according to the interaction between users and items [1]. The goal of item recommendation is to recommend a specific list of items to each user. Most existing methods do not consider the recent behavior of users but recommend items based on the whole purchase history of users. Predicting the user's recent behavior requires recommendation based on the time sequence of the items purchased.

In order to reasonably predict the user's personalized preferences, it is necessary to combine the user's long-term preferences with short-term preferences. The long-term preferences means that the user's preferences are inherently have not changed with time. For example, if the user prefers a certain brand of goods, the purchase behavior of the brand's goods will not change with time. Short-term preferences means that the

© Springer Nature Singapore Pte Ltd. 2019
X. Cheng et al. (Eds.): ICPCSEE 2019, CCIS 1058, pp. 20–31, 2019.
https://doi.org/10.1007/978-981-15-0118-0_2

user's recent behavior will constantly change. For example, the user will purchase a specific item at certain festivals, which will change over time. So short-term preferences are not stable and are easily affected by others. When a user buys a good product, he tends to recommend the product to his friends. In this way, his friends will be influenced by him to buy the goods.

The common challenge of collaborative filtering or other types of recommendation systems is how to make accurate recommendations for massive data in the Internet environment. There are three difficulties [2]. The first is a cold start problem. It is difficult to effectively recommend new items to a new user without history information [3]. The second is data sparse problem. For a large e-commerce platform, because of the large number of products and users, the overlap of different users' purchases is low. At this time, the data is sparse and it is hard to recommend [4]. In addition, the recommendation mode is too singular, the user's long-term preferences can only be mined based on the user's purchase history. The changes in the user's recent behavior can't be captured [5]. Therefore, it cannot meet the existing recommendation requirements.

In this paper, we leverage long-term and short-term preferences for recommendation to make more reasonable predictions. In order to make full use of the sequence information of users to predict short-term preferences, the Interest-Forgotten properties are merged into Markov model. We integrate the trust relationship into the calculation of transition matrix to solve the sparse data problem of personalized Markov model. Besides this, we also integrate the similarity of the items in the calculation of the recommendation results. it represent long-term preference of users. This can improve the recommendation accuracy and further solve the problem of sparse data of the transition matrix.

2 Related Works

The traditional recommendation methods mainly meet the user's preferences according to the user's explicit or implicit feedback. For example, Xu *et al.* [6] exploit a method to find the similarity between users and items through clustering, which can improve the recommendation performance of the algorithm. Ning *et al.* [7] propose a sparse linear method (SLIM) for top-N recommendations that can quickly generate high-quality top-N recommendations. Kabbur *et al.* [8] propose a method for decomposing item similarity (FISM). FISM learns item similarity as the product of two matrices, which enables it to generate high-quality recommendations on sparse data sets.

The traditional method ignores the trust relationship between users. So some scholars have integrated the trust relationship into the recommendation algorithm. For example, Jamali *et al.* [9] propose a model for factorization of social trust relationships. Each user's feature vector depends on the feature vector of his immediate neighbor in the social network. Guo *et al.* [10] exploit a three-element similarity model for item recommendation use social trust. According to the similarity between user-user and item-item, user preferences are found between rated and non-rated items. Lee *et al.* [11] propose a hybrid recommendation algorithm that combines user ratings and social trust for better recommendations, utilize k-nearest neighbors and matrix factorization methods to maximize user ratings and trust information.

Most of the methods only use the user history information to predict the user's long-term preferences and does not utilize the user history with the sequence to predict the user's short-term preference. For example, Rendle et al. [12] propose a sequence recommendation method (FPMC) based on personalized Markov chain. Instead of using the same transition matrix for all users, this method uses a separate transition matrix for each user. Ultimately resulting in a multi-dimensional transition matrix. Chen et al. [13] consider people's Interest-Forgetting when conducting personalized recommendation, and proposed a personalized framework that combines the Interest-Forgetting attribute with the Markov model. Which can significantly improve the accuracy of recommendation. He et al. [14] propose a method of predicting person-alized sequence behavior by combining similarity-based models with Markov chains.

3   Sequence Recommendation

This article mainly uses user history rating and user relationship data with sequential information to recommend. The user rating information is divided into fixed time periods. A period of time is considered a 'basket', similar to the "shopping basket", which was proposed by Rendle et al. [12]. On the FPMC model.

3.1   Next-Basket Markov Model

Before describing the problem, we will introduce the symbols of this article. The user is $U = \{u_1, u_2, \cdots, u_{|U|}\}$. Item set is $X = \{x_1, x_2 \cdots, x_{|X|}\}$. The set of users that user u trusts is F_u. The history basket for each user u is B^u. $B^u := \{B_1^u, B_2^u, \cdots, B_{t-1}^u\}$, $B_t^u \subseteq X$. The time t is not a specific time, but represents the purchase sequence of the user. And the purchase history of all users is $B := \{B^{u_1}, B^{u_2}, \cdots, B^{u_{|U|}}\}$.

The definition of n-order Markov chains is:

$$P(X_t = x_t | X_{t-1} = x_{t-1}, \ldots, X_{t-n} = x_{t-n}) \tag{1}$$

Where $X_t, \cdots, X_{t-n}$ represents any item in X, Because the data in this paper is in the form of a basket, each basket contains multiple items. The use of high-order Markov chains requires a lot of space, so this paper uses a first-order Markov chain to solve the problem.

$$P(B_t | B_{t-1}) \tag{2}$$

Equation 2 represents the transition probability of the entire basket, and the probability of transition between items in the basket is:

$$a_{j,i} := P(i \in B_t | j \in B_{t-1}) \tag{3}$$

For item recommendation, the purchase probability of an item is determined by the last basket item of the user. This can be defined as the average of all transition probabilities from the last basket to the purchase of the item. The probability of an item being purchased is shown in Eq. 4:

$$P(i \in B_t | B_{t-1}) := \frac{1}{|B_{t-1}|} \sum_{j \in B_{t-1}} a_{j,i} \tag{4}$$

Calculation the transition probability in items:

$$
\begin{aligned}
\hat{a}_{j,i} = \hat{P}(i \in B_t | j \in B_{t-1}) &= \frac{\hat{P}(i \in B_t \wedge j \in B_{t-1})}{\hat{P}(j \in B_{t-1})} \\
&= \frac{|(B_t | B_{t-1}) : i \in B_t \wedge j \in B_{t-1}|}{|(B_t | B_{t-1}) : j \in B_{t-1}|}
\end{aligned}
\tag{5}
$$

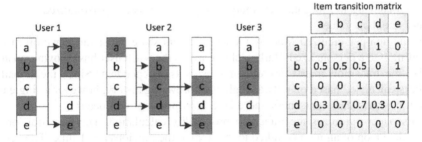

Fig. 1. Non-personalized Markov transition matrix.

As shown in Fig. 1, the transition probability between all items in the transition matrix is initialized to zero. There are five items $\{a, b, c, d, e\}$, three users. In a non-personalized Markov chain, all users share an item transition matrix. The transition probability between items can be calculated using Eq. 5 based on the purchase history of each user. Assuming that all users share the same transition matrix, the transition matrix can be computed by the purchase history of user 1, user 2 and user 3. According to the item of user 3 at the $t - 1$ time, the probability of user 3 purchasing each item at t time is calculated by Eq. 4. For example, compute the probability of user 3 purchasing a:

$$P(a \in B_t | \{b, c, e\}) = \frac{1}{3}(0.5 + 0 + 0) = 0.17$$

By analogy, the probability of user 3 purchasing b, c, d, e can be calculated, and then the corresponding recommendation is made according to the calculation result.

4 Interest-Forgetting Markov Model for Next-Basket Recommendation

4.1 Model Description

Forgetting attribute is an important attribute of people. People's memory of things will be forgotten with time. Similarly, users' preferences are also 'forgotten'. Different users have different forgetting properties and are independent of each other. The user's forgetting rate varies with the length of the interval, and the longer time, the higher the forgetting rate. There are many mathematical representation models for forgetting properties. This paper uses the Hyperbolic model (1978) [15]. The model has two parameters C_u and α_u.

$$\lambda_k^{u,t} = \frac{C_u}{k - \alpha_u}, \ 0 \le \alpha_u < 1, 0 < C_u \le 1 \tag{6}$$

Where α_u is the starting step of compensating for the number of consumption steps and personalizing the forgotten attribute, and C_u is the user u personalized maximum interest retention value.

In the non-personalized Markov model, all users share the same transition matrix. Extending it to a personalized Markov chain means that each user has its own transition matrix, the transition matrix for different users is independent. Since the transition matrix is calculated based on only a single user browsing history, however a single user transition matrix data sparseness and a cold start problem will occur.

This paper optimizes the transition probability calculation. The transition probability calculation of an item is linked to the item at the previous $t - 1$ time. The current item is independent of the transition probability calculations of the previous moments, which will generate $t - 1$ transition probability matrices. In order to measure the influence of the previous $t - 1$ time on different degrees of current time, this paper incorporates the user's Interest-Forgetting attribute in the process of calculating the transition probability. The degree of influence of the previous moments on the transition probability of the current moment is measured by the forgotten attribute. Finally, the joint solution is solved, which can effectively solve these problems. The transition probability between a user item is shown in Eq. 7.

$$a_{j,i,k}^u := P(i \in B_t^u | j \in B_{t-k}^u), k \in \{1, 2 \ldots t - 1\} \tag{7}$$

The probability that a user will purchase an item:

$$P(i \in B_t^u | B_{t-k}^u) \propto \frac{1}{|t - 1|} \sum_{k=1}^{t-1} \frac{1}{|B_{t-k}^u|} \sum_{j \in B_{t-k}^u} \lambda_k^{u,t} a_{j,i,k}^u \tag{8}$$

Personalized item transition probability calculation:

$$\hat{a}_{j,i,k}^u = \hat{P}(i \in B_t^u | j \in B_{t-k}^u) = \frac{\hat{P}(i \in B_t^u \wedge j \in B_{t-k}^u)}{\hat{P}(j \in B_{t-k}^u)}$$

$$= \frac{\left| (B_t^u | B_{t-k}^u) : i \in B_t^u \wedge j \in B_{t-k}^u \right|}{\left| (B_t^u | B_{t-k}^u) : j \in B_{t-k}^u \right|} \tag{9}$$

4.2   Parameter Estimation

The traditional method uses Matrix Factorization to make predictions to solve the data sparse problem. Matrix Factorization methods also have some inherent problems, such as low interpretability and large computational overhead. This paper can effectively solve the data sparse problem by integrating social trust, the multi-step Markov chain of Interest-Forgetting attributes, and the similarity of the fusion items to be introduced later. To reduce computational overhead, the Maximum A Posteriori probability estimation (MAP) method is used herein to estimate the parameters of this paper. Here, Θ is used to represent all the parameters that the MAP is to estimate, assuming that each user's parameters are independent.

$$\underset{\Theta}{\arg\max} = \ln \prod_{u \in U} \prod_{B_{t-k}^u \in B^u} P(i \in B_t^u | B_{t-k}^u)$$

$$= \sum_{u \in U} \sum_{B_{t-k}^u \in B^u} \ln \left(\frac{1}{|t-1|} \sum_{k=1}^{t-1} \frac{1}{|B_{t-k}^u|} \sum_{j \in B_{t-k}^u} \lambda_k^{u,t} a_{j,i,k}^u \right) \tag{10}$$

The gradient of log-likelihood L is relative to the model parameters.

$$\frac{\partial L}{\partial \Theta} = \sum_{u \in U} \sum_{B_{t-k}^u \in B^u} \frac{\frac{1}{|t-1|} \sum_{k=1}^{t-1} \frac{1}{|B_{t-k}^u|} \sum_{j \in B_{t-k}^u} \frac{\partial \lambda_k^{u,t}}{\partial \Theta} a_{j,i,k}^u}{\frac{1}{|t-1|} \sum_{k=1}^{t-1} \frac{1}{|B_{t-k}^u|} \sum_{j \in B_{t-k}^u} \lambda_k^{u,t} a_{j,i,k}^u}$$

$$= \sum_{u \in U} \sum_{B_{t-k}^u \in B^u} \frac{\frac{1}{|t-1|} \sum_{k=1}^{t-1} \frac{1}{|B_{t-k}^u|} \sum_{j \in B_{t-k}^u} \left(\frac{c_u}{(k-\alpha_u)^2} + \frac{1}{k-\alpha_u} \right) a_{j,i,k}^u}{\frac{1}{|t-1|} \sum_{k=1}^{t-1} \frac{1}{|B_{t-k}^u|} \sum_{j \in B_{t-k}^u} \lambda_k^{u,t} a_{j,i,k}^u} \tag{11}$$

For Eq. 11, the gradient descent method can be used to give a training set, and the parameter Θ can be iteratively updated. When the parameters are estimated, the calculation of the recommended results can be performed.

4.3   Fusion Trust Relationship

However, the historical information of a single user is limited. Calculating the transition matrix between items, based on the purchase history of one user may result in data sparseness of the transition matrix. and also have a cold start problem. If a user does not have a purchase history, the user transition matrix cannot be computed.

In order to solve these problems, this paper combines the social trust relationship when calculating the transition matrix. The trusted user of each use u is placed in the set F_u. In order to calculate the similarity between trusted users, this paper uses the Adjusted Cosine Similarity to calculate the similarity of interest between users.

$$sim(u,f) = \frac{\sum_{i \in X} (R_{u,i} - \overline{R_u})(R_{f,i} - \overline{R_f})}{\sqrt{\sum_{i \in X} (R_{u,i} - \overline{R_u})^2} \sqrt{\sum_{i \in X} (R_{f,i} - \overline{R_f})^2}} \tag{12}$$

$R_{u,i}$ and $R_{f,i}$ represent the ratings of user u and user f for item i, and $\overline{R_u}$ represents the average rating of user u for all items. $\overline{R_f}$ represents the average rating of user f for all items. Personalized transition probability calculation assumptions:

$$\eta = \begin{cases} 0, & j \in B_t^u \\ 1, & else \end{cases} \tag{13}$$

The calculation method of transition probability between items in the transition matrix of the fusion trust relationship is as follows:

$$\hat{a}_{j,i,k}^u = \hat{P}(i \in B_t^u | j \in B_{t-k}^u)$$
$$:= (1 - \eta) \frac{\hat{P}(i \in B_t^u \wedge j \in B_{t-k}^u)}{\hat{P}(j \in B_{t-k}^u)} +$$
$$\eta \frac{1}{|F_u|} \sum_{f \in F_u} sim(u,f) \frac{\hat{P}(i \in B_t^f \wedge j \in B_{t-k}^f)}{\hat{P}(j \in B_{t-k}^f)} \tag{14}$$

Where $|F_u|$ represents the number of friends trusted by user u, and Eq. 14 represents that the calculation of the transition matrix of a certain user needs to be jointly computed with the trusted user.

4.4 Fusion Item Similarity

In order to further solve the problem of sparse data of personalized transition matrix, and better improve the accuracy of recommendation. This paper introduces the similarity of the item and calculates the similarity between the items based on the user's rating information on the item. Here, the Adjusted Cosine Similarity is used to calculate the similarity between items.

$$sim(i,l) = \frac{\sum_{u \in U} (R_{u,i} - \overline{R_u})(R_{u,l} - \overline{R_u})}{\sqrt{\sum_{u \in U} (R_{u,i} - \overline{R_u})^2} \sqrt{\sum_{u \in U} (R_{u,l} - \overline{R_u})^2}} \tag{15}$$

$R_{u,i}$ and $R_{u,l}$ indicate that user u rates items i and l, and $\overline{R_u}$ represents the average rate of user u for all items.

Combining the transition matrix with the item similarity, find the item l similar to the i item. When calculating the transition matrix of the item i, and compute the transition probability of the basket item to the item l at all $t - 1$ moments.

$$P_{u,i} = \beta P(i \in B_t^u | B_{t-k}^u) + (1 - \beta)$$
$$\left(\frac{1}{|I|} \sum_{l \in I} sim(i,l) * P(l \in B_t^u | B_{t-k}^u) \right) \quad (16)$$

I represents the top n items with the highest similarity to the item i, parameters $\beta \in [0, 1]$. Schematic diagram of the recommended results are shown in Fig. 2.

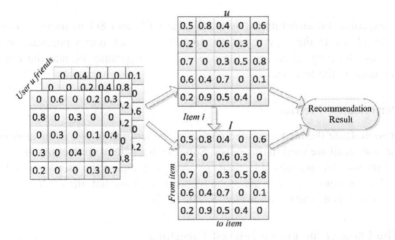

Fig. 2. Schematic diagram of the recommended results.

5 Performance Evaluation

5.1 Dataset and Experimental Environment

In order to evaluate the efficiency and applicability of the proposed algorithm, a series of comparative experiments are designed to verify it.

This paper uses Ciao and Epinions real data as experimental datasets. Ciao is a review site where users can rate and comment on a variety of products. The dataset was captured by Tang *et al.* [16] in 2012 from Ciao's official website. The feedback time was May 2011 and the data set was available online.

Epinions is a popular consumer review site, collected by Zhao *et al.* [17] in 2014. The dataset contains different trust relationships between users. The dataset spans from

January 2001 to November 2013. The statistical information of the dataset can be found in Table 1.

Table 1. Dataset statistics.

	Ciao	Epinions
Users	1708	5621
Items	16485	25996
Feedback	34494	46732
Trusts	35753	23915
Votes/users	20.2	8.88
Trusts/users	20.93	4.55

The experimental environment is a 2.5 GHz CPU and 8G memory, a computer running Windows. In the experiment, the first 80% of each user's purchase sequence was used as a training set to learn the personalization parameters, and the remaining 20% was used as the test set.

5.2 Performance Metrics

In order to evaluate the recommended quality of the algorithm, two evaluation metrics precision and recall are used in this paper. Suppose that N item are recommended to the user, and the user has purchased L, then $@N = L/N$. Suppose the user has purchased a total of M items, some of which are recommended, some are not, then $@N = L/M$. We test the performance when setting N at 5, 10, 15, 20.

5.3 The Effect of the Fusion Interest-Forgetting

In order to verify whether the Markov method of the fusion Interest-Forgetting property can improve the performance of the algorithm. We proposed the method Interest-Forgetting Markov Model for Next-Basket Recommendation (IFNR) and our method that removing the Interest-Forgetting attribute (IFNR-QF) are compared experimentally. To verify whether the Markov method that combing with the Interest-Forgetting property can improve the performance of the algorithm. The experimental results are shown in Fig. 3.

As you can see from Fig. 3, as N gradually increases, precision has some decline, and recall has some rise. Nevertheless, the IFNR recommendation method with the method of Interest-Forgetting attribute is better than the IFNR-QF method without Interest-Forgetting attribute. So the Markov method combining the Interest-Forgetting attribute has a better recommendation effect.

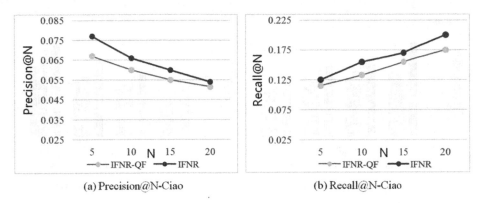

(a) Precision@N-Ciao   (b) Recall@N-Ciao

Fig. 3. Two methods comparison in Ciao dataset.

5.4 Performance Comparison

We compare the performance of IFNR with the following three algorithms.

(1) IFMM algorithm [13]: This algorithm is a method that combines the user's Interest-Forgetting characteristics with the first-order Markov model.
(2) SBPR algorithm [17]: The algorithm is based on the assumption that users and their social circles should have similar interests in objects.
(3) FPMC algorithm [12]: This algorithm is a method of sequence prediction using personalized Markov chain. FPMC is a combination of matrix decomposition and first-order Markov chain.

The precision and recall evaluation metrics were used on the Ciao and Epinions data sets to compare the recommended performance of the proposed algorithm with the three algorithms when setting N of 5, 10, 15, and 20.

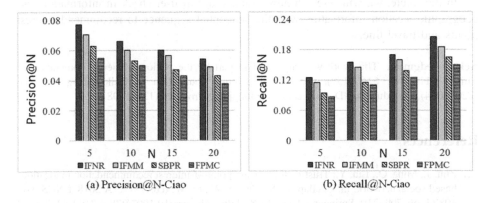

(a) Precision@N-Ciao   (b) Recall@N-Ciao

Fig. 4. Performance compare in Ciao dataset.

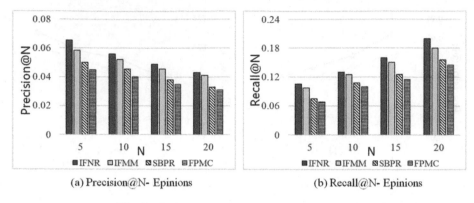

(a) Precision@N- Epinions   (b) Recall@N- Epinions

Fig. 5. Performance compare in Epinions dataset.

As shown in Figs. 4 and 5, with the increase of N, there are some declines in the precision, but IFNR has better performance than SBPR and FPMC methods. IFMM is similar to the IFNR algorithm in some cases, but the recommended performance of IFNR under each evaluation metrics is better than IFMM method. Therefore, the sequence recommendation method IFNR based on the Interest-Forgetting attribute Markov model has a better recommendation effect.

6   Conclusion and Future Work

In this paper, we propose a IFNR model to leverage long-term and short-term preferences for recommendation that integrates social trust, and the Interest-Forgetting attribute into the calculation of personalized Markov transition matrix. It can effectively solve the data sparsity problem of the transition matrix. Moreover, fusing the user's trust relationship and the user's similarity method in the transition matrix calculation can improve the accuracy of the algorithm and further enhance the stability of the algorithm.

In the future, we will research how to make use of user check-in information and geographic location information with the sequential sequence to recommend interest points and travel line.

Acknowledgment. This work was supported in part by the National Science Foundation of China (61100048, 61602159), the Natural Science Foundation of Heilongjiang Province (F2016034), the Education Department of Heilongjiang Province (12531498).

References

1. Zhu, J., Ming, Q., Liu, Y.: Trust-distrust-aware point-of-interest recommendation in location-based social network. In: Chellappan, S., Cheng, W., Li, W. (eds.) WASA 2018. LNCS, vol. 10874, pp. 709–719. Springer, Cham (2018). https://doi.org/10.1007/978-3-319-94268-1_58
2. Wang, H., Wang, N., Yeung, D.Y.: Collaborative deep learning for recommender systems. In: Proceedings of the 15th KDD, pp. 1235–1244 (2015)

3. Jing, L., Wang, P., Yang, L.: Sparse probabilistic matrix factorization by Laplace distribution for collaborative filtering. In: Proceedings of the 24th International Conference on Artificial Intelligence, pp. 1771–1777. AAAI Press (2015)
4. Heckel, R., Vlachos, M., Parnell, T., Duenner, C.: Scalable and interpretable product recommendations via overlapping co-clustering. In: Proceedings of the 33th IEEE International Conference on Data Engineering, pp. 1033–1044 (2017)
5. Gao, H., Tang, J., Hu, X., Liu, H.: Content-aware point of interest recommendation on location-based social networks. In: Proceedings of the Twenty-Ninth AAAI Conference on Artificial Intelligence, pp. 1721–1727. AAAI Press (2015)
6. Xu, B., Bu, J., Chen, C., et al.: An exploration of improving collaborative recommender systems via user-item subgroups, pp. 21–30 (2012)
7. Ning, X., Karypis, G.: SLIM: sparse linear methods for top-N recommender systems. In: Proceedings of the 11th IEEE International Conference on Data Mining, pp. 497–506 (2011)
8. Kabbur, S., Karypis, G.: FISM: factored item similarity models for top-N recommender systems. In: Proceedings of the 19th ACM SIGKDD International Conference on Knowledge Discovery and Data Mining, pp. 659–667 (2013)
9. Jamali, M., Ester, M.: A matrix factorization technique with trust propagation for recommendation in social networks. In: Proceedings of the 2010 ACM Conference on Recommender Systems, pp. 135–142 (2010)
10. Guo, G., Zhang, J., Zhu, F., et al.: Factored similarity models with social trust for top-N item recommendation. Knowl.-Based Syst. 135–142 (2010)
11. Lee, W.P., Ma, C.Y.: Enhancing collaborative recommendation performance by combining user preference and trust-distrust propagation in social networks. Knowl.-Based Syst. **106**, 125–134 (2016)
12. Rendle, S., Freudenthaler, C., Schmidt-Thieme, L.: Factorizing personalized Markov chains for next-basket recommendation. In: Proceedings of the 19th International Conference on World Wide Web, pp. 811–820 (2010). Knowledge-Based Systems
13. Chen, J., Wang, C., Wang, J.: A personalized interest-forgetting Markov model for recommendations. In: Proceedings of the Twenty-Ninth AAAI Conference on Artificial Intelligence, pp. 16–22. AAAI Press (2015). Knowledge-Based Systems
14. He, R., Mcauley, J.: Fusing similarity models with Markov chains for sparse sequential recommendation. In: Proceedings of the IEEE 16th International Conference on Data Mining (2016)
15. Mazur, J.E., Hastie, R.: Learning as accumulation: a reexamination of the learning curve. Psychol. Bull., 1256–1274 (1978)
16. Tang, J., Gao, H., Liu, H.: mTrust: discerning multi-faceted trust in a connected world. In: Proceedings of the 5th ACM International Conference on Web Search and Data Mining, pp. 93–102 (2012)
17. Zhao, T., McAuley, J., King, I.: Leveraging social connections to improve personalized ranking for collaborative filtering. In: Proceedings of the 23rd ACM International Conference on Conference on Information and Knowledge Management, pp. 261–270 (2014)

Attention Neural Network for User Behavior Modeling

Kang Yang and Jinghua Zhu[✉]

School of Computer Science and Technology, Heilongjiang University,
Harbin 150080, China
zhujinghua@hlju.edu.cn

Abstract. The recommendation system can effectively and quickly provide valuable information for users by filtering out massive useless data. User behavior modeling can extract all kinds of aggregated features over the heterogeneous behaviors to help recommendation. However, the existing user behavior modeling method cannot solve the cold-start problem caused by data sparse. Recent recommender systems which exploit reviews for learning representation can alleviate the above problem to a certain extent. Therefore, a user behavior modeling is proposed for recommendation task using attention neural network based on user reviews (AT-UBM). Firstly vanilla attention was used to sample reviews, and then CNN+Pooling method was applied to extract user behavior features. Finally the long-term behavior was combined with short-term behavior in feature spaces. Experimental results on real datasets show that the review-based user behavior model has better prediction accuracy and generalization capability.

Keywords: Recommendation system · Attention neural network · Behavior model · CNN+Pooling

1 Introduction

With the rapid development of Internet technology, e-commerce websites such as Amazon, eBay, and Tmall, generate the majority of data which causes users to face serious information overload problems. How to obtain valuable information quickly and efficiently from complex data has become a key issue in the current development of big data [1]. Therefore, the recommendation system becomes an effective way to filter information and ease information overload problems. By capturing user interest behaviors, it is able to predict the potential needs of users and provides valuable recommended products to users.

The traditional Matrix Factorization methods and collaborative filtering methods have achieved good performance in user rating recommendation, but in the face of sparse data, because MF only considers the scoring relationship between users and products, it is difficult to extract other features. In addition, the collaborative filtering method is a shallow model, which is difficult to model the complex relationship between users and products. At the same time, the cold start problems encountered by these two methods are also hard to solve. As a result, the deep learning method based on neural network has become an effective framework for constructing recommendation systems [2–5]. A user can be represented as what he or she does along the history and the user's historical behavior data also contains massive user features [6]. Recent research on recommendation models based on user behavior modeling proposes ATRank [6], which uses Self-Attention model to learn the behavioral relationship between users and products in the latent semantic space [7]. The ATRank model provides a new perspective for us to study the user behavior prediction model. Nevertheless, ATRank only considers the interaction information between users and products. In the experimental results of the ATRank, it shows that the most recent behavior has the greatest impact on prediction, so the extraction of long-range information is not ideal. For most e-commerce platforms, the review information is not only a product feature but also reflects the user's strong willingness to interact with a product [8]. The information contained in the reviews reflects the user's perception, attitude and experience feedback on a product. The review information can effectively capture a user's preference features or a product's quality features. Through the rich feature semantics of the text, we can extract a long-term depiction model of the user's behavior features and product feedback features. In recent work, the MPCN model based on user review information has been proposed [9]. This method uses the multi-pointer Attention method to model and sample the relevance between users' reviews and products' reviews. Whereas this model's feature extraction layer is still a Matrix Factorization, and it is not possible to extract features based on user behavior, thereby predicting users' future needs.

However, the review information of a user or a product is often extremely rich, which causes excessive computational overhead, and the noise in the review information is also a problem that cannot be ignored. Thus, for the user behavior recommendation system based on review information, there exists the following problem: (1) The feature extraction model of the review information should be independent, instead of simply appending the reviews to the user behavior prediction model. (2) Different review contributes differently to the features of users or products, and the model should be able to dynamically adjust the weight of reviews. (3) The model cannot use all the review information, so it should reduce the computational complexity of the model. (4) The integration of long-term features and short-term features should be within the same feature space, i.e., the two kinds of feature should be distributed within the same feature dimension.

To this end, we propose the AT-UBM model, which is a user behavior modeling for recommendation task using Attention Neural Network based on user reviews. In the part of the long-term feature, it mainly relies on the users' review data, and regard each word as a feature in the review. The part of the long-term feature calculates the affinity matrix, which can represent the users and the products. In the part of the short-term feature, it depends on the historical interaction information between the users and the products that can be depicted as latent feature, then latent feature is embedded into the user's historical interaction sequence. At the end of part, we use CNN+Pooling model to extract features.

The possible contributions of this paper are as follows: (1) We extract users' reviews and integrate them into the user interaction behavior prediction model. (2) We conduct experiments on real data set and the results show that our model can effectively improve the performance of recommendation with an average improvement of about 8 to 10%.

2   Related Works

2.1   User Behavior Modeling

The RNN-based user behavior recommendation system has been intensively researched and discussed by a large number of researchers in recent years [10]. These recommendation systems build model directly on each user interaction behavior [11]. However, the user recommendation system based on the RNN method still confronts plenty of problems. First of all, due to the limitation of RNN theory, it is difficult to implement parallel computing for the user behavior recommendation system based on RNN model, which makes the model only select fewer user feature parameters to boost calculation speed. Secondly, the traditional RNN method relies on converting the input information into an intermediate context vector. The size of intermediate context vector is fixed, which limits capability to of carrying information, therefore, as user interactions features become more complex, the accuracy of the model is difficult to be guaranteed. Finally, RNN-based user recommendation model has limited capability to retain long-term stable features of users, so that the short-term user behavior has an excessive influence on the judgment of the model.

With the development of research of Self-Attention method [7], Self-Attention method provides powerful support for the correlation learning of input data and accelerates the data convergence speed. The ATRank model applies Self-Attention method to the user behavior recommendation system, which is effective. The weight of the context of the user sequence is extracted, which increases the prediction accuracy of the model while reducing the amount of calculation. But the ATRank model relies on the adjustment of product latent features in the learning process and it does not consider the stability of the users' long-term features, and does not consider factors other than the interaction between the user and the product, such as review information. Therefore, the ATRank model still has a lot of progressive space. In the results of the experiment, it suggests that the performance difference between ATRank and CNN +Pooling is comparable, and the accuracy of CNN+Pooling learning method in the later stage exceeds that of the ATRank model.

2.2   User Behavior Modeling

MPCN (Multi-Pointer Co-Attention Networks) is a comment-based Co-Attention Matrix Factorization model [9]. This model considers the contribution of each review to users and products from the level of word vector and selects the most contributing review as the representative of the user or product. Learned representatives are input into the Matrix Factorization model, and the Matrix Factorization model is used to learn the potential relationship between the user and the product. This model uses the gumbel-softmax technology to sample the review information [12], which greatly reduces the amount of calculation on a model which based on the review recommendation. It can also effectively extract useful comment information. In addition, the process can be dynamically performed, which makes the model extremely great flexibility. However, this model is only a model based on Matrix Factorization and cannot be directly applied to the sequence data of user behavior.

3   Attention-Based Behavior Modeling

See Table 1.

Table 1. Symbol description.

Symbol	Description
$W_g, W_u \in \Re^{d \times d}$	Weight parameters of filter layer
$b_g, b_u \in \Re^{n \times n}$	Bias parameters of filter layer
$M \in \Re^{d \times d}$	Weight matrix in Attention
R_A	Total review sets
R_u	User u's review sets
R_i	Item i's review sets
$D(.)$	Dense layer
$\sigma(.)$	Sigmoid function
$f_{emb}(.)$	Embedded function
$f_{gumbel-softmax}(.)$	Gumbel-Softmax function, which can update gradients while hard sampling
$max_{col}(.)$	Max column pooling function
$max_{row}(.)$	Max row pooling function
$randint(.)$	Round function
$equal(.)$	Equate function
$\odot(.)$	Hadamard product
S	Total test sets
S^-	Total negative test sets
S^+	Total positive test sets

3.1   Model Framework

The model of this paper consists of two parts: long-term user behavior model and short-term user behavior model. The input of our model includes the following: the reviews and the historical interaction sequence of the user. The long-term user behavior model needs to input the reviews and the historical interaction sequence of the user sets, while the short-term user behavior model only needs to input user history interaction sequences sets. The sampled data will be extracted through a CNN+Pooling model to extract the long-term and short-term features of the user [13]. Finally, the summation method is used to balance the impact of long-term and short-term on the prediction results.

$$\hat{Y} = f_{cnn+pooling}\left(f_{gumbel-softmax}\left(f_{emb}\left(R_u, R_i, R_A\right)\right)\right) + f_{cnn+pooling}\left(f_{emb}(H_u)\right) \tag{1}$$

3.2 Filtering Input Reviews

Each review vector is represented by the summed average of its word vectors, so each review vector size is $x \in \Re^d$. Reviews often contain majority of noise, and not all reviews contain value information. So, we firstly define a preprocessing gate that can control whether reviews are streamed to the next neural network or discarded directly.

We input each review x_i into the preprocessing gate.

$$\bar{x}_i = \sigma\left(W_g x_i\right) + b_g \odot \tanh\left(W_u x_i + b_u\right) \tag{2}$$

Where σ is the Sigmoid activation function, $\odot$ is a matrix point multiplication symbol, i.e., the product between each value of the two matrices and each value is calculated. x_i is the review vector i in the total review matrix. $W_g, W_u \in \Re^{d \times d}$ and $b_g, b_u \in \Re^{n \times n}$ is the parameter matrix and vector of the preprocessing gate and is the number of total reviews (Fig. 1).

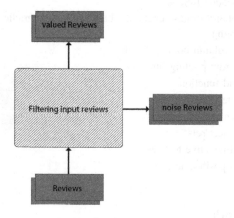

Fig. 1. Review filter gate.

3.3 User-Product Affinity Attention Neural Network

Through the Attention neural network, the most representative reviews for users or products can be extracted to represent the user's long-term behavior and long-term product feedback. Here we use Soft-Attention to learn the relevance of users and product reviews. So that we have user reviews embedded matrix and the product reviews embedded matrix, $u \in \Re^{l_r \times d}$ and $i \in \Re^{l_r \times d}$. In turn, we use the following Attention calculation method to calculate the affinity between the user comment embedded matrix and the product reviews embedded matrix.

$$a_{ij} = D(u_i)^T M D(i_j) \tag{3}$$

$M \in \Re^{d \times d}$ is the weight matrix in Attention, which can adjust the association relation between the user comment embedded matrix and the product reviews embedded matrix. $D(.)$ is a fully connected layer to achieve nonlinear calculation of the model.

We utilize the max-pooling operation to achieve the sampling of the user comment embedding matrix and the product embedding matrix. By sampling the row and column wise maximum of the matrix a, the most representative comment embedding feature vector for each user and each product is obtained. The reason for using the maximum pooling is that this stage of the model needs to extract the most representative reviews for users and products. The above method is defined as follows:

$$u' = \left(G\left(\max_{col}(a) \right) \right) u \qquad (4)$$

$$i' = \left(G\left(\max_{row}(a) \right) \right) i \qquad (5)$$

The $G(.)$ function is to achieve sampling of the reviews. $G(.)$ is similar with the SoftMax function, making normalization processing to sampling information. But the data processed by the SoftMax function is continuous, therefore it can only achieve soft sampling. If the model uses soft sampling, the independence of the reviews will be destroyed along with the learning process. This is why it is not suitable for our model. So, we use the Gumbel-Max trick to hard-sample the reviews embedding vector. By sampling, the long-term features of the user and the product are represented by the reviews, i.e., the behavioral features of users or products do not change, the review representing their feature does not change. Finally, the hard-sampled information is passed to the underlying CNN+Pooling layer to extract long-term features (Fig. 2).

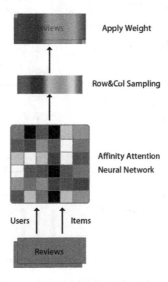

Fig. 2. User-product affinity attention neural network.

3.4   Extraction and Fusion of Long-Term and Short-Term Features

This part is used to learn users' stable long-term preferences from the preference behaviors. Actually, only part of behaviors implies users' long-term preferences. Thus, in this model, we view review information as a user's stable preference for items.

In the short-term behavior model, the model generates a latent space vector for each product and generates a user history behavior matrix composed of product latent space vectors by the user's historical interaction behavior. Finally, the user history behavior matrix is input into the CNN+Pooling layer. It is used to capture the short-term behavior change of users. The historical behavior matrix of the representative users' changes with the training in real time, so the short-term user behavior model can effectively capture the short-term dramatic changes of the user. Therefore, we propose a long-term model based on user reviews and a short-term model based on user interaction. It can be concluded that the long-term model is defined as follows:

$$\hat{Y}_{long-term} = f_{cnn+pooling}\left(f_{gumbel-softmax}\left(f_{emb}(R_u, R_i, R_A)\right)\right) \tag{6}$$

The short-term model is defined as follows:

$$\hat{Y}_{short-trem} = f_{cnn+pooling}\left(f_{emb}(H_u)\right) \tag{7}$$

Finally, we add the long-term prediction and the short-term prediction, and get the objective function of the model (Formula 1). The purpose of the addition is to make the model restrict the weight of effects of long-term and short-term in the simplest way. This method also facilitates the rapid convergence of the model (Fig. 3).

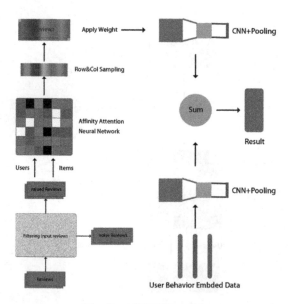

Fig. 3. AT-UBM model.

4 Experimental Evaluation

4.1 Experimental Data Set

The experimental dataset uses Amazon's commodity information[1], which contains reviews on Amazon from May 1996 to July 2014, as well as product meta-information. We generated user-product interaction sequence $(b_1, b_2, b_3, \cdots, b_n)$ in the chronological order of user interactions. At the same time, the user and product reviews index sequence are generated separately by user grouping and product grouping, in other words, they are $(r_{u1}, r_{u2}, r_{u3}, \cdots, r_{un})$ and $(r_{i1}, r_{i2}, r_{i3}, \cdots, r_{in})$ respectively. We randomly shuffle the divided data sets instead of perceiving the $1 \cdots n - 1$ interaction and behavior n as the training set and the test set, so as to prevent the user from being predicted only by the most recent user behavior, thus ignoring the impact of long-term user behavior on user behavior. To reflect the impact of commentary on the user behavior model, we selected users and products with more than 20 reviews into the dataset (Table 2).

Table 2. Dataset.

Dataset	Users	Items	Reviews
Electro.	1406	2768	45968

4.2 Comparative Test

As a comparative test, we selected the CNN+Pooling method as a comparative experiment. The evaluation method of the test uses the following formula:

$$\text{avg}_{\text{HR}} = \frac{1}{|S|} \left(\sum_{x_i \in S^+} equal\left(randint\left(\hat{Y}(x_i), Y\right)\right) + \sum_{x_i \in S^-} equal\left(randint\left(\hat{Y}(x_i), Y\right)\right) \right) \tag{8}$$

As can be notes from the experimental chart, the AT-UBM model performed by the test set was higher 8%–10% than the CNN+Pooling method. Therefore, the long-term user behavior model with reference to user reviews also has a significant improvement in the accuracy of prediction, which enhances the generalization ability of the model to some extent (Fig. 4 and Table 3).

---

[1] http://snap.stanford.edu/data/amazon/productGraph/.

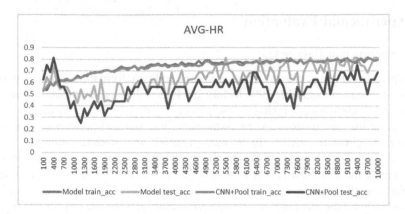

Fig. 4. The average hit rate of the CNN+Pooling and AT-UBM models in the training set and the test set.

Table 3. AVG-HR of different models on the test set.

Model	Electro.
CNN+Max Pooling	0.544875429
ATRank	0.569834235
Our Model	0.637092

5   Experimental Evaluation

This paper proposes prediction model of a long-short-term user behavior based on user reviews and user behavior named AT-UBM model. The AT-UBM model can effectively extract the user's long-term feature behavior from user reviews and combine it with short-term user behavior to achieve performance improvements over previous algorithms. At the same time, our model is based on the Attention framework and the CNN framework, so it has a good parallelization effect. Compared with the traditional RNN framework's sequence prediction model, the AT-UBM model has a natural performance advantage. It also demonstrates in the experiment that after considering the user's review information, the performance of the model in the test set has been greatly improved. In the future, we will focus on models that predict user behavior based on other side information, and try to port the model to the POI recommendation in future work.

Acknowledgment. This work was supported in part by the National Science Foundation of China (61100048, 61602159), the Natural Science Foundation of Heilongjiang Province (F2016034), the Education Department of Heilongjiang Province (12531498).

References

1. He, L., Jiang, B., Lv, S., Liu, Y., Li, D.: A review of research on recommendation systems based on deep learning. Chin. J. Comput. **7**, 11 (2018)
2. He, X., Liao, L., Zhang, H., Nie, L., Hu, X., Chua, T.-S. (eds.): Neural collaborative filtering. In: Proceedings of the 26th International Conference on World Wide Web. International World Wide Web Conferences Steering Committee (2007)
3. Xiao, J., Ye, H., He, X., Zhang, H., Wu, F., Chua, T.-S.: Attentional factorization machines: learning the weight of feature interactions via attention networks. arXiv preprint arXiv:04617 (2017)
4. Xiong, C., Zhong, V., Socher, R.: Dynamic coattention networks for question answering. arXiv preprint arXiv:01604 (2016)
5. Batmaz, Z., Yurekli, A., Bilge, A., Kaleli, C.: A review on deep learning for recommender systems: challenges and remedies. Artif. Intell. Rev., 1–37 (2018)
6. Zhou, C., Bai, J., Song, J., Liu, X., Zhao, Z., Chen, X., et al. (eds.): ATRank: an attention-based user behavior modeling framework for recommendation. In: Thirty-Second AAAI Conference on Artificial Intelligence (2018)
7. Vaswani, A., Shazeer, N., Parmar, N., Uszkoreit, J., Jones, L., Gomez, A.N., et al. (eds.): Attention is all you need. In: Advances in Neural Information Processing Systems (2017)
8. Zheng, L., Noroozi, V., Yu, P.S. (eds.): Joint deep modeling of users and items using reviews for recommendation. In: Proceedings of the Tenth ACM International Conference on Web Search and Data Mining. ACM (2017)
9. Tay, Y., Luu, A.T., Hui, S.C. (eds.): Multi-pointer co-attention networks for recommendation. In: Proceedings of the 24th ACM SIGKDD International Conference on Knowledge Discovery & Data Mining. ACM (2018)
10. Rendle, S., Gantner, Z., Freudenthaler, C., Schmidt-Thieme, L. (eds.): Fast context-aware recommendations with factorization machines. In: Proceedings of the 34th International ACM SIGIR Conference on Research and Development in Information Retrieval. ACM (2011)
11. Hariri, N., Mobasher, B., Burke, R. (eds.): Context-aware music recommendation based on latenttopic sequential patterns. In: Proceedings of the Sixth ACM Conference on Recommender Systems. ACM (2012)
12. Jang, E., Gu, S., Poole, B.: Categorical Reparameterization with Gumbel-Softmax (2016)
13. Gehring, J., Auli, M., Grangier, D., Yarats, D., Dauphin, Y.N. (eds.): Convolutional sequence to sequence learning. In: Proceedings of the 34th International Conference on Machine Learning-Volume 70. JMLR.org (2017)

Method for Extraction and Fusion Based on KL Measure

Zuocong Chen[✉]

College of Computer Science and Technology,
Hainan Tropical Ocean University, Sanya 572022, China
twsf2005@163.com

Abstract. Feature extraction and fusion is important in big data, but the dimension is too big to learn a good representation. To learn a better feature extraction, a method that combines the KL divergence with feature extraction is proposed. Firstly the initial feature was extracted from the primitive data by matrix decomposition. Then the feature was further optimized by using KL divergence, where KL divergence was introduced to the loss function to make the goal function with the shortest KL distance. The experiment is implemented in four datasets such as COIL-20, COIL-100, CBCI 3000 and USPclassifyAL. The result shows that the proposed method outperforms the other four methods in the accuracy when using least number of features.

Keywords: Feature extraction · KL divergence · Samples · Loss function

1 Introduction

It is very known much useful and valuable information is included in the big data [1]. However, the dimensionality of the data is always big, so it is hard to learn from them. Feature extraction is an important research area in pattern recognition and machine learning for big data. Learning important features from these data can not only reduce the computation complexity but also improve the learning ability of the algorithm. A robust feature representation for big data can make the learning model be better obtained.

The traditional methods for feature extraction and fusion always starts at the view of the statistics [2]. With the development of the information technology and network, the feature extraction is attracting more and more attentions. Feature extraction algorithm can be thought as a composite part for the model learning, and it can be divided into three parts: supervised, semi-supervised and unsupervised. The supervised method refers to that all labeled-data are considered and feed to train the model. If some of the labeled-data are used to train the model, we say the method is semi-supervised. No labeled-data are considered in the unsupervised method. All the three methods are preconditioned that the distance among the data represent the similarity property. The distance in feature extraction is always measured by Euclidean distance. However, a large error will be generated when using the Euclidean distance due to the elements of every data have different units.

© Springer Nature Singapore Pte Ltd. 2019
X. Cheng et al. (Eds.): ICPCSEE 2019, CCIS 1058, pp. 42–51, 2019.
https://doi.org/10.1007/978-981-15-0118-0_4

In order to get a better feature extraction representation for big data. We propose a better measure method based on KL distance, and designed a loss function for it. From the experiment, it can be shown that our method has better effect on loss function.

2 Related Work

2.1 Traditional Methods

Traditional feature extraction methods are composed of feature sort method and feature search method. The earliest research on feature sort are mainly based the distance measure methods such as Relief algorithm, the improved ReliefF [3] and the mutual information based DMIFS algorithm [4]. The advantage of the feature sort method is the high execution efficiency. However, two problems hinder the development, one is the number of features should be assigned in advance, and the other is that the feature set may be composed of m optimal features. Furthermore, the feature sort method only evaluates the relevance of the feature itself and the label other than considering the relations among features, so that the redundant feature cannot be distinguished.

For the dataset with high dimensionality, the feature extraction cannot get an effective result because of the redundant features. In order to distinguish the redundant features and then search the optimal feature set, many search algorithms for obtaining the feature subset is proposed. The feature subset search algorithm can be divided into two kinds: comprehensive method and two-phase method. The former one combines the relevance and the redundancy to get a comprehensive factor. The classical algorithms are such as CFS algorithm [5], the mRMR algorithm [6] based on the principle of minimal redundancy and the mRR algorithm based on maximal relevance and clustering technology. The latter algorithms consider the relevance and the redundancy respectively. The representative methods include FCBF algorithm [7] based on uncertainty measurement, the IAMB algorithm [8] based on Markov blanket and their improved methods. Generally speaking, the two-phase analysis method for redundancy and relevance have good quality in feature extraction than the comprehensive method.

2.2 Deep Learning Methods

The deep learning method is one of learning method in machine learning. The learning structure can be divided to shallow one and deep one. Support vector machine is one kind of shallow learning structure [9]. Deep structure proposed by Hinton for the first time in 2006. It is gradually well known by the representative ability in classification and recognition. The main methods in deep learning include restricted Boltzmann machine (RBM), deep belief networks (DBN), convolutional neural network (CNN) and auto encoder (AE). AE and DBN are unsupervised methods, while CNN is a deep supervised method. Then a series of deep network who get much process on the match of Imagenet dataset are appeared. The familiar methods are VGGNet [10], GoogleNet [11], Resnet [12] and DenseNet [13].

Though deep learning methods can learn a good representation for data automatically, it needs much time to train the model. At some times, it only extracts the global feature. Local feature may play a more import role in the specific task.

3   Feature Extraction Based on KL Measure

3.1   Feature Extraction

Let X be a matrix composed by n-dimension, and it can be obtained by normalizing the orthogonal vectors u_j:

$$X = \sum_{j=1}^{\infty} a_j u_j \qquad (1)$$

X can be evaluated as:

$$\bar{X} = \sum_{j=1}^{N} a_j u_j \qquad (2)$$

The mean root square error can be represented as:

$$\zeta = \mathrm{E}[(X - \bar{X})^T (X - \bar{X})] \qquad (4)$$

According to the definition of the orthogonal vector, the following formulas all can be satisfied:

$$u_i^T u_i = \delta_{ij} = \begin{cases} 1, & j = i \\ 0, & j \neq i \end{cases} \qquad (5)$$

$$\zeta = \mathrm{E}[\sum_{j=N+1}^{\infty} a_j^2] \qquad (6)$$

$$a_j = u_j^T X \qquad (7)$$

After feeding the value of a_j to the Eq. (6), we can get:

$$\zeta = \mathrm{E}[\sum_{j=N+1}^{\infty} u_j^T X X^T u_j]$$
$$= \sum_{j=N+1}^{\infty} u_j^T \mathrm{E}(X X^T) u_j \qquad (8)$$

Let R be:

$$\zeta = E[XX^T] \tag{9}$$

Then ζ can be represented by

$$\zeta = E[\sum_{j=N+1}^{\infty} u_j^2 R u_j] \tag{10}$$

where R is auto-correlation matrix.

The goal is to minimize the goal ζ, so the value of u_j should be determined in advance. Therefore, we can construct the Lagrange formula by introducing Lagrange coefficient:

$$g(u_j) = \sum_{j=N+1}^{\infty} u_j^T R u_j - \sum_{j=N+1}^{\infty} \lambda(u_j^T u_j - 1) \tag{11}$$

Then we can get the result by differentiating Eq. (11) with u_j:

$$(R - \lambda_j I)u_j = 0, \quad j = N+1, \ldots, \infty \tag{12}$$

We can get the mean square error form the former N items:

$$\zeta = \sum_{j=N+1}^{\infty} u_j^T R u_j$$
$$= \sum_{j=N+1}^{\infty} tr[u_j R u_j^T] \tag{13}$$
$$= \sum_{j=N+1}^{\infty} \lambda_j$$

From the Eq. (13), we can conclude the feature value λ_j is smaller, the goal value ζ is smaller. They are positive relevance relation between them.

3.2 Feature Generation

The feature extraction process based on KL can be represented as:

(1) The autocorrelation R of X can be denoted as:

$$R = E[XX^T]$$
$$\approx 1/N \sum_{j=1}^{N} X_j X_j^T \tag{14}$$

(2) The diagnosing of matrix R can be denoted as

$$u_i^T R u_j = u_i^T \lambda_j u_j = \lambda_j \delta_{ij} \tag{15}$$

$$E[U^T X X^T U] = U^T R U$$

$$= \begin{pmatrix} \lambda_1 & & 0 \\ & \ddots & \\ 0 & & \lambda_N \end{pmatrix}$$

(3) The eigenvalues of the matrix are sorted as:

$$\lambda_1 \geq \lambda_2 \geq \lambda_3 \dots \lambda_N \tag{16}$$

(4) The largest K eigenvalues are selected to get the eigenvector. The transform matrix U can be obtained by normalizing them.
(5) The primitive matrix X is transformed by K-L measure, the K-dimension matrix X^* can replace the primitive matrix X

$$X^* = U^T X \tag{17}$$

3.3   Loss Function Based on KL Divergence

KL divergence (Kullback-Leibler divergence) is also called KL distance. It is a tool that can describe the distinction between two distributions. The form of KL divergence has the following form:

$$D_{KL}(p||q) = \sum_{x \in \Omega} p(x) \log \frac{p(x)}{q(x)} \tag{18}$$

where p and q are two probability distribution in the probability space Ω.

$D_{KL}(p||q)$ is the KL divergence of p respective to distribution q. It is easy to see from Eq. (17) that $D_{KL}(p||q) \neq D_{KL}(q||p)$, namely, the KL divergence does not satisfy the symmetry. For the KL divergence as the form $D_{KL}(p||q)$, the distribution p is the true distribution and q is the approximate distribution. The value of $D_{KL}(p||q)$ is larger, the difference between real distribution and the approximate distribution is large.

According to the Jensen inequality formula, $D_{KL}(p||q) \geq 0$ is hold, where the equality equation is hold only when $p = q$.

After the features are extracted by (17), we can use it to apply the specific tasks and then compare it with the other methods by KL divergence. If the KL distance of the result is larger than the threshold, then the feature extraction is not good enough, the feature extraction has to be optimized further. And the KL distance will be added to the Eq. (11), shown as:

$$g(u_j) = \sum_{j=N+1}^{\infty} u_j^T R u_j + D_{KL}(p||q) - \sum_{j=N+1}^{\infty} \lambda(u_j^T u_j - 1) \tag{19}$$

The final goal will be that KL distance is nearly 0, which is shown as:

$$D_{KL}(p||q) = 0 \tag{20}$$

4 Experiment Analysis

4.1 Dataset

In order to verify the proposed method, we compare with the other methods. The image dataset in Columbia university COIL-20, COIL-100, CBCI 3000 and USPclassifyALL.

(1) COIL-20 dataset includes 1400 images composed by 20 different goods. Every image has the 32 * 32 resolution.
(2) COIL-100 dataset is composed of 7200 images which attribute to different classifications. Every image is with the resolution 32 * 32.
(3) CBCL3000 dataset is organized by 3000 images, where every image with the resolution 19 * 19.
(4) USPSclassifyAll dataset is composed of 11000 images with the resolution 19 * 19.

The threshold of the KL divergence is set to 0.01 in our method.

4.2 Simulation Result

In order to verify the effectiveness of the proposed, we compare it with the classical method such as SVM, DMIFS, mRMR, VGG, DBN and our method. The first experiment is implemented in the first dataset COIL-20. The features are firstly generated by the five methods. Then, the obtained features are then used to recognize the images. The accuracies obtained with the changing of the number of the features for the five methods are shown as Fig. 1.

It is easy to see that our method has the best accuracy in the five methods. When using just 100-dimension feature, our method can get a accuracy about 93.1%, compared with the best accuracy value 79.2%. When the number of selected features reaches 300, the accuracy in our method does not change anymore. This means that our method can get a very high accuracy at the smallest number of features.

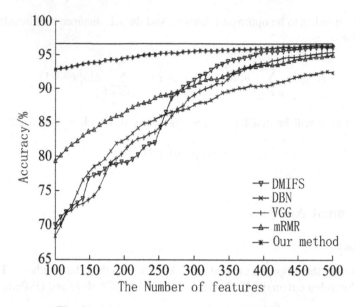

Fig. 1. Accuracy comparison for dataset COIL-20

COIL-100 have 5-times more images than COIL-20. Therefore, the five method are simulated again, and the result is shown in Fig. 2. As the same with the former experiment, our method has the best accuracy in the all five methods. The accuracy seems to be better in this experiment when the number of the features is small. Most of them can obtain an accuracy above 90% when the number of the feature is about 100.

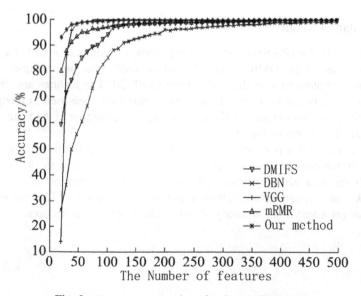

Fig. 2. Accuracy comparison for dataset COIL-100

The method with the poorest performance in accuracy is VGG. This may be caused by that VGG needs a lot of the samples to learn good feature representation. Better than the former experiment, our method obtained the accuracy 96% when the number of the samples is only 50.

The third experiment is implemented in CBCI 3000. The simulation results of five methods are shown in Fig. 3. What we can see is that all five methods have best accuracies in the three experiments. It is clearly to see that our method has the best accuracy when the number of the features is about 9, with the accuracy value of 90%. All the five methods can reach the accuracy 90% except from VGG, when the number of the features are about 10. As the former experiment, VGG perform poorest in the five all.

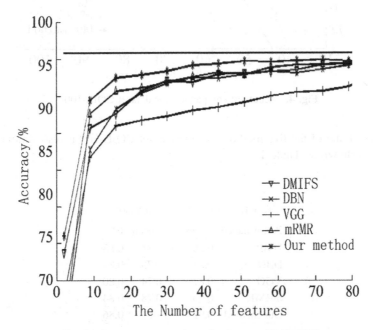

Fig. 3. Accuracy comparison for dataset CBCI 3000

The final experiment is simulated in the dataset USPSclassifyAll. Compared with the former three experiment, this experiment has most samples. Different from the former experiments, the five methods seem to have poorer performance in this experiment. Our method and mRMR can only obtain an accuracy at 80% when the number of the features is nearly 30. DMIFS, VGG and DBN have the accuracy 0 when the number of the samples is 38, 43 and 55. They can get an accuracy of 85% after the number of the samples reaches 80. This may be caused by that the number of the samples are far more than the former experiments. Therefore, it needs more features to represent the samples and then get a good learning model (Fig. 4).

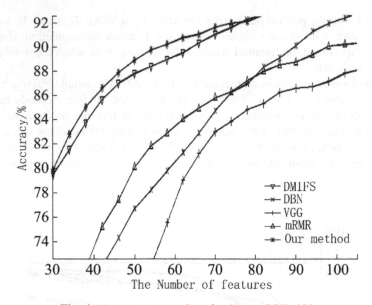

Fig. 4. Accuracy comparison for dataset COIL-100

The recall rate of the five methods for the dataset COIL-100 is also simulated, and the result is shown as Table 1.

Tab 1. Performance comparisons

Performance	Precision	Recall	F1
DMIFS	0.86	0.81	0.83
DBN	0.88	0.75	0.81
VGG	0.87	0.74	0.80
mRMR	0.90	0.78	0.84
Our method	0.92	0.81	0.86

From the Table 1, we can see our method has the best performances on precision rate, recall rate and F1.

5   Conclusion

To improve the feature extraction from the data, we propose a novel method that combines the KL divergence with feature extraction. The features are initially extracted from the primitive data by matrix decomposition. Then the features are optimized by using KL divergence. The KL divergence is introduced to the loss function to make the goal function has the shortest KL distance. The experiment is implemented in four datasets. The result shows that our method has the best performance on accuracy compared with the other methods such as SVM, DMIFS, mRMR, VGG, DBN. Our method always can get a good accuracy when the number of the features is small, which proves that our method has the better ability in extracting better feature.

The next work will be applying our method to more practical applications with enormous samples.

Acknowledgments. This work was financially Project supported by the Education Department of Hai-nan Province, project number: hnjg2017ZD-17. The Hainan Provincial Department of Science and Technology under Grant No. ZDKJ201602.

References

1. Saidi, R., Aridhi, S., Nguifo, E.M., et al.: Feature extraction in protein sequences classification: a new stability measure. Med. Phys. **6**, 683–689 (2018)
2. Chu, Q.C., Wang, Q.G., Qiao, S.B., et al.: Feature analysis and primary causes of pre-flood season "cumulative effect" of torrential rain over South China. Theoret. Appl. Climatol. **131**, 91–100 (2018)
3. Robnik, S.M., Kononenko, I.: Theoretical and empirical analysis of ReliefF and RReliefF. Mach. Learn. **53**, 23–69 (2003)
4. Liu, H., Sun, J., Liu, L., et al.: Feature selection with dynamic mutual information. Pattern Recogn. **42**, 1330–1339 (2009)
5. Hall, M.A.: Correlation-based feature selection for discrete and numeric class machine learning. In: Proceedings of the 7th International Conference on Machine Learning. Morgan Kaufmann, San Francisco, pp. 359–366 (2000)
6. Ding, C., Peng, H.: Minimum redundancy feature selection from microarray gene expression data. In: Proceedings of the IEEE Computer Society Conference on Bioinformatics, pp. 523–528. IEEE Computer Society Press, Washington DC (2003)
7. Yu, L., Liu, H.: Efficient feature selection via analysis of relevance and redundancy. J. Mach. Learn. Res. **5**, 1205–1224 (2004)
8. Tsamardinos, I., Aliferis, C., Statnikov, A.: Local causal and Markov blanket induction for causal discovery and feature selection for classification part I: algorithms and empirical evaluation. J. Mach. Learn. Res. **11**, 171–234 (2010)
9. Yang, L., Wen, K., Gao, Q., et al.: SVM based multi-label learning with missing labels for image annotation. Pattern Recogn. **78**, 307–317 (2018)
10. Simonyan, K., Zisserman, A.: Very deep convolutional networks for large-scale image recognition. arXiv preprint arXiv:1409.1556 (2014)
11. Szegedy, C., Liu, W., Jia, Y., et al.: Going deeper with convolutions. In: IEEE Conference on Computer Vision and Pattern Recognition, pp. 1–9 (2015)
12. He, K., Zhang, X., Ren, S., et al.: Deep residual learning for image recognition. In: IEEE Conference on Computer Vision and Pattern Recognition (2016)
13. Huang, G., Liu, Z., Weinberger, K.Q.: Densely connected convolutional networks. In: IEEE Conference on Computer Vision and Pattern Recognition (2017)
14. Han, Q.L., Liang, S., Zhang, H.L.: Mobile cloud sensing, big data, and 5G networks make an intelligent and smart world. IEEE Netw. **29**(2), 40–45 (2015)
15. Song, Y., Wang, H., Li, J., Gao, H.: MapReduce for big data analysis: benefits, limitations and extensions. In: Che, W., et al. (eds.) ICYCSEE 2016. CCIS, vol. 623, pp. 453–457. Springer, Singapore (2016). https://doi.org/10.1007/978-981-10-2053-7_40

Evaluation of Scholar's Contribution to Team Based on Weighted Co-author Network

Xinmeng Zhang[1,2,3(✉)], Xinguang Li[1], Shengyi Jiang[2,3], Xia Li[2,3],
and Bolin Xie[2,3]

[1] Laboratory of Language Engineering and Computing,
Guangdong University of Foreign Studies, Guangzhou 510006,
Guangdong, China
xmzhang@mail.gdufs.edu.cn

[2] Non-universal Language Intelligent Processing Laboratory,
Guangdong University of Foreign Studies, Guangzhou 510006,
Guangdong, China

[3] School of Information Science and Technology,
Guangdong University of Foreign Studies,
Guangzhou 510006, Guangdong, China

Abstract. The contributions of scientific researchers include personal influence and talent training achievements. In this paper, using 9964 high-quality co-author scientific papers in English teaching research from China citation database from 1997 to 2016, a weighted coauthor network with variety factors is constructed. A model was proposed to calculate the author's contribution to the research team by combining personal and network characteristics. The results reveal a variety of characteristics of the co-author networks in English teaching research field, including statistical properties, community features, and authors' contribution to teams in this discipline.

Keywords: Social network analysis · Co-author network · Research team · Academic contribution

1 Introduction

Scientific collaboration is a very important method for scholars to trigger innovations. The amount of scientific research cooperation, cooperation rate and cooperation scope show an increasing trend year by year. A scientific collaboration networks is a set of scientists which has partnership to some of the others, two scientists are considered connected if they have coauthored one or more papers together [1]. Many scientists attempt to study the nature, structure and laws of scientific cooperation networks from various angles. Newman [2] conducted a detailed study of the cooperative networks in the four databases in SCI from the perspective of complex networks, and revealed the small world phenomenon and the phenomenon of scale-free phenomena in each cooperative network. Acedo [3] carried out an exploratory analysis of co-authorships in the field of management, results show a growing tendency of the co-authored papers in the field of management, similar to what can be observed in other disciplines. Velden

© Springer Nature Singapore Pte Ltd. 2019
X. Cheng et al. (Eds.): ICPCSEE 2019, CCIS 1058, pp. 52–61, 2019.
https://doi.org/10.1007/978-981-15-0118-0_5

[4] studied the relationship between the co-author network from mesoscopic level. They combine qualitative methods (participant interviews) with quantitative methods (network analysis) and demonstrate the application and value of their approach in a case study comparing three research fields in chemistry.

Many scholars use a collaborative network to evaluate scientific research teams and researchers. Jiancheng [5] present an improved DEA model to evaluate the quality of research team. Ordóñez-Matamoros [6] studied the International Scientific Collaboration team performance of in Colombia. Liu [7] proposed enhanced network-based approach to increase discrimination in data envelopment analysis. Tone [8] proposed a dynamic DEA model involving network structure in each period within the framework of a slacks-based measure approach. The influence analysis of scientific research scholars is more concerned. Chen [9] confirmed in the coauthor network that the higher the node aggregation coefficient, the smaller the influence of the node. Xiang [10] studied the relationship between authoritativeness and influence of users in the co-author network, and introduced prior knowledge to improve PageRank's measurement of user authority.

At present, the evaluation of the contributions of scientific researchers is basically from the academic point of view. However, the achievements of scientific researchers in talent training are ignored. Based on the traditional influence analysis, we introduce the analysis of the achievements of the partners through the co-author network, and comprehensively evaluate the contribution of the researchers to the team. This paper proposes a weighted co-author network model based on the number of co-authors, author rankings, and citation frequency, and builds a weighted co-authors network based on English teaching research papers in the China Citation Database from 1997 to 2016. We study a variety of statistical properties of the weighted co-authors networks, and propose a model for individual contribution to the team based on personal characteristics and network characteristics. We reveal the characteristics of the author's collaborative network in the field of English teaching research and discover researchers who have played an important role in this research field.

2   The Weighted Co-author Network

2.1   Data Source

We collected high quality cited papers in English teaching research field from 1997 to 2016 in Chinese citation database. The raw data contains a list of papers, including the author's name and other information such as abstract, title, publish date, unit, and more. A total of 33,843 papers involving 29,173 authors and 4,994 scientific research units, of which 23,879 papers have the sole author and 9,964 papers have one or more authors. Only about 34% of papers have co-authors, which means that the authors in the field of English teaching have a low level of scientific cooperation. We calculate the distribution p_k of numbers k of papers per author, and the results show that the distribution followed a power law, as shown in Fig. 1.

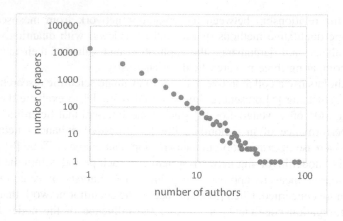

Fig. 1. Histograms of the numbers of papers per author in English teaching research area

The distribution shows that many authors have only published a small number of papers, and the "Fattail" made up of a handful of authors who have published a large number of papers. More than 10,000 authors only publish one paper. It shows that most Chinese authors have weak research ability in the field of English teaching research.

The distribution of total number of citations per author is shown in Fig. 2. The x-axis represents the total number of citations, and the y-axis represents the number of authors. It shows that only a small number of authors have a high number of citations. The authors with the highest number of citations have been cited more than 10000 times.

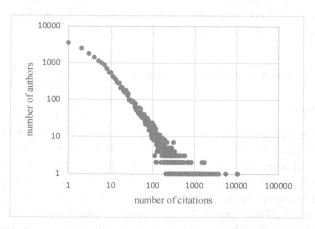

Fig. 2. Distribution of the number of citations per author in English teaching research area

2.2 Weighted Weighted Co-author Network

Of the 33,843 papers, 23,879 papers were independently completed by the only author, so a network was constructed based on other 9,964 papers. Taking the Author as the vertex, edges are added between each pair of authors on each paper. A total of 29,173 authors, of whom 14,831 have not collaborated with other authors in their papers, other 14342 authors have 12106 edges of cooperation.

Newman [1] calculates the weight between authors based on the number of cooperation and the number of partners. Börner [11] calculated edge weights based on the number of co-authors and the number of citations, The calculation model of the edge weight between node i and node j is as follow

$$w_{ij} = \sum \frac{1 + c_p}{n_p * (n_p - 1)} \tag{1}$$

Where n_p is the number of authors of a collaborative paper, c_p is the number of citations of the paper. But the weighted model ignores the author's ranking, and the number of citations affects the weight of edges too much.

In this paper, we propose a weighted network model that takes into account the number of authors, the author's ranking, and the number of citations, while using logarithm to reduce the excessive impact of the number of citations. The calculation formula is as follow

$$w_{ij} = \sum \frac{2 * n_p + 2 - t_i\_t_j}{n_p * (n_p + 1)} \ln(c_p + e) \tag{2}$$

Where n_p is the number of authors of a collaborative paper, c_p is the number of citations of the paper, t_i is the author i ranked in the paper, t_j is the author j ranked in the paper, e is natural number.

We studied some basic statistical features of the weighted co-author network, the results are shown in Table 1. Obviously, the weighted co-author network is a sparse one.

Table 1. The characteristics of weighted co-author network.

Characteristic	Value
Number of vertices	14342
Number of edges	12106
Average degree:	1.688
Average weighted degree	1.914
Average clustering coefficient	0.719
Total triangles	3587
Number of weakly connected components	4861
The size of max connected components	91
Maximum degree	27
Maximum weighted degree	50

3   The Contribution of Scholar to Team

3.1   Community Structure Characteristics

This coauthor network is a non-connected network graph, which consists of many connected sub-networks, showing a high degree of disconnection. It shows that the research teams in the field of English teaching research are scattered and lack of enough interlinkages between teams. We use community detection algorithm [12] to identify the communities in the coauthor network. The coauthor network includes a total of 4861 communities. The team size distribution is shown in Fig. 3.

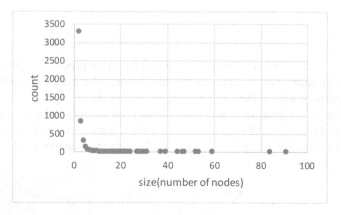

Fig. 3. Team size distribution

The distribution of community size presents power-law distribution characteristics, and most communities are very small, especially two-thirds of them have only two vertices. The largest community is from Beijing Foreign Studies University, with a total of 91 members and 129 edges.

The community structure of the research team presents a hybrid structure with a mesh inside and a star outside. The core members of research team cooperate closely with each other and form a mesh structure, and the core members cooperate with other members to form a star structure. The tight mesh structure indicates the scientific research cooperation between researchers, and the star shape is generally the relationship between teachers and students. The structure of the largest community and the second largest community are shown in Figs. 4 and 5.

Fig. 4. The largest community is from Beijing Foreign Studies University, with a total of 91 members and 129 edges.

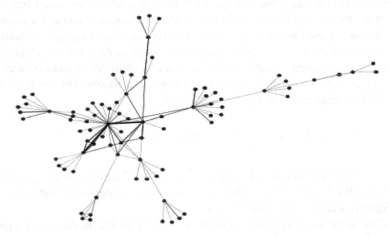

Fig. 5. The second largest community is from Tsinghua University, with a total of 84 members and 119 edges.

3.2 Scholar's Contribution Calculation Model

Leaders of scientific research play an important role in the development of scientific research teams. There are many ways to assess a scholar's contribution, including personal characteristics, network topology, relationships between individuals, etc. [13]. But each method has its advantages and disadvantages. For example, the influence measurement method based on random walk uses neighbor nodes to describe the influence of the nodes, avoiding noise, but ignoring the nature of the nodes themselves [10, 14]. The method based on individual characteristics only considers individual local characteristics and ignores neighbor's network location. Therefore, we evaluate the author's influence and contribution to the team from multiple factors such as the author's paper, collaboration, and network location. We not only consider the influence of individual scholars, but also the contribution of scholar to the team. How to evaluate the scholars' effectiveness in cultivating other members of the team? We evaluate the contribution of leading scholars by the quality of collaborative papers. The calculation formula is as follow

$$b_{ij} = \frac{c_{ij}}{c_j} \tag{3}$$

Where b_{ij} indicates the benefit of author j from author i, c_{ij} is the total number of citations of papers co-authored by author i and author j, c_j is the total number of citations of papers of author j.

The total benefit of author i to team is as follow

$$b_i = \sum_{j \in CN(i)} b_{ij} \tag{4}$$

Where $CN(i)$ is the neighbors of node i in the community that node i belongs to.

According to the weighted co-author network, we propose a model for calculating the contribution of authors by synthesizing various methods. Pagerank considers the global structure of the network and is suitable for large-scale networks. Degree is the most direct representation of the author's influence in team. Total number of citations of author is the representative of the quality of personal papers. The mode for calculating scholar's contribution to team is as follows

$$contribution_i = \frac{rank_i}{max_{rank}} * \ln(d_i + e) * \ln(c_i + e) * b_i \tag{5}$$

Where $rank_i$ is the weighted PageRank [15] value of author i, max_{rank} is the max weighted PageRank value of all authors, d_i is the weighted degree of author i, c_i is the total number of citations of author i's papers.

The distribution of contribution of authors is shown in Fig. 6, the x-axis represents the value of contribution, and the y-axis represents the count of the contribution value. The distribution conforms to the power law characteristic, and only a few authors have high contribution value.

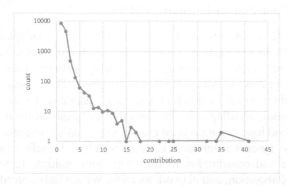

Fig. 6. Distribution of the contribution of authors in English teaching research area

Tables 2 and 3 show the top ten citation authors and the top 10 high contributors, where d is the weighted degree, Nc is the number of citations, Pr is value of pagerank, b is the total benefit of author to team, C_1 is the contribution value without considering the benefit of author to team in non-weighted Co-Author Network, C_2 is the contribution value according to the formula 5 in weighted Co-Author Network. This article discusses the author's contribution to the team, so if the author works in different units at different times and belongs to different research teams, we calculate the author's contribution value to different teams. Interestingly, the first author is the same person as the sixth author in Table 3. In fact, she worked at Nanjing University before 2005 and entered Beijing Foreign Studies University after 2005. She played a important role in both research teams. Some scholars have extraordinary personal achievements, but they are relatively weak in the construction of scientific research teams, so the contribution to the team is relatively low. For example, Jigang Cai's paper is ranked second in total citations, but only six of his authors have worked with him, so his contribution to the team is lower. Although some authors have high citations, there are too few collaborators and the location of the network is not important, so their contribution to the team is low. In Table 3, Liu Yongbing has only 733 quotations, but his weighed degree is very high, which indicates that he works closely with other authors. The higher b value indicates that he is of great help to improve the quality of collaborators' papers. Only 4 of the 10 authors with the highest number of citations appear in Table 3.

Table 2. Top 10 high citation authors in English teaching research.

Author	Work unit	d	wd	Nc	Pr	b	C_1	C_2
QiuFang Wen	NJU	14	103	15736	0.00037	9.1	14.15	213.1
Jigang Cai	Fudan	6	32	11111	0.00023	1.8	6.58	19.4
Lifei Wang	UIBE	17	106	6232	0.00058	10.1	21.33	338.8
QiuFang Wen	BFSU	21	99	6130	0.00062	10.2	24.38	363
Weidong Dai	SISU	10	84	5193	0.0003	6.5	9.26	105.6
Ruiqing Liu	BFSU	3	22	4874	9.4E−05	1.9	1.97	6.9
Dingfang Su	SISU	10	44	3963	0.00034	5.6	10.02	84.8
Wenyu Wang	NJU	8	55	3902	0.00019	3.2	5.36	29.3
Chuming Wang	gdufs	4	20	3859	0.00015	2.1	3.43	11.8
Yan Zhou	BFSU	8	50	3856	0.00023	4	6.27	41.9

Table 3. Top 10 high contributing authors in English teaching research.

Author	Work unit	d	wd	Nc	Pr	b	C_1	C_2
QiuFang Wen	BFSU	21	99	6130	0.00062	10.2	24.38	363.0
Tongshun Wang	SJTU	22	94	1683	0.00071	10.3	23.85	350.1
Lifei Wang	UIBE	17	106	6232	0.00058	10.1	21.33	338.8
LiSheng Luo	Tsinghua	24	118	1619	0.00066	9.9	22.59	326.3
Jinfen Xu	HUST	18	131	3788	0.00054	9.5	19.06	292.4
Qiufang Wen	NJU	14	103	15736	0.00037	9.1	14.15	213.1
Wenbin Wang	NBU	16	65	2070	0.00051	9.1	16.15	211.5
Kaibao Hu	SJTU	16	58	1029	0.00053	9.2	15.17	195.6
Yongbing Liu	NENU	12	82	733	0.0004	8.4	10.11	140.2
Weidong Dai	SISU	10	84	5193	0.0003	6.5	9.26	105.6

4 Conclusion

Through the analysis of the weighted co-author network in english teaching research area, we draw the following conclusions.

- We propose a weighted co-author network model based on multiple factors. The co-author network is a non-connected graph, and scientific research cooperation in the field of English teaching research is not enough.
- We propose a computational model of individual contribution to team research based on individual characteristics and network structure characteristics. We find that the probability distribution of personal contribution presents a power-law distribution feature. Only by publishing a high-quality paper and working closely with team members can we obtain high contribution values.

Only by strengthening team building, adjusting research and production relations, and forming a close teamwork relationship can we promote the development of scientific research productivity.

The parameters in our model need to be further improved, and more evidences need to be found to evaluate the experimental results.

Acknowledgments. This research is supported by the National Natural Science Foundation of China (No. 61402119).

References

1. Newman, M.E.J.: Scientific collaboration networks I: network construction and fundamental results. Phys. Rev. E **64**(1), 016131 (2001)
2. Newman, M.E.J.: The structure and function of complex networks. Soc. Ind. Appl. Math. Rev. **45**(2), 167–256 (2003)
3. Acedo, F.J., Barroso, C., Casanueva, C., et al.: Co-authorship in management and organizational studies: an empirical and network analysis. J. Manag. Stud. **43**(5), 957–983 (2006)
4. Velden, T., Haque, A., Lagoze, C.: A new approach to analyzing patterns of collaboration in co-authorship networks: mesoscopic analysis and interpretation. Scientometrics **85**(1), 219–242 (2010)
5. Jiancheng, G., Junxia, W.: Evaluation and interpretation of knowledge production efficiency. Scientometrics **59**(1), 131–155 (2004)
6. Ordóñez-Matamoros, H.G., Cozzens, S.E., Garcia, M.: International co-authorship and research team performance in Colombia. Rev. Policy Res. **27**(4), 17 (2010)
7. Liu, J.S., Lu, W.M.: DEA and ranking with the network-based approach: a case of R&D performance. Omega **38**(6), 453–464 (2010)
8. Tone, K., Tsutsui, M.: Dynamic DEA with network structure: a slacks-based measure approach. Omega **42**(1), 124–131 (2014)
9. Chen, D.B., Gao, H., Lü, L.Y., Zhou, T.: Identifying influential nodes in large-scale directed networks: the role of clustering. PLoS One **8**(10), e77455 (2012)

10. Xiang, B., Liu, Q., Chen, E., Xiong, H., Zheng, Y., Yang, Y.: PageRank with priors: an influence propagation perspective. In: Proceedings of the 23rd International Joint Conference on Artificial Intelligence (IJCAI 2013), pp. 2740–2746. AAAI Press, Menlo Park (2013)
11. Börner, K., Dall'Asta, L., Ke, W., et al.: Studying the emerging global brain: analyzing and visualizing the impact of co-authorship teams. Complexity **10**(4), 57–67 (2005)
12. Blondel, V.D., Guillaume, J.-L., Lambiotte, R., Lefebvre, E.: Fast unfolding of communities in large networks. J. Stat. Mech. Theory Exp. (10), P1000 (2008)
13. Han, Z.M., Chen, Y., Liu, W., Yuan, B.H., Li, M.Q., Duan, D.G.: Research on node influence analysis in social networks. Ruan Jian Xue Bao/J. Softw. **28**(1), 84–104 (2017)
14. Li, Q., Zhou, T., Lü, L., Chen, D.: Identifying influential spreaders by weighted LeaderRank. Phys. A **404**, 47–55 (2014)
15. Ding, Y.: Applying weighted PageRank to author citation networks. J. Assoc. Inf. Sci. Technol. **62**(2), 236–245 (2011)

Community Evolution Based
on Tensor Decomposition

Yuxuan Liu[✉], Guanghui Yan, Jianyun Ye, and Zongren Li

School of Electronic and Information Engineering, Lanzhou Jiaotong University,
Lanzhou 730070, China
61050049@qq.com

Abstract. With the high dimensionality of data, the method of tensor decomposition has attracted much attention in the field of data research and analysis. The tensor decomposition is well reflected in the study of high-dimensional data. The existing research uses the results of tensor decomposition to conduct community discovery. Based on the existing research, this paper presents a method to study community evolution using the results of tensor decomposition. The feature matrix obtained by the tensor decomposition algorithm was analyzed, and the real-time activity of the community with the feature matrix with time slice direction was studied to obtain the event process of community evolution. Experimental results in real data sets show that this method can well analyze dynamic events in the dataset and community evolution events.

Keywords: Tensor · CP decomposition · Community discovery ·
Community evolution · Common theme

1 Overview

All The rapid development of computers and networks has led to an increase in the amount of online information and data, forming a social system for online social networks. Any structure that contains a large number of individual units and individuals can be studied as a complex network. Watts and Strogatz [1] first proposed the small world theory, and Barabasi and Albert [2] proposed the scale-free network (BA) model, which became the founder of complex network research. The community structure [3] in complex networks is one of the basic characteristics of the network. Therefore, mining communities has become one of the important research directions of complex network analysis.

Most networks in the real world, such as common social networks, blogs, Twitter, WeChat, etc., as well as common social relationships, such as literature citation networks, movie evaluation networks, community media networks, etc. due to frequent changes and some of them Frequent activities, community structure will gradually evolve in the network, these networks were defined by Newman in 2003 as a dynamic network [4]. In a dynamic network, communities may grow, shrink, form, or disappear, and the community ownership of individual nodes may also change frequently. In the current research, in the community detection research of social networks, the division and analysis of static networks is the mainstream direction [5]. This method obviously

© Springer Nature Singapore Pte Ltd. 2019
X. Cheng et al. (Eds.): ICPCSEE 2019, CCIS 1058, pp. 62–75, 2019.
https://doi.org/10.1007/978-981-15-0118-0_6

does not take into account the changes of nodes and the evolution of communities between time and time. The research of dynamic networks has only attracted the attention of researchers in recent years [6]. Dynamic networks are a three-dimensional structure composed of static networks of several time slices. In the existing research, tensor as a multi-dimensional structure Express this data and can be used for research in community discovery [7].

As a form of saving data, tensor can preserve the unique structural characteristics of the data and the high dimensional relationship of the data itself. In the research of dynamic networks, the dimensions of data are generally high. The method of tensor decomposition is similar to the method of matrix decomposition. It can map difficult-to-handle high-dimensional data into lower data space, and achieve dimensionality reduction and compression data. The purpose of algorithm complexity. Tensor decomposition exploits the main idea of approximating the original tensor by a low-rank decomposition, extending many methods of tensor decomposition. Among them, the widely used methods are CP decomposition and Tucker decomposition [8]. The CP tensor decomposition is a general term for two equivalent models of Parallel Factor-ization (PARAFAC) and Canonical Decomposition (CANDECOMP) proposed by Carroll and Chang [9] by Harshman [10] et al.

The method of tensor decomposition has been widely used in many aspects in today's research, such as signal data processing, visual computer research, data mining statistics, image analysis, etc., and various research directions in data mining have different applications. Such as personalized recommendation of multi-dimensional networks, node importance ranking and community discovery [11]. However, there is no good research method in the study of community evolution. On the basis of community discovery, it is a new research point to study the changes of communities at different moments through the results and methods of community discovery and the results of tensor decomposition.

In this paper, the data is degraded by CP tensor decomposition, and different feature matrices are obtained. The tensor decomposition results in three directions of the data set are used to analyze the feature matrix with time slice direction and give the eigenvalue based on tensor. Divide the criteria for the community, calculate the activity of each community over time, and analyze the communities and nodes with higher activity to get the main community of the network evolution within the dynamic community. This approach can get information about changing communities. In the study of community evolution, the complex methods of tracking each node and community are eliminated, greatly simplifying the complexity of community evolution research. This article uses 2 real data and evaluates. The events of community evolution (maintenance, disappearance, integration, division, addition, reduction) are judged by studying the similarity changes in the active time of the active community.

2 Experimental Related

Tensor is a general term for high-dimensional arrays. A common one-dimensional array is called a vector, and a matrix is a two-dimensional array. When the dimension is three-dimensional and above, this high-dimensional array is called a high-order tensor (Fig. 1).

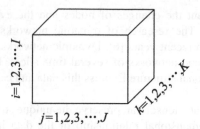

Fig. 1. Third-order tensor $x \in \mathbb{R}^{I \times J \times K}$

In Fig. 2, the modulo-n matrix tensor is called the tensor matrix expansion form (Matricization) [8], which divides the tensor according to the method of recombining into different matrices. For example, a $2 \times 3 \times 2$ third-order tensor x can be expanded to obtain three matrices, which are respectively 2×6 modulo-1 matrix $X_{(1)}$, 3×4 modulo-2 matrix $X_{(2)}$ and 2×6. Mode-3 matrix $X_{(3)}$.

Definition: The result of the nth dimensional $x \in \mathbb{R}^{I_1 \times I_2 \times \cdots \times I_N}$ expansion of the Nth order tensor can be expressed as Eq. (1)

$$X_{(n)} \in \mathbb{R}^{I_n \times (I_1 \cdots I_{n-1} I_{n+1} \cdots I_N)} \tag{1}$$

$$X_{(1)} = \begin{bmatrix} 1 & 2 & 3 & 7 & 8 & 9 \\ 4 & 5 & 6 & 10 & 11 & 12 \end{bmatrix}$$

$$X_{(2)} = \begin{bmatrix} 1 & 4 & 7 & 10 \\ 2 & 5 & 8 & 11 \\ 3 & 6 & 9 & 12 \end{bmatrix}$$

$$X_{(3)} = \begin{bmatrix} 1 & 4 & 2 & 5 & 3 & 6 \\ 7 & 10 & 8 & 11 & 9 & 12 \end{bmatrix}$$

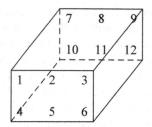

Fig. 2. The Matricization of $x \in \mathbb{R}^{2 \times 3 \times 2}$

The Fig. 2 is a way of expressing the product of the tensor and the matrix. The result of the tensor-n product is still expressed as a tensor.

Definition: If the Nth order tensor $x \in \mathbb{R}^{I_1 \times I_2 \times \ldots \times I_N}$ and the matrix $U \in \mathbb{R}^{J \times I_n}$ have a modulo-n product, it will be expressed as follows:

$$y = x \times_n U \qquad (2)$$

In Eq. (2) $y \in \mathbb{R}^{I_n \times (I_1 \ldots I_{n-1} J I_{n+1} \ldots I_N)}$, it can also be expressed in the form of a matrix as Eq. (3):

$$Y_{(n)} = U X_{(n)} \qquad (3)$$

The tensor CP decomposition (CANDECOMP/PARAFAC) mainly performs the decomposition tensor by decomposing a number of order tensors to obtain a finite number of rank-one tensor sums [8] as Fig. 3.

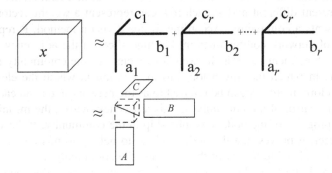

Fig. 3. The CP decomposition of a 3-order tensor

For example, the third-order tensor $x \in \mathbb{R}^{I \times J \times K}$, the decomposition of the tensor is expressed as Eq. (4):

$$x \approx \sum_{r=1}^{R} a_r \circ b_r \circ c_r \qquad (4)$$

Shown in the above formula is the outer product of the rank one, and the rank of the tensor is expressed as R, where $a_r \in \mathbb{R}^I$, $b_r \in \mathbb{R}^J$, $c_r \in \mathbb{R}^K$, $r = 1, \ldots, R$. Can be expressed as Eq. (5):

$$x_{ijk} \approx \sum_{r=1}^{R} a_{ir} b_{jr} c_{kr} \qquad (5)$$

If the vectors in the rank one can be respectively composed into different matrix forms, the matrix can be called a factor matrix of tensors. For example, $a_r \in \mathbb{R}^I$ a factor matrix $A = [a_1, a_2, \ldots, a_R] \in \mathbb{R}^{I \times R}$ that can form a target tensor. Similarly, a third-order tensor can obtain two other factor matrices $B \in \mathbb{R}^{J \times R}$, $C \in \mathbb{R}^{K \times R}$.

If the normalization factor matrix is normalized, the normalization coefficient can be obtained $\lambda \in \mathbb{R}^R$. In summary, the tensor CP decomposition can be expressed as Eq. (6):

$$x \approx \sum_{r=1}^{R} \lambda_r a_r \circ b_r \circ c_r = \|\lambda; A, B, C\| \tag{6}$$

In a dynamic network with time slices, the results are obtained by tensor CP decomposition $A = [a_1, a_2, \ldots, a_R] \in \mathbb{R}^{I \times R}$, $B \in \mathbb{R}^{J \times R}$ $C \in \mathbb{R}^{K \times R}$, and the three feature matrices have different meanings [12]. In the undirected graph, the adjacency matrix of each time slice network in the dynamic network with time slices has a symmetrical structure. In general, the obtained adjacency matrix A = B. In matrix A, the elements of each row represent different nodes, each row can represent a weight vector, and each column represents a community. According to this division method, R represents the total number of network communities in this time slice. In the adjacency matrix, the meaning of each element value is the weight or weight of the community represented by the column in which the row belongs to the column in which the element value belongs. Therefore, matrix A can be referred to as the network node and each network. A membership matrix of a community. In this adjacency matrix, the meaning of each weight is the progress of the node's membership in the community. If the value of the element is larger, it proves that the node is likely to belong to this community. After obtaining the membership matrix of the node and the community, it is next to find the column with the largest weight of each node in the row, and then the node can belong to the node of the community, or use the aggregation under the premise of obtaining the membership matrix. The class algorithm can use the membership matrix as the feature matrix of the node in the network when clustering. The algorithm can use some classical clustering algorithms, such as K-means algorithm, to cluster and get the result of community division. In the research of this paper, matrix C is mainly used. The rows in matrix C represent the network layer of each time slice. Each column represents the community. In the C matrix, the meaning of the element value is in the network layer. The evolutionary events generated at this point in time contribute to the activity of the community or to the changes in the community. Find the most active community on the film at a certain time (that is, the maximum value on each column), and compare the time-slice similarity of the community to get the event of community evolution and analyze it.

A_{ij} Indicates the i-th row and j-th column elements in the matrix A obtained by the tensor CP decomposition. Since the matrix A and the matrix B are completely equal in the result of the tensor CP decomposition in the network with time slices, A_{ij} also represent The elements of the i-row j-column of the feature matrix B. Similarly, C_{ij} the i-th row and the j-th column element in the feature matrix C are represented.

In the network with time slices, the application and improvement of modularity formulas provide powerful help for the study of community evolution of dynamic networks, community discovery, and provide evaluation indicators for multi-dimensional community division and community evolution issues such as dynamic networks. On this basis, Zhang et al. [13] proposed a further derivation of modularity in the research of multi-layer networks. Sarzynska et al. [14] scholars proposed a zero model.

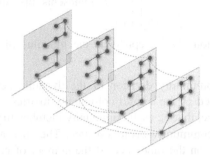

Fig. 4. Module degree diagram of multi-chip network

According to the principle of modularity formula of static network in Fig. 4, Mucha et al. [12] proposed a modularity formula suitable for dynamic networks or networks with high latitude as Eq. (7).

$$Q_{multislice} = \frac{1}{2\mu} \sum_{ijsr} \left[\left(A_{ijs} - \gamma_s \frac{k_{is}k_{js}}{2m_s} \right) \delta_{sr} + C_{jsr}\delta_{ij} \right] \delta(g_{is}, g_{jr}) \tag{7}$$

If the multi-dimensional network is regarded as a dynamic network, A_{ijs} is the connection matrix formed by the snapshot of the network at time s, K_{is} indicating the degree of node i (s time), indicating the number of connected edges (s time) shared in the network, C_{jsr} indicating the node Whether j and r have a connection relationship (s time), as shown in the above figure, the dotted line indicates the connection relationship between the two nodes, g_{is} indicating the community to which the node i belongs (s time), if $\delta_{sr}\, r = s$ is satisfied, $\delta_{sr} = 1$, otherwise $\delta_{sr} = 0$. The above modularity formula can be divided into two parts. The first part is basically similar to the modularity calculation formula in the static network, mainly to ensure that each time slice in the multi-dimensional network and the dynamic network is divided into distinct community structures, the second part Mainly to show the connection between dynamic networks on-chip at different times.

It can be obtained from the tensor decomposition formula, because the factor matrices A, B, and C are composed of modulo-n matrices in three directions, and the singular value vectors obtained by tensor singular value decomposition are sequentially composed, so these three matrices are The relationship of each vector is orthogonal to each other. Available from A = B:

$$Q_m = \sum_{i=1}^{R} a_i u_i^T U \times \hat{\lambda}_s \times U^T \sum_{j=1}^{R} a_j u_j = \sum_{i=1}^{R} a_j u_j = \sum_{i=1}^{R} a_i^2 \left\| \hat{\lambda}_s \right\|_i, \tag{8}$$

In Eq. (8), $\hat{\lambda}_s = \begin{bmatrix} \lambda_1 & & \\ & \ddots & \\ & & \lambda_R \end{bmatrix} \begin{bmatrix} t_{s1} \\ \vdots \\ t_{sR} \end{bmatrix}$ represents the normalization coefficient

after tensor decomposition, $\left\| \hat{\lambda}_s \right\|_i$ represents the modulus of the ith column of the coefficient matrix $\hat{\lambda}_s$.

Since the overall community division of the dynamic network is obtained, the divided results are defined as the overall community metrics, and each community is analyzed C_{ij} by time slice, that is, each community membership value j is traversed to obtain the most active community evolution time. The slice analyzes the number of nodes in the community on the time slice. If the number of nodes is greater than the number of metric nodes, the community performs a growth event, and similarly, the event can be reduced to form an event and a disappearance event.

Based on the above relationship and formula, this paper proposes a dynamic network module degree division based on the results obtained by tensor CP decomposition in dynamic networks. The community is found based on the activity change of each community along the time axis. Evolution events, tracking the community at the current time, and comparing the communities based on modular tensor discovery to determine whether the community evolution event is one of growth, reduction, formation, or disappearance.

If the time series network is $G = \{G_{(1)}, G_{(2)}, \ldots, G_{(T)}\}$, $G_{(T)}$ is the network snapshot at time t, all the adjacency matrices on each time slice are combined into tensors $g \in \mathbb{R}^{N \times N \times T}$.

Dynamic network community and time slice community similarities as Eq. (9):

$$sim(C_i, C_{ij}) = \begin{cases} \dfrac{|V_i \cap V_{ij}|}{\max(|V_i|, |V_{ij}|)}, & if \ \dfrac{|V_i \cap V_{ij}|}{\max(|V_i|, |V_{ij}|)} \geq k \\ 0, otherwise \end{cases} \tag{9}$$

Community similarity is divided by different communities in the same community, and the number of members is used to determine whether dynamic communities and a certain time slice community are similar. For example, a dynamic community shares the same 25-node membership with the same community of a time slice, and if the similarity threshold k is less than 0.2, the two communities are similar communities (e.g., 25/100 > 0.25). The selection of the threshold k is based on the relevant characteristics of different networks. The threshold of the similarity k of the dynamic society is different. The evolution of the community over time and the number of evolved communities will be different.

Active communities C_{ij} in time j. Comparing dynamic community-based counterparts, they experience different evolutionary events.

Formation: If a community C_{ij} at an active time point does not find a matching community in the dynamically divided community, then an event occurs in the community at time j.

Dissolution: If the community dynamic division results C_i can not find a similar community C_{ij} on the j time film, then the community dissolution event occurred at time j.

Survival: If no active community is found on the J time slice, there is no evolution event on the time slice. The time slice community is basically consistent with the dynamically divided community results, and the community remains alive.

Splitting: If the ratio of several communities at an active time point, and the nodes that dynamically divide the community is at least k, then the community splits into a group of communities at j time. In order to prevent the members from leaving the network, the community gathers. The ratio of members to and should be greater than k.

Merger: If the community results are dynamically divided, and the proportion of active community members in several communities and j time slices is at least k, then the active community at j time is synthesized by the community, and also to prevent most of the members from previously not existing. In the case, the ratio of community members to active community members should be greater than k.

Growth and attenuation: If the community at the active time and the dynamic division correspond to the community, the proportion of the common members is at least k, when the number of active community nodes is greater than the number of dynamic community nodes, instant $N_{ij} > N_i$, community growth, and vice versa $N_{ij} < N_i$ The community decays at the point in time.

3 Experiment and Analysis

This section experiments through two real data sets to verify the feasibility and efficiency of the algorithm. The two real data sets verified are the PSTN data set [15] and the Facebook wall posts data set.

The PSTN (primary school temporal network) data set is the information data of the contact network generated by the activities of the students of 10 classes of a school within 8 days. The contact information data is recorded once every 20 s through the sensor on the student. The time when the information is generated is included in the record of the information. If there is a certain relationship between the two students, their id and the class to which the two students belong will be recorded. The data was obtained by observing contact information among 242 students in 8 days, and the total number of recorded data generated was 125,733. The information of the data set is shown in the figure, and each record is (time, student 1 id, student 2 id, student 1 class, student 2 class). The part of the data are in Fig. 5.

"31220	1560	1570	3B	3B"
"31220	1567	1574	3B	3B"
"31220	1632	1818	4B	4B"
"31220	1632	1866	4B	4B"
"31220	1673	1698	1B	1B"
"31220	1819	1836	4B	4B"
"31220	1819	1866	4B	4B"
"31240	1558	1567	3B	3B"
"31240	1567	1574	3B	3B"
"31240	1632	1818	4B	4B"
"31240	1632	1866	4B	4B"
"31240	1673	1698	1B	1B"
"31240	1741	1820	3A	3A"
"31240	1809	1822	3A	3A"
"31240	1819	1836	4B	4B"
"31260	1558	1564	3B	3B"
"31260	1558	1567	3B	3B"
"31260	1560	1570	3B	3B"
"31260	1564	1567	3B	3B"
"31260	1564	1574	3B	3B"
"31260	1567	1574	3B	3B"

Fig. 5. PSTN part of the data

The Facebook wall posts dataset is a directional network of small posts on the Facebook friend's wall. The nodes of the network are Facebook users, each directed edge represents a post, and the user who composes the post is linked to the user who wrote the wall of the friend with the post. Since users can write multiple posts on a friend's wall, the network allows a single node to connect multiple edges. Since the user can post on their own wall, the network contains loops.

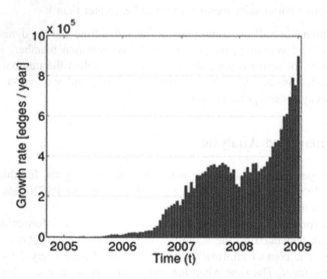

Fig. 6. Facebook friend wall posting time distribution

Figure 6 shows the time distribution growth table of Facebook's friend wall posting. It can be seen that the number of posts tends to increase due to the change of time. The data set contains 46,592 users, 876,993 edges, that is, the number of posts posted on the wall of the friend's user. The experiment intercepts the 2009 post data and divides the time slice into 12 months for experimentation.

In the experiment of discovering the PSTN dataset community, the rank of the tensor is selected to be 20 in the CP tensor decomposition. When the factor matrix obtained by the decomposition is used to divide the community, the number of clusters in the k-means algorithm is 10, so 10 is obtained. A real community, the number of members in each community is around 20.

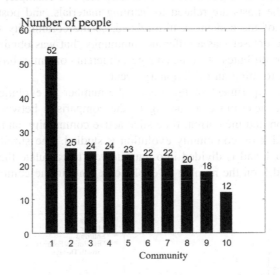

Fig. 7. PSTN dataset community population distribution

Figure 7 above shows the number of communities in the PSTN dataset obtained by the tensor decomposition discovery community algorithm. The above experimental results show that because the activities between members in class 3A and class 4B are relatively frequent, when clustering, the two are easily combined into one community.

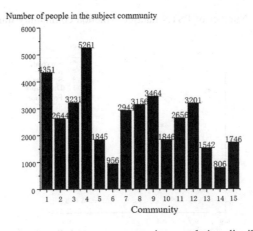

Fig. 8. Friend wall data part community population distribution

Figure 8 above shows the experiment on the Facebook wall posts dataset, when the factor matrix obtained by CP tensor decomposition is used for community division, the number of clusters obtained by k-means algorithm is 65, so 65 real community divisions are obtained. More 15 communities and their numbers. Due to the nature of the members of each community and the nature of posting, the content is different. Each community represents a neighborhood's post attribute. For example, in the first community, most of the posts are related to learning materials, and posts in the second community. Most of the content is film-related, so the community that was experimented with in the dataset was a different community that was obtained according to different content or attributes. As time changes, different content published by different users will change, forming an evolutionary event.

In the data set experiment of this paper, the number of evolution events of the community can be determined according to the comparison between the results of community division and the current time slice active community, and the influence of similarity threshold k on community evolution is studied. The similarity threshold k ranges from 0.1 to 1 and is divided into ten intervals to calculate the impact of each similarity threshold k on the number of evolution events in the community.

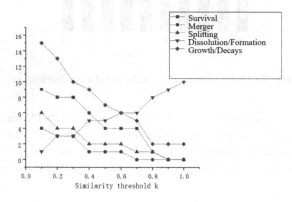

Fig. 9. Number of PSTN dataset evolution community transitions

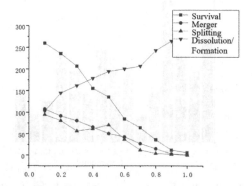

Fig. 10. Facebook wall posts dataset evolution events

It can be clearly seen from Figs. 9 and 10 that the selection of the similarity threshold k has a significant influence on the number of community evolutions. The formation and dissolution events increase with the increase of k. The formation and dissolution are reciprocal, so they are represented by a line. On the contrary, survival, merger and split events basically decrease with the increase of k. When the k value is lower, more communities will be matched, so the number of observed community evolution survival events is larger. Relatively when the threshold k is high, the matching community evolution events are relatively more conservative, and the evolutionary community life span is relatively shorter.

In order to evaluate the optimal similarity threshold k in the evolution of the community, a keyword extraction algorithm [16] is used here. This algorithm generates a series of keywords that match the keywords of each community theme in the information or posts in the community, take the 10 most frequently. The goal is to expect a community with multiple time periods to discuss or conduct the same topic. We refer to the situation in which a community on a single time slice and a dynamically divided community remain the same topic. The observed data set is in different similarity thresholds k, and the average number of common themes between any two surviving communities is shown in the following figure:

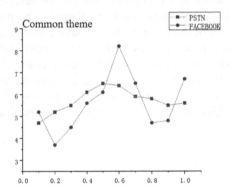

Fig. 11. Facebook wall posts dataset evolution events

It can be observed from Fig. 11 that in the change of the common subject number of the two data sets, the PSTN data set has a smoother effect on the similarity threshold due to the data characteristics. In the process of selecting the similarity threshold, only the higher average common subject number is selected. However, there is a significant fluctuation in the Facebook wall posts dataset because the number of topics discussed in this dataset is very large, and the keywords of the theme are also very different. At nearly 0.6, the number of common themes has increased significantly. There is a significant decrease in the after greater than 0.6. On the PSTN dataset, set k = 0.5, the highest average common subject number in the living community, and 0.6 as the optimal similarity threshold k on the Facebook wall posts dataset. The reason for this stability and volatility depends on the characteristics of the two data sets. The PSTN data set has a very stable community structure, and the large number of members in the data set fluctuate less, and the content price is single, but on the dataset of the social networking site. There is a huge amount of content information, and there are many fields involved in each member, so this kind of fluctuation will occur.

4 Conclusions

Based on the dynamic community tensor decomposition research, this paper proposes to divide the community with time domain network based on tensor decomposition, and then compare the community nodes and members of the current active community time slice to discover the evolution of the community. It mainly utilizes the characteristics of the three feature matrices obtained by tensor CP decomposition to divide the community and find active communities at a certain point in time. In the comparison of community similarity in community evolution, a keyword extraction algorithm is used to determine the similarity threshold, so that the optimal similarity threshold k can be selected to determine the number of time of the final community evolution for the analysis of community evolution. The community discovery algorithm based on tensor decomposition and finding active communities on each time slice can greatly reduce the work of community evolution in each time slice for the comparison of adjacent practice articles. Only some active communities can be selected to observe evolutionary events. The active community discovery algorithm proposed in this paper can target the evolution events of the community and save a lot of redundant work. In the future work, the method is planned to expand from the active community to the active nodes tracking the active community, capture the active change nodes of the dynamic social network, and analyze the significance of the active points in the community. It is also possible to capture the evolutionary abstraction patterns of dynamic community networks in order to predict the structural changes of communities with time slices in future time slices.

References

1. Watts, D.J., Strogatz, S.H.: Collective dynamics of 'small-world' networks. Nature **393** (6684), 440–442 (1998)
2. Barabasi, A.L., Albert, R.: Emergence of scaling in random networks. Science **286**(5439), 509–512 (1999)
3. Shen, H., Cheng, X., Cai, K., et al.: Detect overlapping and hierarchical community structure in networks. Phys. A: Stat. Mech. Appl. (2008)
4. Newman, M.E.J., Park, J.: Why social networks are different from other types of networks. Phys. Rev. E **68**(3), 036122 (2003)
5. Wan, L., Liao, J., Zhu, X.: CDPM: finding and evaluating community structure in social networks. In: Tang, C., Ling, C.X., Zhou, X., Cercone, N.J., Li, X. (eds.) ADMA 2008. LNCS (LNAI), vol. 5139, pp. 620–627. Springer, Heidelberg (2008). https://doi.org/10.1007/978-3-540-88192-6_64
6. Sun, J.M., Philip, S.Y., Faloutsos, C., et al.: GraphScope: parameter-free mining of large time-evolving graphs. In: Proceedings of KDD 2007. ACM Press, New York (2007)

7. Mørup, M.: Applications of tensor (multiway array) factorizations and decompositions in data mining. Wiley Interdisc. Rev. Data Min. Knowl. Discovery **1**(1), 24–40 (2011)
8. Kolda, T.G., Bader, B.W.: Tensor decompositions and applications. SIAM Rev. **51**(3), 455–500 (2009)
9. Carroll, J.D., Pruzansky, S., Kruskal, J.B.: A general approach to multi dimension analysis of many-way arrays with linear constraints on parameters. Psychometrika **45**(1), 3–24 (1980)
10. Harshman, R.A.: Foundations of The PARAFAC procedure: model and Conditions for An 'explanatory' multi-mode factor analysis. UCLA Working Papers in Phonetics, vol. 16, pp. 1–84 (1970)
11. Kolda, T.G., Bader, B.W., Kenny, J.P.: Higher-order web link analysis using multilinear algebra. In: Proceedings of the 5th IEEE International Conference on Data Mining, Texas, USA, pp. 1–8. IEEE (2005)
12. Mucha, P.J., Richardson, T., Macon, K., et al.: Community structure in time-dependent, multiscale, and multiplex networks. Science **328**(5980), 876 (2009)
13. Zhang, H., Wang, C.D., Lai, J.H., et al.: Modularity in complex multilayer networks with multiple aspects: a static perspective. Appl. Inform. **4**(1), 7–31 (2017)
14. Sarzynska, M., Leicht, E.A., Chowell, G., et al.: Null models for community detection in spatially embedded, temporal networks. J. Complex Netw. **4**(3), 363–406 (2018)
15. Lathauwer, L.D., Moor, B.D., Vandewalle, J.: A multilinear singular value decomposition. SIAM J. Matrix Anal. Appl. **21**(4), 1253–1278 (2000)
16. Witten, I.H., Paynter, G.W., Frank, E., et al.: KEA: practical automatic keyphrase extraction. In: Proceedings of the 4th ACM Conference on Digital Libraries, Berkeley, USA [s. n.] (1999)

A Content-Based Recommendation Framework for Judicial Cases

Zichen Guo, Tieke He, Zemin Qin, Zicong Xie, and Jia Liu[✉]

State Key Laboratory for Novel Software Technology, Nanjing University,
Nanjing 210093, China
liujia@nju.edu.cn

Abstract. Under the background of the Judicial Reform of China, big data of judicial cases are widely used to solve the problem of judicial research. Similarity analysis of judicial cases is the basis of wisdom judicature. In view of the necessity of getting rid of the ineffective information and extracting useful rules and conditions from the descriptive document, the analysis of Chinese judicial cases with a certain format is a big challenge. Hence, we propose a method that focuses on producing recommendations that are based on the content of judicial cases. Considering the particularity of Chinese language, we use "jieba" text segmentation to preprocess the cases. In view of the lack of labels of user interest and behavior, the proposed method considers the content information via adopting TF-IDF combined with LDA topic model, as opposed to the traditional methods such as CF (Collaborative Filtering Recommendations). Users are recommended to compute cosine similarity of cases in the same topic. In the experiments, we evaluate the performance of the proposed model on a given dataset of nearly 200,000 judicial cases. The experimental result reveals when the number of topics is around 80, the proposed method gets the best performance.

Keywords: Recommendation · Content-based · LDA · Cosine similarity

1 Introduction

With the development of computer science, it has been a very common ways to solute some difficult problems in reality by simulating with computer. Meanwhile, with the advancement of artificial intelligence, judicial judgement is getting closer to the justice of law with the aid of big data analysis. It is worth noting that the similarity analysis of judicial cases is the basis of wisdom judicature. A formative judicial case contains the court, the accuser and the accused, the fact, and the result of the case. In order to give credibility within a community, jury trials must take all these complicated factors into consideration with reference to similar cases. With the explosion in the number of judicial cases, it is difficult to consider similar cases without omission. Because of this, we seek to provide a novel recommendation method to assist judicial processing.

© Springer Nature Singapore Pte Ltd. 2019
X. Cheng et al. (Eds.): ICPCSEE 2019, CCIS 1058, pp. 76–88, 2019.
https://doi.org/10.1007/978-981-15-0118-0_7

Starting with the study of Becker [1], researchers focus on what factors influence the optimal amount of enforcement, like the cost of catching criminals, the subjective decisions that affect the result. However, in practice, these factors are affected by political, moral and many other subjective constraints. Our main purpose is to make use of the objective factors among judicial cases.

Despite the fact that judicial study has gained some achievements in many aspects, such as legal word embeddings [2], inferring of the penalty [3] and judicial data standard [4], a recommender system is needed to deal with the large volume problem of judicial cases. In general, three filtering techniques such as content-based [5], collaborative [6,7] and hybrid filtering [8,9] are presented in the recommender system literature to filter records and identify the relevant information. Some of the progressive collaborative filtering algorithm [10,11] take cold start into consideration on the situation of lack of users or users' behaviours. In the meantime, it is challenging in judicial area because there exist many one-time users.

In view of the current situation, we propose an effective way to get recommendations, which is to collect the judicial cases a certain user put in. Our primary focus is to explore the judicial cases that are used to capture semantic similarities among text snippets. As mentioned above, given the cases that user input, the proposed model can return a recommended list of the relevant cases. We proposed our framework of content-based judicial case recommendation, as shown in a flow chart, Fig. 1.

In summary, we do the following work in this paper.

- We propose a content-based recommendation method for judicial cases.
- We develop a co-training process with TF-IDF and LDA to gain a plausible performance.
- We conduct an extensive experiments to test the performance of our proposed method, and the result reveals when the number of topic is around 80, our proposed method shows best performance.

The rest of this paper is organized as follows. Section 2 first describes relevant background of the models and algorithms, then sets out the proposed model and theoretical basis. Section 3 presents the experimental results and Sect. 4 summarizes this paper.

2   Methodology

2.1   Background

In this part, we provide detailed background of the models and algorithms used in this paper.

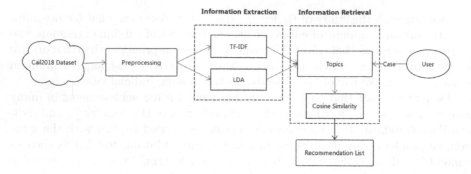

Fig. 1. Framework of content-based judicial case recommendation

Recommender Systems. Recommendation systems recommend items that specific users may be interested in books, news, movies, etc. At present, the methods of recommender systems are mainly based on collaborative filtering [12], association rules [13], content or hybrid algorithm [14]. LDA-based recommendation belongs to content-based recommendation.

Cold-starting is taken into consideration on the situation of lack of users or user behaviours and can be proved efficiently in many real projects. It also calls for attention in judicial field because existing a mass of one-time users.

Content-Based Recommendation with LDA. In natural language processing field, topic modeling is a kind of modeling for discovering the abstract "topics" that occur in a collection of documents. The LDA(Latent Dirichlet Allocation) model proposed by Blei in 2003 [19] has set the topic model on fire. The so-called generation model indicates that we think that every word in a document is achieved through the process of selecting a topic with a certain probability.

It's been a long time that LDA has been used to study user interests and build a system to recommend more friends with the same or similar user interests [17]. However, considering the lack of label of user interest and behavior among judicial cases, it is difficult to focus on user-generated content. We seek to turn to a new direction, which is to analyze and classify judicial cases input by users as content instead of user-generated content. In addition, TF-IDF is another reasonable algorithm in case recommendation.

TF-IDF Algorithm. TF-IDF is a commonly used weighting technology for information retrieval and data mining. TF means word frequency, IDF means inverse document frequency. TF-IDF proved useful and effective in stop-word filtering in various subject fields including text summarization and classification [18].

- TF Score (Term Frequency) considers documents as bag of words, agnostic to order of words. A document with 10 occurrences of the term is more relevant than a document with term frequency 1.
- We also want to use the frequency of the term in the collection for weighting and ranking. Rare terms are more informative than frequent terms. We want low positive weights for frequent terms and high weights for rare terms.

2.2 Preliminaries

For convenience, we define the custom data formats and definitions used in Table 1.

Table 1. Notations

Symbol	Description
M	Number of judicial cases
m	Index of a judicial case
N	Number of words in judicial case m
K	Number of topics
k	Index of a topic
R_m	Collection of words in judicial case m
c	Cause of action of a judicial case
q	Quantified data of a judicial case
l	Location of a judicial case
p	People involved of a judicial case
W_m	Collection of words in judicial case m
θ_m	Topic distribution of law case m
φ_k	Word distribution for topic k
$Z_{m,n}$	Topic assignment for $w_{m,n}$
$w_{m,n}$	The n-th word in case W_m

Definition 1 Judicial Case. A judicial case consists of a collection $R_m(c, q, l, p)$, which means that judicial case m is made up of the collections of words R_m with four elements c, q, l, p.

Definition 2 Topic. LDA defines each topic as a bag of words. Given a dataset of cases, topics maximize the posterior probability of the observed corpus.

2.3 Data Preprocessing

In light of the difference between Chinese and Romance languages, we use "jieba" text segmentation to get word sequences from dataset. For each judicial m in the dataset, we get the collection $R_m(c, q, l, p)$. Also, a special filter is set up to filter out key data and sensitive vocabulary in the cases to remove interferences. We make a transformation $R_m(c, q, l, p) \rightarrow W_m(c)$ to get filtered collection of words in judicial case m.

2.4   Information Extraction

TF-IDF and LDA are trained to constitute the recommendation knowledge together in this part.

First, in order to smooth frequency of words in preprocessed data of M judicial cases, we use TF-IDF to obtain new corpus for the following training. TF-IDF assumes that if a word is important for a document, it would repeatedly appear in that document whereas it would be relatively rare in other documents. The TF is associated with the former assumption and the IDF is associated with the latter. TF-IDF is defined as

$$tfidf(t, d, D) = tf(t, d) \times idf(t, D)$$

where $f_{d(t)}$ is the normalized frequency of term $t \in w$ Therefore, it is defined as:

$$tf(t, d) = \frac{f_{d(t)}}{\max_{w \in d} f_{d(w)}}$$

In document d, $f_{d(t)}$ is the frequency of term t and w is an existing word. Also, $idf(t, D)$ shows the IDF t, which is defined as

$$idf(t, D) = \log_2 \left(\frac{|p|}{|(d \in D, t \in d)|} \right)$$

where $|D|$ indicates the total number of documents in the corpus, and $|(d \in D, t \subset d)|$ is the number of documents in which the term t appears.

The remaining words were filtered by frequency using the TF-IDF score. TF-IDF measures the importance of a word in a corpus as seen above. It increases with the number of occurrences in the document and decreases with the frequency in the corpus. We compute TF-IDF for each word of each document-plot in the corpus and keep a certain number of words with the highest score to optimize the corpus.

Although LDA assumes the documents to be in bag of words (bow) representation. We find success when using TF-IDF representation as it can be considered a weighted bag of words. It changes θ_m and φ_k in LDA model, as shown in Fig. 2.

We describe the LDA process of a judicial case data set in formal language, as shown below. Dirichlet() represents Dirichlet distribution and Multi() represents multinomial distribution.

1. For each topic $k \in 1, \ldots, K$, draw $\varphi_k \sim$ Dirichlet(β), denoting the specific word distribution for topic k.

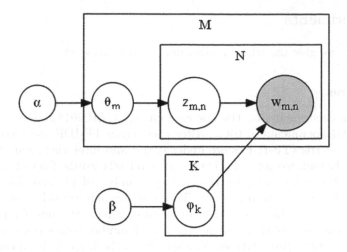

Fig. 2. Graphical representation of LDA model

2. For each judicial case $m \in 1, \dots, M$:
   - Draw $\theta_m \sim \text{Dirichlet}(\alpha)$, indicating the distribution of topics embedded in judicial case m;
   - For the n-th word in case m, $n \in 1, \dots, N$, draw a W Multi(ϕ_z) for each word $w \in W_{m,n}(c)$.

The progress above can be used to gain knowledge among different kind of judicial cases. In order to generate recommendations for uses, we also need to do information retrieval from the topic distribution.

2.5 Information Retrieval

For each judicial case $m \in 1, \dots, M$, we can get a vector of K topic distribution via information extraction, which is defined as

$$m = (s_1, \dots, s_k)$$

where we seek s_i referring to the maximum among $s_1, \dots, s_k$. On this occasion, i is the topic we regarded as the classification of case S. On account of two cases are similar if they contain similar topic contribution, similarity between cases is measured by cosine angle between vectors. Given a judicial case s input by user, which belongs to classification i, for each judicial case $t \in 1, \dots, M_i$, we get $\text{Sim}(s, t)$, which is defined as:

$$\text{Sim}(s, t) = \cos(s, t) = \frac{s \cdot t}{\|s\| \times \|t\|}$$

Recommendation list is composed of Top 5 cases of $\text{Sim}(s, t)$.

3   Experiments

In this part, we give the whole realization of our framework.

3.1   Dataset

We perform experiments on the law case dataset CAIL2018_Small, which contains 204, 231 documents in total. After conducting TF-IDF, we retrieve a list of low value words (TF-IDF score under 0.025) and filter them out of the dictionary. In the end, we get a dictionary with 311, 024 words. Considering actual processing of judicial cases, we take a large number of judicial cases without manual labeling results into account. Therefore, we only consider using the fact description label in this dataset. In order to eliminate the interference items, we add the screening of time, place, person and number before data preprocessing, so as to get the final dataset. The specific methods for judicial cases are as follows:

- Regular expressions are used to match time keywords that appear in the cases.
- Regular expressions are used to match location keywords that appear in the cases, such as 'province', 'city', 'district'.
- Characters in the format of "XXX" are replaced by "PERSON" fields.
- For the regular matching of measurement units, the size of money is judged and divided into seven grades and marked as follows (Table 2):

Table 2. Measurement labels

Money range	Label
$[0, 10)$	$m1$
$[10, 100)$	$m2$
$[100, 1000)$	$m3$
$[1000, 10, 000)$	$m4$
$[10, 000, 100, 000)$	$m5$
$[100, 000, 1000, 000)$	$m6$
$> 1000, 000$	$m7$

To analyze the dataset as a whole, we give the statistics of money in the dataset, as shown in Fig. 3. Among the whole dataset, the proportion of Small-money criminal cases is very high, while the cases involving large amounts of money are very low. In all, the amount of m7-level criminal cases is 0. This figure reflects the case characteristics of CAIL2018_Small dataset from aspect of money. And the timeline of CAIL2018_Small dataset shows in Fig. 4.

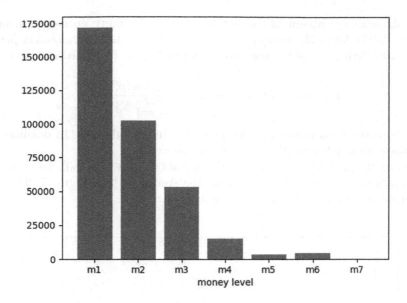

Fig. 3. Statistics of money

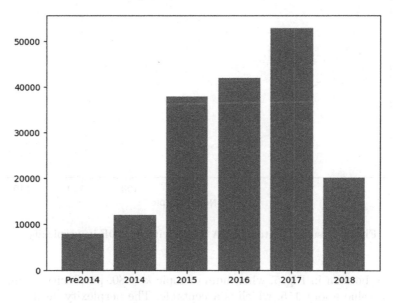

Fig. 4. Statistics of time

3.2 Experimental Results

We implement perplexity as the indicator [19]. Perplexity is a statistical measure of how well a probability model predicts a sample. In information theory, perplexity is the probability that the test data is monotonically decreasing, which is the algebraic equivalent of the inverse of the probability geometric mean of each word. The lower the complexity score, the better the generalization performance [20]. Perplexity of the untrained dataset (D_{test}) is defined as follows:

$$\text{perplexity}(D_{\text{test}}) = \exp(\frac{-\sum_{d=1}^{M} \log(p(w_d))}{\sum_{d=1}^{M} N_d})$$

where M is the total number of documents in judicial dataset. In document d, W_d represents words and N_d is the number of words.

Among the primary setting, for each num of topic $k \in [10, 150]$, we set hyperparameters $\alpha = \frac{50}{k}$, $\beta = 0.01$, following the studies of [21]. Figure 5 illustrates the perplexity figures with different numbers of topic k.

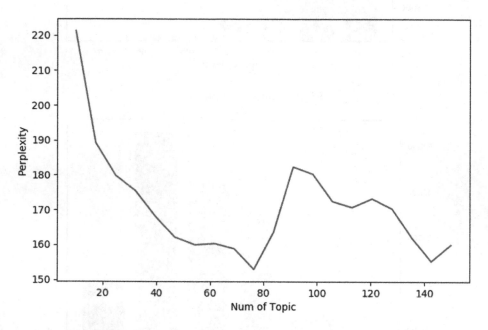

Fig. 5. Results of k-topic LDA model with TF-IDF in perplexity

As can be seen in Fig. 5, when num of topic $k \simeq 80$, perplexity requires the minimum value about 155, which is acceptable. The perplexity declines significantly when $k \in [10, 50]$, and are in an upward trend when $k \in [80, 95]$, but also generally falls for $k > 95$ in the process.

Next we figure out exactly the value of k, we reduce the scope and choose $k = 75, 76, 77, 78, 79, 80$, then calculate the perplexity as showing in Fig. 6.

As shown in Fig. 6, when $k = 78$, perplexity achieves the minimum value nearly 154. In all, we choose $k = 78$ as ideal topic number. We display the top 30 words with TF-IDF value in the model with $k = 78$, as shown in Fig. 7.

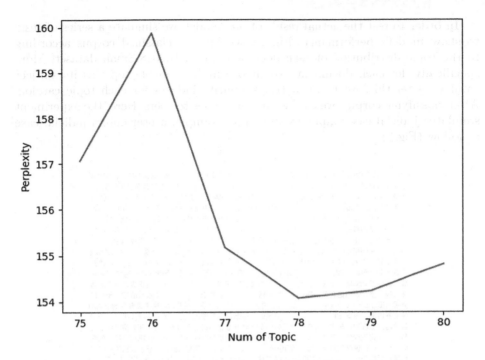

Fig. 6. Results of perplexity $k \in [75, 80]$

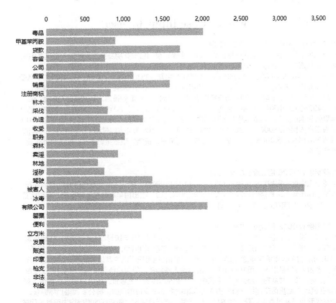

Fig. 7. TOP 30 WORDS

In order to test the actual result of our model, we simulate a series of tests to show model's performance. Firstly, we build a classified corpus according to the topic distribution of each document in CAIL2018_Small dataset. More specifically, for each document, we choose most probable topic as its subject catalog. After this, we build matrix similarity indexes for each topic catalog. After classifying corpus, we can recommend cases to users. Here, the experiment simulates judicial cases input by user. For example, a user enters judicial case as follow (Fig. 8):

2017年10月中旬，被告人于伟仁联系刘海明欲购买甲基苯丙胺（冰毒）2000克，刘海明遂联系李伟寻找毒品卖家。10月20日左右，李伟、王少成（刑拘在逃）分别联系王少廷购买甲基苯丙胺，其中王少廷联系购买1000克，李伟联系购买2000克。王少廷遂联系范敬仰购买甲基苯丙胺3000克，范敬仰又联系张高阳购买3000克甲基苯丙胺。后经联系约定，范敬仰出售给王少廷甲基苯丙胺每克100元，王少廷出售甲基苯丙胺每克110元，刘海明告知于伟仁甲基苯丙胺每克125元。10月22日早上6时许，张高阳、范敬仰、王少廷、李伟、刘海明、于伟仁、王少成等人相互电话联系询问毒资情况并约定进行毒品交易。李伟电话告知王少廷让刘海明和于伟仁带着毒资去姜寨镇，上午10点左右到姜寨镇见面交易。王少成联系王少廷于10点左右到庞营韩老家集上取毒资，后王少廷找到王少成拿到毒资10万元，两人约定剩余1万元毒资等毒品出售后再补足，王少廷携带该10万元毒资回到姜寨镇玉楼村与前来购买毒品的刘海明、于伟仁见面，于伟仁称毒资在随身携带的银行卡上，问王少廷是否能转账付钱，王少廷电话联系范敬仰后不同意转账交易，刘海明、于伟仁便去银行取款，于伟仁到银行取出59000元现金。王少廷找到范敬仰交10万元毒资，后范敬仰携带该毒资到张高阳家交易毒品，张高阳将一包毒品交给范敬仰。范敬仰携带该毒品回到家中交给王少廷。王少廷将毒品放在出租车内，开车带着范敬仰返回姜寨准备将毒品交给王少成。在姜寨镇玉楼村，民警将王少廷、范敬仰抓获，当场从其驾驶的出租车后排座上查获一包嫌疑毒品。刘海明、于伟仁计议返回临泉县城继续用银行卡取款，在前往临泉县城的路上被公安机关查获，民警从后排座上查获现金59000元。经称量，嫌疑毒品净重994.61克；经鉴定检出甲基苯丙胺成分，含量为79.6%。侦查机关当日将张高阳抓获，12月13日将李伟抓获。

Fig. 8. Case input by user

推荐案例170536 相似度：0.961252
公诉机关指控：被告人范某某与张某（另案处理）系同居男女朋友关系，被告人范某某明知张某实施贩卖毒品行为，仍用自己的银行账户，帮助张某保管毒赃、提取毒资。具体事实如下：1、2015年7月底至11月期间，张某雇佣余某（另案处理）去广东购买1000克甲基苯丙胺毒品用于贩卖，后在温岭市、台州市椒江区等地将该1000克甲基苯丙胺卖给曹某、王某1等人，其中王某1于2016年9月初至10月期间，先后汇到被告人范某某持有的账户尾号为4925的农村信用社储蓄卡毒赃人民币共计47500元。2、2015年10月4日，被告人范某某受张某指使，至中国农业银行玉环县清港支行给帮助张某购买毒品的余某控制的银行账户汇去毒资人民币35000元。2015年11月28日上午，温岭市公安局禁毒大队民警在玉环县清港镇下潥村八片区0066号出租房抓获被告人范某某。被告人范某某归案后，如实供述主要涉案事实。

推荐案例137726 相似度：0.949229
上海市宝山区人民检察院指控：2015年8月13日19时许，被告人牟某某携带18张他人银行卡至上海市闸北区秣陵路XXX号火车站邮政支局门口，欲出售牟利时被民警抓获，同时民警缴获其随身携带的上述银行卡。被告人牟某某到案后如实供述上述犯罪事实。

推荐案例160012 相似度：0.946199
湖北省武汉市武昌区人民检察院指控：被告人张1某于2016年10月21日16时许，应周某购买毒品的要求，在武汉市武昌区武泰闸栅栏口与周某相约见面，收取周某的购毒款人民币1100元，并与周某约定稍后在武汉市武昌区南湖社区雅安街南嘉宾馆向周某交付毒品。随后，被告人张1某以人民币1000元的价格向他人购得塑料袋装白色晶体颗粒毒品1包及红色片剂毒品3颗，并从白色晶体颗粒毒品中分装少许留作其个人吸食。当日17时许，被告人张1某在武汉市武昌区南湖社区雅安街南嘉宾馆欲向周某交付毒品时，被公安民警当场抓获。公安民警当场从其身上查获上述毒品。经称量、取样及鉴定，所查获的毒品均为甲基苯丙胺，共计净重10.37克。指控上述事实，公诉机关提供有下列证据予以证实：1、公安机关出具的抓获经过、破案经过；2、物证（扣押清单及照片）；3、证人周某的证言；4、被告人张1某的供述与辩解；5、称量、取样笔录及鉴定材料。公诉机关认为，被告人张1某违反国家毒品××法规，以牟利为目的为他人代购毒品，其行为触犯了《中华人民共和国刑法》×××、××的规定，已构成贩卖毒品罪。提请本院依法追究其刑事责任。

Fig. 9. Recommended case

Then we load the topic index, calculate the similarity between the input case and each cases in the indexcatalog by cosine similarity. We select the top 5 cases of similarity as the recommendation judicial cases to present to the user. Top three judicial cases is shown in Fig. 9 and the cosine similarities are 0.9613, 0.9492, 0.9462.

4 Conclusion

In this paper, we present a content-based method of judicial case recommendation to address the problem of how to help user better understand judicial cases in depth. Specifically, we develop a co-training process with TF-IDF and LDA to gain a plausible model performance. Given LDA is an unsupervised learning algorithm, we conduct experiments to evaluate the performance of the proposed recommender system. The results show the optimal number of topic. Our recommendation method still has some room for improvement. Putting state-of-the-art algorithms into practice with good performance is always a critical problem, which we will focus on in the future.

Acknowledgment. The work is supported in part by the National Key Research and Development Program of China (2016YFC0800805) and the National Natural Science Foundation of China (61772014).

References

1. Becker, G.S., Landes, W.M.: Essays in the Economics of Crime and Punishment, Number 3 in Human Behavior and Social Institutions. National Bureau of Economic Research: Distributed by Columbia University Press
2. He, T., Lian, H., Qin, Z., Zou, Z., Luo, B.: Word embedding based document similarity for the inferring of penalty. In: Meng, X., Li, R., Wang, K., Niu, B., Wang, X., Zhao, G. (eds.) WISA 2018. LNCS, vol. 11242, pp. 240–251. Springer, Cham (2018). https://doi.org/10.1007/978-3-030-02934-0_22
3. He, T.-K., Lian, H., Qin, Z.-M., Chen, Z.-Y., Luo, B.: PTM: a topic model for the inferring of the penalty. J. Comput. Sci. Technol. **33**(4), 756–767 (2018)
4. Qin, Z., He, T., Lian, II., Tian, Y., Liu, J.: Research on judicial data standard. In: 2018 IEEE International Conference on Software Quality, Reliability and Security Companion (QRS-C), pp. 175–177. IEEE (2018)
5. Balabanovic, M., Shoham, Y.: Fab: content-based, collaborative recommendation. Commun. ACM **40**, 66–72 (1997)
6. Linden, G., Smith, B., York, J.: Amazon.com recommendations: item-to-item collaborative filtering
7. Das, A.S., Datar, M., Garg, A., Rajaram, S.: Google news personalization: scalable online collaborative filtering. In: Proceedings of the 16th International Conference on World Wide Web - WWW 2007, p. 271. ACM Press (2007)
8. Badaro, G., Hajj, H., El-Hajj, W., Nachman, L.: A hybrid approach with collaborative filtering for recommender systems. In: 2013 9th International Wireless Communications and Mobile Computing Conference (IWCMC), pp. 349–354, July 2013

9. Strub, F., Mary, J., Gaudel, R.: Hybrid collaborative filtering with autoencoders (2016)
10. Ahn, H.J.: A new similarity measure for collaborative filtering to alleviate the new user cold-starting problem. Inf. Sci. **178**(1), 37–51 (2008)
11. Patra, B.Kr., Launonen, R., Ollikainen, V., Nandi, S.: A new similarity measure using Bhattacharyya coefficient for collaborative filtering in sparse data. Knowl.-Based Syst. **82**(C), 163–177 (2015)
12. Ekstrand, M.D.: Collaborative filtering recommender systems **4**(2), 81–173
13. Lin, W., Alvarez, S.A., Ruiz, C.: Efficient adaptive-support association rule mining for recommender systems. Data Min. Knowl. Disc. **6**(1), 83–105 (2002)
14. Kardan, A.A., Ebrahimi, M.: A novel approach to hybrid recommendation systems based on association rules mining for content recommendation in asynchronous discussion groups. Inf. Sci. **219**, 93–110 (2013)
15. Nagori, R., Aghila, G.: LDA based integrated document recommendation model for e-learning systems, pp. 230–233, April 2011
16. Luostarinen, T., Kohonen, O.: Using topic models in content-based news recommender systems. In: Proceedings of the 19th Nordic Conference of Computational Linguistics (NODALIDA 2013), pp. 239–251. Linköping University Electronic Press, Sweden (2013)
17. Pennacchiotti, M., Gurumurthy, S.: Investigating topic models for social media user recommendation. In: Proceedings of the 20th International Conference Companion on World Wide Web, WWW 2011, pp. 101–102. ACM, New York (2011)
18. Ramos, J.: Using TF-IDF to determine word relevance in document queries
19. Blei, D.M.: Latent Dirichlet allocation, p. 30
20. Arora, K.: Contrastive perplexity: a new evaluation metric for sentence level language models. CoRR, abs/1601.00248 (2016)
21. Yin, H., Sun, Y., Cui, B., Hu, Z., Chen, L.: LCARS: a location-content-aware recommender system, pp. 221–229, August 2013

Collaboration Filtering Recommendation Algorithm Based on the Latent Factor Model and Improved Spectral Clustering

Xiaolan Xie[1,2] and Mengnan Qiu[1(✉)]

[1] College of Information Science and Engineering, Guilin University of Technology, Guilin, Guangxi Zhuang Autonomous Region, China
837315716@qq.com
[2] Guangxi Universities Key Laboratory of Embedded Technology and Intelligent System, Guilin University of Technology, Guilin, China

Abstract. Due to the development of E-Commerce, collaboration filtering (CF) recommendation algorithm becomes popular in recent years. It has some limitations such as cold start, data sparseness and low operation efficiency. In this paper, a CF recommendation algorithm is propose based on the latent factor model and improved spectral clustering (CFRALFMISC) to improve the forecasting precision. The latent factor model was firstly adopted to predict the missing score. Then, the cluster validity index was used to determine the number of clusters. Finally, the spectral clustering was improved by using the FCM algorithm to replace the K-means in the spectral clustering. The simulation results show that CFRALFMISC can effectively improve the recommendation precision compared with other algorithms.

Keywords: Collaboration filtering · Recommendation algorithm · Latent Factor Model · Cluster validity index · Spectral clustering

1 Introduction

In recent years, with the rapid development of technologies such as cloud computing, big data, and Internet of Things, various applications in the Internet space have emerged the explosive growth of data size [1]. Big data contains rich value and great potential, which will bring about transformative development to human society, but it also brings serious problem of information overload. How to quickly and effectively obtain valuable information from complicated data has become a key problem in the current development of big data. As an effective method to solve the problem of information overload [2], the recommendation system has become a hot topic in academia and industry and has been widely used. The recommendation system mines interesting items (such as information, services, articles, etc.) from the massive data through the recommendation algorithm according to the user's needs, interests, etc., and recommends the results to the user in the form of a personalized list. Traditional recommendation methods mainly include collaborative filtering, content-based recommendation methods and hybrid recommendation methods. Among them, the most

© Springer Nature Singapore Pte Ltd. 2019
X. Cheng et al. (Eds.): ICPCSEE 2019, CCIS 1058, pp. 89–100, 2019.
https://doi.org/10.1007/978-981-15-0118-0_8

classic algorithm is collaborative filtering, such as matrix factorization, which uses the interaction information between users and items to recommend for users. Collaborative filtering is currently the most widely used recommendation algorithm. It has been repeatedly won in the Netflix Grand Prix in recent years., but it also encountered severe problem of data sparseness (a user-rated item only accounts for a very small portion of the total number of items) and cold start (new users and new items often do not have score data) [3].

In order to resolve the above problems, some scholars have made the related improvements on CF recommendation and obtained some achievements. Sarwar et al. [4] solved the scalability problem of the traditional collaborative filtering recommendation algorithm by using the clustering algorithm. The algorithm improves the recommendation accuracy and reduces the time complexity of the algorithm. Xu et al. [5] proposed a multi-clustering algorithm, which aggregates all users and items into several user-item subgroups. Each subgroup contains some data of users and items. The experiment shows that the accuracy of the recommendation results is better than the original algorithm. Gong et al. [6, 7] proposed a combination of fuzzy clustering and collaborative filtering recommendation algorithm. The experiment shows that the algorithm is better than the traditional collaborative filtering recommendation algorithm; Bo et al. [8, 9] mined the user's interest in the item category according to item score of the user and information of the item category, and constructed the user-item preference to improve similarity calculation method. This way improves the accuracy of the recommendation result.

The above improvements of recommendation are built on a single independent data space, while in big data environment, a large number of new data have grown explosively in many fields, such as movies, music, books, news, e-commerce and social networks [10]. We need a lot of identified training data to apply conventional machine learning approaches in new fields, but it will cost a lot of manpower and material resources. Although the users from different websites are not the same group, they are subsets sampled from all social members, and users will reflect the similar interest distributions on movies, books, news and others. For example, people who like adventure novels may also like adventure movies and related news. Therefore, in the case of no explicit correspondence between users and items, the users or items in the same category can also have similar potential features [11, 12].

In this paper, we propose the CF recommendation algorithm based on the Latent Factor Model and improved spectral clustering. The processes of CFRALFMISC are as follows: CFRALFMISC firstly adopts the Latent Factor Model to predict the missing score. Then, CFRALFMISC uses the cluster validity index to determine the number of clusters. Finally, CFRALFMISC improves the spectral clustering by using the FCM algorithm to replace the K-means in the spectral clustering. We make simulation experiments on the data set of MovieLens (100 KB) to prove the effectiveness of CFRALFMISC.

The rest of this paper is organized as follows. Section 2 describes the CF recommendation and the standard spectral clustering algorithm. Section 3 presents our research approach based on the Latent Factor Model and improved spectral clustering. Section 4 presents the experimental results and makes comprehensive comparison. Section 5 concludes the paper.

2 Related Work

2.1 Collaborative Filtering Recommendation

Collaborative filtering recommendation is originated from the process of word-of-mouth in real life. Collaborative filtering uses the methods that there are similar interest preferences among similar users to discover potential user preferences for items. It mainly includes heuristics and model-based collaborative filtering. The heuristic method first calculates the similarity between users (or items) by the user's historical score difference, and then calculates the utility value according to the user's historical score and the similarity between users. The model-based approach primarily predicts a user's potential preference for a item by constructing a user preference model. Collaborative filtering only needs to use the user's historical score data, so it is simple and effective, and it is the most successful recommendation method. However, since the user's rating data of the item is much smaller than the total number of items, the problem of sparse data is often encountered. In addition, for new users or projects, since there is no scoring data and it is impossible to recommend, there is the problem of cold start [3].

The collaborative filtering recommendation algorithm is mainly divided into the following three steps.

(1) Similarity calculation between users (items). The calculation methods of similarity mainly include Euclidean distance, cosine similarity and Pearson correlation similarity.
(2) Find the nearest neighbor. The method of the nearest neighbor selection can be divided into neighbors of the fixed number and neighbors based on the similarity threshold, specifically selecting users with the top K preference similarities as neighbor users or setting users that its preference similarity threshold is greater than a certain threshold as similar users.
(3) Produce recommendations. The current user's rating of the unrated items is predicted by finding the preference information of the nearest neighbor users.

2.2 Standard Spectral Clustering Algorithm

Spectral clustering is an algorithm that is evolved from the graph theory and has been widely used in clustering. Its main idea is to treat all the data as points in space, which can be connected by edges. The edge weight value between two distant points is lower, and the edge weight value between two closer points is higher. By cutting the graph that is composed of all the data points, the weights between different subgraphs after cutting graph are as low as possible, and the weights in subgraphs are as high as possible, thus achieving the purpose of clustering.

The most commonly used similarity matrix is generated by the full connection of Gaussian kernel distance. The most commonly used cut mode is Ncut. The last commonly used clustering method is K-Means. The following summarizes the spectral clustering algorithm flow with Ncut.

Input: the sample set $D = (x_1, x_2, \ldots, x_n)$, the generation method of the similarity matrix, dimension k_1 after dimension reduction, the clustering method, dimension k_2 after clustering.

Output: the cluster division $C(c_1, c_2, \ldots, c_{k_2})$.

(1) Construct the similarity matrix S of the sample according to the generation method of the input similarity matrix
(2) Construct the adjacency matrix W and the degree matrix D according to the similarity matrix S
(3) Calculate the Laplacian matrix L
(4) Construct a normalized Laplacian matrix $D^{-1/2}LD^{-1/2}$
(5) Calculate the eigenvector f corresponding to the minimum k_1 eigenvalues of $D^{-1/2}LD^{-1/2}$
(6) Normalize the matrix composed of the corresponding feature vectors f, and finally form the feature matrix F of $n \times k_1$ dimension
(7) For each row in F as a sample of k_1 dimension, it is a total of n samples that are clustered by the input clustering method. The clustering dimension is k_2.
(8) Get the cluster division $C(c_1, c_2, \ldots, c_{k_2})$.

3   CF Recommendation Based on the Latent Factor Model and Improved Spectral Clustering

3.1   Latent Factor Model

Latent Factor Model can be seen as the improved SVD model. Because the SVD model fills a sparse matrix, which requires a large amount of storage space and the computational complexity is high, it is difficult to apply. In order to solve the shortcomings of the SVD, Simon Funk proposed the Latent Factor Model in the Netflix Prize [13]. The core idea of the model is to connect user interests and items through implicit features. The algorithm decomposes the scoring matrix into two low-dimensional matrices through matrix decomposition [14]. The one is the user implicit feature matrix, and the other is the item implicit feature matrix. Then multiplying the low-dimensional implicit matrices to represent the predicted score. Suppose user u scores item i as $\bar{r}_{ui}$, the formula is as follows.

$$\bar{r}_{ui} = p_{uf}^T q_{if} \tag{1}$$

p_{uf} represents the user implicit feature matrix. q_{if} represents the item implicit feature matrix. p_{uf} and q_{if} can be learned by minimizing the RMSE through the observations in the training set.

$C(p, q)$ is called the loss function and is used to estimate the parameters. $\lambda \left(\|p_u\|^2 + \|q_i\|^2 \right)$ is a regular term added to prevent overfitting. The loss function described above can be optimized by using the random gradient descent method [15].

The fastest direction of descent can be found by calculating the partial derivative of the parameter. What you can get is as follows.

$$\begin{cases} \frac{\partial C}{\partial p_{uf}} = -2q_{ik} + 2\lambda p_{uk} \\ \frac{\partial C}{\partial p_{if}} = -2p_{uk} + 2\lambda q_{ik} \end{cases} \tag{2}$$

Move forward along the fastest descent direction according to the stochastic gradient descent method. The following recursive formula can be obtained:

$$\begin{cases} p_{uf} = p_{uf} + \alpha(q_{ik} - \lambda p_{uk}) \\ q_{if} = q_{if} + \alpha(p_{uk} - \lambda q_{ik}) \end{cases} \tag{3}$$

In the formula, α is the learning rate, and its value can be obtained through repeated experiments. The termination condition of the iteration is that the loss function reaches a minimum. Since the single eigenvector elements are partially derived by the gradient descent method, the items related to the remaining eigenvectors are all 0, which avoids the problem that the general SVD algorithm needs to iterate or interpolate the unknown elements in the matrix. In addition, it can be seen that the entire training process only needs to store the memory space of the feature vector when a scoring data is used for each iteration, so it is especially suitable for cases with very large data sets [16].

3.2   User Spectral Clustering

3.2.1   Improved Spectral Clustering

The formation of the nearest neighbor of the user affects the mining of the user interest, which affects the accuracy of the recommendation. Because of the sparseness of the scoring matrix, finding the nearest neighbor in the global scope and recommending will increase the computational cost, so the clustering algorithm is needed to improve [17–20]. Traditional clustering algorithms are sensitive to the spatial distribution of datasets, and the clustering effect is not ideal [21]. Therefore, this paper selects the spectral clustering algorithm that is often used in the field of image segmentation to identify arbitrary shape spaces and converges to the global optimal solution. Considering that users actually have different preferences for different types of movies, such as a user likes love and war movies, which should belong to different categories of user clusters, but the hard clustering algorithm can only classify the user's preferences as the class of love or war, which is not in line with reality. Therefore, the FCM (fuzzy C-means) algorithm is used to replace the K-means in the spectral clustering, and the user's degree of membership function for each class makes the clustering result more relevant to the fact [24]. The procedure is as follows.

(a) Input a scoring matrix R. Construct a user similarity matrix W and an undirected weighting graph G(V, E) from Eq. (1), wherein vertex V represents a user in the matrix R, and edge E represents a similarity between users. Then the unnormalized Laplacian matrix L of the graph G is

$$L = D - W \tag{4}$$

In the formula, D is the degree matrix, and D_{ii} is the sum of the elements of the i-th row of the matrix W.

(b) Normalize the unnormalized Laplacian matrix L:

$$L_{sym} = D^{-1/2}LD^{-1/2} = I - D^{-1/2}WD^{-1/2} \tag{5}$$

In the formula, I is the unit matrix.

(c) Calculate the top k minimum eigenvalues of the matrix L_{sym} and their corresponding eigenvectors $V_1, V_2, \ldots, V_k$ constitute the matrix $V = [V_1, V_2, \ldots, V_k]$.

(d) The matrix V is clustered by the FCM algorithm.

3.2.2   Cluster Validity Index

The number of clusters will have a direct impact on the quality of recommendation [22]. Too many numbers will reduce the nearest neighbors of the user and reduce the accuracy of recommendation. Too few numbers will increase the search cost of the nearest neighbors [18]. Previous algorithms [17–20] lack the determination of the number of clusters. This paper determines the number of user clusters based on the degree of membership matrix after the FCM clustering.

Chen et al. [23] proposed the validity index V_{cs} based on the degree of membership rather than distance, thus it can overcome the influence of noise points or abnormal points on the cluster center. The index V_{cs} consists of two parts: compactness and dispersion. The compactness is defined as follows.

$$C = \sum_{i=1}^{c} \sum_{j=1}^{n} C_{ij} \tag{6}$$

In the formula, when $u_{ij} \geq 1/c$, $C_{ij} = u_{ij}^2$; $u_{ij} < 1/c$, $C_{ij} = 0$. c is the number of clusters. u_{ij} is the degree of membership that the j-th object to the i-th class, C_{ij} is the compactness between the i-class and j-class samples, and the larger the C_{ij}, the more compact the clustering result.

The dispersion between the i-th class and the j-th class is defined as

$$S_{ij} = \min(u_{ik}, u_{jk}) \, k = 1, 2, \ldots, n \tag{7}$$

The overall dispersion is defined as $S = \max_{i=1, j=1, i \neq j} S_{ij}$, the smaller the dispersion, the more clear the division between the two classes.

In summary, the cluster validity index based on the degree of membership is defined as

$$V_{cs} = C/S \tag{8}$$

In the paper, the index V_{cs} is used to determine the number of clusters. The larger the V_{cs} value, the better the clustering result.

3.3 Collaboration Filtering Recommendation Algorithm Based on the Latent Factor Model and Improved Spectral Clustering

There are many ideas for combining recommendation algorithms. In the paper, a combination recommendation algorithm is designed by two collaborative filtering recommendation algorithms. The Latent Factor Model is used to evaluate the user's prediction scores for some unrated items firstly, and the score value is filled into the user-item scoring matrix. Then after solving the problem of data sparsity, the improved spectral clustering algorithm is used to obtain better recommendation results based on the filled matrix.

Steps of recommended algorithm for combining the Latent Factor Model and the improved spectral clustering model are as follows.

Step1. Fill the missing data matrix to obtain the whole user-item matrix R by using the Latent Factor Model.

Step2. The scoring matrix R is clustered by the improved spectral clustering algorithm to aggregate users into different k clusters. The distance function uses the modified cosine similarity, and the formula of the modified cosine similarity [25, 26] is as follows.

$$sim(u_i, u_j) = \frac{\sum_{c \in I_{i,j}} (r_{i,c} - \overline{r}_i)(r_{j,c} - \overline{r}_j)}{\sqrt{\sum_{c \in I_i} (r_{i,c} - \overline{r}_i)^2} \sqrt{\sum_{c \in I_j} (r_{j,c} - \overline{r}_j)^2}} \tag{9}$$

In the formula, $r_{i,c}$ and $r_{j,c}$ respectively represent the score of the user u_i and the user u_j on item i_c. $\overline{r}_i$ and $\overline{r}_i$ respectively represent the average score of the user u_i and user u_j on item. Let the item set of the user u_i and the user u_j's public score as $I_{i,j}$. I_i and I_j respectively represent the item set of the user u_i and the user u_j's score.

Step3. For the active user a, the distance between a and k class centers is calculated, and a is designated as the nearest cluster to the center.

Step4. In clustered clusters, the user similarity $sim(U_a, U_i)$ is calculated, and N most similar users are selected as the nearest neighbors.

Step5. According to the scoring data of the nearest neighbor user set, The current user a's scoring of non-scored items is weightedly predicted.

Step6. Select the first N predictive scoring to output.

4 Experimental Analysis and Discussion

In this section, we investigate the performance of CFRALFMISC. First, we describe experimental data sets and evaluation metrics. Then, we describe the optimization experiment of the cluster number. Finally, we conduct our experiments on different collaboration recommendation algorithms and analyze experimental results.

4.1   Experimental Data Sets and Evaluation Metrics

The experiment used film data collected by the GroupLens team in the university of Minnesota. There are data sets of MovieLens (100 KB), MovieLens (1 MB) and MovieLens (10 MB). Take a 1 MB data set as an example. MovieLens (1M) contains 1 million ratings from 6000 users on 4000 movies. The rating level is 1–5. The data set used in the experiment is a 100 KB dataset that contains 100,000 ratings from 1000 users on 1700 movies, and the data set has a sparsity of 93.7%. Sparse data refers to data in which most values in the data frame are missing or zero. In the experiment, 80% of the data was used as the training set and 20% of the data was used as the test set.

Root mean square error (RMSE) is the most commonly used evaluation metric of recommendation quality. The smaller the RMSE, the more accurate the prediction results. The paper adopts the RMSE as the evaluation metric of the recommendation performance. The formula of the RMSE is as follows.

$$\text{RMSE} = \sqrt{\frac{\sum_{i,j} |R_{ij} - \overline{R_{ij}}|^2}{N}} \tag{10}$$

In the formula, R_{ij} is the true score in the test data, $\overline{R_{ij}}$ is the score predicted by the recommendation algorithm, and N is the number of the test score data.

The formula of the accuracy is as follows.

$$\text{precision} = \frac{\sum_{u \in U} |R(u) \cap T(u)|}{\sum_{u \in U} |R(u)|} \tag{11}$$

R(u) represents a list of recommendations made by the algorithm for the user on the training set. T(u) represents the list of behaviors of the user on the test set. I represents a collection of all items. The higher the value of the accuracy, the better the effect of the recommendation.

4.2   The Optimization Experiment of the Cluster Number

The clustering algorithm based on the FCM is used to cluster users. The degree of membership matrix after clustering and the V_{cs} index are used to determine the optimal number of clusters. The fuzzy clustering index in the FCM algorithm is selected as 2. As shown in Fig. 1, when the number of clusters is 6, the cluster validity index V_{cs} reaches the maximum, so the user is selected to be clustered into six categories. Table 1 is the distribution of the number of users in each category after clustering.

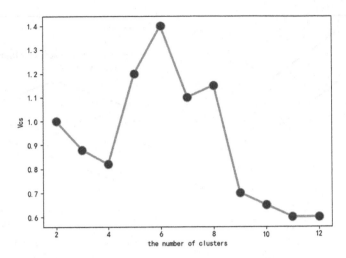

Fig. 1. Determine the number of the user cluster

Table 1. The distribution of the number of users in each category

	C_1	C_2	C_3	C_4	C_5	C_6
The number of users in each category	141	170	185	158	165	124

4.3 The Results of the Experiment

In order to test the recommended effect of the CFRALFMISC, this paper adopts the measure of the RMSE to compare the recommended effects with UserCF, ItemCF and the basic spectral clustering, and this paper also adopts the measure of the precision to compare the recommended effects with the basic spectral clustering, the LFM and the CFRALFMISC. In the Latent Factor Model, $\alpha = 0.01$, F = 40, $\lambda = 0.03$. In the spectral clustering model, the number of clusters are selected as 6. The experimental results are shown in Fig. 2. It can be seen from Fig. 2 that the proposed algorithm improves the recommendation accuracy compared with other algorithms and effectively improves the recommendation performance. It can be seen from Fig. 3 that the CFRALFMISC outperforms the basic spectral clustering in the precision. When the number of the nearest neighbors is more than 17, the performance of the CFRALFMISC is better than that of the LFM.

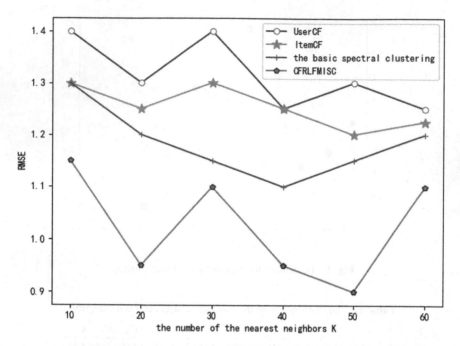

Fig. 2. The comparison of RMSE in different algorithms

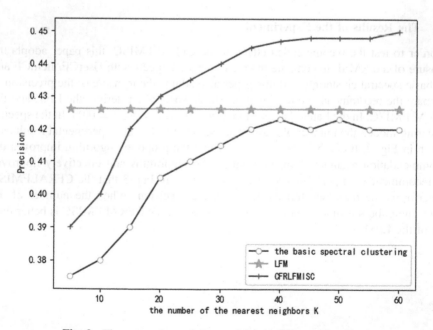

Fig. 3. The comparison of the precision in different algorithms

5 Conclusion

CF algorithm has been widely used in the field of recommender systems and has got better recommendation results, but it is affected largely by the sparse rating matrix and other problems. Combining the advantages of spectral clustering, the Latent Factor Model, the cluster validity index and the FCM algorithm, CFRALFMISC can achieve better recommendation results. In the future, we will try to use other clustering algorithms instead of the spectral clustering algorithm to further improve the clustering effect of the CFRALFMISC algorithm.

Acknowledgements. This research work was supported by the National Natural Science Foundation of China (Grant No. 61762031), Guangxi Key Research and Development Plan (Gui Science AB17195029, Gui Science AB18126006), Guangxi key Laboratory Fund of Embedded Technology and Intelligent System, 2017 Innovation Project of Guangxi Graduate Education (No. YCSW2017156), 2018 Innovation Project of Guangxi Graduate Education (No. YCSW2018157), Subsidies for the Project of Promoting the Ability of Young and Middle-aged Scientific Research in Universities and Colleges of Guangxi (KY2016YB184), 2016 Guilin Science and Technology Project (Gui Science 2016010202).

References

1. Marz, N., Warren, J.: Big Data: Principles and Best Practices of Scalable Realtime Data Systems. Manning Publications Co., Greenwich (2015)
2. Adomavicius, G., Tuzhilin, A.: Toward the next generation of recommender systems: a survey of the state-of-the-art and possible extensions. IEEE Trans. Knowl. Data Eng. **17**(6), 734–749 (2015)
3. Huang, L.W., Jiang, B.T., Lv, S.Y., Liu, Y.B., Li, D.Y.: Survey on deep learning based recommender systems. Chin. J. Comput. **41**(07), 1619–1647 (2018)
4. Sarwar, B.M., Karypis, G., Konstan, J., et al.: Recommender systems for large-scale e-commerce: scalable neighborhood formation using clustering. In: Proceedings of the 5th International Conference on Computer and Information Technology (2002)
5. Xu, B., Bu, J., Chen, C., et al.: An exploration of improving collaborative recommender systems via user-item subgroups. In: Proceedings of the 21st International Conference on World Wide Web, pp. 21–30 (2012)
6. Gong, S.J.: The collaborative filtering recommendation based on similar-priority and fuzzy clustering. In: Proceedings of Power Electronics and Intelligent Transportation System (PEIT 2008), pp. 248–251 (2008)
7. Li, H., Zhang, Y., Sun, J.H.: Research on collaborative filtering recommendation based on user fuzzy clustering. Comput. Sci. **39**(12), 83–86 (2012)
8. Bo, D., Jiang, J.L.: Collaborative filtering recommendation algorithm based on score difference level and user preference. J. Comput. Appl. **36**(04), 1050–1053+1065 (2016)
9. Wang, X.J.: Employing item attribute and preference to enhance the collaborative filtering recommendation. J. Beijing Univ. Posts Telecommun. **37**(06), 68–71+76 (2014)
10. Meng, X.F., Ci, X.: Big data management: concepts, techniques and challenges. J. Comput. Res. Dev. **50**(1), 146–169 (2013)
11. Zhao, S., Cao, Q., Chen, J.: A multi-atl method for transfer learning across multiple domains with arbitrarily different distribution. Knowl. Based. Syst. **94**, 60–69 (2016)

12. Saha, B., Gupta, S., Dinh, P.: Multiple task transfer learning with small sample sizes. Knowl. Inf. Syst. **46**(2), 315–342 (2016)
13. Piatetsky, G.: Interview with Simon Funk. ACM SIGKDD Explor. Newsl. **9**(1), 38–40 (2007)
14. Koren, Y., Bell, R., Volinsky, C.: Matrix factorization techniques for recommender systems. IEEE Comput. J. **42**(8), 30–37 (2009)
15. Koren, Y.: Factor in the neighbors: scalable and accurate collaborative filtering. ACM Trans. Knowl. Discov. Data (TKDD) **4**(1), 1–11 (2010)
16. Qiao, P., Cao, Y., Ren, Z.: A recommendation algorithm on fusion latent factor model and K-meansplus clustering model. Comput. Digit. Eng. **46**(06), 1108–1111 (2018)
17. Deng, A., Zuo, Z., Zhu, Y.: Collaborative filtering recommendation algorithm based on item clustering. MINI-MICRO Syst. **09**, 1665–1670 (2004)
18. Moradi, P., Ahmadian, S., Akhlaghian, F.: An effective trust-based recommendation method using a novel graph clustering algorithm. Phys. A: Stat. Mech. Appl. **436**(10), 462–481 (2015)
19. Mingjia, W.A.N.G., Jingti, H.A.N., Songqiao, H.A.N.: Collaborative filtering algorithm based on fuzzy clustering. Comput. Eng. **38**(24), 50–52 (2012)
20. Alper, B., Huseyin, P.: A comparison of clustering-based privacy-preserving collaborative filtering schemes. Appl. Soft Comput. **13**(5), 2478–2489 (2013)
21. Li, Z., Guiqiong, X., Zha, J.: Social recommendation algorithm based on spectral clustering with Nystrom extension. Appl. Res. Comput. **32**(11), 3238–3241 (2015)
22. Arbelaitz, O., Gurrutxaga, I., Muguerza, J., et al.: An extensive comparative study of cluster validity indices. Pattern Recogn. **46**(1), 243–256 (2013)
23. Chen, J., Pi, D.: A cluster validity index for fuzzy clustering based on non-distance. In: Proceedings of the 5th International Conference on Computational and Information Science, pp. 880–883 (2013)
24. Li, Q., Li, S., Xu, G.: Collaborative filtering recommendation algorithm based on spectral clustering and fusion of multiple factors. Appl. Res. Comput. **34**(10), 2905–2908 (2017)
25. Qian, H., Xia, Q., An, D., Zhang, Q.: Application of heuristic drift reduction algorithm in data processing for MEMS gyroscope. Transducer Microsyst. Technol. **29**(03), 109–111 +114 (2010)
26. Shi, Z., Han, B., Xu, Y., Ping, L.I.: Horizontal attitudes determination based on gravity adaptive complementary. Chin. J. Sens. Actuators **22**(07), 993–996 (2009)

Database

Hadoop + Spark Platform Based on Big Data System Design of Agricultural Product Price Analysis and Prediction by HoltWinters

Yun Deng[1,2(✉)], Yan Zhu[1,2], Qingjun Zhang[1,2], and Xiaohui Cheng[1,2]

[1] Guangxi Key Laboratory of Embedded Technology and Intelligent System, Guilin University of Technology, Guilin 541004, China
woshidengyun@sina.cmn
[2] Department of Information Science and Engineering, Guilin University of Technology, Guilin 541004, China

Abstract. In the market of agricultural products, the price of agricultural products is affected by production cost, market supply and other factors. In order to obtain the market information of agricultural products, the price fluctuation can be analyzed and predicted. A distributed big data software platform based on Hadoop, Hive and Spark is proposed to analyze and forecast agricultural price data. Firstly, Hadoop, Hive and Spark big data frameworks were built to store the data information of agricultural products crawled into MYSQL. Secondly, the information of agricultural products crawled from MYSQL was exported to a text file, uploaded to HDFS, and mapped to spark SQL database. The data was cleaned and improved by Holt-Winters (three times exponential smoothing method) model to predict the price of agricultural products in the future. The data cleaned by spark SQL was imported and predicted by improved Holt-Winters into MYSQL database. The technologies of pringMVC, Ajax and Echarts were used to visualize the data.

Keywords: Hadoop · Spark · Big data ·
Analysis and forecast of agricultural product prices · Holt-Winters

1 Introduction

In the agricultural products market, the price of agricultural products is affected by various factors, including producing cost, market supply, market demand, currency and internation. As enjoying vast territory and abundant resources, China has fairly large amounts of production as well as market information of the agricultural products annually. There are so many advantages with using big data technology to count and predict the market information of agricultural products: first, integrating and sharing data; second, accessing to the market information of agricultural products from all aspects; third, making advance prejudgment and analysis as well as prediction.

This paper is to establish big data platforms such as Hadoop and spark, through storing and cleaning the acquired data on the platform for many times and then using the improved Holt-Winters index sliding method, to forecast agricultural products price

X. Cheng et al. (Eds.): ICPCSEE 2019, CCIS 1058, pp. 103–123, 2019.
https://doi.org/10.1007/978-981-15-0118-0_9

in a future period. Finally, the cleaned data and the predicted price are input into the MYSQL database to visualize the data. The algorithmic development and expected results on this platform will enrich the research results based on price-forecasting of agricultural products at home and abroad and will be of more significant theoretical meaning and study value at present.

2   System Construction

2.1   A Subsection Sample

The technical route of the work done in this paper is divided into the following steps: The first step is the foundation layer–establishing the standby big data platforms such as Hadoop, spark and so on; The second step is the acquisition of data, that is, collecting national agricultural data of authoritative web pages and parsing and processing the HTML page with utilizing beautiful Soup; The third step is data storage–inputting the parsed data into distributed file management of HDFS; The fourth step is data cleaning–mapping the data that are input into the cluster of HDFS to the database of Spark-sql, using HQL programming to clean data; The fifth step acts as the price forecast of agricultural products in a future period based on the upgraded Holt-Winters index sliding method; The sixth step is to channel the cleaned data and predicted price into the MYSQL database for data visualization. The process is shown in the Fig. 1.

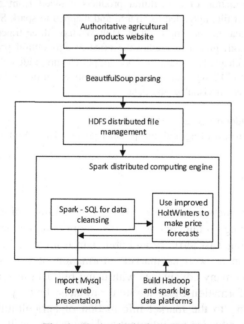

Fig. 1. Technological road map.

The quality of the data hardware equipment of the machine plays a considerable role. Therefore, the platform primarily selects two HP 360PGEN81U servers to install the Linux operating system, of which big data software, ranging from the mysql of version 5.7.18, the hive of version 2.1.1, the spark of version 2.1.1, to the jdk of version 8u131-linux-x64, to build a cluster of a major node and a affiliated node. This platform deploys the software platforms of big data, covering Hadoop, Hive, and Spark, of which the architecture is shown as the Fig. 2.

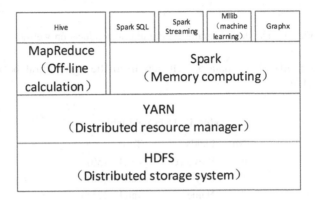

Fig. 2. Architecture diagram.

This big data platform mainly includes the components of HDFS, YARN, MapReduce, Spark and Hive. The functions of each component are listed as follows:

(1) HDFS system is mainly used to store huge data sets to increase the throughput of data access;
(2) MapReduce is primarily for the data set on HDFS operation;
(3) YARN is principally utilized to provide the upper application with unified resource management and scheduling;
(4) Hive is the infrastructure of data warehouse on the Hadoop, which is primarily applied for storing, inquiring and analyzing large-scale data stored in the Hadoop;
(5) Spark starts its memory distribution data sets mainly to speedily perform the large-scale data.

3 The Acquisition and Cleaning of Data

3.1 The Acquisition of Data

3.1.1 Detailed Design of Data Source Module

The data here all stem from an authoritative agricultural websites that are similar to the agriculturally benefit network (http://www.cnhnb.com). Hence, this paper takes one of the agriculturally benefit networks as an example. And the data of Shanyao (shanyao is also named as yam) are shown in the Fig. 3.

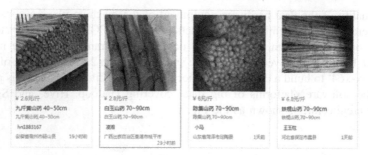

Fig. 3. The data diagram of agricultural products on website.

The various fields of the data are chosen from the agricultural benefit network, shown in the Table 1.

Table 1. Data field table.

Num	Fields	Type
0	Province	varchar(50)
1	City	varchar(50)
2	Name	varchar(255)
3	Price	float
4	Price_unit	varchar(50)
5	Year	varchar(50)
6	Month	varchar(50)
7	Day	varchar(50)
8	Time_release	datetime
9	Variety	varchar(255)

3.1.2   Detailed Design of the Chosen Data

This article utilizes the language Python3 to select the data on the website. The specific process for edit codes is listed in the Fig. 4 below.

In the whole process of choosing data, the steps are as follows:

(1) Obtaining information of the entire Web page according to the given URL;
(2) Using BeautifulSoup to parse the entire Web page to get the number of pages following the page;
(3) Setting the initial value of the page number as 1, and utilizing a "for circulation" to load all the URL into a list of "produce details";
(4) Establishing an empty list data "produce_total" to use as containing all the data of agricultural products, and establishing a dictionary "result" to involve all the data of agricultural products;
(5) In the two-layer "for circulation", adopting "select" to capture the release time that is similar to "a few months ago", "how many days ago", "several hours" and "a few minutes ago", and working out the time that is similar to 2018, 2018-03, 2018-03-21 based on the current date and then putting these dates into the dictionary "result";

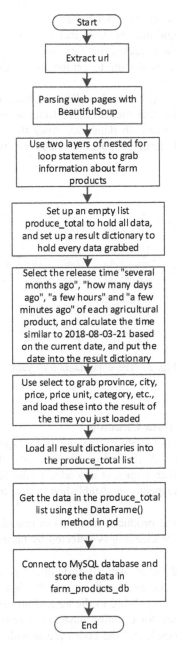

Fig. 4. The flow chart of acquiring.

(6) Employing "select" to select the factors, involving provinces, cities, prices, price units, categories, etc., and putting these factors into the "result" loading time just now. After a page finished, adding the data stored in the dictionary "result" to the list "produce_total" and continuing to get the data of the unfinished page until all the pages are completed;

(7) Utilizing the DataFrame () method in PD to obtain the data stored in the list Produce_total list, connecting the MYSQL database, and storing all the obtained data into the farm_products_db database. After the work done, all the data is already in the MYSQL farm_products_db database, which are presented as the Fig. 5.

河南	焦作温县 铁棍山药 4 7.5	元/斤	2018	2018-01	2018-01-05	2018-01-05 10:5! 铁棍山药
河南	焦作温县 铁棍山药 5 8	元/斤	2018	2018-01	2018-01-05	2018-01-05 10:5! 铁棍山药
河南	焦作武陟 铁棍山药 5 8	元/斤	2018	2018-01	2018-01-05	2018-01-05 10:3! 铁棍山药
河南	焦作武陟 铁棍山药 7 12	元/斤	2018	2018-01	2018-01-05	2018-01-05 10:3! 铁棍山药
广东	韶关翁源 桂淮2号淮 2.3	元/斤	2018	2018-01	2018-01-05	2018-01-05 10:3; 桂淮2号淮山药
河南	焦作温县 铁棍山药 2 5	元/斤	2018	2018-01	2018-01-05	2018-01-05 10:3(铁棍山药
广西	来宾兴宾 紫玉淮山山 3.5	元/斤	2018	2018-01	2018-01-05	2018-01-05 10:2(紫玉淮山山药
云南	文山文山 紫玉淮山山 4	元/斤	2018	2018-01	2018-01-05	2018-01-05 10:2; 紫玉淮山山药
云南	文山文山 紫玉淮山山 4	元/斤	2018	2018-01	2018-01-05	2018-01-05 10:2; 紫玉淮山山药
云南	文山文山 紫玉淮山山 6	元/斤	2018	2018-01	2018-01-05	2018-01-05 10:2; 紫玉淮山山药
河南	郑州二七 铁棍山药 4 7.5	元/斤	2018	2018-01	2018-01-05	2018-01-05 10:1! 铁棍山药
河南	焦作温县 铁棍山药 4 7.5	元/斤	2018	2018-01	2018-01-05	2018-01-05 10:1! 铁棍山药
河南	焦作温县 铁棍山药 3 7	元/斤	2018	2018-01	2018-01-05	2018-01-05 10:1(铁棍山药
河南	焦作温县 铁棍山药 5 5	元/斤	2018	2018-01	2018-01-05	2018-01-05 10:1(铁棍山药
河南	焦作温县 铁棍山药 3 4	元/斤	2018	2018-01	2018-01-05	2018-01-05 10:0! 铁棍山药
河南	焦作温县 铁棍山药 7 6	元/斤	2018	2018-01	2018-01-05	2018-01-05 10:0! 铁棍山药

Fig. 5. The data diagram of yam.

3.2   Data Cleaning

3.2.1   Detailed Design of Data Cleaning

Firstly, the work of preparation before cleaning is to export the selected data from MYSQL by the way of text and file first of all. Secondly, the data of the agricultural products, chosen from the Web pages, is uploaded to the HDFS after simple prepossessing. Finally, folders are respectively established to save the data in the future on the Hadoop cluster.

The number of agricultural products, involved in this platform, is relatively large, and the general situation of its cleaning is reflected by taking the yam as an example, shown in the Fig. 6.

The article has uploaded the all data of agricultural products to the HDFS previously, but they're still not in the Spark-sql. Hence, the first thing to do is to create an external table to map the data involving in the agricultural products to the database of SPARK-SQL, so as to prepare for cleaning data of the next step (yam is taken as the example of its cleaning process), and the codes are as follows:

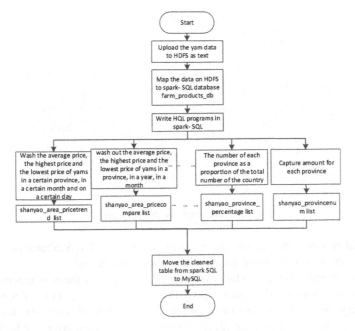

Fig. 6. Overview of yam cleaning.

-The constructed external table of original data of yam
DROP TABLE IF EXISTS farm_products_db.zongshanyao;
CREATE EXTERNAL TABLE zongshanyao(
province string,
city string,
name string,
price float,
price_unit string,
time_release string,
year string,
month string,
day string)
ROW FORMAT DELIMITED FIELDS TERMINATED BY '\t'
LOCATION '/user/glhz/data/farm_products/shanyao';

Following operating the program written by the above HQL and after the shift between map and reduce, the data of the agricultural products on the HDFS has been saved in the database farm_products_db of the Spark-sql, as shown in the Fig. 7.

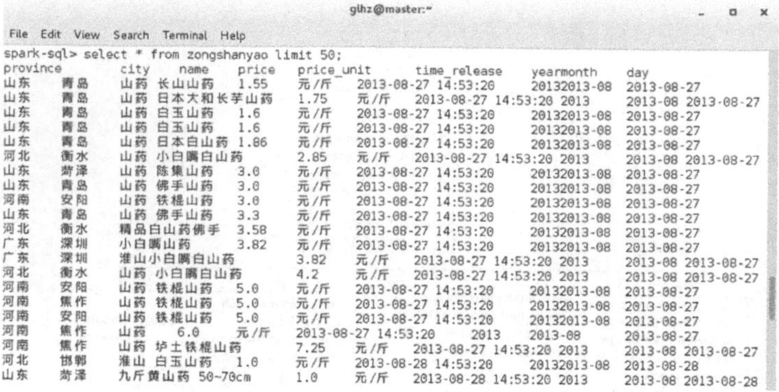

```
glhz@master:~
File Edit View Search Terminal Help
spark-sql> select * from zongshanyao limit 50;
province      city   name   price   price_unit   time_release   yearmonth   day
山东  青岛  山药 长山山药  1.55  元/斤  2013-08-27 14:53:20  20132013-08  2013-08-27
山东  青岛  山药 日本大和长芋山药  1.75  元/斤  2013-08-27 14:53:20 2013  2013-08 2013-08-27
山东  青岛  山药 白玉山药  1.6  元/斤  2013-08-27 14:53:20  20132013-08  2013-08-27
山东  青岛  山药 白玉山药  1.6  元/斤  2013-08-27 14:53:20  20132013-08  2013-08-27
山东  青岛  山药 日本白山药 1.86  元/斤  2013-08-27 14:53:20  20132013-08  2013-08-27
河北  衡水  山药 小白嘴白山药  2.85  元/斤  2013-08-27 14:53:20 2013  2013-08 2013-08-27
山东  菏泽  山药 陈集山药  3.0  元/斤  2013-08-27 14:53:20  20132013-08  2013-08-27
山东  青岛  山药 佛手山药  3.0  元/斤  2013-08-27 14:53:20  20132013-08  2013-08-27
河南  安阳  山药 铁棍山药  3.0  元/斤  2013-08-27 14:53:20  20132013-08  2013-08-27
山东  青岛  山药 佛手山药  3.3  元/斤  2013-08-27 14:53:20  20132013-08  2013-08-27
河北  衡水  精品白山药佛手  3.58  元/斤  2013-08-27 14:53:20  20132013-08  2013-08-27
广东  深圳  小白嘴山药  3.82  元/斤  2013-08-27 14:53:20  20132013-08  2013-08-27
广东  深圳  淮山小白嘴白山药  3.82  元/斤  2013-08-27 14:53:20 2013  2013-08 2013-08-27
河北  衡水  山药 小白嘴白山药  4.2  元/斤  2013-08-27 14:53:20 2013  2013-08 2013-08-27
河南  安阳  山药 铁棍山药  5.0  元/斤  2013-08-27 14:53:20  20132013-08  2013-08-27
河南  焦作  山药 铁棍山药  5.0  元/斤  2013-08-27 14:53:20  20132013-08  2013-08-27
河南  安阳  山药 铁棍山药  5.0  元/斤  2013-08-27 14:53:20  20132013-08  2013-08-27
河南  焦作  山药  6.0  元/斤  2013-08-27 14:53:20  2013  2013-08  2013-08-27
河南  焦作  山药 垆土铁棍山药  7.25  元/斤  2013-08-27 14:53:20 2013  2013-08 2013-08-27
河北  邯郸  淮山 白玉山药  1.0  元/斤  2013-08-28 14:53:20  20132013-08  2013-08-28
山东  菏泽  九斤黄山药 50~70cm  1.0  元/斤  2013-08-28 14:53:20 2013  2013-08 2013-08-28
```

Fig. 7. Data graph in the yam table.

It should be noted that due to there are more species of agricultural products, the yam, one of them, is only listed to introduce above and below.

The data of agricultural products is mapped to spark-sql through external tables. And the specific demands demonstrated by this platform are as: (1) the demonstration of the average price, the highest price and the lowest price of some species in every month in a certain year of an area; (2) the display of the average price, the highest price and the lowest price of some species in some month in a certain year of a certain area; (3) the demonstration of the price fluctuations of some agricultural product in the recent half of a month; (4) the collecting amounts of the data of agricultural products price in the national agricultural market, etc. As the concrete cleaning terms are so many, here several varieties are introduced, and their codes are as the follows:

a. Regional price comparisons

```
DROP TABLE IF EXISTS farm_products_db. shan-
yao_area_pricecompare;
CREATE TABLE  shanyao_area_pricecompare AS
SELECT data_2.year,
split(data_2.month,'-')[1] AS month,
split(data_2.day,'-')[2] AS day,
AVG(data_2.price) AS average_price,
MAX(data_2.price) AS max_price,
MIN(data_2.price) AS min_price
FROM
(SELECT year,month,day,price
FROM farm_products_db.zongshanyao
)AS data_2
GROUP BY year,month,day
ORDER BY year,month,day;
```

The cleaned data results have already been in the table Shanyao_area_pricecompare of the database farm_products_db from the Spark-sql, as shown in the Fig. 8.

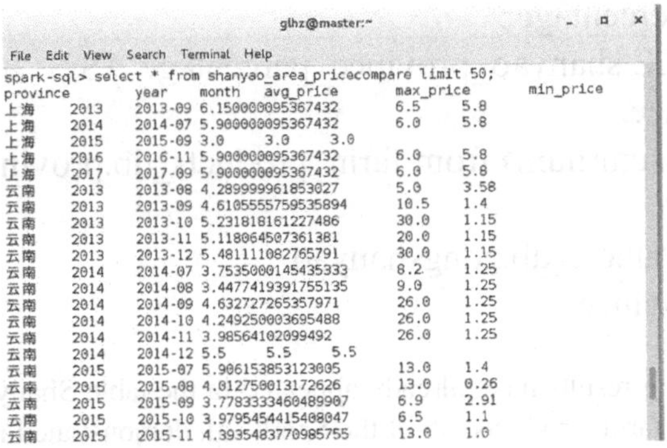

```
glhz@master:~
File Edit View Search Terminal Help
spark-sql> select * from shanyao_area_pricecompare limit 50;
province        year    month   avg_price               max_price       min_price
上海    2013    2013-09 6.150000095367432       6.5     5.8
上海    2014    2014-07 5.900000095367432       6.0     5.8
上海    2015    2015-09 3.0     3.0     3.0
上海    2016    2016-11 5.900000095367432       6.0     5.8
上海    2017    2017-09 5.900000095367432       6.0     5.8
云南    2013    2013-08 4.289999961853027       5.0     3.58
云南    2013    2013-09 4.610555557595358 94    10.5    1.4
云南    2013    2013-10 5.231818161227486       30.0    1.15
云南    2013    2013-11 5.118064507361381       20.0    1.15
云南    2013    2013-12 5.481111082765791       30.0    1.15
云南    2014    2014-07 3.7535000145435333      8.2     1.25
云南    2014    2014-08 3.4477419391755135      9.0     1.25
云南    2014    2014-09 4.632727265357971       26.0    1.25
云南    2014    2014-10 4.249250003695488       26.0    1.25
云南    2014    2014-11 3.98564102099492        26.0    1.25
云南    2014    2014-12 5.5     5.5     5.5
云南    2015    2015-07 5.906153853123005       13.0    1.4
云南    2015    2015-08 4.012750013172626       13.0    0.26
云南    2015    2015-09 3.7783333460489907      6.9     2.91
云南    2015    2015-10 3.9795454415408047      6.5     1.1
云南    2015    2015-11 4.3935483770985755      13.0    1.0
```

Fig. 8. The data graph in the shanyao_area_pricecompare table.

b. Regional price trends

```
DROP TABLE IF EXISTS
farm_products_db.shanyao_area_pricetrend;
CREATE TABLE shanyao_area_pricetrend AS
SELECT province,month,day,
AVG(price) AS avg_price
FROM farm_products_db.zongshanyao
GROUP BY province,month,day
ORDER BY province,month,day;
```

The finished results have existed in the table Shanyao_area_pricetrend of the database farm_products_db from Spark-sql, as shown in the Fig. 9.

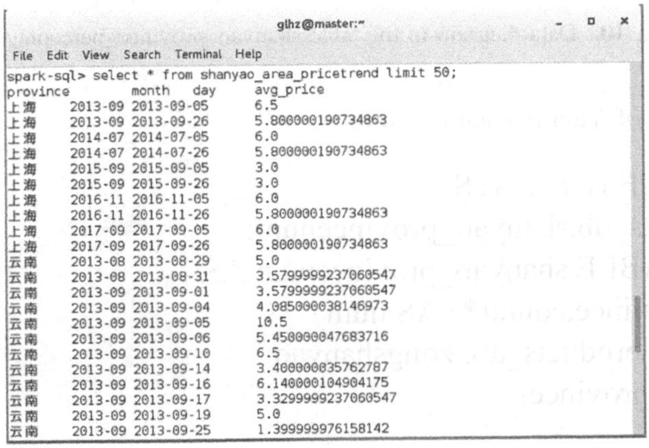

```
glhz@master:~
File Edit View Search Terminal Help
spark-sql> select * from shanyao_area_pricetrend limit 50;
province        month   day             avg_price
上海    2013-09 2013-09-05      6.5
上海    2013-09 2013-09-26      5.800000190734863
上海    2014-07 2014-07-05      6.0
上海    2014-07 2014-07-26      5.800000190734863
上海    2015-09 2015-09-05      3.0
上海    2015-09 2015-09-26      3.0
上海    2016-11 2016-11-05      6.0
上海    2016-11 2016-11-26      5.800000190734863
上海    2017-09 2017-09-05      6.0
上海    2017-09 2017-09-26      5.800000190734863
云南    2013-08 2013-08-29      5.0
云南    2013-08 2013-08-31      3.5799999237060547
云南    2013-09 2013-09-01      3.5799999237060547
云南    2013-09 2013-09-04      4.085000038146973
云南    2013-09 2013-09-05      10.5
云南    2013-09 2013-09-06      5.450000047683716
云南    2013-09 2013-09-10      6.5
云南    2013-09 2013-09-14      3.400000035762787
云南    2013-09 2013-09-16      6.140000104904175
云南    2013-09 2013-09-17      3.3299999237060547
云南    2013-09 2013-09-19      5.0
云南    2013-09 2013-09-25      1.399999976158142
```

Fig. 9. The data graph of the table shanyao_area_pricecompare.

c. The proportion of yam to the national total quantity in each province

DROP TABLE IF EXISTS farm_products_db. shan-
yao_province_percentage;
CREATE TABLE shanyao_province_percentage AS
SELECT province,
count(*)/(select sum(num) from farm_products_db.provincenum)
AS percentage
FROM farm_products_db.zongshanyao
GROUP BY province;

The data-cleaning results have already appeared in the table Shanyao_province_per-
centage of database farm_products_db of the Spark-sql, demonstrated in the Fig. 10.

Fig. 10. Data diagram in the table shanyao_province_percentage.

d. The number of Yam in each province

DROP TABLE IF EXISTS
farm_products_db.shanyao_provincenum;
CREATE TABLE shanyao_provincenum AS
SELECT province,count(*) AS num
FROM farm_products_db. zongshanyao
GROUP BY province;

The data-cleaning results have been already in the table Shanyao_provincenum of the farm_products_db stored in the Spark-sql database, listed in the Fig. 11.

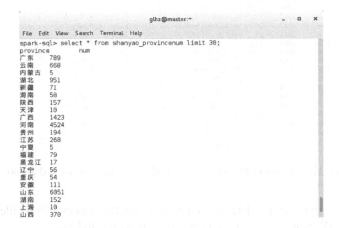

Fig. 11. Data diagram in the table shanyao_provincenum.

3.2.2   Transferring Data from Spark-sql to MYSQL

After cleaning the data in the spark-sql, the technologies such as SpringMVC, Ajax, Echarts, and MYSQL can't be directly invoked to visual display. Therefore, the data must be transferred into the farm_products_db of MYSQL from the farm_products_db of the database Sspark-sql so as to display the data in the interface with the Web tables. Since there are too many cleaned tables, here is only a description of the data imported into MYSQL of the table Shanyao_area_pricecompare of the cleaned yam.

The data in the Spark-sql is directed from the Spark-sql to the local/home/glhz/output/folder in the form of a script, concretely listed as the follows:

insert overwrite local directory '/home/glhz/output/'
row format delimited
fields terminated by '\t'
select * from shanyao_area_pricecompare;

After functioning the script, contents appears in the Shanyao_area_pricecompare table of the database farm_products_db of Spark-sql in the text/home/glhz/output/000000_0, as shown in Fig. 12.

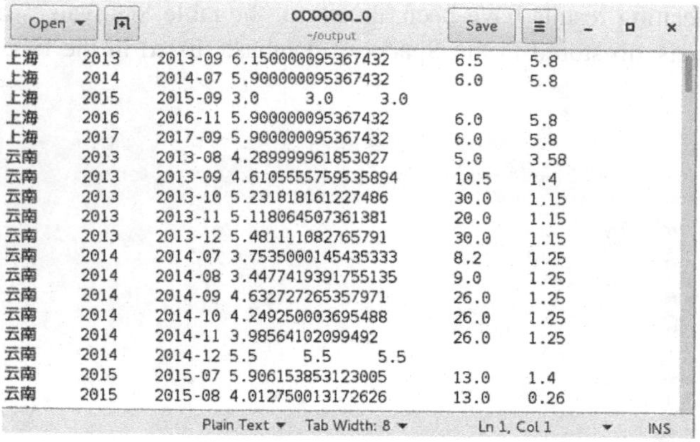

Fig. 12. Importing the data of the spark- SQL leads to local diagram.

Following importing the data into a local text file, the data in the text file should be put in to the MYSQL. The specific operations are as the follows;

(1) Constructing a table that matches these data in the database farm_products_db of MYSQL, with the following specific scripts:

CREATE TABLE `shanyao_area_pricecompare` (
  `province` varchar(50) NOT NULL,
  `year` varchar(50) NOT NULL,
  `month` varchar(50) NOT NULL,
  `avg_price` double NOT NULL,
  `max_price` double NOT NULL,
  `min_price` double NOT NULL
) ENGINE=InnoDB DEFAULT CHARSET=utf8;

(2) Importing the data in text to the table Shanyao_area_pricecompare of farm_products_db of the MYSQL, as follows:

Load data local infile '/home/glhz/output/000000_0' into table shanyao_area_pricecompare;

The data in the table Shanyao_area_pricecompareof farm_products_db in MYSQL is shown in the Fig. 13.

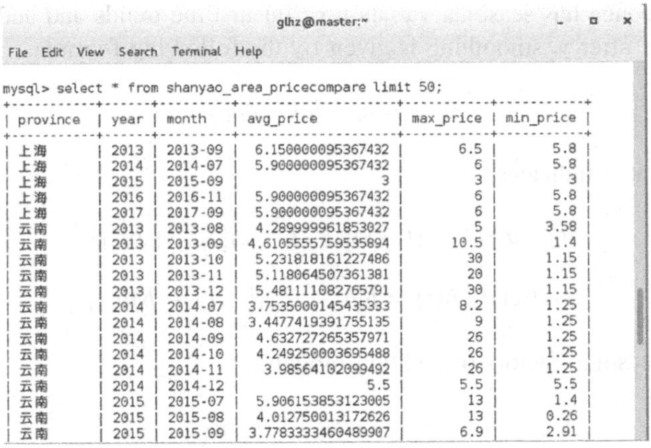
glhz@master:~

File Edit View Search Terminal Help

mysql> select * from shanyao_area_pricecompare limit 50;

province	year	month	avg_price	max_price	min_price
上海	2013	2013-09	6.150000095367432	6.5	5.8
上海	2014	2014-07	5.900000095367432	6	5.8
上海	2015	2015-09	3	3	3
上海	2016	2016-11	5.900000095367432	6	5.8
上海	2017	2017-09	5.900000095367432	6	5.8
云南	2013	2013-08	4.289999961853027	5	3.58
云南	2013	2013-09	4.6105555759535894	10.5	1.4
云南	2013	2013-10	5.231818161227486	30	1.15
云南	2013	2013-11	5.118064507361381	20	1.15
云南	2013	2013-12	5.481111082765791	30	1.15
云南	2014	2014-07	3.7535000145435333	8.2	1.25
云南	2014	2014-08	3.4477419391755135	9	1.25
云南	2014	2014-09	4.632727265357971	26	1.25
云南	2014	2014-10	4.249250003695488	26	1.25
云南	2014	2014-11	3.98564102099492	26	1.25
云南	2014	2014-12	5.5	5.5	5.5
云南	2015	2015-07	5.906153853123005	13	1.4
云南	2015	2015-08	4.012750013172626	13	0.26
云南	2015	2015-09	3.7783333460489907	6.9	2.91

Fig. 13. The data diagram in the table shanyao_area_pricecompare table of mysql.

4 The Price Prediction of Agricultural Product

4.1 Detailed Description of the Price Prediction

4.1.1 Holt-Winters Index Smoothing Method

The Holt-winters Index smoothing method can be used in the three following situations, and they are:

Under the situation of linear time trends without seasonal variations, similar to the double exponential smoothing method, this method serves as a sequence which is a predictable linear trend and does not include seasonal components. The only difference is that this smoothing method adopt two parameters, while the double smoothing method only employs one parameter. $\widehat{y_t}$ is the sequence after y_t smoothing, given by the following formula:

$$\hat{y}_{t+k} = a(t) + b(t)k \tag{1}$$

The intercept is represented by a and the trend is represented by b. These two parameters are defined by the following recursive formulas:

$$a(t) = \alpha y_t + (1 - \alpha)[a(t-1) + b(t-1)] \tag{2}$$

$$b(t) = \beta[a(t) - a(t-1)] + (1 - \beta)b(t-1) \tag{3}$$

α, β is between 0 and 1, which is an exponential smoothing method with two parameters. The predicted values are calculated as follows:

$$y_{t+k} = a(T) + b(T)k \tag{4}$$

These predicted values feature in a linear trend, of which the intercept is $a(T)$ and the slope is $b(T)$.

As the sequence has seasonal variation of linear time trends and addition models, the sequence $\hat{y}_t$ after y_t smoothing is given by the following formula:

$$\hat{y}_{t+k} = a_t + b_t k + c_{t+k} \tag{5}$$

Among then: a indicates:

$$a_t = \alpha[y_t - c_t(t-s)] + (1-\alpha)[a_{t-1} + b_{t-1}]$$

$$B \text{ is trend}: b(t) = \beta[a_t - a_{t-1}] + (1-\beta)b_{t-1}$$

c_t acts as seasonal factors of addition model:

$$c_t = \gamma[y_t - a_t] + (1-\gamma)c_{t-s}$$

α, β, γ, among 0 and 1, are damping factors. s is determined as seasonal frequency, and the predicted value is calculated by the following formula:

$$\hat{y}_{T+k} = a_T + b_T k + c_{T+k-s}$$

The estimation of seasonal factors uses the s the last period.

The smooth sequence $\hat{y}_t$ of y_t, which is of seasonal variations with linear time trends and multiplication models, is given by the following formula:

$$y_{t+k} = (a(t) + b(t)k)c_{t+k} \tag{6}$$

Of then: a(t) is referred to:

$$a(t) = \alpha\frac{y_t}{c_{t-s}} + (1-\alpha)[a(t-1) + b(t-1)]$$

$b(t)$ is trend:

$$b(t) = \beta[a(t) - a(t-1)] + (1-\beta)b(t-1)$$

c_t indicates seasonal factors of multiplication model:

$$c_t(t) = \gamma\frac{y_t}{a(t)} + (1-\gamma)c_{t-s}$$

Among them, the damping factors α, β, γ are during 0 and 1. s is specified as the seasonal frequency of the Cycle for Season, s is just like the monthly data of 12. And the predicted value is calculated by the following formula:

$$\hat{y}_{T+k} = [a(T) + b(T)k]c_{T+k-s} \tag{7}$$

The seasonal factor is estimated by s of the last period.

4.1.2   The Construction of the Model

The time model used here is upgraded based on the original Holt-Winters model, which can take one more province than the original model into consideration. As for the price predicting, the fact that accounting for not only the only single time series as well as adding a regional series is helpful for comprehensively searching for some hidden information renders price forecast of agricultural products be more accurate. The specific program flow is shown in the Fig. 14.

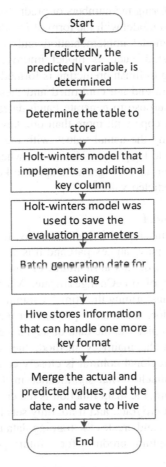

Fig. 14.   Flow chart of modeling construction.

The implementation process of the concrete model: (1) Setting crucial parameters to predict how many values there are after the original sequence; (2) Establishing and training a new model based on the original Holt-Winters model to enable it to address one more key series and to set its RDD format as (Holt-Winters Model, vector); (3) Constructing the vectors of the posteriori values established previously and then importing them into the Forecast method of the Holt-Winters model for prediction; (4) The predicted numbers subtract the intended predicted numbers set previously, and if the difference is less than 0, the prediction should be continued. Otherwise, the prediction will be ended; (5) To test whether the predicted values meet the requirements, these values are measured by the evaluation criteria of Holt-Winters model, mainly for the measurement evaluation of predictive values, including quantity, extreme difference, maximum, minimum, mean, variance, and standard deviation. Then, vector transposition–storing the numbers of prediction and the numbers obtained after the end of the evaluation model with the form of Dataframe. The whole set model, in addition to the specific model implementation and conduct. For example, other operations need to be assisted, such as model improvement and implementation. (1) The dates from beginning to finish predicting is formed and its format is set; Then, the training dates of start and end are confirmed and these two dates' difference are worked out to achieve the purpose to generate batch dates of specific day to save it. (2) The information that can cope with more than one key series format is saved into the table Hive. (3) Combining the actual values with the predicted ones with a date forms Ataframe (Date, Data) to keep in the Hive table, when the processed data format creates one more key column, and designates a specific key for output, with the real values being prior to the predicted values.

4.1.3   The Training of Model

Following the finished building model is built, the next task is to train the model, of which the training procedure plays an key part and it is also related to the final results. The training model algorithm of this agricultural product price prediction is mainly composed of three methods of timeChangeToDate, Main and mainKey. Of them, timeChangeToDate is that transforming the "time" column of the data into a fixed time format, while the mainKey does not mainly prepare for the null value, but effectively delete the null value, so as to pick out the data of keyName. Due to all the parameters are taken into account with the mainKey method, the region of this keyName is involved. Besides, and this regional column is closely connected with the search for hidden information, which attaches vital significance in the price prediction of agricultural products. Further more, the main function is the entry of the whole program as well as the specific execution program, and it not only sets the environmental variables and parameters, but also reads and creates the training data to apply them to the model, predicting the price of agricultural products for a future period. Its specific program flow is shown in the Fig. 15.

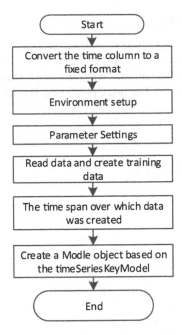

Fig. 15. The training flow diagram of model.

The training process is implemented after the success of the model building. Firstly, the three aspects, including transforming the date specific to day into a fixed time format, generating the concrete pattern required here according to the character string, and transforming the generated model into the format dataframe, are all conducive to the subsequent operations. Secondly, the surrounding and parameters are set, in which the main parameters contain: the data sheet in the Hive, the column names waiting for processing in the Hive, the starting time and ending time selected as the data reference, the number of predicted data, the table name stored, the key column of the output and the seasonal parameters. After that, all these, including reading the data in the Hive, omitting the null value and the null value of the key series according to the mainKey, and initializing all the parameters, covering the dtIndex (the formats are: yyy-MM-dd) that is specific to the day, and supplementing the missing date, set the RDD of the training data as TimeSeriesRDD ((key, DenseVector(series)) and complement the missing values. These can change the time span in the data into: start date + end date + increment. The next step is to do the critical action, that is, instantiating the previous models, creating and training the Holt-Winters model of one more province column. And after the entire process has run successfully, merging the actual values and predicted values and adding the date creates the form of Dataframe (Date, Data) to save it.

4.2   Experimental Analysis

This platform mainly analyzes yam, and analyzes its price forecast results for the next 15 days within Guangxi province. The predicted price is shown in Fig. 16. (Note: (1) This platform models with national agricultural products to predict the price of Guangxi in the next 15 days; (2) The price prediction for this platform is on December 7, 2017).

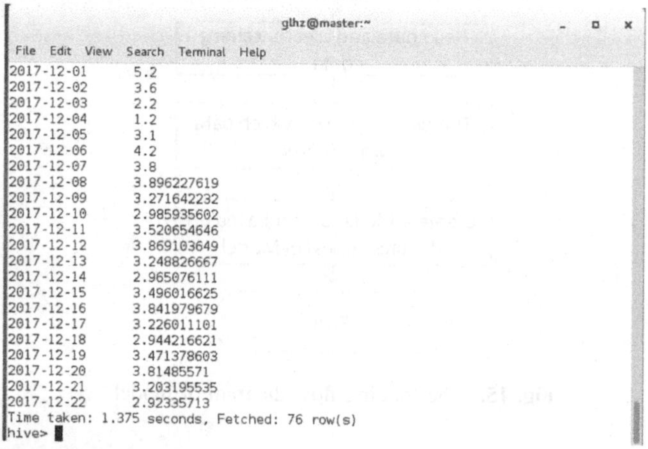

```
                              glhz@master:~                    _  □  x
 File  Edit  View  Search  Terminal  Help
2017-12-01        5.2
2017-12-02        3.6
2017-12-03        2.2
2017-12-04        1.2
2017-12-05        3.1
2017-12-06        4.2
2017-12-07        3.8
2017-12-08        3.896227619
2017-12-09        3.271642232
2017-12-10        2.985935602
2017-12-11        3.520654646
2017-12-12        3.869103649
2017-12-13        3.248826667
2017-12-14        2.965076111
2017-12-15        3.496016625
2017-12-16        3.841979679
2017-12-17        3.226011101
2017-12-18        2.944216621
2017-12-19        3.471378603
2017-12-20        3.81485571
2017-12-21        3.203195535
2017-12-22        2.92335713
Time taken: 1.375 seconds, Fetched: 76 row(s)
hive> █
```

Fig. 16. Data diagram of yam prediction in guangxi.

(3.8962276, 3.2716422, 2.9859357, 3.5206547, 3.8691037, 3.2488267, 2.9650762, 3.4960167, 3.8419797, 3.226011, 2.9442167, 3.4713786, 3.8148558, 3.2031956, 2.9233572). These value above is used to predict the price of yam in Guangxi for the next 15 days, which assesses the measurement through the Holt-Winters model: mean, variance, standard deviation, the maximum, the minimum, extreme difference, quantity, as shown in the Table 2.

Table 2. Measurement evaluating table of Holt-winters model.

Mean	Variance	Standard deviation	Max	Min	Range	Count
3.3785652	0.1256857	0.3545218	3.8962276	2.9233571	0.9728705	15

4.3   Visualization Design

The visualized data in this platform is in majority. Several aspects of the yam are listed in the following:

(1) The changing trend of historical data and price is shown in Fig. 17.

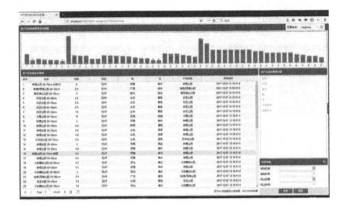

Fig. 17. The change trend chart of historical data and price data.

The module provides queries not only about historical data, but also about price changes of agricultural products. Through the chart, we can fully learn about not only the information of agricultural products, but also the trend of price changes in time.

(2) The price trend is shown in Fig. 18.

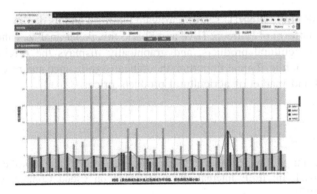

Fig. 18. Price trend chart.

The module provides a demonstration of the highest prices, average prices and the lowest prices of a certain agricultural output season monthly in the approaching years, in order to observe its market price tendency in some province.

(3) Price forecast is shown in Fig. 19.

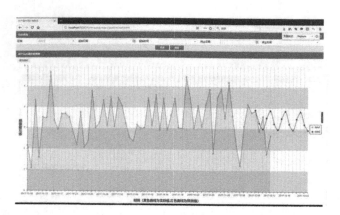

Fig. 19. Price trend chart.

The module mainly estimates the price level in the future based on the law of price movement with making use of the historical and current prices and the demand of provincial supply.

5   Conclusions

Based on the HP 360PGEN81U server hardware platform and its deploy distributed big data software platform of Hadoop, Hive and Spark, this paper analyzes and predicts the price data of agricultural products, and graphically displays the original price and forecast price, which is of great important theoretical significance and research value on the price trend and market of agricultural products price.

References

1. Yang, Y.M., Yu, H., Sun, Z.: Aircraft failure rate forecasting method based on Holt-Winters seasonal model. In: IEEE International Conference on Cloud Computing & Big Data Analysis (2017)
2. Assis, M.V.O.D., Hamamoto, A.H., Abrão, T., et al.: A game theoretical based system using Holt-Winters and genetic algorithm with fuzzy logic for DoS/DDoS mitigation on SDN Networks. IEEE Access **5**(99), 9485–9496 (2017)
3. Li, G., Shuang, Y.: Study on construction of agricultural product supply chain and benefit distribution based on agricultural industrialization poverty relief project: taking Rosa sterilis processing industry as the example. In: International Conference on Service Systems & Service Management (2017)
4. Mahdian, E., Karazhian, R.: Effects of fat replacers and stabilizers on rheological, physicochemical and sensory properties of reduced-fat ice cream. J. Agric. Sci. Technol. **15** (6), 1163–1174 (2018)

5. Yang, Z., Hu, C., Zhu, X.: The data analysis and the database construction for Pleurotus Eryngii product traceability system. In: International Conference on Information Science & Control Engineering (2017)
6. Yan, J.P., Huang, Y.F., Chua, K.C., et al.: River catchment rainfall series analysis using additive Holt-Winters method. J. Earth Syst. Sci. 125(2), 269–283 (2016)
7. Yan, Z., Fei, W., Jian, X.: Research on energy saving method of IDC CRAC system based on prediction of working load. In: International Conference on Smart & Sustainable City & Big Data (2016)
8. Rodriguez, H., Puig, V., Flores, J.J., et al.: Combined holt-winters and GA trained ANN approach for sensor validation and reconstruction: application to water demand flowmeters. In: Control & Fault-tolerant Systems (2016)
9. Razali, S.N.A.M., Rusiman, M.S., Zawawi, N.I., et al.: Forecasting of water consumptions expenditure using Holt-Winter's and ARIMA. J. Phys. Conf. Ser. (2018)
10. Wu, L., Gao, X., Xiao, Y., et al.: Using grey Holt-Winters model to predict the air quality index for cities in China. Nat. Hazards 88(2), 1–10 (2017)
11. Dasgupta, S.S., Mahanta, P., Roy, R., et al.: Forecasting industry big data with Holt Winters method from a perspective of in-memory paradigm. In: Meersman, R., et al. (eds.) On the Move to Meaningful Internet Systems: OTM 2014 Workshops 2014. LNCS, vol. 8842, pp. 80–85. Springer, Heidelberg (2014). https://doi.org/10.1007/978-3-662-45550-0_11
12. Fang, D., Yin, Z., Spicher, K.: Forecasting accuracy analysis based on two new heuristic methods and Holt-Winters-Method. In: IEEE International Conference on Big Data Analysis (2016)
13. Salamah, M., Kuswanto, H., Rachmi, A., et al.: Simulation study to construct the prediction interval for double seasonal Holt-Winters. Int. J. Appl. Math. Stat.™ 52(7), 43–50 (2014)

Oracle Bone Inscriptions Big Knowledge Management and Service Platform

Jing Xiong[1,2]($\boxtimes$) (iD), Qingju Jiao[1,2] (iD), Guoying Liu[1,2], and Yongge Liu[1,3]

[1] Anyang Normal University, Anyang 455000, Henan, China
jingxiong125@gmail.com
[2] Key Laboratory of OBI Information Processing, Ministry of Education, Anyang 455000, Henan, China
[3] Collaborative Innovation Center of International Dissemination of Chinese Language Henan Province, Anyang 455000, Henan, China

Abstract. Oracle bone inscriptions (OBI) has important historical and cultural values. The traditional OBI research methods are at a choke point, and the research method using computer science and information technology has opened up the way of OBI information processing. However, it faces some problems, such as large but not centralized knowledge system, long learning cycle, learning difficulties, being difficult to obtain resources, inconsistent format, low retrieval accuracy, and low knowledge sharing and reuse. To solve these problems, it designs an OBI big data research resource management and intelligent knowledge service platform. Its goal is to make full use of OBI big data characteristics and use the knowledge engineering technology to build an OBI knowledge ecosystem. The experiment results show that the platform proposed provides effective one-stop knowledge management and knowledge service.

Keywords: Oracle bone inscriptions · Big Knowledge · Knowledge graph

1 Introduction

Oracle Bone Inscriptions (OBI) is regarded as the primary form of modern Chinese characters. They are the oldest mature characters that exist in China. They are the treasures of the Chinese nation and have important historical value and scientific significance [1]. Because of the cultural relic characteristics of OBI, most researchers cannot directly touch the bones or tortoise shells and can only conduct research based on the OBI literature and documents. In addition, worldwide OBI fragments are gradually aging, damaged and disappearing, so the digital protection of OBI is imperative. After OBI has been included in the UNESCO Memory of the World Register, it attracted more and more researchers' attention. At the same time the problems OBI faces in digital sorting, basic data management and knowledge service are increasingly prominent.

The research of OBI must rely on a large number of literature and documents [2], and it must also rely on the background information of the bones and tortoise shells

© Springer Nature Singapore Pte Ltd. 2019
X. Cheng et al. (Eds.): ICPCSEE 2019, CCIS 1058, pp. 124–136, 2019.
https://doi.org/10.1007/978-981-15-0118-0_10

unearthed. Archaeological excavation can fully verify the reliability of OBI interpretation and research, and point out the direction for them. How to manage the massive archaeological information effectively and make full use of the information for research? How to find the required information in the massive literature? How to provide clues for OBI experts through intelligent information retrieval? These all need to rely on the literature database. Song Zhenhao, a famous OBI expert, pointed out that what OBI scholars need most is a comprehensive, authoritative and functional literature database. Many experts and scholars engaged in OBI agree with this view. Therefore, it is necessary and urgent to build the database of OBI literature.

Due to historical reasons, OBI bones or tortoise shells are scattered all over the world, so the researchers can only rely on OBI rubbings, pictures, descriptions, academic literature and other resources. Therefore, it is particularly necessary to establish a comprehensive and rich OBI literature database by collecting and organizing existing oracle bones. In addition, OBI and other cultural relics will eventually be damaged and disappear with the passage of time. Therefore, the OBI digitization is an effective measure for the rescue and protection of cultural relics. Therefore, it is urgent to effectively manage these digital OBI data and provide intelligent knowledge services based on them.

Now the accumulated OBI research data has gradually reflected the characteristics of big data [1]. Since the discovery of OBI in 1899, a large number of research materials have been gradually digitized. These materials include OBI archaeological data, images, videos, animation, papers and records, network data, etc. At present, about 150,000 pieces of oracle bones or tortoise shells have been found. Digital resources for these bones and tortoise shells include pictures, rubbings and facsimile copy.

The fast growing OBI data include the digital calligraphy, document digitization, online review information, Chinese character education resources, and application of ancient Chinese characters based on OBI. OBI data can be divided into structured data, semi-structured data and unstructured data, including text, image, video, animation, font barcode and other storage formats. In addition, OBI glyph generated automatically based on the dynamic description library [3], OBI fragment conjugation [4], and OBI visual input method [5] have also contributed a large number of heterogeneous data in various forms to OBI research resources.

The basic data of OBI have the feature of low value density. At present, there are about 5000 OBI characters found, among which about 1500 characters can be recognized. Moreover, a large number of images of OBI are not clear or even have no information. It is difficult for computers to deal with these problems, and they must be intervened with the experience and knowledge of OBI experts. However, it is impossible for human experts to memorize thousands of bones and tortoise shells information and literature. Therefore, it is necessary to use big data knowledge engineering to assist experts in the interpretation of OBI.

2   ObigK Platform

In order to solve the problem of knowledge management in OBI information processing, we put forward the construction plan of OBI Big Knowledge Management and Service Platform, ObigK for short. Its goal is to provide massive OBI data sharing and big data knowledge services for OBI researchers. The framework structure of the platform is shown in Fig. 1.

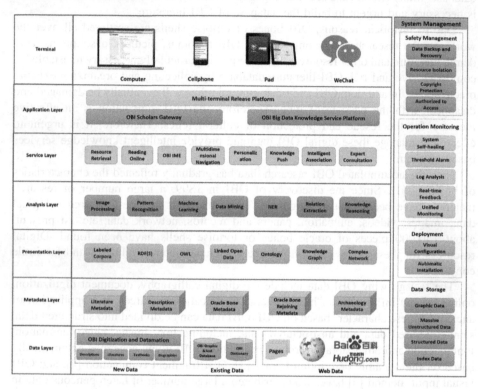

Fig. 1. The framework structure of OBI Big Knowledge Management and Service Platform (ObigK)

As can be seen from Fig. 1, ObigK platform covers all processes of OBI digitalization, data storage, data mining and analysis, data release and sharing, data security management and intelligent knowledge service. From bottom to top, it is divided into data layer, metadata layer, presentation layer, analysis layer, service layer and application layer, and supports multi-terminal user access.

2.1 ObigK Layers

The data layer contains multi-source and heterogeneous data. The objects of OBI research are all kinds of materials, some examples are shown in Fig. 2.

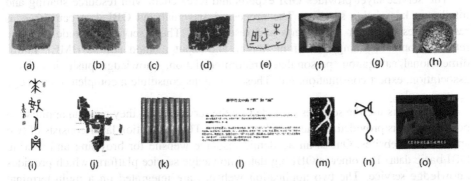

(a)   (b)   (c)   (d)   (e)   (f)   (g)   (h)

(i)   (j)   (k)   (l)   (m)   (n)   (o)

Fig. 2. OBI data including but not limited to the following types: (a) OBI picture (front); (b) OBI picture (back); (c) OBI picture (profile); (d) OBI rubbing; (e) OBI facsimile, it is an image according to a piece of bone or tortoise shell or its rubbing from artificial copying; (f) OBI 3D model; (g) OBI DNA identification data; (h) OBI archaeological data; (i) OBI calligraphy; (j) conjugate data of OBI fragments; (k) OBI descriptions; (l) OBI papers; (m) OBI characters cut off from the rubbings, it is also an image format; (n) artificial copy of the OBI characters, they are very clear, but will lose some feature information. They are also images; (o) OBI video.

For ObigK data layer, the data collection needs to fully integrate the existing data resources and constantly expand new data. The newly created data is mainly completed by manual operation to ensure the data quality. Their sources are authoritative OBI descriptions, literature, textbooks, biographies, etc. Network data is mainly completed through web mining to improve work efficiency. Finally, the data of web mining needs to be supplemented by manual validation in order to obtain high-quality data.

Since OBI data come from different sources and take various forms, it is necessary to establish metadata in order to realize effective discovery, query, integrated organization, effective management and sharing of data resources. OBI metadata includes literature metadata, description metadata, oracle bone and tortoise shell metadata, oracle bone and tortoise shell rejoining metadata, archaeological metadata, etc.

The knowledge representation layer is based on the metadata layer. Different representations are designed for different levels of requirements, including Scheme, RDF(S), OWL, Linked Open Data, ontology, knowledge graph and complex network. These knowledge representations correspond to different granularity of OBI data resources. For example, the complex network is constructed with each OBI character on the oracle bone or tortoise shell as the node. The OBI knowledge graph includes the above metadata, and its nodes can be refined to each OBI character.

In the analysis layer, various methods are used to make statistics and analysis of OBI data and metadata, so as to find the correlation between data resources and realize the mining of hidden relationships among them. The objects of analysis include not only OBI basic data such as description and rubbing, but also knowledge representation layer objects such as ontology, knowledge graph and complex network. Therefore, analysis methods include image processing and pattern recognition, machine learning and data mining, knowledge reasoning, etc.

The service layer provides OBI experts and researchers with resource sharing and intelligent knowledge services. The processed and analyzed OBI data can serve as research resources for the experts and researchers. These services include resource retrieval, online reading, OBI font library and input method editors (IME), multidimensional navigation, personalized recommendation, knowledge push, intelligent association, expert consultation, etc. These functions constitute a complete knowledge service ecology.

When users use the services provided by the service layer, they need an application portal, which is provided by the application layer. The application layer consists of two application websites. One is an academic resource website for browsing and sharing OBI basic data. The other is OBI big data knowledge service platform which provides knowledge service. The two application websites are integrated on a multi-terminal publishing platform and accessed through different links.

ObigK platform users can access the multi-terminal publishing platform of application layer through PC, mobile phone, Pad and WeChat.

2.2 ObigK System Management

ObigK system management includes data storage, installation and deployment, operation monitoring and security management. It meets the basic requirements of a general enterprise-level knowledge services solution. Compared with other knowledge service platforms, ObigK platform is unique in two aspects: (1) for data storage, it integrates structured data and unstructured data, and OBI image and text coexist in distributed storage of graphic and textual data. (2) the platform is planned to be open to all OBI researchers in a public benefit, but the intellectual property rights of the OBI data providers must be protected at the same time. Therefore, authorization access method is adopted in security management, and different levels of permissions are provided. For copyright protection, it allows users to browse OBI basic data online and provides restricted download services. If a user needs to download the restricted data, he can send a resource request via E-mail. After the application and the authorization of the data provider, these resources can be transmitted through a bilateral negotiation, instead of widely available download on the platform.

3 Key Technologies

3.1 OBI Big Data Management

The key technologies involved in OBI big data management include efficient storage technology, specific and multi-modal data preprocessing technology, OBI character

IME technology and OBI character granularity precision search technology. Big data is the basis of ObigK platform. Therefore, it is necessary to design an integrated knowledge management framework of OBI, including literature, descriptions, rubbings, glyphs, radicals and semantic components. So that it can provide convenient knowledge services for OBI experts. The integrated knowledge management framework is named *Three Libraries and One Platform*. It is shown in Fig. 3.

Fig. 3. The Three Libraries and One Platform framework. Three Libraries means OBI font library, OBI description library and OBI literature library. One Platform means the ObigK platform. Three Libraries are the backbones of the One Platform.

The construction of OBI font library needs to collect and sort out all the OBI characters under the guidance of experts, including the heteromorphic writing of the characters, and classify and mark them according to the font, meaning and radical of the characters. In order to process OBI characters correctly on computers, it is necessary to design the editing and coding methods and develop an effective OBI IME for OBI experts.

The construction of OBI description library needs to collect and sort out the currently published catalogues and digitize them. This is a difficult task, because there are many descriptions, such as the earliest one *Tieyun Canggui* (铁云藏龟), which was written by Liu E in 1903. Moreover, many pieces of bone and tortoise shell were collected and published abroad. Therefore, the collection of OBI descriptions needs the support and help of experts. The bones and tortoise shells recorded in OBI descriptions, including rubbings, pictures, facsimiles, three-dimensional models and other forms, need to be sorted out and classified. The properties of each piece of bone or tortoise shell, such as the place where it was unearthed, the place where it is stored, its material, its size and its conjugation, etc., need to be labeled and managed by database. Finally we will build a high quality OBI description library with abundant resources.

The construction of OBI literature library needs to collect, sort out and process all the literature published since the first discovery of OBI in 1899. Then realize the digitization and datamation of these literature in order to provide a full text search service. Compared with other literature, the OBI characters which cannot be typeset are embedded in the articles using pictures, and quite a few literature are all handwritten. These problems bring great challenges to the full text retrieval of OBI literature.

Based on the above three libraries, the ObigK platform can be built. At the data level, this platform can provide OBI experts all over the world with complete, authoritative, accurate and convenient data access. At the level of knowledge services, OBI experts are provided with more comprehensive and intelligent knowledge services such as semantic retrieval, image retrieval and personalized recommendation by means of artificial intelligence technology. It is shown in Fig. 4.

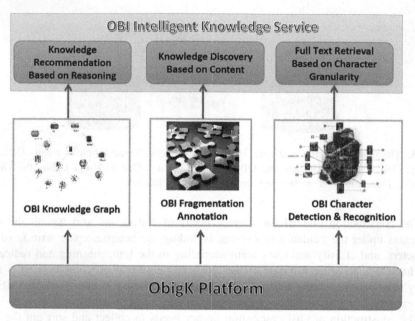

Fig. 4. Intelligent OBI knowledge service based on ObigK platform. It provides full-text retrieval, knowledge discovery and knowledge recommendation.

ObigK platform collects academic materials such as OBI descriptions, books, literature, rubbings, pictures, facsimiles and 3D modeling data, realizes unified resource management through document fragmentation annotations, and provides intelligence knowledge services. Different from traditional OBI document retrieval, ObigK platform can provide OBI character granularity-based, content-based and image-based full-text retrieval, as well as knowledge reasoning and personalized recommendation based on knowledge graph.

3.2　OBI Characters Detection and Recognition

The basic data of OBI mainly exist in the form of images, including pictures, rubbings, facsimiles and 3D models. The detection work of OBI character is to find out which areas of the image are characters, and the work of recognition is to determine which OBI characters are in the detected area. If those OBI characters have been examined and interpreted, it is necessary to give the corresponding modern Chinese characters. The tasks of detection and recognition are shown in Fig. 5.

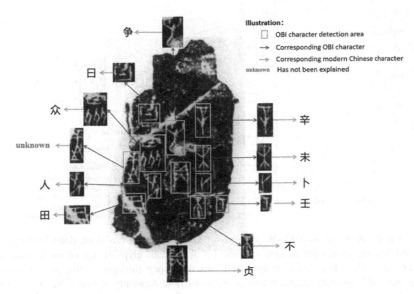

Fig. 5. OBI characters detection and recognition. If an OBI character has been explained, the corresponding modern Chinese character is pointed out.

It is very challenging to use computer to detect and recognize OBI characters automatically. Specifically, the background noise of OBI rubbings is high, which is mainly caused by the corrosion, excavation damage and the texture of the bones or tortoise shells. Shi et al. [6] studied the detection method of OBI characters on rubbings based on threshold segmentation and morphology. Shaotong [7] studied the OBI font recognition method based on topological registration. Ying et al. [8] proposed an OBI image segmentation algorithm based on wavelet transform and FCM fuzzy clustering. Liu et al. [9] studied the OBI character image recognition by using the SVM classification technology. Kappa coefficient and recognition accuracy were used for evaluation, and the accuracy rate reached 88%. Although these methods have achieved some beneficial results, there are many expert interventions in the process of data preprocessing, and the results are not satisfactory when identifying OBI characters on rubbings with large noise or incomplete parts. They cannot deal with uninterpreted OBI characters. The multi-modal recognition method which combines text and image may be a feasible solution. In addition, the deep learning method also shows advantages in processing large-scale corpus of ancient Chinese [10]. A cross-modal method using deep neural network to unify semantic representation of images and texts of ancient Chinese characters such as OBI and bronze inscription is shown in Fig. 6.

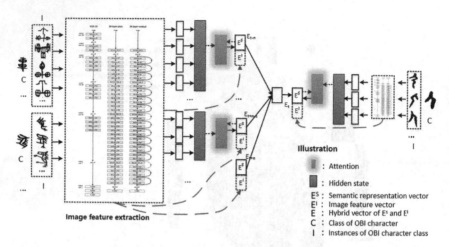

Fig. 6. The cross-modal recognition method for OBI characters based on deep learning. The dotted red line represents the feature information of the font glyph image of each instance of character class. This deep learning approach was developed through a joint project between Anyang Normal University, Xiamen University and China Academy of Social Sciences. (Color figure online)

This method needs to collect large scale images and text data of similar ancient Chinese characters such as OBI and bronze inscriptions and make full use of the similarity of images and text semantic similarity to realize OBI characters recognition. The image data is divided into characters first, and then the image and text are integrated by deep neural network to form a unified semantic representation. Finally, the end-to-end deep neural network based on that unified semantic representation is used to recognize oracle bones.

3.3 OBI Document Fragmentation Annotation

Since most of the digitized OBI resources are images, they cannot be searched accurately by existing search engines. And most OBI characters or some other ancient characters cannot be directly input into the literature at the time of their publication, the place where OBI characters need to be displayed in the text are usually replaced by pictures, as shown in Fig. 7.

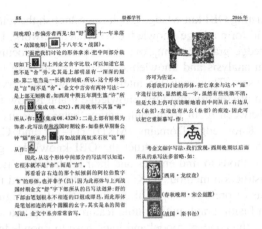

Fig. 7. In literature, OBI or other ancient characters that cannot be input are usually replaced by pictures of the glyph. Please look at the part marked by the red box. (Color figure online)

In order to solve these problems, the fragmentation processing and annotation methods were adopted. In this way, we can realize the full text retrieval based on OBI glyph images. The core of fragmentation annotation is to split OBI documents according to four levels of article, chapter, section and paragraph on the basis of catalog editing. After the document catalog editing is completed, the specific contents of each chapter can be separated through human-computer interaction. Split by paragraph is the basic unit, in which pictures and tables are split by image format and inserted into corresponding positions. For the OBI glyph images, the positions in the document and the corresponding document metadata are marked. The result are XML annotated documents. The fragmentation annotation process is shown in Fig. 8.

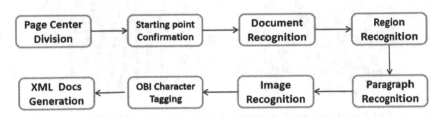

Fig. 8. The fragmentation processing and annotation flow of OBI literature.

3.4 OBI Knowledge Graph Construction

We already know that there is an outstanding contradiction in OBI information processing. It is between high dependence and low sharing and reuse of OBI experts' knowledge and existing research achievements. The main reason is that the researchers separated the other discipline knowledge related to OBI and neglected its evolution of knowledge. In order to solve this contradiction, the OBI knowledge graph construction is proposed. Constructing OBI knowledge graph can establish the relation between OBI knowledge nodes and form a huge knowledge associated network. Using the semantic advantage of knowledge graph can improve the efficiency of document retrieval, and based on association analysis and knowledge reasoning can push more relevant document resources to users. Thus realizing intelligent knowledge service.

The key problem of constructing OBI knowledge graph is how to find entities and semantic relations between them from multi-source heterogeneous data. The combination of Mapping Knowledge Domains (MKD) [11–13] and Google Knowledge Graph (KG) [14, 15] was used to construct the OBI knowledge graph. First, using MKD construction methods to find entities and their relationships from OBI literature. Second, use the construction methods of KG to extract entities and their relations, and integrating the results of previous step. Third, after entity disambiguation and relation fusion, it will gain a fusion graph with double features of map and spectrum. Because ontology can provide the concept model and logic base to knowledge mapping, it can mine implied semantic relations by using ontology and rules based reasoning. Then the fusion graph can be enriched. At last, we will build the OBI knowledge graph. The implementation process will make full use of the resources of the knowledge representation layer in the ObigK platform shown in Fig. 1. A fragment of the OBI knowledge graph is shown in Fig. 9.

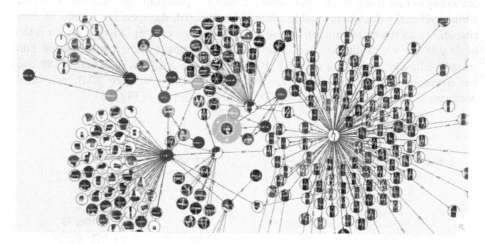

Fig. 9. The OBI knowledge graph fragment. Entity extraction and relation extraction from multi-source heterogeneous data sets are realized.

4   Conclusion

It is of great significance to build the ObigK platform. This knowledge service platform has important social benefits for inheriting and carrying forward Chinese traditional culture, popularizing OBI knowledge to the public, lowering the threshold of OBI learning, promoting the sharing and reuse of OBI knowledge, and even changing the existing primary school students' literacy and enlightenment education. The construction of this platform is of great significance to promote the development of OBI research and fully excavate the treasure of the Chinese nation and reproduce the original appearance of ancient society and history. The promotion and application of this platform can not only develop the theory and technology of oracle information processing, but also promote the development of other disciplines such as history, archaeology, history of ancient science, philology and geography.

At present, we have collated the key data for ObigK platform as follows: the digitization of 239 kinds of OBI descriptions, the digitization of 13315 OBI literature, OBI ontology (373 concepts, 116 attributes and 8403 instances), OBI character complex network (6199 nodes and 160964 edges), OBI knowledge graph (142106 entities and 273068 relations). Based on these organized data we intend to complete the development of OBI academic resource website and OBI intelligent knowledge service platform, so as to provide high-quality and efficient knowledge management and knowledge services for oracle researchers.

Acknowledgement. This work is supported by the National Natural Science Foundation of China (No. U1504612, 61806007, U1804153), the National Language Committee scientific research projects of China (No. YWZ-J023, YB135-50), the Development Projects of Henan Province Science and Technology (No. 182102310039), the Science and Technology Key Project of Henan Province Education Department (No. 17A520002).

References

1. Jing, X., Feng, G., Wu, Q.: Research on semantic mining for large-scale Oracle bone inscriptions foundation data. New Technol. Libr. Inf. Serv. **31**(2), 7–14 (2015)
2. Xiong, J., Luo, Z., Wang, A.M.: Research on entity relation discovery for Oracle Bone Inscriptions knowledge mapping construction. Comput. Eng. Sci. **37**(11), 2188–2194 (2015)
3. Li, Q., Wu, Q., Yang, Y.: Dynamic description library for Jiaguwen characters and the research of the characters processing. Acta Scientiarum Naturalium Universitatis Pekinensis **49**(1), 61–67 (2013)
4. Wang, A., Ge, W., Zhao, Z., et al.: Research on computer matching of inscriptions on tortoise fragments. Comput. Eng. Des. **32**(10), 3570–3573 (2011)
5. Liu, Y., Li, Q.: Design and implementation of visual input method of oracular inscriptions on tortoise shells and bones. Comput. Eng. Appl. **40**(17), 139–140 (2004)
6. Shi, X.S., Huang, Y.J., Liu, Y.G.: Location method of rubbing text of oracle bone inscriptions based on threshold segmentation and morphology. J. Beijing Inf. Sci. Technol. Univ. **29**(6), 7–10 (2014)
7. Shaotong, G.: Identification of Oracle-bone script fonts based on topological registration. Comput. Digit. Eng. **44**(10), 2001–2006 (2016)
8. Ying, H.E., Xiaoju, H.E., Gang, Z.: Image segmentation of JiaGuWen image based on wavelet transform and FCM. J. Tianjin Univ. Sci. Technol. **33**(6), 62–66 (2018)
9. Liu, Y., Liu, G.: Oracle bone inscription recognition based on SVM. J. Anyang Normal Univ. **19**(2), 54–56 (2017)
10. Wang, B., Shi, X., Tan, Z., et al.: A sentence segmentation method for ancient Chinese texts based on NNLM. Acta Scientiarum Naturalium Universitatis Pekinensis **53**(2), 255–261 (2017)

11. Hu, Z., Sun, J., Wu, Y.: Research review on application of knowledge mapping in China. Libr. Inf. Serv. **57**(3), 131–137 (2013)
12. Changjiang, Q., Hanqing, H.: Mapping knowledge domain-a new field of information management and knowledge management. J. Acad. Libr. **27**(1), 30–37 (2009)
13. Lee, Y., Chen, C., Tsai, X.: Visualizing the knowledge domain of nanoparticle drug delivery technologies: a scientometric review. Appl. Sci. **6**(1), 11 (2016)
14. Liu, Q., Li, Y., Duan, H., et al.: Knowledge graph construction techniques. J. Comput. Res. Dev. **53**(3), 582–600 (2016)
15. The Google Knowledge Graph: Information gatekeeper or a force to be reckoned with? Strateg. Dir. **30**(4), 15–17 (2014)

Meteorological Sensor Data Storage Mechanism Based on TimescaleDB and Kafka

Liqun Shen[1], Yuansheng Lou[1], Yong Chen[2], Ming Lu[3], and Feng Ye[1,2(✉)]

[1] College of Computer and Information, Hohai University,
Nanjing 211100, China
yefeng1022@hhu.edu.cn
[2] Postdoctoral Centre, Nanjing Longyuan Micro-Electronic Company,
Nanjing 211106, China
[3] Jiangsu Water Resources Department, Nanjing, China

Abstract. The scale of meteorological sensor data increases at TB level every week. Traditional relational database is inefficient in storing and processing such data and cannot satisfy many soft requirements. However, the heterogeneity and diversity of the numerous existing NoSQL systems impede the well-informed comparison and selection of a data store appropriate for a given application context. Implementing a meteorological sensor data storage mechanism is a key challenge. Therefore, a meteorological sensor data storage mechanism based on TimescaleDB and Kafka is proposed. In this solution, meteorological sensor data was acquired and transmitted by Kafka and was sent to TimescaleDB for storage and analysis. Based on simulated meteorological sensor dataset, it compared the solution with other NoSQL stores such as Redis, MongoDB, Cassandra, HBase and Riak TS. The experimental results show that the storage mechanism proposed is superior in the storage and processing of massive meteorological sensor data.

Keywords: NoSQL · Meteorological sensor data · Kafka · TimescaleDB

1 Introduction

In recent years, due to continuous development of meteorological monitoring technology, the observation coverage rate, observation frequency and the number of observation stations have been greatly improved, which makes the scale of meteorological sensor data sustainably growing and shows the "3V" characteristics of big data [1]: large scale, various types and fast growth rate. There are two main types of meteorological sensor data in common use [2]: (1) structured data, it is always stored, analyzed and calculated by traditional relational databases, such as Oracle and MySQL. Station precision is a good example. (2) Semi-structured/unstructured data, such as video monitoring information, is stored and processed in NoSQL databases in the majority of cases.

However, the National Weather Forecasting Platform MICAPS version 3.0 has begun to need coping with the increasingly severe storage pressure and processing

© Springer Nature Singapore Pte Ltd. 2019
X. Cheng et al. (Eds.): ICPCSEE 2019, CCIS 1058, pp. 137–147, 2019.
https://doi.org/10.1007/978-981-15-0118-0_11

performance since 2010 [3], mainly reflected in: (1) there are challenges in the analyzing and writing of massive meteorological data. (2) Massive meteorological sensor data is faced with the hard problem of slow access and query speed. These pressures pose great difficulty to the traditional methods of meteorological data storage and processing, such as big data storage, real-time processing, and fast respond speed, etc. Meteorological data not only require satisfying rapid storage, but also need to be analyzed in millisecond time. Traditional relational databases do not have the ability to meet the efficient performance requirements of meteorological sensor data storage and access. Fortunately, NoSQL databases can achieve high storage and access performance of large scale meteorological sensor data. Therefore, on the basis of comparing various NoSQL databases, this paper proposes a storage mechanism based on TimescaleDB and Kafka to implement the storage, analysis and calculation of meteorological data.

Firstly, this paper introduces the related work of meteorological sensor data, and briefly describes the key-value database Redis [4], document database MongoDB [5], column database HBase [6], Cassandra [7] and time series database Riak TS [8] which belong to NoSQL databases, and then elaborates TimescaleDB technology and Kafka message engine system. After that the storage mechanism and architecture are described in detail. Finally, the experimental process and results are discussed, which verifies that our solution has better work efficiency and workload performance in storing and processing meteorological sensor data than other NoSQL databases.

2 Related Work

Natural disasters may endanger human life and lead to property losses. Teng [9] studied a method of multi-classifier collaborative mining of regional meteorological data to predict the occurrence of natural disasters. The existing large meteorological data storage system has poor real-time performance, opaque transmission of data processing, and poor self-adaptation. To solve these problems, Shao Lei and others studied a Hadoop meteorological cloud platform to improve the performance requirements of storage, management and reading of meteorological data [10, 11]. According to the multi-dimensional model of meteorological data and user behavior, Wang and Huang [3] designed and implemented a high-performance massive data storage system where applied a non-relational distributed key value database. Xu, et al. [12] proposed an index-based query optimization framework for structured meteorological data that called HBase4M (HBase for Meteorology) according to the storage characteristics of HBase. Wang [13] used TimescaleDB database to store retrieved rain cells in the research of convective rain cell database based on high resolution radar images.

2.1 Overview of NoSQL and TimescaleDB

NoSQL is a collection of a variety of non-relational databases [14]. According to different application scenarios, NoSQL databases can be summarized into four types: (1) NoSQL databases for high performance of reading and writing, such as Memcached and Redis, are commonly used in the construction of large-scale network platforms.

(2) Document-oriented databases, such as MongoDB and CouchDB, are generally stored in JSON format. (3) NoSQL databases for distributed computing, such as Cassandra and Voldemort, have good lateral scalability. (4) Time-oriented NoSQL databases, such as TimescaleDB and Riak TS, have high performance in storing and processing time series data.

TimescaleDB is an open source time series database optimized for fast reading and complex queries [15]. It is implemented as an extension of PostgreSQL. Due to supporting a fully SQL interface, TimescaleDB can be used as easily as traditional relational databases and its expansion model is the same as existing NoSQL databases. Compared with traditional relational databases, TimescaleDB has the following advantages: (1) TimescaleDB has a high data reading rate for large-scale data; (2) TimescaleDB has higher query performance; (3) the time-oriented feature can help TimescaleDB provide a huge performance improvement for the time-oriented query performance. Compared with common NoSQL databases, such as MongoDB and Cassandra, TimescaleDB has the following advantages: (1) TimescaleDB can query data with standard SQL statements, while most NoSQL databases need to learn a new query language; (2) Simple operation makes the users quickly grasp the use of TimescaleDB if you are familiar with using PostgreSQL (a traditional relation database); (3) TimescaleDB supports any third tool that uses SQL, etc.

From the user's point of view, TimescaleDB exposes content which is similar to a single table (called HyperTable). HyperTable actually contains many abstract or virtual views (called chunks) of data. TimescaleDB has high performance in writing and querying because of its interaction with HyperTables. Though HyperTable behaves like a regular table, TimescaleDB can maintain high performance even when the database storage is extended to a degree where the scale of data is beyond the bearing capacity of the storage. Internally, TimescaleDB automatically divides HyperTables into chunks, each of which corresponds to a two-dimensional space determined by the specified time interval and the location of the partitioning key, thus TimescaleDB can maintain high performance in query parsing. TimescaleDB that community released currently only supports single-node deployment, and cluster deployment is still under test, so the experiment in this paper is based on single-node deployment of TimescaleDB.

2.2 Kafka Stream Processing

Stream processing is one of the main application scenarios of Kafka [16]. Since the version 0.10.0.0, Kafka has launched a new stream processing component that called Kafka Streams. Intrinsically, stream processing is a processing mode or a data processing engine whose main purpose is to process an infinite number of data sets. Kafka Streams is a client API library, which relies on the functions that provided by Kafka core to process and analyze the message data stored in Kafka topic. Kafka Streams is rooted in the idea of stream processing and with the help of Kafka transactions and producers, It can not only realize the precise processing semantics, but also provide tools for deriving event time and processing time. Event time refers to the time when the message or event actually occurred, and it also be known as the creation time. Processing time refers to the time when the message or event was observed. At the same time, Kafka Streams provides fully support for windowing, and can also realize persistent storage and real-time query of application state.

In conclusion, a meteorological sensor data storage mechanism based on TimescaleDB and Kafka is proposed.

3 The Proposed Mechanism

The architecture of meteorological sensor data storage mechanism based on TimescaleDB and Kafka is shown in Fig. 1. Meteorological sensors receive real-time data from various stations and transmit the data to the Kafka platform. Kafka platform sends received data to the TimescaleDB database in the form of stream. TimescaleDB stores, calculates and analyses the received data, then transmits it to Kafka Streams for processing. Kafka Streams sends the results to users in the form of stream, so that users can get the required meteorological data. Similarly, in the process of testing Redis, MongoDB, Cassandra, HBase and Riak TS databases, the Kafka platform sends the received data in the form of stream through clusters to these databases, and then the same operations will be carried out on these databases.

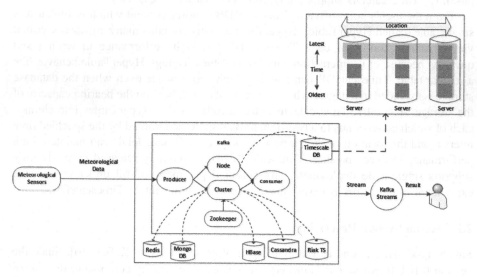

Fig. 1. Architecture of meteorological data storage scheme based on TimescaleDB and Kafka.

The main function of producers is to send messages to Kafka, that is, Producer is responsible for writing data to Kafka. In this architecture, producers send data received by meteorological sensors to Kafka. Consumer's function is mainly to read information from Kafka. In this architecture, consumers read meteorological data from Kafka and send data to database. TimescaleDB automatically partitions the received meteorological data according to time and geographical space. Kafka Streams uses stream processors to process the received data.

To sum up, the producer sends real-time meteorological data from each station that received by meteorological sensors to Kafka. Consumers get asked information from Kafka and send it to various databases. Kafka acts as a data transmission pipe line intrinsically. Each database calculates and analyses the received data, and then sends the results to users in form of stream through Kafka Streams. Therefore, users can get real-time rainfall, rainfall at a specific time and other information.

4 Experiment and Analysis

4.1 Experimental Data and Environment

The experiment uses simulated meteorological sensor data that generated by batch data generation tool. There are 69 stations with simulated meteorological sensors, and each station receives meteorological data every 5 min. The time span is from January 1, 2015 to July 24, 2017, with a total of 18 million pieces of data. The meteorological sensor data contains five fields, including primary key number (ID), station number (STCD), time (TM), rainfall (RF) and station name (RFROM), as shown in Table 1.

Table 1. T examples of meteorological sensor data.

ID	STCD	TM	RF	RFROM
1	62916750	2015-01-01 00:05:00	14.61	swj

The experimental environment is Ubuntu 64 bit operating system. On account of the current version of TimescaleDB just supports single-node deployment, only one machine is used in the test of TimescaleDB. Because other databases support clustering, four machines with the same configuration are used to deploy the cluster in the process of testing them. These machines are configured as follows: The main board is ASUS PRIME B350-PLUS, the processor is AMD Ryzen 7 1700X Eight-Core Processor 8-core, the memory is 32 GB DDR4 3000 MHz, and the main hard disk is Samsung MZVLW 128HEGR-000L2 (128 GB/SSD). The workload tool for monitoring NoSQL databases mentioned above is Ganglia.

TimescaleDB, Redis, MongoDB, Cassandra, HBase and Riak TS are installed in the four machines. The four machines are all equipped with Ubuntu 16.04 system, Zookeeper and Kafka. The version of Kafka is 1.0.0. Zookeeper is a tool to provide coordination services for Kafka.

4.2 Experiments and Analysis

This experiment will use TimescaleDB to insert meteorological sensor data and query those data in different scenarios, then record the corresponding execution time and CPU utilization of four used machines of each operation. At the same time, Redis, MongoDB, Cassandra, HBase, Riak TS are used to perform the same operation on meteorological data and record corresponding experimental data. Comparing the experimental data can verify the high efficiency of the meteorological sensor data storage mechanism based on TimescaleDB.

Scenario 1 (Data Import Operation). We Insert 1,000, 10,000, 100,000 and 250,000 data into the tables created by each database respectively. This paper uses JDBC to add, delete and modify the data of each database. The execution time of each database is shown in Fig. 2 and the CPU utilization of each database is shown in Fig. 3.

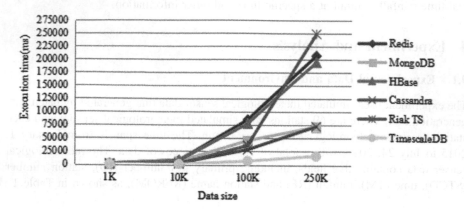

Fig. 2. The execution time of scenario 1.

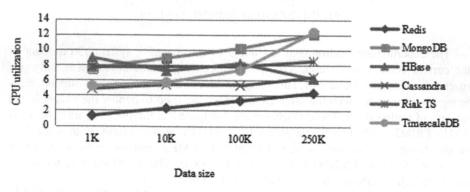

Fig. 3. CPU utilization of scenario 1.

From Fig. 2, the execution time of inserting 1,000 and 10,000 data into each database is nearly the same, but it can still be seen that the execution time of TimescaleDB is less than that of other databases. In the test of inserting 100,000 data into various databases, we can see that TimescaleDB takes less than 1/10 of the execution time of Redis database. In the experiment of inserting 250,000 data into each database, the execution time of TimescaleDB database does not exceed 1/15 of Riak TS which execution time is the longest among these databases and 1/4 of the execution time of MongoDB which execution time is only more than TimescaleDB.

From Fig. 3, the CPU utilization of TimescaleDB for data import operation increases with the rise of data scale. When the data scale reaches 250,000, the CPU utilization of TimescaleDB exceed that of other databases.

Scenario 2 (Query Data by Time Zone). We query the data from stations for one day (287), one week (2011), one month (8845) and one year (134047), and record the execution time and CPU utilization of each NoSQL database. The execution time and CPU utilization of each database selecting station information according to a time zone are shown in Figs. 4 and 5 respectively (Note: Redis, HBase, Cassandra cannot implement this function directly).

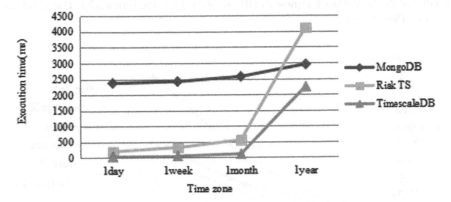

Fig. 4. The execution time of scenario 2.

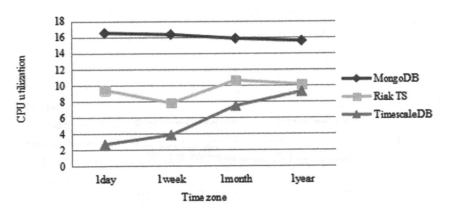

Fig. 5. CPU utilization of scenario 2.

From Fig. 4, the execution time of TimescaleDB is about 1/100 of MongoDB and no more than 1/8 of Riak TS when querying the data volume for one day. When querying the data for one week and one month, TimescaleDB's execution time is much less than that of other databases. The execution time of TimescaleDB increases dramatically when querying the data volume for one year, and is about 20 times that of querying data volume for one month, but the execution time is still less than the other two databases.

From Fig. 5, the CPU utilization of MongoDB is the highest, and its CPU utilization is substantially unchanged with the growth of data scale. Although the CPU utilization of TimescaleDB increases with the augment of time interval, it is lower than that of other NoSQL databases.

Scenario 3 (Indexing Query). Each database queries 10000,100000 and 250000 data items by index, and records the execution time and CPU utilization of each database correspondingly. The index of Redis, MongoDB, HBase, and Cassandra is the primary key (id), while the index of TimescaleDB and Riak TS is the timestamp. The results are shown in Figs. 6 and 7.

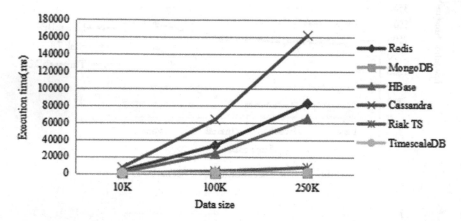

Fig. 6. The execution time of scenario 3.

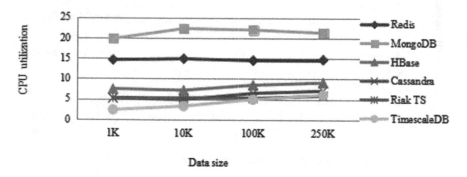

Fig. 7. CPU utilization of scenario 3.

From Fig. 6, the execution time of TimescaleDB query data by index is much less than that of other databases. At a data size of 250,000, TimescaleDB takes nearly 1/332 of the execution time of Cassandra, while Riak TS takes more than 13 times the execution time of TimescaleDB.

As you can see from Fig. 7, the CPU utilization of each database has a little change with the increase of data size. Compared with other databases, the CPU utilization of TimescaleDB is relatively stable and low.

Scenario 4 (Increase the Number of Queries While Insert Data into Database).
While constantly insert data into various databases through Kafka, the data volume of query is increasing correspondingly. The number of query is increased by 1,000, 10,000, 100,000 and 250,000. The execution time and CPU utilization of each database are shown in Figs. 8 and 9 respectively.

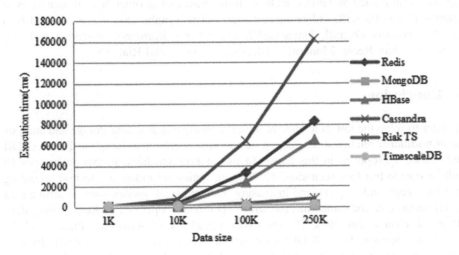

Fig. 8. The execution time of scenario 4.

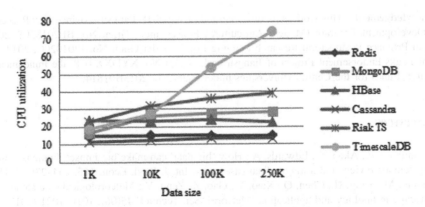

Fig. 9. CPU utilization of scenario 4.

As can be seen from Fig. 8, the time consumed by TimescaleDB in the process of continuously inserting data increases with the growth of the number of queries, but the degree of changes is not large. From Scenario 1 and Scenario 2, it is easy to infer that in this scenario, the execution time of TimescaleDB mostly result from data insertion. TimescaleDB runs much less time in this scenario than Redis, HBase, Cassandra, and Riak TS, and there is little difference in execution time between TimescaleDB and MongoDB.

From Fig. 9, we can see the CPU utilization of TimescaleDB grows with the increasing number of queries in the process of data insertion, and the growth rate is the highest in all databases. TimescaleDB has the highest CPU utilization when the query volume grows to 100, 000 and 250,000.

From the experimental results of the four scenarios mentioned above, it can be seen that the performance of TimescaleDB is higher than that of other NoSQL databases in general, though the CPU utilization of TimescaleDB is higher than that of MongoDB in specific scenario. Overall, TimescaleDB has better performance in storing and analyzing data than Redis, MongoDB, HBase, Cassandra, and Riak TS.

5 Conclusion

In order to solve the low efficiency of massive meteorological data storage and analysis using traditional relational database, a data storage mechanism based on TimescaleDB and Kafka is proposed in this paper. In this storage mechanism, this paper uses 18 million data to test four scenarios: data import, station information selection according to a time zone, index query and increasing the number of queries while inserting data into databases. At the same time, the same experiment is operated on Redis, MongoDB, HBase, Cassandra and Riak TS. The experimental results verify the efficiency of the storage mechanism. We will take study on the performance of TimescaleDB for data storage and analysis in the environment of cluster deployment possibly after the TimescaleDB community releases the version of TimescaleDB supporting cluster.

Acknowledgement. This work is partly supported by the 2018 Jiangsu Province Key Research and Development Program (Modern Agriculture) Project under Grant No. BE2018301, 2017 Jiangsu Province Postdoctoral Research Funding Project under Grant No. 1701020C, 2017 Six Talent Peaks Endorsement Project of Jiangsu under Grant No. XYDXX-078, the Fundamental Research Funds for the Central Universities under Grant No. 2013B01814.

References

1. Wamba, S.F., Akter, S., Edwards, A.: How 'big data' can make big impact: findings from a systematic review and a longitudinal case study. Int. J. Prod. Econ. **165**, 234–236 (2015)
2. Yang, M., Yang, H., Chen, Q., Xiao, Y., Gao, Z., Zeng, Y.: Meteorological data cloud data storage technology and application. Meteorol. Sci. Technol. **45**(06), 1017–1021 (2017)
3. Wang, R., Huang, X., Zhang, B., Wang, J., Luo, B.: Design and implementation of real-time analysis and storage system for massive meteorological data. Comput. Eng. Sci. **37**(11), 2045–2054 (2015)
4. Redis. https://redis.io/. Accessed 22 Jan 2019
5. Jiang, H., Shen, F., Chen, S., Li, K.C., Jeong, Y.S.: A secure and scalable storage system for aggregate data in IoT. Future Gener. Comput. Syst. **49**, 133–141 (2015)

6. Apache HBase. http://hbase.apache.org/. Accessed 22 Jan 2019
7. Apache Cassandra. http://cassandra.apache.org/. Accessed 22 Jan 2019
8. Riak TS. http://basho.com/products/riak-ts/. Accessed 22 Jan 2019
9. Teng, S., et al.: A cooperative multi-classifier method for local area meteorological data mining. In: Proceedings of the 2014 IEEE 18th International Conference on Computer Supported [::Cooperative::] Work in Design (CSCWD), pp. 435–440
10. Shao, L., Liu, J., Dong, G., Mu, Y., Guo, P.: The establishment and data mining of meteorological data warehouse. In: 2014 IEEE International Conference on Mechatronics and Automation, pp. 2049–2052 (2014)
11. Jiang, X., Chen, W., Wang, Y.: The adaptive research of data layout in large-scale meteorological data storage system. In: 2013 IEEE Third International Conference on Information Science and Technology (ICIST), pp. 1016–1020 (2013)
12. Xu, X., Yang, Z., Ma, T.: Query optimization of meteorological structured data based on HBase. Comput. Eng. Appl. (9), 80–84(2017)
13. Wang, L.P., Munoz Lopez, C., Homg, T.C., et al.: A convective rain cell database based upon high-resolution radar images: unravelling convection patterns. In: EGU General Assembly Conference Abstracts, vol. 20, p. 362 (2018)
14. Chandra, G.: Deka: BASE analysis of NoSQL database. Future Gener. Comput. Syst. **52**, 13–21 (2015)
15. TimescaleDB. https://docs.timescale.com/v1.1/introduction. Accessed 28 Jan 2019
16. Apache Kafka. http://kafka.apache.org/. Accessed 28 Jan 2019

Cost Optimization Strategy for Long-Term Storage of Scientific Workflow

Zaibing Lv, Cheng Zhang$^{(\boxtimes)}$, and Futian Wang

School of Computer Science and Technology, Anhui University, Hefei, China
cheng.zhang@ahu.edu.cn

Abstract. With the rapid development of cloud environment, the capabilities of systems have been promoted with powerful computing and storage. But for the characteristic of "pay-as-you-go" of cloud resources, it is necessary to consider the different data storage cost. Especially for processing of "old data" in long-term storage, an appropriate strategy is needed to reduce users' cost. Considering the characteristics of price stratification in the current commercial cloud environment, a three-level price stratified storage strategy is proposed based on the CTT-SP algorithm, which stores part of the "old data" on relatively inexpensive secondary and tertiary storage, and ensures that the time delay caused by three-level storage does not exceed the deadline. Compared with other storage methods, the experimental result shows the strategy proposed can guarantee the time delay while reducing the cost of users significantly in long-term storage.

Keywords: CTT-SP algorithm · Price stratification · Long-term storage

1 Introduction

In recent years, with the arrival of the era of big data, more and more scientists pay attention to the processing of data-intensive applications with scientific workflows. Such as astronomy [1], high-energy physics [2] and bioinformatics [3]. Due to the requirement of data-intensive processing, the cloud computing as a new parallel and distributed computing model can provide secure and reliable flexible computing services according to users' needs. Now scientists no longer need to purchase expensive hardware and software infrastructure, nor need to consider how to operate and maintain, so that they can concentrate more on their own scientific research.

Currently several researches have been performed on the data-intensive scientific applications. Li et al. [4–6] considers putting high correlation data blocks together to reduce transmission costs. Xu et al. [7] establishes the data model, the data nodes in scientific workflow are sorted according to the importance of the model, and stored in turn until reaching the lowest cost. Dong et al. [8–10] propose a cost transitive tournament shortest path (CTT-SP) algorithm, which is based on the minimum cost benchmark. The optimal balance between computation and storage can be found by using the shortest path algorithm dijkstra. Sha [11] aims to improve CTT-SP algorithm. The main focus is to reduce the complexity and improve the accuracy of the original algorithm by constructing critical paths.

© Springer Nature Singapore Pte Ltd. 2019
X. Cheng et al. (Eds.): ICPCSEE 2019, CCIS 1058, pp. 148–157, 2019.
https://doi.org/10.1007/978-981-15-0118-0_12

To sum up, the previous researches focus on two main areas, namely, the data placement and the storage of intermediate data. For the data placement, it is the collaboration between multiple cloud environment service providers. The data nodes with high interdependence are concentrated in the same data center, which can reduce the unnecessary transmission cost on the whole. For the storage of intermediate data, some optimal methods are considered in a single cloud environment. The shortest path can be found by using dijkstra algorithm after constructing CTT [10]. This algorithm has an excellent effect on linear workflow [12]. But they only consider the current data in workflow, such as Chinese FAST engineering, which produces 10 to 100 TB of data per day and requires that all the data should be processed on that day. However, in the context of long-term storage, the data in workflow is no longer One-off, but becomes "old data" over time. The "old data" with its own characteristics is not the synonym of the "useless data". On the one hand, it is usually much lower than the current "new data" in the frequency of using, making it unreasonable to equate it with the current data processing; On the other hand, the "old data" is also very important. For example, in the patrol system, the problem has been gradual rather than abrupt. Therefore, the analysis of historical data can achieve early warning well. In this scenario, although the access frequency of "old data" is not high, once hit, and we do not store it, we need to regenerate them [10], which will make the whole workflow stop waiting for data generation, and then affect the progress of the whole process. So we need a reasonable strategy to deal with the "old data".

Nowadays, cloud service providers also provide a variety of combinations according to users' needs. As far as cloud storage resources are concerned, there are mainly three echelons. We can reduce costs by placing some "old data" in relatively inexpensive secondary and tertiary storage, but we need to pay attention to the extra "seeking" time.

The remainder of this paper is organized as follows: Sect. 2 states the basic relevant definitions, introduces the CTT-SP algorithm and current cloud storage strategies of cloud. Section 3 proposes a novel evolutionary strategy, and describes how price is stratified on "storage" nodes. In Sect. 4, the experiments are designed and conducted, in which the results are given and analyzed. Finally, we draw conclusions in Sect. 5.

2   Relevant Premises and Definitions

2.1   Cloud Storage Data

Data stored on the cloud can be divided into two categories: raw data and intermediate generated data. Because raw data costs are fixed, so we only focus on the generated data optimization problem.

We call the data generated during the execution of the scientific workflow as intermediate generated data, and the decision of whether to store or delete is up to the users themselves. If stored, the storage fee from the starting date will be counted. Correspondingly, we can use the data directly when we need them, without seeking time. Conversely, if users choose to delete the data, they must regenerate them from the prior and stored node sets as needed. This process involves paying for CPU computing in the cloud environment.

2.2   CTT-SP Algorithm

Dong [10] provided an optimal storage strategy of intermediate data in a single cloud environment. They use a data dependency graph (DDG) to represent the dependencies between data. Simply put, a data dependency graph is a directed acyclic graph and records the precursor and successor relationships of task nodes in a scientific workflow so that any intermediate data can be regenerated.

For example, Fig. 1 is a simple linear DDG example. The precursor of D recorded in DDG is {A, B, C}. If the D user chooses to delete but needs to be used at present, the system needs DDG to get its first precursor and stored node. If C is stored, then C is used. If C is still deleted, the first precursor B of C is found again, and the analogy is made until D can be regenerated.

Fig. 1. A simple DDG example.

After constructing DDG, CTT-SP algorithm can be used to find the shortest path. The first step of CTT-SP is to add two virtual nodes, namely, one virtual node ds before the first node and one virtual node de after the tail node. And their sizes and frequencies are both 0. The second step is to add edges to transform IDG graph of scientific workflow into corresponding CTT graph, thus transforming the optimization problem of intermediate data storage of scientific workflow into the shortest path problem corresponding to CTT graph. In the third step, dijkstra algorithm is used to solve the shortest path of CTT drawing. The point in the shortest path is the point that needs to be stored.

But, because the frequency of data usage changes with time, especially when the data becomes long-term storage, this phenomenon becomes more obvious. We assume the trend of the data as shown in Fig. 2, which may not be fully consistent with the actual situation, but the overall trend is gradually declining. And the size of the data in the period is assumed to be unchanged.

In the context of long-term storage, scientific workflow has emerged some new problems: (1) How to store the "old data" which have low using frequency? (2) In the long-term workflow, if the shortest path constructed by CTT-SP algorithm changes due to the change of data usage frequency, how to deal with these "old data"?

So, the original CTT-SP algorithm is no longer suitable for long-term storage environment, a new strategy is necessary for this. The method adopted in this paper is to stratify the single price and store part of the "old data" on relatively cheap storage resources. In this way, when the frequency of data decreases, we can directly transfer it from the first level of storage to the second and third levels of storage resources, and consequently reduce the overall cost.

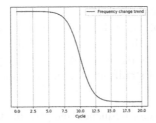

Fig. 2. Frequency trend of data in long-term storage.

2.3 Three Level Storage

The cloud operators provide a variety of combinations according to users' requirements. The three-tier storage strategy has been applied by some IT giants [13–15]. The characteristics of the three-tier storage strategy are as follows:

1. First-tier storage costs the most, but its response time is at the millisecond level, and there is no seeking cost.
2. The second-tier storage costs are relatively high, and the response time is basically the same as that of first-tier storage, but it has the cost of data retrieve. So this resources are more suitable for storing the accessed infrequently data.
3. Third-tier storage costs are the lowest, but the response time is longer. Each cloud provider is different, ranging from minutes to hours. And it has a higher seeking cost, which is more suitable for low-frequency archiving data storage.

3 Evolution Strategy

3.1 Hierarchical Storage Model and Optimizing Ideas

We can get the required allocation strategy in multi-price when we only use the original CTT-SP algorithm. The specific approach is to expand each storage node into three layers, which means to choose storage from one price to three, with full connection before and after.

Figure 1 is a simple DDG example which is only using one price, while Fig. 3 is a multi-price model.

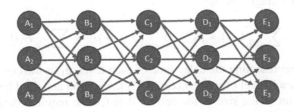

Fig. 3. Extended to the three-level storage model

Assuming that there are five nodes in our workflow, when we expand the price to three levels, the complexity of path selection changes from linear $O(1)$ to exponential $O(3^{n=5})$. Therefore, in practice, the original CTT-SP algorithm is too complex for the three-tier price stratification idea. But if we take the whole process apart, we can get the storage point at first, and then stratify on the basis of the storage point. The specific process is as follows:

a. Defining two functions: $f(t)$ for frequency and $g(t)$ for size. The formula 1 and 2 can be used to obtain the number of times used and the amount of data generated in the $[t1, t2]$ period.
b. Discovering the required storage points. In order to prevent inconsistencies caused by the previous single-price storage and then reach three price stratification, we bring all three storage prices into the CTT-SP algorithm at the time of construction and then intersect each result set.
c. Multi-objective optimization. The hierarchical price is introduced into the set obtained in the second step, and the multi-objective algorithm is used to ensure that the solution is non-dominated sets.
d. User choices. Take the sets of solutions from step 3 back to the multi-objective algorithm again for getting final solution set, and the user is given the choice.

$$F(t) = \int_{t1}^{t2} f(t)dt \tag{1}$$

$$G(t) = \int_{t1}^{t2} g(t)dt \tag{2}$$

User choice is mainly due to the introduction of price stratification, the overall workflow has more "time" cost than the original process. The "seeking" delay time of data storage in the multilevel storage, which is several minutes multiplied by the frequency of use during that period, that is:

$$seek = delay\_t * F(t) \tag{3}$$

'delay_t' is fixed by cloud operators and usually 1–5 min, so, the total seeking time is:

$$Seek\_time = \sum_{i=1}^{3} seek_i * h_i \tag{4}$$

And the total storage cost is:

$$Storage\ cost = \sum_{i=1}^{3} G(t) * p_{1i} + \sum_{i=1}^{3} F(t) * p_{2i} + \sum_{i=2}^{3} F(t) * s * p_{3i} \tag{5}$$

h_i are constants, representing the delay time of each level, p_{1i} is the cost of storage cost per level, p_{2i} for each level of request cost, p_{3i} is the second and third level seeking fee. And s is a constant of the size of the current data (or the required data set).

3.2   Long-Term Storage Strategy

In the previous section, it can be clearly found that cost and seek_time are a pair of conflicting objective functions. The reduction of cost must be accompanied by more use of secondary and tertiary storage resources, but this inevitably increases the "seeking" time of the whole process.

On the other hand, assuming that we separate three segments representing three prices on a workflow with a period of n, we require that the second segment point be greater than or equal to the first segment point, so that the total solution domain is as follows:

$$(n+1) * n * (n-1) * \dots * 1 = (n+1)! \qquad (6)$$

The solution set chosen in this way is too large to be enumerated completely.

So, multi-objective genetic algorithm NSGAII is selected, cost and time are selected as two objective functions, and an appropriate strategy is obtained by iteration to "place old data" in three-level storage. So our model becomes:

$$\min(f) = \min(\text{Cost,Time}) \qquad (7)$$

Formula 7 is a multi-objective function for long-term storage problem in cloud environment. It contains two sub-goals: Cost and Time.

$$\min(\text{Cost}) = \min(\text{Computing Cost} + \text{Storage cost}) \qquad (8)$$

Formula 8 shows that Cost includes two aspects: calculation cost and storage cost, requiring the minimum value as far as possible.

$$\min(\text{Time}) = \min(\text{Seek\_time}) = \min(\sum\nolimits_{i=1}^{3} seek_i * h_i) \qquad (9)$$

Formula 9 represents the delay time of seeking, requiring that the delay time of the system should not exceed the deadline of the user.

4   Simulation Experiment

4.1   Hierarchical Price Comparison

To verify the correctness of the scheme, we assume that a linear workflow consists of five nodes: a, b, c, d, e (Fig. 1) and two virtual nodes: ds, de, and random generation of size (100G–1T), rebuilding time (1–10 h) and frequency of use (3–30/month). Specifically as Table 1 shows.

Through CTT-SP algorithm, we can get the storage points are 'C', 'E'.

We combine the real Amazon cloud resources [13] and the parameters [10] to get the CPU and Table 2 parameters:

CPU Computing Cost: $0.1 per hour.

Table 1. Node parameter 1

	Size	Rebuilding time	Frequency
a	226	1	6
b	298	1	16
c	478	7	22
d	402	8	9
e	455	9	18

Table 2. Price stratification parameters

	Standard storage	Low frequency storage	Archive storage
Storage pricing	0.023	0.0125	0.004
Seeking cost	0	0.01	0.03
Request cost	0.000005	0.00001	0.01

The results for a quarter (three months) are as follows:

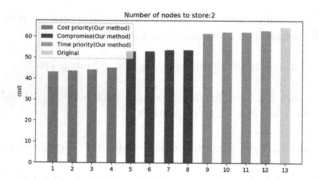

Fig. 4. An example of cost map for storing two points

We divide the non-dominant set into three groups and plot it. Each group contains three legends, such as 1, 2, 3 … 12, 13.

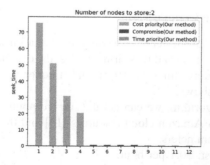

Fig. 5. An example of time map for storing two points

Figure 4 shows that, compared with the original CTT-SP algorithm, the cost savings changes from 30% at the "cost priority" (legend 1) to 5% at the "time priority" (legend 9). The change range is about 25%.

But Fig. 5 shows that, when choosing "cost priority", the time is about 75, but then it drops rapidly, and it is almost zero when choosing "time priority". That is to say, when "time priority" is chosen, the seeking time of the whole workflow can be neglected.

Then we test stored four nodes of a five-node workflow storage system with the Table 3 parameters:

Table 3. Node parameter 2

	Size	Rebuilding time	Frequency
a	804	8	3
b	565	8	16
c	340	7	23
d	404	9	28
e	169	10	11

The points need to store are 'E', 'D', 'B', 'C'.

Other invariable, the results for a quarter (three months) are as Figs. 6, 7 shows. As we can see from the two figures, the more "storage points" a workflow needs to store, the better cost saving of our strategy, cost savings range from Fig. 4 approximate 30% to Fig. 6 approximate 60% if we choose "cost priority" first.

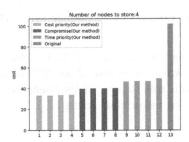

Fig. 6. An example of cost map for storing four points

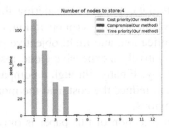

Fig. 7. An example of time map for storing four points

4.2 Long-Term Storage Comparison

In Fig. 8, we conclude that "cost priority" is the lowest in overall workflow progress. The "time priority" strategy is basically the same as the original CTT-SP algorithm, which is due to the full consideration of the time cost, thus reducing the cost of optimization. In addition, the black dotted line strategy is full storage, which cost fee is the most, but its time overhead is zero and there is no waiting time when we need to use any data. Blue dotted line is a non-storage strategy, which has the fastest cost growth and the longest delay. Without special requirements, other strategies are generally recommended.

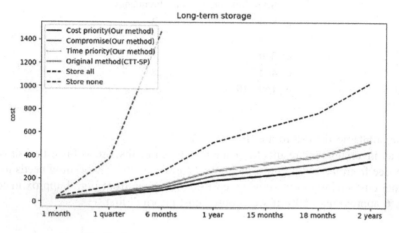

Fig. 8. Long-term storage (Color figure online)

Therefore, our method has strong flexibility. If it is insensitive to time parameters, we choose "cost priority", which saves cost about 30% of the original strategy while 65% savings compared with full storage. If time-sensitive, choose "time priority", which saves cost about 3% of the original strategy.

5   Conclusion

Aiming at the data-intensive cloud workflow, this paper proposes a method that considers long-term intermediate data storage. Considering that there are a lot of useful intermediate data in workflow due to long-term operation, our method combines CTT-SP algorithm with price stratification, and multi-objective genetic algorithm is used to move part of "old data" to relatively inexpensive cloud resources, while guaranteeing the overall workflow time delay. Finally, through a comparison experiment, it is verified that hierarchical price can reduce the cost, and the more storage points, the effect of saving storage costs is obvious.

Because the optimization direction of this paper is only for storage costs. For the future research the optimization of computing costs will be considered in real scenarios.

Acknowledgment. This work is supported by Anhui Natural Science Foundation 1908085MF206 and National Natural Science Foundation of China (NO. 61402007, 61573022), the Scientific Research Foundation for the Returned Overseas Chinese Scholars, State Education Ministry.

References

1. Deelman, E., et al.: Pegasus: mapping scientific workflows onto the grid. In: Grid Computing, Second European Across Grids Conference, Axgrids, Nicosia, Cyprus, January, Revised Papers (2004)
2. Ludäscher, B., et al.: Scientific workflow management and the Kepler system: research articles. Concurr. Comput.: Pract. Exp. **18**, 1039–1065 (2006)
3. Oinn, T., et al.: Taverna: a tool for the composition and enactment of bioinformatics workflows. Bioinformatics **20**, 3045–3054 (2004)
4. Li, X., et al.: A novel workflow-level data placement strategy for data-sharing scientific cloud workflows. IEEE Trans. Serv. Comput. (1939)
5. Ikken, S., Renault, E., Barkat, A., Kechadi, M.T., Tari, A.: Efficient intermediate data placement in federated cloud data centers storage. In: Boumerdassi, S., Renault, É., Bouzefrane, S. (eds.) MSPN 2016. LNCS, vol. 10026, pp. 1–15. Springer, Cham (2016). https://doi.org/10.1007/978-3-319-50463-6_1
6. Zhao, Q., Xiong, C., Zhao, X., Yu, C., Xiao, J.: A data placement strategy for data-intensive scientific workflows in cloud. In: IEEE/ACM International Symposium on Cluster (2015)
7. Xu, R., Zhao, K., Zhang, P., Dong, Y., Yun, Y.: A novel data set importance based cost-effective and computation-efficient storage strategy in the cloud. In: IEEE International Conference on Web Services (2017)
8. Dong, Y., Cui, L., Li, W., Xiao, L., Yun, Y.: An algorithm for finding the minimum cost of storing and regenerating datasets in multiple clouds. IEEE Trans. Cloud Comput. (2016)
9. Dong, Y., Yun, Y., Xiao, L., Chen, J.: A cost-effective strategy for intermediate data storage in scientific cloud workflows, pp. 1–12 (2010)
10. Dong, Y., Yun, Y., Xiao, L., Chen, J.: On-demand minimum cost benchmarking for intermediate dataset storage in scientific cloud workflow systems. J. Parallel Distrib. Comput. **71**, 316–332 (2011)
11. Lei, F., Sha, M., Liu, X., Liang, Y.: Improved CTT-SP algorithm with critical path method for massive data storage in scientific workflow systems. Int. J. Pattern Recognit. Artif. Intell. **30** (2016)
12. Lei, F., Sha, M., Liang, Y., Liu, X.: Experimental analysis on CTT-SP algorithm for intermediate data storage in scientific workflow systems. In: International Conference on Computational Intelligence & Security (2016)
13. Amazon EC2 Pricing. https://amazonaws-china.com/cn/s3/pricing/. Accessed 1 Mar 2019
14. Baidu Cloud Pricing. https://cloud.baidu.com/product/bos.html. Accessed 16 Feb 2019
15. Tencent Cloud Pricing. https://cloud.tencent.com/product/cos. Accessed 16 Feb 2019

Optimal Overbooking Mechanism in Data Plan Sharing

Yaxin Zhao, Guangsheng Feng[⊠], Bingyang Li, Chengbo Wang,
and Haibin Lv

Harbin Engineering University, Harbin 150080, China
fengguangsheng@hrbeu.edu.cn

Abstract. For mobile devices users, the inconvenience caused by the fixed data plans becomes a popular issue. With the help of personal hotspot (pH), however, the smartphone users can share network connection with other users nearby. In this paper a data plan sharing platform with reservation mechanism is proposed. The users on this platform can sell their surplus data plan traffic to make profit. A Markov decision process model was formulated and analyzed for data plan sharing platform revenue management through overbooking mechanism. The results of experiment and simulation show that the platform revenue has a considerable growth using this model.

Keywords: Data-plan sharing · Markov decision process ·
Revenue management · Overbooking reservation mechanism ·
Dynamic programming

1 Introduction

Getting the internet connection becomes increasingly important to people these days. You can hear "Is Wi-Fi available here?" more and more often. However, many places do not provide Wi-Fi access. Or the Wi-Fi signal is not good enough to surf online fluently. Then we have to use data plan which we have selected from all data-plan options provided by Internet Service Providers. We must note that data-plan services provided by ISP have many shortcomings. However, there are two typical issues which we are interested in.

First, the unlimited data-plan is not an option provided by ISPs. Therefore, If the Internet users want to avoid the expensive excess fee, they must try to make sure the data is not overused.

Secondly, to all the internet users, the data using status is different every month. Meanwhile they cannot change their data plan forward. So sometimes they will have to pay the overused data cost which is really expensive. But on the other hand, sometimes at the end of month there still a lot data remain which is wasted.

Nowadays, the technology is mature enough to support people share the internet connection via Bluetooth or personal hotspots. The internet users can download data packets cooperatively on the premise that the users are all interested in the same data source. But the preference of the internet users is not so highly unified in real life. The users of smartphone which supported by this technology can also share internet connection to other devices nearby. In daily life, people just share their personal hotspots to their family, friends or acquaintances. Therefore, a reliable and reasonable platform is urgently needed. Through this platform people can sell their surplus data to acquire economic profits. And to whom the data plan is out of use can obtain data at low cost.

There are several papers have been proposed to address the data plan sharing problem. These works mainly focus on how to motivate the internet users to cooperate with each other or the economic benefits of data buyers. Some works studies the pricing schemes in data sharing market. And as far as we are acknowledged of, there is no work study the platform with reservation mechanism to supervise this transaction, which provides high reliability and security. Besides, it also allows the data buyers have different preference of QoS (Such as the speed of internet connection). And little attention has been paid to the revenue management in this area.

Motivated by this, this paper first proposes a platform for data plan sharing with reservation mechanism. It is not only capable of guaranteeing the users both can achieve economic benefits with high reliability and security, but also the data buyers can choose different speed of internet connection according to his or her time delay tolerance.

There is a problem comes along with reservation mechanism. After the customers make the reservation, the business will save the deal for them. If they cancel the reservation or even do not show up, there is no punishment economically or just very low cost. The risk is only on the business' shoulders. The situation can be improved via overbooking. The airline company usually accepts more reservation than the actual accommodation. It can guarantee the flight is nearly full even though some customers cancel the reservation or do not show up, which lower the risk at some level.

Meanwhile the platform can optimize its profits using dynamic programing.

Our contributions can be summarized as follows:

- We propose a data plan sharing platform with reservation mechanism, that allows a user to set diverse demands of quality of service (QoS) according to his or her preference.
- We formulate and analyze a Markov decision process model for data plan sharing platform. Considering real life circumstances, we allow cancellation, no showing up, and overbooking.
- We draw the overbooking into the platform revenue management. And we solve the problem by using dynamic programming.

The rest of the paper is organized as follows. Section 2 reviews related work and Sect. 3 presents the system model and the problem formulation. Section 4 shows several numerical examples, and we analyze the results. And we conclude the paper in Sect. 5.

2   Related Work

There are several works focus on how to motivate the internet users to participate data plan sharing. An incentive mechanism named RAP is proposed in [1] to encourage participation. And the users in their system can manage their surplus data plan more efficiently while a high-speed download rate can be achieved. This paper [2] studies how to motivate such pH-enabled data-plan sharing between local users and travelers in the ever-growing roaming markets by proposing pricing incentive for a data-plan buyer to reward surrounding pH sellers (if any). Another pricing incentive model [3] is for a secondary data buyer (typically, a traveler) to opportunistically demand and reward pHs (if any) in the vicinity to reach a win-win situation. This paper [8] analyzes and compares different incentive mechanisms for a master to motivate the collaboration of smartphone users on both data acquisition and distributed computing applications.

And some works studies the pricing in data plan sharing. They [4] derive the optimal prices and amount of data that different buyers and sellers are willing to bid in this market and then propose an algorithm for ISPs to match buyers and sellers. They [7] study the economic interests of a wireless access point owner and his paying client, and model their interaction as a dynamic game. This paper [9] analyzes two pricing schemes commonly used in Wi-Fi markets: the flat-rate and the usage-based pricing. A platform is proposed in [10] which facilitates the trade by matching the supply and demand. We propose a data plan sharing platform with reservation mechanism with the consideration of QoS.

The overbooking is an effective method to increase the revenue of the airline, rail or shipping company. Even some customers miss the trip or even do not show up, it can still keep the seats are highly used in most runs. They [12] formulate and analyze a Markov decision process model for airline sear allocation on a single-leg flight with multiple fare classes. We formulate and analyze a Markov decision process model for data plan sharing platform. And we draw the overbooking into the platform revenue management. The cancellation, no showing up, and overbooking are allowed under the consideration of the daily life circumstances.

3   System Model and Problem Formulation

3.1   The Data Plan Sharing Platform

The data buyers should have been reserved the order earlier enough before the beginning of data plan sharing process. As to the data sellers, after they upload the time and the location they can serve as a personal hotspot. The platform will pay them before the deal according to the analysis of the data buyers' data usage habits at different time in a day. However, if the data sellers cannot finish the deal, there is a penalty they should pay to the platform enough to cover the loss so the platform can buy cellular data from ISPs to serve the data buyers. In the data buyers side, the platform collects the reservation information of them a fixed period of time earlier, and then decides whether accept the reservation aims to maximize the profits of the platform.

3.2   The Discrete-Time Model

Between the beginning of the data plan sharing process and the time when the platform starts the reservation, there are N time periods. The value decreases as time goes, $t = T, T - 1, \ldots, 1, 0$. Easily to know, T represents the time when the customer can make the reservation. And 0 is the beginning of the data plan sharing process. In our assumption, only one event happens at each time period. There are three kinds of situation: (1) A customer makes a reservation for a data plan sharing time unit. (2) A customer cancels the reservation he or she makes before. (3) Nothing happens.

3.3   The Probabilities and Distribution in Data Plan Sharing

In this model, the probability that the customers cancel the reservation and the probability of customers' no showing up are irrelevant with time. So in the following discussion, we can use a one-dimensional state variable.

Let p_{it} denote the probability that at the time period t a customer makes a reservation for a data sharing time unit. Similarly, let $q_t(x)$ denote the probability that a customer cancels his or her reservation in time period t. And $p_{0t}(x)$ denote the probability that nothing happens in time period t. Where x is the number which the platform has accepted the reservation until the time period t.

In our assumption, $q_t(x)$ is a function of x. Its value does not decrease as the x increasing when t is the same. At every time period t, we assume that besides nothing happens, otherwise there will be at least one customer make or cancel a reservation. Then we can get,

$$\sum_i^T p_{it} + q_t(x) + p_{0t}(x) = 1, \forall x, t \geq 1. \tag{1}$$

At time period t, there is a customer make a reservation for a data plan sharing unit. Then the platform manager will check the service level which the customer has chosen and the reserved data plan sharing unit number. The customer will choose different service level to obtain different internet speed. Aiming to maximize the benefits, the platform manager will decide whether to accept the reservation or not. If nothing happens or a customer cancels the reservation, the decision is no need to make then. At time period t, the platform manager accepts a data plan sharing reservation. The platform will obtain economic benefits $b_{it} \geq 0$.

There still is another situation. The customer who has made reservation do not show up at the data plan sharing process. We assume that probability is β. Then the probability of showing up is $1 - \beta$. Let $f(x)$ denote at the beginning of data plan sharing process, the number of whom have the reservation and show up. x is the current number which the platform manager has accepted reservation. It is easily to be seen that $x - f(x)$ represents the number of no showing up. The $f(x)$ follows a $(x, 1 - \beta)$ binomial distribution. Let M denote the total number of the data plan sharing unit provided by the data sellers. When the data plan sharing process starts, if $f(x) = y$. we create a convex function $\gamma(x)$ that represents the platform's economic loss from overbooking. As y ($y \geq 0$) increases, the value of $\gamma(x)$ is non-decreasing. When $y \leq M$, $\gamma(x) = 0$. We assume that $\gamma(x)$ is a convex function which is non-decreasing. It is reasonable.

From a realistic view, we can know that the reservation for a data plan sharing unit is irrelevant with the number of the data plan sharing units which the platform already booked. Meanwhile the probabilities of the cancellation of the reservation and no showing up for the reservation are depends on the current number of booked data plan sharing units. And in our assumption the probabilities are the same in different service levels. $q_t(x)$ is a non-decreasing function of x. The number of the booked data plan sharing units gets bigger, the probability of the reservation cancellation is higher.

3.4   The Economic Benefit Maximization Problem

From time period T to 0, maximizing the economic benefit of the platform is our objective. At the beginning, the x is equal to 0 which means no data plan sharing units have been booked. We can transform our objective the economic benefit maximization problem into a Markov Decision Process (MDP) problem with the discrete time model in which the time period $t = T, T - 1, \ldots, 0$. Where the state x is the current number of booked data plan sharing time units. The platform can overbook and the customers can cancel the reservation or even do not show up at all in our model. Due to the existence of the overbooking, the x does not have to lower than M. At the beginning (time period = N), there is no data plan sharing unit has been booked. So as long as the customer makes a reservation, the platform manager will accept it probably. However, because we start with no data sharing time units booked at stage N and, at most, one data unit request can be accepted at each stage (because, at most, one arrives), $x \leq T - t$ is satisfied in every time period t. We introduce another variable o to make the computation easier. Let o denote the overbooking number which allowed by the platform manager. Then the constraint is $x \leq M + o$.

let $S_t(x)$ denote the maximum economic benefit the data plan sharing platform expect to obtain from time period t to 0. It is a function of x. S_t as the expecting benefit maximization, can get by

$$S_t(x) = \sum_{i=1}^{n} p_{it} \max\{b_{it} + S_{t-1}(x+1), S_{t-1}(x)\} + q_t(x)S_{t-1}(x-1) + p_{0t}(x)S_{t-1}(x),$$

$$(2)$$

where $0 \leq x \leq T - t, t \geq 1, f(x) \sim Bin(x, 1 - \beta)$.

This problem can be seen as a queueing optimal problem.

We can get $p_t = \sum_{i=1}^{n} p_{it}$, B_t as the economic benefit of the platform is also a random variable with time period t. And we can get

$$P\{B_t = b_{it}\} = \frac{p_{it}}{p_t}, \quad i = 1, \ldots, n. \ t \geq 1. \tag{3}$$

Let $A_t(x, b) = \max\{b + S_{t-1}(x+1), S_{t-1}(x)\}$, x as an integer, $x \in [0, T - t]$. Where b is a real number which gets from B_t. The B_t follow more arbitrary distribution in other common queue optimal problem while the bound is $[0, \infty)$. But in this problem, the value of b gets from the $b = b_{it}$ when for some i in i = 1, …, n the probability is positive. Then we can get the equal optimal equation,

$$S_t(x) = p_t E[A_t(x, B_t)] + q_t(x)S_{t-1}(x-1) + (1 - p_t - q_t(x))U_{t-1}(x) \tag{4}$$

$$A_t(x) = \max\{b - S_{t-1}(x) - S_{t-1}(x+1), 0\} + S_{t-1}(x) \tag{5}$$

where $0 \le x \le T - t, t \ge 1$, and we have

$$S_0(x) = E[-\gamma(f(x))], 0 \le x \le T \tag{6}$$

In our assumption, $\gamma(\cdot)$ is convex and non-decreasing. And $S_0(\cdot)$ and S_t are both concave and non-increasing. The value of $S_t(x) - S_t(x+1)$ is the expected economic loss of the platform could suffer when the platform accepts a reservation in time period $t + 1$.

We introduce the variable o to make the computation easier. Let o denote the overbooking number which allowed by the platform manager. Then the constraint is $0 \le x \le M + o$ for each single time period t.

At time period $t - 1$, we define the value functions,

$$S_{t-1}(x) = S_{t-1}(M+o) - Z(x-M-o), \quad x \ge M+o \tag{7}$$

where Z satisfies the condition $Z \ge max_{i,t}\{b_{it}\}$. As we can see, when the number booked equals $M + o$ at time period t, the platform manager will reject any reservation of the customers from all service levels.

Due to the optimal policy, the platform manager does not accept reservation of the customers from all service levels to avoid the economic loss when the booked number is $M + o$. we define the opportunity cost functions as follows: $S_{t-1}(x) - S_{t-1}(x) - S_{t-1}(x+1)$, for all x, $0 \le x \le M + o$. Then the optimality Eq. 2 can be simplified to

$$S_t(x) = \sum_{i=1}^{n} p_{it} \max\{b_{it} - S_{t-1}(x), 0\} + xq_t S_{t-1}(x-1) + (1 - xq_t)S_{t-1}(x) \tag{8}$$

where $0 \le x \le M + o - 1$, $t \ge 1$,

$$S_t(M+o) = (M+o)q_t S_{t-1}(M+o-1) + (1 - (M+o)q_t)S_{t-1}(M+o), t \ge 1 \tag{9}$$

$$S_0(x) = E[-\gamma(f(x))], 0 \le x \le M+o \tag{10}$$

In the next section, we will apply these equations in the numerical examples to get the optimal result.

4 Numerical Examples Measurement and Evaluation

In this section, we report the results of numerical solution of several examples.

4.1   Example 1

In Example 1, We consider two booking service levels with economic benefits $b_1 = 0.5$ and $b_2 = 0.1$ The refund amounts are $c_{in} = b_i, i = 1, 2$ The maximum overbooking level is $o = 2$ The economic loss from overbooking is given as 0.3 per data plan sharing time period overbooked. We assume all the customers show up in this example. The values of the parameters $p_{it}, i = 1, 2$ and q_t are shown in Table 1.

Table 1. Parameters for Example 1

Parameters/Time period n	p_{1t}	p_{2t}	q_t
10–9	0.182	0.318	0
8–7	0.086	0.2	0.014
6–5	0.143	0.143	0.014
4–3	0.241	0	0.017
2–0	0.241	0	0.0167

Figure 1 plots the economic benefits, with respect to t for the same amount of data plan sharing units ($M = 8$) and the various final values of x. $x = 10$ and $x = 8$. As the Fig. 1 shows, the former plot is always above the latter. And the benefits functions are monotone (non-increasing) in t.

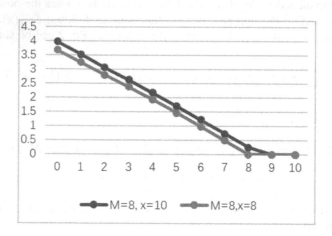

Fig. 1. The economic benefits versus t for different x values for Example 1

4.2   Example 2

In Example 2, We consider booking service levels with economic benefits $b_1 = 0.5$ and $b_2 = 0.3$. The refund amounts are $c_{in} = b_i, i = 1, 2$. The maximum overbooking level is $o = 5$. The economic loss from overbooking is given as 0.3 per data plan sharing time period overbooked. With no consideration of the no showing up situation. The values of the parameters $p_{it}, i = 1, 2$ and q_t are shown in Table 1.

Figure 2 plots the benefits, with respect to t for the different amounts of data plan sharing units (M = 5 and M = 8) and the same overbooking limit of o. (o = 2). As the Fig. 2 shows, the former plot is always above the latter. And the benefits functions are monotone (non-increasing) in n.

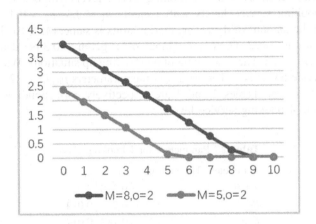

Fig. 2. The economic benefits versus t for different M with same overbooking limit for Example 2.

And we take random probability to do 100 groups of experiments, the results are still same. In Fig. 3 plots the profits, with respect to n for the different amounts of data plan sharing units (M = 49, 51, 53 and M = 55) and the same overbooking limit of o. (o = 5).

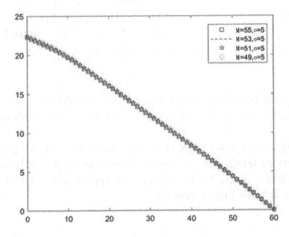

Fig. 3. The economic benefits versus t for different M with same overbooking limit for Example 2.

So when the number of the data buyers is large enough, the benefit will grow with the amount of data plan sharing units the platform purchased.

4.3 Example 3

In Example 3, We consider three cases with different numbers of booking service levels. There are 1, 2, 4 kinds of booking service levels. We consider the 1 booking service level with benefit $b_1 = 0.5$, 2 booking service levels with benefits $b_1 = 0.5$ and $b_2 = 0.1$, 4 booking service levels with benefits $b_1 = 0.5$, $b_2 = 0.4$, $b_3 = 0.3$ and $b_4 = 0.1$. And the amount of data plan sharing units is same. ($M = 8$). The maximum overbooking limit is $o = 2$. The overbooking cost is given as 0.3 per data plan sharing time period overbooked. There are without no showing up situations. When the number of booking service levels is 2, The values of the parameters $p_{it}, i = 1, 2, 3, 4$ and q_t are shown in Table 1. The Tables 2 and 3 are respectively corresponding to 1 booking service level and 4 booking service levels.

Table 2. Parameters for 1 booking class

Parameters/Time period t	p_{1t}	q_t
10–9	0.318	0
8–7	0.2	0.014
6–5	0.182	0.014
4–3	0.143	0.017
2–0	0.157	0.017

Table 3. Parameters for 4 booking classes

Parameters/Time period t	p_{1t}	p_{2t}	p_{3t}	p_{4t}	p_{5t}
10–9	0.182	0.182	0.318	0.318	0
8–7	0.086	0.086	0.2	0.2	0.014
6–5	0.143	0.143	0.143	0.143	0.014
4–3	0.241	0.241	0	0	0.017
2–0	0.25	0.25	0	0	0.0167

Figure 4 plots the benefits, with respect to n for the different amounts of booking classes and the same data plan sharing units with the same overbooking limit of o. (o = 2).

The 2 booking service levels plot is always above the 1 booking service level plot and 4 booking service levels plot. 4 booking service levels plot is between the 2 and 1 booking service levels plots. So it indicates an appropriate number of the booking service levels can achieve higher benefits.

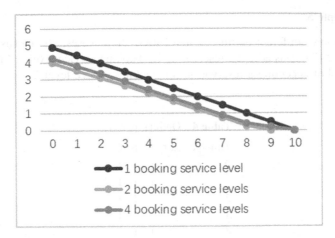

Fig. 4. The economic benefits versus t for different number of the booking service levels for Example 3.

4.4 Example 4

In Example 4, We consider four cases with different overbooking limits. $o_1 = 2$, $o_2 = 3$, $o_3 = 4$ *and* $o_4 = 5$. We consider the benefits $b_1 = 0.5$ and $b_2 = 0.3$. And the amount of data plan sharing units is same. (M = 55). The overbooking cost is given as 0.3 per data plan sharing time period overbooked. There are no no-shows. The values of the parameters p_{it}, $i = 1, 2$ and q_t are randomly obtained. We proceed the experiment for 100 times.

As shown in Fig. 5, the three plots are slightly overlapped at the time closing to the beginning of the data plan sharing process. But at the beginning, the overbooking limit is lower, the benefit gets higher.

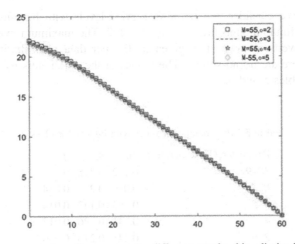

Fig. 5. The economic benefits versus t for different overbooking limits for Example 4.

4.5   Example 5

In Table 4 shows different optimal policies under three cases, "0" means accept one cancellation. "1" means accept a reservation for booking service level 1. Likewise, "2" is the symbol of accepting a reservation for booking service level 2.

Table 4. Optimal policy

Time period t	A	B	C
0	1	1	1
1	1	1	1
2	1	1	1
3	1	1	1
4	1	1	1
5	1	0	1
6	1	2	1
7	0	1	0
8	2	1	0
9	2	1	2
10	1	1	1

We can see from the Table 4. For the reason of the benefits of different booking service levels are different. The benefit the platform can get from booking service level 1 is higher than the booking service level 2. So when both booking service levels upload the reservation requests. The platform will give priority to the reservation for booking service level 1.

4.6   Example 6

In Example 6, We consider two booking service levels with benefits $b_1 = 0.5$ and $b_2 = 0.1$. The refund amounts are $c_{in} = r_i$, $i = 1, 2$. The maximum overbooking limit is $o = 2$. The overbooking cost is given as 0.3 per data plan sharing time period overbooked. There are no no-shows. The values of the parameters $p_{it}, i = 1, 2$ and q_t are shown in Tables 5 and 6.

Table 5. Parameters for more data buyers for class 2

Parameters/Time period t	p_{1t}	p_{2t}	q_t
10–9	0.182	0.318	0
8–7	0.086	0.2	0.014
6–5	0.143	0.143	0.014
4–3	0.241	0.276	0.017
2–0	0.241	0.276	0.0167

Table 6. Parameters for more data buyers for class 1

Parameters/Time period t	p_{1t}	p_{2t}	q_t
10–9	0.318	0.182	0
8–7	0.2	0.086	0.014
6–5	0.143	0.143	0.014
4–3	0.276	0.241	0.017
2–0	0.276	0.241	0.0167

Figure 6 plots the benefits, with respect to t for the different distributions of booking service levels. The two plots are almost overlapped. So we check the optimal policies for both situations.

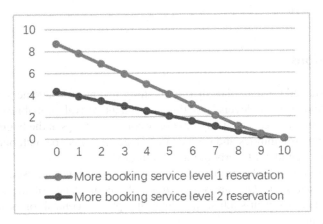

Fig. 6. The benefits versus n for different distributions of multiple booking service levels reservation

In Table 7 shows different optimal policies under three cases, "0" means accept one cancellation. "1" means accept a reservation for booking service level 1. Likewise, "2" is the symbol of accepting a reservation for booking service level 2. In this example, the optimal policies are exactly same under both situations.

We can see from the Table 7. For the reason of the profits of different booking service levels are different. The benefit the platform can get from booking service level 1 is higher than the booking service level 2. So when both booking service levels upload the reservation requests. The platform will give priority to the reservation for booking service level 1. However, when the probability of booking service level 2 reservation is higher than the booking service level 1. The platform will still give priority to the reservation for booking service level 1. Even though there is the over-booking risk, the platform will take it for the booking service level 1 reservation. Because the benefit from the booking service level 1 reservation is big enough to cover the overbooking economic loss.

Table 7. Optimal policy for different distributions of multiple classes reservation

Time period t	More booking service level 2 reservation	More booking service level 1 reservation
0	1	1
1	1	1
2	1	1
3	1	1
4	1	1
5	1	1
6	1	1
7	0	0
8	0	0
9	0	0
10	1	1

5 Conclusions

We have considered the data plan sharing platform revenue management problem with multiple booking service levels and time-dependent arrival probabilities. Currently booked customers may cancel at any time or become no-shows at the beginning time of the service, with probabilities and refunds that, in general, may be both booking service level and time dependent. Overbooking is permitted.

We have formulated the problem as a discrete-time MDP and used dynamic programming to analyze it. Our analysis exploits the equivalence of the revenue management problem to well-studied problem of optimal control of admission to a queueing system.

When the cancellation and no-shows probabilities are independent of the booking service levels, we have used a transformation to convert the problem to an equivalent problem with a one-dimensional state variable; the total number of data plan sharing units booked.

In our numerical examples, we have demonstrated that an optimal booking policy can have counterintuitive properties when cancellations and no-shows are included.

References

1. Yu, T., et al.: INDAPSON: an incentive data plan sharing system based on self-organizing network. In: IEEE INFOCOM 2014 - IEEE Conference on Computer Communications. IEEE (2014)
2. Wang, F., Duan, L., Niu, J.: Optimal pricing of user-initiated data-plan sharing in a roaming market. IEEE Trans. Wirel. Commun. 17(9), 5929–5944 (2018)
3. Wang, F., Duan, L., Niu, J.: Pricing for opportunistic data sharing via personal hotspot. In: GLOBECOM 2017–2017 IEEE Global Communications Conference. IEEE (2017)

4. Zhang, Y., et al.: Incentive mechanism for mobile crowdsourcing using an optimized tournament model. IEEE J. Sel. Areas Commun. **35**(4), 880–892 (2017)
5. Wang, X., Duan, L., Zhang, R.: User-initiated data plan trading via a personal hotspot market. IEEE Trans. Wirel. Commun. **15**(11), 7885–7898 (2016)
6. Musacchio, J., Walrand, J.: WiFi access point pricing as a dynamic game. IEEE/ACM Trans. Netw. **14**(2), 289–301 (2006)
7. Duan, L., et al.: Motivating smartphone collaboration in data acquisition and distributed computing. IEEE Trans. Mob. Comput. **13**(10), 2320–2333 (2014)
8. Duan, L., Huang, J., Shou, B.: Pricing for local and global Wi-Fi markets. IEEE Trans. Mob. Comput. **14**(5), 1056–1070 (2015)
9. Yu, J., et al.: Mobile data trading: a behavioral economics perspective. In: 2015 13th International Symposium on Modeling and Optimization in Mobile, Ad Hoc, and Wireless Networks (WiOpt). IEEE (2015)
10. Zheng, L., et al.: Secondary markets for mobile data: feasibility and benefits of traded data plans. In: 2015 IEEE Conference on Computer Communications (INFOCOM). IEEE (2015)
11. Rothstein, M.: OR forum—OR and the airline overbooking problem. Oper. Res. **33**(2), 237–248 (1985)
12. Subramanian, J., Stidham Jr., S., Lautenbacher, C.J.: Airline yield management with overbooking, cancellations, and no-shows. Transp. Sci. **33**(2), 147–167 (1999)

Survey of Data Value Evaluation Methods Based on Open Source Scientific and Technological Information

Xiaolin Wang, Cheng Dong, Wen Zeng$^{(\boxtimes)}$, Zhen Xu,
and Junsheng Zhang

Institute of Scientific and Technical Information of China, Beijing 100038, China
zengw@istic.ac.cn

Abstract. It is important to effectively identify the data value of open source scientific and technological information and to help intelligence analysts select high-value data from a large number of open-source scientific and technological information. The data value evaluation methods of scientific and technological information is proposed in the open source environment. According to the characteristics of the methods, the data value evaluation methods were divided into the following three aspects: research on data value evaluation methods based on information metrology, research on data value evaluation methods based on economic perspective and research on data value assessment methods based on text analysis. For each method, it indicated the main ideas, application scenarios, advantages and disadvantages.

Keywords: Data value evaluation ·
Open-source scientific and technological information · Information metrology ·
Economic perspective · Text analysis

1 Introduction

The emergence of the Internet has led to a sharp increase in the amount of scientific and technological information in the open source environment. From the GB level at the end of the 20th century to the EB level in 2009 to the ZB level in recent years, the data volume has been continuously expanding [1], and a lot of data can be obtained directly from the open network channels. In this case, the data has the characteristics of large scale, various forms, unpredictable quality, and complicated calculation. In the face of the intermingled open source scientific and technological information, if there is no effective method, really valuable intelligence may be flooded by massive invalid information, so how to achieve the collection and processing of the data from the massive open source data, It is an important problem to be solved urgently to obtain high information value data to meet the needs of users. The choice of data value evaluation method is one of the methods to solve this problem. That is, according to the actual needs, the data value assessment method is used to evaluate the value of the data, and perceive the high-value data that meets the actual needs. This paper will summarize the data value evaluation methods of scientific and technological information in the open source environment. According to the characteristics of the methods, the data value evaluation methods are divided into the

© Springer Nature Singapore Pte Ltd. 2019
X. Cheng et al. (Eds.): ICPCSEE 2019, CCIS 1058, pp. 172–185, 2019.
https://doi.org/10.1007/978-981-15-0118-0_14

following three aspects: research on data value evaluation methods based on information metrology; research on data value evaluation methods based on economic perspective; research on data value assessment methods based on text analysis.

2 Research on Data Value Evaluation Methods Based on Information Metrology

2.1 Qualitative Evaluation Method

In the early stage, the main method of data value evaluation is qualitative evaluation, which is carried out in the form of expert evaluation and questionnaire survey. According to the needs of users, a targeted evaluation index is established.

The earliest data value assessment was the "10C" principle proposed by Betsy RichIllon in 1991, namely content, credibility, critical thinking, copyright, citation, continuity, censorship, connectivity, comparability, context [2]. The presentation of these 10 qualitative indicators has had a great influence on the follow-up research. After that, various qualitative indicators have been proposed and applied to the value evaluation of open source information. GL Wilkinson proposes 11 indicators for information and website quality evaluation: site access and availability; resource identification and documentation; author identity; author authority; information structure and design; content relevance and scope; content effectiveness; accuracy and balance of content; navigation within documents; link quality; aesthetic and emotional aspects [3]. Cooke and Stoke provide eight data value evaluation indicators for library information workers, namely information source, scope and discussion (discipline scope, revision method, accuracy, timeliness, readership, etc.), technology factors, text format, information organization, price and availability, user support system, authority [4]. Harris has proposed a well-known "CARS inspection system" for network resources, namely credibility, accuracy, reasonableness and support [5]. Batini pointed out that the quality dimensions of data standards include timeliness, accuracy, integrity and consistency [6].

In addition, database developers have also formulated qualitative evaluation indicators for academic databases and related websites, which focus on the content and quality of network information resources. For example, ISI Web of Knowledge relies on expert advice and tools to evaluate academic websites. The main indicators include authority, accuracy, timeliness, navigation system and page design, content and practicality, scope, readers' level and text level, etc. [7].

Qualitative assessment mainly establishes different value evaluation indicators according to different needs. Some of these indicators can directly evaluate the value of information data, such as source authority, content accuracy, timeliness, etc. In addition, some indicators are used to evaluate the value of data indirectly, such as text format, availability, technical factors, etc. These indicators are indirectly evaluated by the external characteristics of the information. The evaluation of data value of open source information begins with qualitative evaluation, and the advantage is that more detailed and comprehensive evaluation of all aspects of data can be done by artificial means. But there are also shortcomings. Firstly, the artificial qualitative evaluation has the subjective concept of each evaluation subject, and the objectivity is insufficient.

Secondly, due to the huge amount of open source information, the workload of qualitative assessment methods alone will be large and the feasibility is insufficient.

2.2 Quantitative Assessment Method

The quantitative evaluation method is to obtain a more objective evaluation of various information collected through mathematical statistics. Because of the wide variety of open source information, different quantitative statistical methods are adopted for different types of information.

For academic information such as papers and patents, quantitative analysis is mainly carried out from the perspective of literature. For example, in addition to the cost method and the revenue method, the evaluation of patent value is mainly based on the reference relationship of the patent [8]. Inspired by Rousseau's idea of "document centripetal citation network" [9], Hu proposed a structured patent value evaluation study based on "patent centripetal citation network", and obtained a set of structured indicators, including the outgrowth index of patent (OIP), the technical interest index (TII), the technology span index (TSI), the technology active degree (TAD), technology extension impact relative index (The CC-index), technology extension base relative index (The RC-index), technology congest basis relative index (The RR-index) seven quantization parameters, using these seven quantifiable calculation indicators to complete the evaluation of the patent value [10]. Anthony proposed to apply new technology to directly visualize the citation relationship of non-patent references (SNPRs), map the landscape around SNPRs through bibliographic coupling and co-citation analysis, and derive the conceptual environment of SNPR through keyword co-occurrence analysis to evaluate the patent value [11]. In addition, the scholar Yang started from the analysis of the characteristics of the cited papers in the journal papers. On this basis, the author puts forward the academic value evaluation method of the low-cited journal papers based on the cited frequency, mainly using the three parameters of citation frequency, mathematical expectation of citation frequency of periodical papers and publication time of papers, and combined with linear regression analysis method to evaluate the value of papers [12].

For network information resources, the early quantitative evaluation is mainly based on simple and intuitive data such as the number of website landings, visits and other simple and intuitive data to measure. With the development of network information metrology, in 1996, McKiernan first used the word "Situation" in the link relationship between networks [13]. Rousseau, a bibliometrics expert, studies the link relationship between networks, and believes that the link relationship between web pages is similar to the citation of the literature. The webpages with more citations had higher value, and "Sitation" was used in academic papers for the first time, which marked the emergence of network link analysis method [14]. The network link analysis method has been well used in information retrieval. For example, the PageRank algorithm proposed by Brin and Page, the founder of the second-generation search engine Google, thinks that if a web page is directed by a lot of pages, web page is considered to be of higher value [15]. Liu provides a focused study on understanding PageRank and the relationship between PageRank and social impact analysis, and along this line, presents a linear social impact model. It also reveals that the model promotes the PageRank-based privilege calculation by introducing some constraints [16]. In the process of developing Clever search engine by IBM, Kleinberg proposed HITS, a new algorithm for ranking web pages. According to the HITS algorithm, when evaluating the value of web pages, it should not only consider the number of inbound links (Hub weight), but also consider the number of outbound links (Authority weight), and finally use these two parameters to measure the value of web pages [17].

Whether it is the citation measurement analysis of academic literature or the web link analysis method of network information resources, it lays a foundation for the quantitative research of data value evaluation of open source technology information, and has achieved certain success in practice. However, there are still some shortcomings: First, for such "cited" measurement methods, it takes time to accumulate, and has good measurement results for old documents and old web pages, but for the evaluation of open source information, new resources tend to It will be more valuable; secondly, the "cited" analysis does not involve the evaluation of the value of the text. Therefore, the applicability of evaluating open source information only based on quantitative evaluation method is not strong.

3 Research on Data Value Evaluation Methods Based on Economic Perspective

For the data value evaluation methods of open source information, many of them are based on the economic perspective, which is mainly aimed at data with market value records, such as open source government information, open source enterprise information, patent information and so on. They use economic methods to evaluate the data and perceive the value of the data.

3.1 Value Evaluation Method Based on Open Source Government Data

On the one hand, value assessment based on open source government data can adjust the direction of future development and optimize the effective allocation of government information resources. On the other hand, it can present the value of open government data to the public. Two evaluation methods are used to evaluate the value of open source government data. One is based on the top-down macroeconomic method, the other is based on the bottom-up microeconomic method.

The top-down macro-economic approach evaluates the value of data from the point of view of the generation and end-use of data resources [18]. For example, PIRA, an internationally renowned energy consulting firm, uses top-down macroeconomic methods and case studies to evaluate the value of data on the supply side and on the demand side. The evaluation of the supply side is based on the cost of investment when collecting data. The evaluation of the demand side is to use the expenditure of the user to measure the data [19]. This method is mainly used to determine the future trend of open source government data, because it evaluates the total value of various industries that use open source government data as investment, and shows the total value of open source government data to the economy. But there are also shortcomings. This method exaggerates economic value and over-emphasizes causality, so it produces biased estimates.

The bottom-up microeconomic approach is mainly to evaluate the data value by collecting business data, case studies, and using productivity analysis to calculate the net economic value. In other words, it is calculated by using the willingness to pay open source data minus the collection of open data and production costs. Productivity analysis of open source government data takes into account the inputs and outputs of individuals and companies in the use of government data. This analysis is only effective in a bottom-up approach to microeconomics. If the goal of the assessment is to figure out why the government should spend money on open source government data, then a bottom-up approach would be more appropriate. However, the bottom-up approach also has its drawbacks. It does not take into account the broader economic impact. If the relevant inputs are not correctly identified, the evaluation results will not be credible, and it is difficult to extend the evaluation from the individual industry assessments to the entire economy [18].

3.2 Value Evaluation Method Based on Open Source Enterprise and Patent Data

For the evaluation of the value of open source enterprise information and patent information, the main economic methods are traditional cost method, market method, income method, real option method, and econometric method.

1. Cost method
   The cost method refers to the evaluation of the data value based on the current acquisition cost of the data. For the enterprise information, it is mainly used for the value perception of the internal data of the enterprise. The cost mainly includes the initial equipment cost and the data collection cost. This method avoids the subjective influence, but ignore the data to bring profit to the company, so the perceived value will be lower than the actual value. For patent information, the cost method uses the cost-related information of the patent to evaluate the value of the patent. however, because the patent is an intangible asset, when using this method to evaluate the intangible asset, as tangible assets cannot be reflected in a clear market price, the current costing approach tends to use potential costs to assess patent value from a side-by-side basis. For example, based on the way of reverse thinking, Zhan used patent infringement judgment and evading the potential cost of protection of the design as the basis for the initial evaluation of patent value [20].
2. Market method
   The market method mainly compares the value of other data in the market with the value of the data being evaluated, and takes the other factors that influence the price of market data as the control variables, and compares it in a more mature trading market, and finally perceives the value of the data. Market method is easy to be recognized by everyone, but because the patent market in China is not perfect and the patent information as reference is not comprehensive, the method also has some limitations [21].
3. Income method
   The income method refers to the evaluation of the expected value of the data within the time limit, and discounts the value according to a certain discount rate to obtains

the current value assessment [22]. For example, Dong chose to use the comprehensive evaluation method to determine the discount rate, technology sharing rate and remaining economic life of patent rights, and to evaluate the patent value of electronic information industry [23]. This method has obvious defects. Since the determination of information such as discount rate, technology sharing rate and residual economic life is subjective to some extent, the estimation of expected patent revenue is also subjective and uncertain.

4. Real option method

Real option refers to the decision-making and choice rights that investors have in the whole process of investing in real assets. The holder of option has the right to choose and decide the quantity of real assets purchased or sold by negotiation price when he holds all kinds of investment information. For example, in 1986, Pakes proposed that patent is an option. The option pricing model is used to evaluate the value of patent. The option pricing model takes into account the high risk of future revenue of patent better, and incorporates the unique "technology shock" factor of patent into the consideration of value evaluation [24]. Tang proposed the idea of applying the real option method to evaluate the value of railway patents in China. Taking a railway patent as an example, based on a detailed analysis of its legal value, technical value and economic value, the patent option was applied to the preliminary evaluation of the patent [25]. The assumptions of the real option method of patent value evaluation are closer to the actual situation, and also take into account various uncertain factors affecting the value of patents. However, the model of this method is relatively complex, and the comprehensibility of the model is poor, and the estimation of various uncertain factors is relatively difficult, which affects the practicability of the method.

5. Econometric method

Generally, econometric economic method takes the estimated value of data as dependent variable and the influencing factors of data value as independent variable, selects samples of the same quality as the data to be evaluated, and carries on multivariate regression analysis with historical data. On this basis, establish the data value evaluation model, and then use the model to carry on the data value evaluation calculation. Professor Harhoff of the University of Munich, Professor Scherer of Harvard University, and Professor Vopel of the Centre for European Economic Studies conducted a practical study of this method using a large number of patent value data [26]. Considering the overall patent value of the enterprise, Zhang constructs the pyramid of the quantification of the patent value of the enterprise, analyzes the factors affecting the patent value, and puts forward to determine the contribution rate of the patent revenue by the method of classification, and set up a quantitative model between patent value and enterprise income [27]. Econometric model method is easy to understand, but also has obvious defects. Firstly, it is difficult to get a large number of samples of homogeneous data value to carry out regression analysis, thus affecting the establishment of the model; Secondly, this kind of methods often assume that the data value and the influencing factors are between Linearity, this assumption itself may have certain limitations that affect the accuracy of the model.

In summary, for the data value evaluation method based on economics perspective, the main data subject is the patent, government or enterprise open source data and other data with market value records, because of the high requirement on the content of the data itself, the application is limited. In addition, this method has uncertainty and subjectivity for the evaluation of data value, and its accuracy will be affected to a certain extent.

4 Research on Data Value Assessment Methods Based on Text Analysis

For the evaluation of the data value of open source technology information, it is necessary to evaluate the value of the text content of the data to a large extent. The text analysis methods for information retrieval and information filtering are all part of the data value assessment, which can solve the problem of overloading the amount of open source information.

4.1 Text Analysis Method for Information Retrieval

Information retrieval is a technology or process that finds, collects, and organizes information according to user needs and in a certain way. Its development has improved the efficiency of information searching and organizing to a certain extent. It is more suitable for searching and organizing life and entertainment information, but it is not comprehensive and scientific for academic information. In a sense, information retrieval is also a part of the evaluation of the value of open-source scientific and technological information data if the demand point of scientific researchers is regarded as the key word to evaluate the value of open-source scientific and technological information data [28].

Information search based on user behavior and preferences can solve the problem of knowledge overload caused by common search engines to a certain extent. For example, Information search based on user behavior and preferences can solve the problem of knowledge overload caused by common search engines to a certain extent. For example, Price proposes a semantic component model for expert user needs for information retrieval, which divides a collection of documents into different classes, each of which has a set of associated semantic components. Each semantic component instance consists of a piece of text about a particular aspect of the document subject and may not correspond to a structural element in the document. The semantic component model represents the content of the document in a way that complements the full text and keyword index, and the semantic component can improve the ranking of documents for precision search [29]. Based on the prior knowledge of the target field, Liu Fang uses the ontology prior knowledge to describe, so as to search the actual needs of the user from the conceptual words of the ontology, and finally calculate the similarity between the concept words and the documents, and the importance of the text data is sorted [30]. Hiren Karodiya presents a new algorithm for searching results using web page categories. The search ranking technique is used to classify users and pages by analyzing user behavior, page content, and URL order of visits [31].

To sum up, information retrieval based on user behavior and preference can solve the problem of overloading of open source scientific and technological information to a certain extent, at the same time, it also provides some reference significance for the evaluation of data value of open source information, but there are also some shortcomings. The technology needs to be designed according to the specific requirements, and the portability is poor. If the user has specific scene requirements, then the deeper semantic analysis of the data can be used to improve the data intelligence value and perceive the high intelligence value information.

4.2 Text Analysis Method for Information Filtering

Information filtering and information retrieval are proposed solutions to solve the overloading of open source information. The difference between them is that information retrieval will perform important sorting of information. However, information filtering only returns the results to "keep" or "delete", does not sort the importance of information, information filtering is more like a two-classification problem. Information filtering is divided into two types: one is content-based filtering, the other is collaborative filtering. Because this paper is based on the open source technology information data value evaluation method research, so it is mainly discussed from the content-based filtering.

Content-based information filtering methods are mostly text categorization methods, such as Bayesian networks, feature extraction, support vector machines (SVM), and K-nearest neighbor classification algorithms. For example, Chang mainly uses Bayesian network to construct information filtering model, and incorporates user preferences and needs into the model to achieve the purpose of information filtering [32]. Bing proposed an information filtering algorithm based on feature vectors. It is believed that feature frequency selection should incorporate word frequency statistics, text attributes and grammatical features of words, combined with text representation and statistical principle to extract feature vectors, and finally complete information filtering [33]. Gu proposed an information filtering method based on Support Vector Machine (SVM) algorithm, selects appropriate kernel functions and parameters, constructs SVM classifier, and uses empirical research of simulation experiments to complete information filtering [34]. Tang established an information filtering model based on K-Nearest Neighbor Classification (KNN) algorithm, from the first filtering of crawling topics to the second filtering of keyword matching, and then to the third filtering of semantic analysis. Through this whole process, information filtering was completed [35].

Information filtering technology is one of the methods to solve the problem of information overload. Because most of the information filtering methods use user interest model to filter information, this method needs to use a large number of data sets for model training, and then evaluate the validity of the model through manual annotation results. In this process, it is very labor-intensive to construct manual labeled training set.

5 Summary

The research on the data value evaluation method of scientific and technological information in the open source environment provides a reference for solving the problem of information overload and efficient application of data. This paper systematically summarizes the data value evaluation methods of the main open source technology information, and points out the main ideas, application scenarios, advantages and disadvantages of each method. As shown in Table 1 below.

Table 1. Comparison of three methods

Method		Application scenario	Advantage	Shortcoming
Data value assessment method based on information metrology	Qualitative evaluation method	• Website management optimization • Providing assistance to library information workers • Database developers evaluate academic databases and related websites	All aspects of the data can be evaluated in a more detailed and comprehensive manner	• The artificial qualitative evaluation has the subjective concept of each evaluation subject, and the objectivity is insufficient • Due to the huge amount of open source information, the workload of qualitative assessment methods alone will be large and the feasibility is insufficient
	Quantitative assessment method	• For academic information such as papers and patents, quantitative analysis is mainly carried out from the perspective of literature The network link analysis method has been well used in information retrieval	It has strong objectivity and has achieved certain success in practice	• For such "cited" measurement methods, it takes time to accumulate, and has good measurement results for old documents and old web pages, but for the evaluation of open source information, new resources tend to It will be more valuable • The "cited" analysis does not involve the evaluation of the value of the text

(*continued*)

Table 1. (*continued*)

Method		Application scenario	Advantage	Shortcoming
Data value evaluation method based on economics	Value evaluation method based on open source government data	• The top-down macro-economic approach: it is used to adjust the future development direction and optimize the effective allocation of government information resources • The bottom-up microeconomic approach: it can present the value of open government data to the public	• The top-down macro-economic approach: it evaluates the total value of various industries that use open source government data as investment, and shows the total value of open source government data to the economy • The bottom-up microeconomic approach: productivity analysis of open source government data takes into account the inputs and outputs of individuals and companies in the use of government data. This analysis is only effective in a bottom-up approach to microeconomics	• The top-down macro-economic approach: this method exaggerates economic value and over-emphasizes causality, so it produces biased estimates • The bottom-up microeconomic approach: it does not take into account the broader economic impact. If the relevant inputs are not correctly identified, the evaluation results will not be credible, and it is difficult to extend the evaluation from the individual industry assessments to the entire economy
	Value evaluation method based on open source enterprise and patent data	• Cost method: for enterprise information, it is mainly used for the value perception of internal data of the enterprise. For patent information, there is currently a tendency to evaluate the value of patents from the	• Cost method: avoid subjective influences • Market method: it is easy to be recognized by the public. • Income method: take into account the value implications of future expectations • Real option method: the	• Cost method: ignore the data to bring profit to the company, so the perceived value will be lower than the actual value • Market method: because the patent market in China is not perfect and the patent information as reference is not comprehensive,

(*continued*)

Table 1. (*continued*)

Method	Application scenario	Advantage	Shortcoming
	side using potential costs • Market method: applying in a mature trade market to evaluate • Income method: used in the context of understanding market information such as discount rate, technology share rate, and remaining economic life • Real option method: used in the context of understanding relevant investment market information • Econometric method: applied to scenarios with large amounts of sample data	hypothetical conditions are closer to the actual situation, and various uncertain factors affecting the value of patents are also considered • Econometric method: it is easy to be recognized by the public	the method also has some limitations • Income method: it has certain subjectivity to determine information such as discount rate, technology share rate and residual economic life • Real option method: the model of this method is relatively complex, and the comprehensibility of the model is poor, and the estimation of various uncertain factors is relatively difficult, which affects the practicability of the method • Econometric method: firstly, it is difficult to get a large number of samples of homogeneous data value to carry out regression analysis, thus affecting the establishment of the model; Secondly, this kind of methods often assume that the data value and the influencing factors are between

(*continued*)

Table 1. (*continued*)

Method	Application scenario	Advantage	Shortcoming
			linearity, this assumption itself may have certain limitations that affect the accuracy of the model
Data value assessment method based on text analysis	Text analysis method for information retrieval	Applied to information retrieval based on user behavior and preferences	To a certain extent, it can solve the problem of open source technology information data overload, and also provide some reference for the data value evaluation of open source information	The technology needs to be designed according to the specific requirements, and the portability is poor
	Text analysis method for information filtering	Applied to information filtering in the context of a large number of labeled training sets	To a certain extent, it can solve the problem of open source technology information data overload	This method needs to use a large number of data sets for model training, and then evaluate the validity of the model through manual annotation results. In this process, it is very labor-intensive to construct manual labeled training set

Finally, it is considered that the evaluation of the data value of the scientific and technological information in the open source environment needs to integrate the characteristics of each level of the open source scientific and technological information to evaluate the value of the data. First of all, the external basic characteristics of the open source scientific and technological information (for example: source authority, information attention, Information impact, etc.) should be evaluated. Secondly, it is necessary to analyze the content features of open source scientific and technological information deeply. Finally, according to the specific needs, the paper points out the actual demand characteristics of information, and makes a directional evaluation of the demand characteristics. In this way, the data value of the scientific and technological information in the open source environment can be truly and effectively evaluated, and finally the purpose of discovering high-value data is achieved.

Acknowledgment. This research is supported by the Project of ISTIC (MS2019-04).

References

1. Liu, W.: Research on methods for web—based open source intelligence and analytics. National University of Defense Technology (2014)
2. Betsy, R.: Ten C'S for evaluating internet sources [EB/OL] (1991-03-17). http://www.doc88.com/p-7186950426397.html. Accessed 10 Feb 2019
3. Wilkinson, G.L., Bennett, L.T., Oliver, K.M.: Evaluating the quality of internet information sources. Counselor Educ. Supervision **34**(4), 369–387 (1997)
4. Stoker, D., Cooke, A.: Evaluation of networked information sources. Publ.-Essen Univ. Libr. **1995**(18), 287 (1995)
5. Robert, H.: Evaluating internet research sources. Virtual Salt Saylor Found. (2010). [EB/OL], 15 June 2007. https://www.rtsd.org/cms/lib/PA01000218/Centricity/ModuleInstance/2137/Evaluating_Internet_Research_Sources.pdf. Accessed 10 Feb 2019
6. Batini, C., Cappiello, C., Francalanci, C., et al.: Methodologies for data quality assessment and improvement. ACM Comput. Surv. **41**(3), 16 (2009)
7. Liu, Yu., Li, G.: Collection of information resources on ISI web of knowledge(R). Chin. J. Sci. Tech. Periodicals **14**(z1), 732–736 (2003)
8. Silverberg, G., Verapagen, H.H.G.: The size distribution of innovations revisited: an application of extreme value statistics to citation and value measures of patent significance. J. Econometrics **139**(2), 318–339 (2004)
9. Rousseau, R., Hu, X.: An outgrow index. Sci. Focus **57**(3), 287–290 (2011)
10. Hu, X.Y., Chen, J.: An investigation on indicators for patent value based on structured data of patent documents. Stud. Sci. Sci. **32**(03), 343–351 (2014)
11. Raan, A.: Patent citations analysis and its value in research evaluation: a review and a new approach to map, technology-relevant research. J. Data Inf. Sci. **2**, 50 (2017)
12. Yang, L., Wan, X.Y.: Research on the evaluation methods of low cited periodical thesis academic value. Inf. Stud.: Theory Appl **37**(7), 95–101 (2014)
13. McKiernan, G.: Cited sites (SM): citation indexing of web resources, 1996(11):2007
14. Yang, Z., Lei, Z.: Review of web information resources evaluation. J. Libr. Sci. China **36**(9), 75–89 (2010)
15. Qiao, S.J., Peng, J., Li, T.R., et al.: Hybrid page scoring algorithm based on centrality and PageRank. J. Southwest Jiao Tong Univ. **46**(3), 456–460 (2011)
16. Liu, Q., Xiang, Y.N.J., et al.: An influence propagation view of PageRank. ACM Trans. Knowl. Discov. Data **11**(3), 1–30 (2017)
17. Kleinberg, J.M.: Authoritative sources in a hyperlinked environment. J. ACM (JACM) **46**(5), 604–632 (1999)
18. Mei, C.: Study on the value evaluation of open government data: progress and enlightenment. J. Intell. **36**(11), 92–98 (2017)
19. CPIRA: Commercial exploitation of Europe's public sector information [EB/OL] (2000-01-01). https://www.researchgate.net/publication/221522328_e-Government_and_the_Re-use_of_Public_Sector_Information. Accessed 11 Feb 2019
20. Zhan, Y.J., Wang, C.W., Xiong, B., Peng, W.F.: Research on patent value evaluation based on the potential cost of protecting intellectual property. Sci. Technol. Manag. Res. **38**(13), 170–174 (2018)
21. Yang, X., Li, Y.Q., Guo, J.: Calculation method of patent value and practice in electric power industry. China Invent. Patent **14**(02), 51–54 (2017)

22. Li, Y.H., Li, J.N.: Research on the method of evaluating the value of Internet enterprise data assets. Econ. Res. Guide **14**, 104–107 (2017)
23. Dong, S.W.: Research on the application of income approach in the evaluation of patent value of electronic information industry. Hangzhou University (2017)
24. Pakes, A.: Patents as options: some estimates of the value of holding european patent stocks. Econometrica **54**(4), 755–784 (1986)
25. Tang, Y.Z., Qin, Q.: Railway patent value evaluation in China based on real option method. China Railway **04**, 33–38 (2018)
26. Scherer, F.M.: Citations, family size, opposition, and the value of patent rights: evidence from Germany. Res. Policy **32**(8), 1343–1363 (2003)
27. Zhang, Y., Zhang, W.: Empirical study on the quantitative evaluation model of the patents value of enterprise. J. Intell. **29**(2), 51–54 (2010)
28. Zou, Y.M., Zhang, Z.Q.: The intelligence value identifying methods of science & technology web resource: a review. J. Intell. **5**, 25–30, 59 (2014)
29. Price, S.L., Nielsen, M.L., Delcambre, L.M.L., et al.: Using semantic components to search for domain-specific documents: an evaluation from the system perspective and the user perspective. Inf. Syst. **34**(8), 724–752 (2009)
30. Liu, F., Yu, C., Meng, W.: Personalized web search for improving retrieval effectiveness. IEEE Trans. Knowl. Data Eng. **16**(1), 28–40 (2004)
31. Kardiya, H., Singh, A.P.D.K.: User specific search ranking technique. Int. Res. J. Comput. Sci. Eng. Appl. **2**(1), 212–215 (2013)
32. Chang, L.J., Yu, J.X., Qin, L.: Context-sensitive document ranking. J. Comput. Sci. Technol. **25**(3), 444–457 (2010)
33. Bing, R.: Information filtering algorithm based on feature vector. In: International Conference on Intelligence Science & Information Engineering, pp. 468–471. IEEE (2011)
34. Gu, W.C., Chai, A.R., Han, J.S.: Filtering methods against junk messages based on support vector machine. Trans. Beijing Inst. Technol. **33**(10), 1062–1066 (2013)
35. Tang, H., Yang, X.J., Wang, J., Huang, W.: Research on the risk information filtering model based on K-nearest neighbor classification algorithm. J. Intell. **37**(03), 64–70 (2018)

Network

Credible Routing Scheme of SDN-Based Cloud Using Blockchain

Qin Qiao[1,2,3], Xinghua Li[1,2,3]([✉]), Yunwei Wang[1,2,3], Bin Luo[1,2,3],
Yanbing Ren[1,2,3], and Jianfeng Ma[1,2,3]

[1] School of Cyber Engineering, Xidian University, Xi'an 710071, Shaanxi,
People's Republic of China
xhlil@mail.xidian.edu.cn
[2] State Key Laboratory of Integrated Services Networks, Xidian University,
Xi'an 710071, Shaanxi, People's Republic of China
[3] Shaanxi Key Laboratory of Network and System Security, Xidian University,
Xi'an 710071, Shaanxi, People's Republic of China

Abstract. Software-defined networks (SDN) have been widely used in Cloud Data Centers in recent years. With the development of cloud technologies, different organizations need to share network resources to achieve common business goals, which requires distributed SDN controllers to collaboratively manage cloud networks and realize cross-domain routing. However, existing distributed controller cooperative routing schemes require a third-party trust center to establish trusted relationships for controllers. Since both trust centers and certified entities are vulnerable to various attacks and security risks, the existing works cannot effectively ensure cross-domain routing is credible. To address this problem, Blockchain is employed to establish trusted relationships between distributed controllers, then a cross-domain routing mechanism was devised based on the trusted relationships. Security analysis and experiments indicate that the proposed scheme can establish trust relationships and provide credible cross-domain routing cooperation for distributed SDN controllers. Besides, the required overhead of storage and bandwidth are very limited, which implies good practicability.

Keywords: SDN · Cross-domain routing · Blockchain · Cloud

1 Introduction

With the development of Cloud computing, Cloud data centers have become widely used as a new network model. To realize the management of Cloud data center, Software Defined Networks (SDN) is introduced due to its high efficiency and flexibility. SDN abstracts the network into a control layer composed of controllers and a data layer composed of forwarding devices and manages the network through controllers. Since Cloud computing enables cross-organizational collaboration, and each Cloud data center can share resources to achieve common business goals [1, 2], SDN controllers are required to manage cross-domain routing collaboratively. Therefore, how to achieve an efficient and secure collaboration of distributed SDN controllers has attracted extensive attention of researchers.

© Springer Nature Singapore Pte Ltd. 2019
X. Cheng et al. (Eds.): ICPCSEE 2019, CCIS 1058, pp. 189–206, 2019.
https://doi.org/10.1007/978-981-15-0118-0_15

The horizontal controller architecture is the mainstream solution for distributed controller collaboration because of its greater scalability [3–7], as shown in Fig. 1. According to different implementations of inter-domain routing, the existing schemes are divided into two categories: BGP-based schemes and New schemes respectively. In the former the controller autonomously manages the intra-domain network, and the inter-domain routing is completed by the border gateway protocol (BGP) [8–11]; and in the latter, the local domain controllers share local network state information, and then each controller can calculate routing paths according to the global information and send cooperative routing requests to other domains [6, 12].

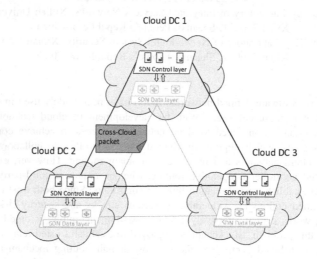

Fig. 1. Network scenario of the SDN horizontal control architecture across the Cloud data centers. The data layer and the control layer of the network between the Cloud data centers are all connected. The controllers of each organization collaborate to generate routing paths for Cross-Cloud packets, then the data layer forwards the packet.

Organizations usually presume there exist trust relationships among them, which will not always be guaranteed in reality, and further lead that the cooperation to be compromised by malicious collaborators. Thus, the premise of inter-organizational controllers' collaboration is to ensure that the trust relationships are established so that packets can be forwarded across domains correctly. However, both the aforementioned schemes require a trusted third party (such as a CA) to establish trusted relationships between the controllers. There are three problems: (1) It is difficult to find a third party that is trusted by all organizations. (2) Trust based on the central authentication system is very fragile. If the trust center is compromised, situations such as controller identity counterfeiting may occur, which brings serious security problems to SDN. (3) Risk of Single Point of Failure. Controllers usually are virtual in the cloud. Both virtual machines and controllers are at risk of being compromised even if it is certified, which may lead to malicious routing rules.

To address the problems above, we introduce Blockchain to build trust relationships, realize state consensus for the controllers among different organizations. Because Blockchain can build trust relationships between entities without the necessity of a third trust center. The main contributions of this paper are summarized as follows:

(1) We establish trusted relationships between controllers. According to the scenario in Fig. 1, by utilizing Consortium Blockchain, we design the horizontal distributed controller architecture and define the contents stored in the block. Moreover, combining with the sliding window, the Blockchain on-demand update mechanism is also presented for information sharing between controllers.

(2) Based on the proposed trusted relationships between controllers, we give a cooperative routing scheme. It enables nodes of each organization to calculate routes by consistent network state information stored in the Blockchain, so that realizes credible cross-domain routing.

(3) Security analysis and experiments indicate that the proposed scheme can effectively establish trusted relationships for controllers in different Cloud data centers and the proposed cooperation routing scheme is credible, while the overhead of storage and bandwidth required are very limited.

The rest of this paper is organized as follows. We review related work in Sect. 2. In Sect. 3, the scheme is presented in detail, and its security analysis is given in Sect. 4. The performance analysis and experimental verification were carried out in Sect. 5. Finally, the work of this paper is concluded in Sect. 6.

2 Related Work

The horizontal controller architecture is the mainstream solution for distributed controller collaboration currently. Due to the equal status of the controllers, this architecture is well suited for communication between different organizations. And the horizontal architecture is more scalable than the hierarchical architecture, which has a performance bottleneck due to centralization. The existing collaborative routing scheme of the horizontal controller architecture is mainly divided into the following two categories.

2.1 Schemes Using BGP for Inter-domain Routing

Since SDN only changes the way of intra-domain routing and does not change the packet itself, BGP can be used for inter-domain routing. The idea of Route Flow [8] is to transfer control functions to virtual machines and run standard routing protocols such as BGP, OSPF, and RIP on virtual machines. The routing information is exchanged through the control plane formed by the virtual machine. The disadvantage is that one hardware device corresponds to one virtual machine, so that the control plane is too complicated, which brings stability and scalability problems. Hydrogen [9] in the Open Daylight project uses the extended BGP-LS protocol and defines the data structure of Type/Length/Value to describe the link status. But the ability of this new format to carry information and the flexibility of the protocol when applied between domains has yet to be verified. SDNi [10] is a multi-domain heterogeneous SDN controllers communication draft proposed by Huawei, which defines the concept of SDN domain, the message, and communication mechanism of the SDNi protocol. The data exchange of SDNi can be implemented based on SIP or BGP. However, the draft does not give a specific implementation. DISCO [11] supports collaboration between SDN networks in multiple domains and uses BGP or OSPF packets as information carriers. It uses AMQP to realize the communication between the inter-domain controllers, sends the BGP or OSPF message to the subscribed controller, thereby achieving the cooperation between the domains.

As a traditional network protocol, the routing policy of BGP is forwarding packets based on the destination address. However, SDN can formulate routing strategies based on multiple fields. Therefore, some researchers have tried to improve the deficiencies of BGP and proposed to use the advantages of SDN for fine-grained routing between domains.

2.2 New Inter-domain Routing Schemes with SDN Features

The second type of schemes apply SDN to the inter-domain routing based on the characteristic of multi-field matching and provide more flexible forwarding strategies than BGP. Nick et al. proposed SDX [12] (SDN exchange point), trying to take advantage of the Internet Exchange Point (IXP) [13]. They replace the original switching network in the IXP with the OpenFlow switch and replace the original Route Server with the SDX controller. Then each domain can implement the inter-domain routing policy by sending a flow table to the OpenFlow switch. However, the number of IXPs in the world is limited. For domains that do not have access to IXP, SDX will fail, so it is not universal. WestEast-Bridge [6] is an SDN east-west interface protocol proposed by Tsinghua University. It realizes the cooperation of heterogeneous controllers without relying on the BGP. Controllers in different domains need to implement P2P network connections. Controllers can exchange and synchronize information based on the information structure proposed by the solution. After calculating the global path, the route calculation application on the controller will send a path fragment installation request to the domain related to the path, and then the corresponding controller generates a flow entry according to the request. It uses the X.509 digital certificate to authenticate controllers.

The existing schemes all adopt a trust center to establish trusted relationships for controllers. However, it's difficult to find a trusted institution that each organization believes. Besides, the trust center and the controllers of each organization also have the risk of being compromised, which will bring serious security problems to SDN.

3 The Proposed Scheme

The proposed SDN architecture based on Blockchain is shown in Fig. 2. Small-scale networks of multiple organizations constitute a large-scale cloud alliance. The data layer network is connected, and the control layer adopts a point-to-point network architecture. Entities within an organization trust each other and inter-organizational trust relationships need to be established through Blockchains

Physical machine: Each physical machine runs a controller and a Blockchain node. Controller: Controller is responsible for managing the local domain network and cooperating with other controllers to handle global events. Controllers can read and write data on the Blockchain through the API.

Node: Blockchain nodes have multiple roles. The most common is the Peer node, but a peer node can serve multiple roles at the same time. Some Peer nodes act as the endorsement node, which will simulate executing the transaction and endorse the execution result. There is also a part of the Peer nodes acting as the Leader in the organization, responsible for obtaining the latest block from the Order service nodes and broadcasting it to other nodes within the organization. The Order service node is a special type of Blockchain node, which is responsible for sorting the transactions, generating blocks and broadcasting the blocks that have reached consensus to Leaders.

MSP: Membership Service Providers issue a certificate for each entity in the network as its identity. Different from the traditional authentication center, the nodes in our scheme do not rely on the identity authentication provided by the MSP to trust other nodes.

Cloud: Cloud is the network of a Cloud data center, including a control layer composed of controllers and a data layer formed of forwarding devices.

Blockchain: Blockchain is a chain structure connected by blocks, which stores the transaction information of the network.

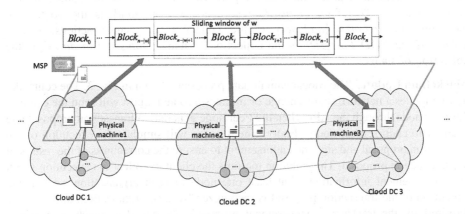

Fig. 2. System model

3.1   Establishment of Trusted Relationships of Distributed Controllers Based on Consortium Blockchain

Building a Blockchain. According to the degree of disclosure and usage scenarios, Blockchains can be divided into three categories: Public, Consortium, and Private. The Consortium Blockchain is typically deployed across multiple organizations to provide trust for information sharing across administrative domains. In this solution, the software-defined Cloud data center network is built based on a Consortium Blockchain. Because the verification efficiency of the Consortium Blockchain is higher than that of the Public Blockchain, and it's more suitable for the scene of multi-organization alliance than Private Blockchain. Block is the structure for storing data, including block header and block body. In this scheme, the block header includes the version number, timestamp, hash value of the previous block, and transaction summary. The transaction information is stored in the block body in the form of a Merkle tree. There are two types of transactions in this solution, one is cross-domain routing, and the other is network state information updating. The content of the cross-domain routing transaction is routing request information, including destination information and source information, and routing rules. The content of the transaction for network status information updating is different at the different stage of an update. The transaction content in the initialization phase is the local network status information acquired by the controller. Considering the speed of reading data, we use the key-value pair to store network status information: {Node_ID: IP_add, is_edge, type; Link_ID: Node_ID_src, Port_ID_src, Node_ID_dst, Port_ID_dst, is_crossdomain; Port_ID: Node_ID, Port_-MAC, is_active, is_edge, Vlan_ID; Reachability: IP_prefixes/IP_add, length}. In the update phase, start or stop of the network entity, connected or disconnected of the link, and activation or deactivation of the port will cause the network status information to change. Therefore, $StatusChange = \{action, item, state\}$ is the content of the transaction. $action = \{add, del\}$ is the collection of change actions, $item = \{Node\_ID, Link\_ID, Port\_ID\}$ is the collection of items, $status = \{up, down\}$ is the collection of status.

Blockchain Update. The Blockchain update process is shown in Fig. 3. The controller that generates a network status change or initiates a route request will send the event to the Peer node through the API. Then the Peer node generates a transaction proposal and sends it to endorsement nodes. Endorsement nodes will simulate the execution of the transaction proposal and sign the endorsement to ensure the correctness of the results of the transaction proposal. This solution requires the client to collect the execution result and endorsement information of at least one node of each organization, so that the execution of the transaction proposal is recognized by each organization, and the time required for the transaction endorsement process is shortened as much as possible. After the client collects the endorsement of the transaction proposal, the transaction proposal content and the endorsement information will be combined into one transaction and sent to the Order service nodes. Order service nodes will sort all transactions by time and pack them into a block. After Order service nodes reach a consensus to ensure the consistency of the uplink data, the block will be broadcasted to every Leader. Leaders synchronize blocks to all Peer nodes in its organization, and finally, the Peer nodes update the local ledger. In the process, the sender needs to sign the message, and the receiver needs to verify the signature. At this point, the Blockchain has completed an update.

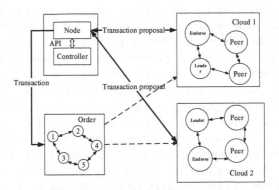

Fig. 3. Blockchain update process

To improve the performance of the Blockchain update process, we propose two optimization methods in this paper. One is the sliding window mechanism, which is used to reduce the storage overhead of the system; the other is the Leader election algorithm, which is used to improve the Blockchain update efficiency.

Sliding Window Mechanism. As the system runs over time, a large number of status change events will be saved in the ledger. To save storage resources, we introduce a sliding window mechanism. The checkout operation will be performed every w block, the global network state information will be updated according to the data in blocks in the window. And as the sliding window moves to the right, the obsolete blocks will be deleted. Thus, the number of network status change events that the ledger needs to save is only equal to the window size, as shown in Fig. 4. Since the information in the block has been effectively utilized, the deletion will not affect the existing network global status information.

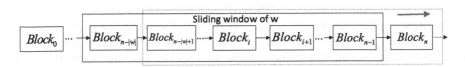

Fig. 4. Sliding window mechanism: Suppose that the checkout operation is performed when $Block_n$ is generated, then $Block_n$ will store global network state information calculated based on $Block_{n-|w|}$ to $Block_{n-1}$. After $Block_n$ is generated, the sliding window will move to the right and $Block_{n-|w|}$ will be deleted.

Leader Election Algorithm. The role of Leader is to connect to Order service, get the latest block and synchronize the block within the organization. When multiple nodes declared that they are Leaders at the same time, the existing election algorithm sorts the nodes in alphabetical order according to the PKI-ID, and all nodes will determine that the node with the highest ranking is the Leader. Since the Leader performs more work than the normal Peer node, in our solution, the node which can efficiently complete the task is more promising to be selected. We consider it from two aspects: (1) Propagation delay T_i. (2) Overload. The workload difference of the SDN controller causes the node load difference, which has two factors. The first one is A_i, the frequency of Asynchronous message. This kind of message carries the request sent by Switches to the controller. A_i depicts the degree of how much the network is dependent on the controller. The more Asynchronous messages, the higher the workload the controller has. The other one is C_i, the frequency of Controller-to-switch message. This kind of message is used by the controller to manage and maintain the switch. A larger C_i indicates that the controller participates in more network activities, which also means that the controller has a larger workload. Based on the above factors, the estimated delay ED_i of the node can be obtained. The estimated delay is used to reflect performance differences between different nodes, rather than specific delay estimates. As shown in Eq. (1), α, β are parameters determined according to the specific network conditions.

$$ED_i = \alpha T_i + \beta(A_i + C_i) \tag{1}$$

During the election process, the node will broadcast a proposal message containing ED_i to each node in the organization and sets a timeout period. Each node receives the message and stores ED_i. If the Leader does not appear until timeout, the node with the lowest ED_i among the nodes participating in the election will send a declaration message to the other nodes, and declare itself to be the Leader. If multiple declarations occur at the same time, the node with the lowest ED_i will become the Leader, and the other nodes will abandon the election.

3.2 Controller Cooperative Routing Scheme

Network Status Update Mechanism. The view of a single controller is limited. To implement cross-domain routing of SDN, it is necessary for the controllers to achieve the consistency of the global view of the network, which prevents the controller from generating conflicting flow rules based on different network information. Since the initialization phase is different from the update phase, a two-stage network status update mechanism is proposed below.

Initialization phase: The controller uses the Link Layer Discovery Protocol to discover the links between the switches in the domain. However, controllers cannot discover cross-domain links through LLDP. It is proposed in [6] to add the "LLDP extension" module. When the controller receives the LLDP packet and finds that the source switch does not belong to the local domain, it determines that the LLDP packet comes from another domain. Then the controller will establish an inter-domain link based on the ports of the source switch and the switch that received the packet. After the topology discovery is completed, the controller submits the obtained network state information of the domain as the transaction content to the Blockchain. After the Blockchain update, each controller can acquire global network status information from the local Peer node.

Status update phase: When the controller detects the network status change event mentioned in 3-B in the local domain, it constructs a status change transaction $SC = \{StatusChange, seq, ID_{C_{req}}\}$, seq is the change event number, and $ID_{C_{req}}$ is the unique identifier of the controller. The controller publishes the status change transaction to the Blockchain. After the Blockchain updating, each node can refresh the global status according to the status change information.

When a new node joins the Blockchain network, the network state information in the new domain needs to be shared with other nodes. The new node submits the local network status information as transaction content to the Blockchain and shares the local network status information according to the process in phase one. And the new node just needs to copies all the blocks in the sliding window from the current Blockchain to the local ledger to synchronize the latest global network status information.

The Cross-Domain Collaborative Routing Mechanism. When the switch receives a new packet, if there is no matching entry in the flow table, the switch will deliver the packet to the controller of the domain. The controller parses the packet to determine whether the destination address is in the organization. If not, the data packet needs to be routed across the organizations, then the controller sends the event to the Peer node to initiate a cross-domain routing transaction. The specific process is shown in Fig. 5:

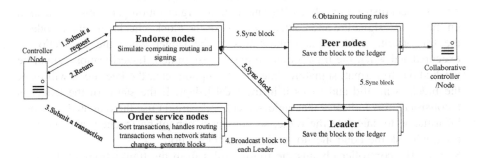

Fig. 5. Flow chart of the routing mechanism

Step 1. The controller plays the role of the client and constructs a Cross-domain Routing Request $Request_{C_{req}} = \{ID_{C_{req}}, seq, m\}$, $ID_{C_{req}}$ is the unique identifier of the controller, seq is the request number, and m is the request content, including the information of source and destination. The controller C_{req} sends the request to the Peer node through the API, then the Peer generates a transaction proposal, signs it and sends it to the endorsement nodes.

Step 2. The endorsement node checks the format, signature, written permission of the request and whether it is a replay request. If the request passes the verification, the endorsement node will simulate the execution of the request, that is, calculating the routing rule based on the global status information, then sign the execution result and return it to the controller. The shortest path algorithm can be used to calculate the routing rules, such as the Dijkstra algorithm. Other algorithms can also be selected as needed.

Step 3. The controller collects the results returned by the endorsement nodes according to the requirements of the endorsement strategy and verifies the signatures. As long as one organization does not endorse the transaction, the transaction will be invalid. After collecting all the endorsements and confirming that the flow rules returned by all endorsement nodes are the same, the controller submits the endorsements and execution results as a transaction to Order service nodes.

Step 4. The task of Order service nodes is to arrange the transactions in chronological order, generate a block and broadcast it to Leaders of each organization. However, in this solution, if the network status changes when the routing request is sent, the routing calculation result will be affected. So if an original transaction sequence is $\{R_1, R_2, R_3, S_1, R_4, S_2, R_5, R_6\}$ (R is the transaction containing the routing request, and S is the transaction containing the status change event), Order service nodes will return the routing request transaction to the controller for re-entry, only sort transactions for status change requests in chronological order: $\{S_1, S_2\}$. The transaction is eventually packaged into a block and broadcasted to Leaders of each organization.

Step 5. Leader performs verification on the format, signature of the block, and then synchronizes the block within the organization through Gossip-based peer-to-peer network data distribution. Each Peer node in the organization performs verification on the format, signature and content integrity of the block, checks the chain code of the transaction, and compares the status of the ledger when the transaction is executed and when the transaction proposal is submitted. If verification cannot be verified due to format, signature, integrity corruption, etc., the Peer node will mark the block as invalid and saves it in the local ledger. If the status of the ledger is inconsistent, this indicates that endorsement nodes might use the outdated topology information to calculate the route, so consensus cannot be reached, and the Peer node will ask the controller to resubmit the transaction.

Step 6. The controller obtains the routing rules from the transactions in the local ledger. As the flow rules include the unique identifier of the switch, the controller can filter the partial flow rules related to the local domain, and send the flow rules to the switches. If the packet does not pass through the domain, which means there is no flow rule associated with the domain, then the controller does not need to do more processing.

In this way, the controllers cooperate to complete the cross-domain routing of the packet.

4   Security Analysis

The security of this solution is divided into the establishment of the trusted environment and the security of the routing mechanism.

This scheme provides a distributed credible environment that tolerates Byzantine nodes for SDN controllers. It can resist tampering and replay attacks during storage and update process, ensuring that trusted relationships between controllers are established.

Analysis: (1) *Establishment of a credible environment.* In the traditional scheme, the trust between nodes is based on identity authentication (for example, using CA to establish trusted relationships), but a node with identity information may also have malicious behavior. Our scheme builds trust on the consensus mechanism. All controllers rely on the data on the Blockchain to make decisions, instead of relying on one or several entities. This scheme adopts the PBFT consensus protocol [14]. Its core idea is that nodes exchange information with each other after receiving the message to ensure the consistency of the information. In the process of node consensus, the node that sends the forged information to destroy the consensus is called the Byzantine node. Compared with Kafka, Raft, Paxos and other protocols that can only tolerate node failure, PBFT can not only tolerate node failure but also ensure that the final consensus will not be affected by Byzantine nodes when Byzantine nodes are less than 1/3 of the total number of nodes. There is a detailed mathematical proof of the relationship between the number of Byzantine nodes and the safety of consensus in [15]. (2) *Tamper protection.* The network status information stored in the ledger of the Peer node is the base information of the cross-domain routing. If it is tampered with, the node will figure out a wrong routing rule and the consensus will fail. This scheme can prevent tampering from two dimensions, one is cryptography, and the other is distributed storage. First, if an attacker attempts to tamper with the information in the ledger, he needs to obtain the private key of all nodes associated with the information, thereby modifying the signature information of the message. However, the amount of computation required to crack the private key is huge, and the Blockchain is the structure in which the blocks are linked by hashing. If the attacker wants to modify a block, he needs to modify every block after it. Second, even if the attacker tampers with a ledger of one node, the distributed storage method makes it hard to affect the consensus. (3) *Prevent replay attacks.* If an attacker intercepts a piece of outdated topology information and shares it with other nodes, it may cause the node to calculate the route based on the wrong topology information. However, $Block_n$ contains the hash value of its parent block $Block_{n-1}$. Suppose the attacker replays a historical block, the hash value of the parent block in the block is not equal to the hash value of the last block in the current Blockchain so that the forward hash chain relationship cannot be satisfied. Therefore, the historical block will not be received.

In summary, our solution extends identity-based trust to consensus-based trust and uses the consensus mechanism of the Blockchain to ensure the state of the nodes is consistent. At the same time, it can effectively prevent malicious tampering and replay attack. It establishes trusted relationships for distributed SDN controllers base on Consortium Blockchain.

The cross-domain collaborative routing mechanism is credible.

Analysis: (1) The source of network status information. In traditional schemes, controllers get global network state information from neighbor controllers. But in our scheme, controllers get information from the Blockchain, which ensures that the information obtained by controllers in whole network is consistent and not determined by individuals. (2) The architecture of the software-defined network allows the development of third-party applications. Applications bring flexibility to SDN, but it also puts pressure on application management. If there is a vulnerability in the source controller and the attacker uses it to install a malicious program, the malicious routing rules will be generated. In traditional schemes, controllers in other domains trust the controller and cooperate with forwarding packets, which may forward the packet to the malicious eavesdropper. But in our mechanism, once a node is attacked and an incorrect route is generated, the route will be marked as invalid because all nodes cannot reach a consensus. The routing rules are jointly calculated by organizations in our mechanism. Therefore, Single Point Failure is prevented by the authentication of the transaction by multiple nodes.

In summary, the cross-domain collaborative routing mechanism between distributed controllers is credible.

5 Performance Analysis and Experiments

To verify the effectiveness and practicability of the proposed scheme, we conducted experiments and analysis from three aspects: cross-domain routing efficiency, storage overhead, and bandwidth overhead. The experiment in this paper uses Go language, the experimental platform is 3.30 GHz Core i5-4590CPU, 8 GB, and the operating system is Windows 10–64 bit.

5.1 Cross-Domain Routing Efficiency

According to the process of the routing in Sect. 3, the time required from the initiation of a route request to the routing rule being installed on the switch is $t_{endorse} + t_{con} + t_{broadcast} + t_{deal}$. $t_{endorse}$ is the time required from initiating a routing request until all the endorsement information is collected. t_{con} is the time required for the transaction to reach a consensus. $t_{broadcast}$ is the time required for Leader to get the block and broadcast it to Peer nodes. The time required for the controller to install the routing flow table to the switch is t_{deal}. Compared with the existing solution, the work of installing the flow table is indistinguishable and t_{deal} does not cause a gap. Therefore, we optimize the endorsement strategy to reduce $t_{endorse}$, adopt PBFT, an efficient consensus protocol, to reduce t_{con}, and optimize the Leader election scheme to reduce $t_{broadcast}$. The optimization effect of the scheme is proved by experiments below.

The impact of endorsement strategies on $t_{endorse}$. Our endorsement strategy requires the controller to collect endorsements from each organization. There is no need to collect endorsements for all nodes of all organizations. Through experimental simulation, we compare the time spent in the endorsement phase under our strategy and the strategy that all nodes need to endorse. In the experiment, there are four organizations, each organization has ten nodes, and each node's endorsement speed is related to the real-time load of the node and the state of the communication link between the node and the controller. We sent 100 transaction proposals, as shown in Fig. 6, the All-node endorsement strategy generally takes longer than our strategy. The experimental data shows that a total of 89 transactions takes longer time under all-node endorsement strategy. Therefore, our endorsement strategy can effectively reduce $t_{endorse}$.

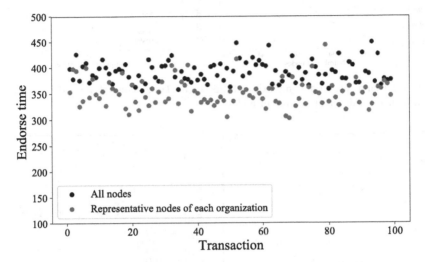

Fig. 6. Time-consuming comparison of different endorsement strategies

The impact of the consensus mechanism on t_{con}. POW is the most well-known consensus algorithm in the Blockchain and is used in Bitcoin. After a block is generated, at least five blocks must be verified before the block can be confirmed. Our scheme adopts PBFT, which is an algorithm with strong consensus. The block is immediately confirmed when stored in the ledger, so the consensus efficiency is higher than POW. However, since the communication complexity is $O(N^2)$, the relationship between the traffic and the number of consensus nodes is shown in Fig. 7(a). We can see with the number of consensus nodes increases, the consensus efficiency will decrease. We can introduce the Sharding Pattern [15] to divide the consensus nodes into different sets, and each set reaches consensus in parallel, which can effectively cope with the efficiency drop caused by the increase of nodes. Through experiments, we compared the time required to reach a consensus when there are a different number of nodes in the network. When there are 4, 10, 15, and 20 nodes in the network (4 is the minimum number of nodes required by PBFT), the agreed average time are 0.977 s, 1.132 s, 1.197 s, and 1.443 s respectively. As shown in Fig. 7(b), the fewer nodes, the shorter the time required to reach a consensus. It can be seen that the Sharding Pattern effectively improves consensus efficiency.

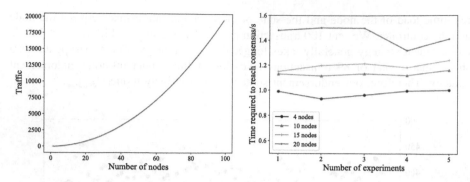

Fig. 7. (a) Communication complexity of PBFT (b) Time required for PBFT to reach a consensus under a different number of nodes

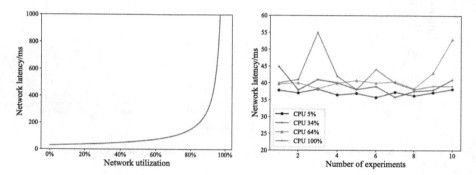

Fig. 8. (a) Assume that the idle network delay is 30 ms, the network delay changes with network utilization. (b) The delay of the Leader under different CPU utilization

In the Leader election scheme, we select the propagation delay between nodes and the load of the node as the influencing factor of the estimated delay of the Leader. In this section, we prove the influence of the above factors on the delay through experiments and analysis. And we testify that the Leader elected by our scheme has fewer estimated delay. The effect of propagation delay on the delay of the Leader is apparent, so we mainly introduce the load of the node. As is well known, greater load leading in the greater network utilization and the higher CPU utilization. The relationship between network utilization and network delay is shown in Fig. 8(a). If the idle network delay is $D0$ and the current utilization of the network is U, then the current network delay $D = \frac{D0}{1-U}$. We can figure out that the higher the network utilization, the greater the delay. As for the impact of CPU utilization on the network delay, we conducted an experiment. We set the CPU utilization as 5%, 34%, 64%, 100% and test the communication delay between the peer node and the Order service node for ten times at each CPU utilization. The average communication delay is 37.143 ms, 39.356 ms, 41.349 ms, 41.663 ms respectively. From Fig. 8(b), we can see that high CPU utilization can cause large delay in network event processing.

In summary, the propagation delay and the load of the node can be used as the influencing factors of the estimated delay of nodes.

In the experiment to verify that the Leader selected by this scheme has less estimated delay, we set up fifteen nodes in each organization, use GT-ITM to randomly
generate topology, simulate random traffic injection network and calculate the estimated delay of each node. We use the original and our Leader selection algorithm to
select the Leader for ten times to compare the difference between the original algorithm
and our proposed one. The experimental results are shown in Fig. 9. Our proposed
scheme has a smaller estimated delay. In the experimental context, the algorithm of this
paper saves the time of the synchronization block more than the original algorithm.

5.2 Storage Overhead

To avoid excessive accumulation of network status change information and storage
resources being occupied, we introduce the sliding window mechanism to limit the
amount of status change information stored in the ledger. In the experiment, we set the
size of the global status information to 2 KB, and the status change information size to
about 300 bytes.

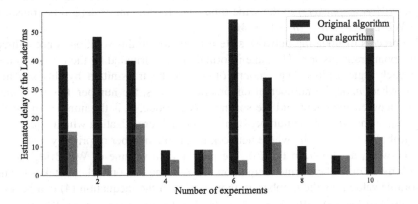

Fig. 9. Estimated delay of the Leader between two Leader election algorithms

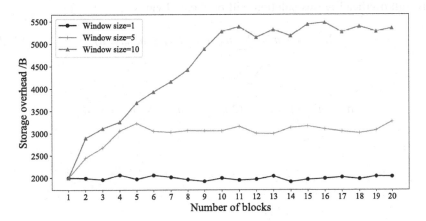

Fig. 10. Effect of sliding window mechanism on storage overhead

As shown in Fig. 10, we tested the storage overhead of the Blockchain when the sliding window size is 1, 5, and 10. If the window size is 1, every time nodes store a new block, they will delete the previous block, so that storage over-head is kept within a range of size of a block containing global network information. If the window size is 5, when the node receives the sixth block, the first block will be deleted. And so on when the window size is 10, the storage overhead will remain in the range of 10 blocks. We can see the sliding window mechanism can effectively limit the storage overhead.

5.3   Bandwidth Overhead

We compared our scheme with two pieces of related work: DHT (Distributed Hash Table), a distributed hash table, which is the most common and successful distributed storage method in distributed SDN networks. WE-Bridge, a well-known solution for horizontal distributed SDN architecture, which is better than DHT in some performance indicators.

Assuming the network state information of the same size M is shared at the same frequency, the amount of data to be transmitted by the DHT is $n * (n - 1) * (M + sizeof(requstpacket))$ in the same time, and the amount of data to be transmitted by the WE-Bridge is $n * (n - 1) * M$.

The process of updating network state information in this solution is not a directly point-to-point transmission. The state information is distributed by the Leader to nodes within each organization. The amount of data to be transmitted by this scheme is Eq. (2), where n_e is the number of endorsement nodes, the number 1 is the communication between the client and the sorting service node, n_l is the number of Leaders, and $\sum n_{org}$ is the sum of the number of nodes in each organization which equals to the total number of nodes in the whole network minus the number of primary nodes in the whole network, as shown in Eq. (3). We put the data volume of WE-Bridge and the data volume of our scheme into inequation (4). Assume that our scheme takes the most unfavorable value n as the number of endorsers, then the inequation (4) can be solved by combining Eq. (3). When $n > 2 + \sqrt{5}$, bandwidth overhead of WE-Bridge is greater than this solution. As the network scale is larger, the advantage of saving bandwidth overhead in this solution will be more obvious.

$$(2n_e + 1 + n_l + \sum n_{org}) * M (n_e \leq n, n_o < n) \tag{2}$$

$$\sum n_{org} = n - n_l \tag{3}$$

$$n * (n - 1) * M > (2n_e + 1 + n_l + \sum n_{org}) * M \tag{4}$$

6 Conclusion

The existing SDN distributed controller collaboration solutions cannot guarantee credible routing in Cloud data centers environment of a multi-organization alliance. The reason for this problem is that the trust relationship between controllers depends on trust center. The vulnerability of the trust center and the authenticated entities may lead to malicious routes in the network. To address this problem, this paper builds trusted relationships based on the Consortium Blockchain for distributed controllers, defines the content of the block, establishes the on-demand update mechanism of the Consortium Blockchain in combination with the sliding window. Subsequently, a routing mechanism was proposed to enable the controller to cooperatively route relying on the consensus data. Security analysis and experiments show that this scheme can establish decentralized trust relationships for controllers and effectively ensure cross-domain routing is credible. Further, the proposed scheme has good practicability.

Acknowledgements. This work was supported by National Natural Science Foundation of China (Grant Nos. U1708262, U1736203, 61772173, 61672413), National Key R&D Program of China (2017YFB0801805), the Fundamental Research Funds for the Central Universities and the Innovation Fund of Xidian University.

References

1. Alansari, S., Paci, F., Sassone, V.: A distributed access control system for cloud federations. In: 2017 IEEE 37th International Conference on Distributed Computing Systems (ICDCS) [Internet]. IEEE, June 2017. https://doi.org/10.1109/icdcs.2017.241
2. Yan, Q., Yu, R., Gong, Q., et al.: Software-defined networking (SDN) and distributed denial of service (DDoS) attacks in cloud computing environments: a survey, some research issues, and challenges. IEEE Commun. Surv. Tutor. 1 (2015). https://doi.org/10.1109/comst.2015.2487361
3. Tootoonchian, A., Ganjali, Y.: HyperFlow: a distributed control plane for OpenFlow. In: Internet Network Management Conference on Research on Enterprise NETWORKING, p. 3. USENIX Association (2011)
4. Koponen, T., Casado, M., Gude, N., et al.: Onix: a distributed control platform for large-scale production networks. In: Usenix Conference on Operating Systems Design and Implementation, pp. 351–364. USENIX Association (2010)
5. Berde, P., Hart, J., et al.: ONOS: towards an open, distributed SDN OS. In: The Workshop on Hot Topics in Software Defined NETWORKING, pp. 1–6. ACM (2014). https://doi.org/10.1145/2620728.2620744
6. Lin, P., Bi, J., Wang, Y.: WEBridge: west-east bridge for distributed heterogeneous SDN NOSes peering. Secur. Commun. Netw. **8**(10), 1926–1942 (2015). https://doi.org/10.1002/sec.1030
7. Medved, J., Varga, R., Tkacik, A., et al.: OpenDaylight: towards a model-driven SDN controller architecture. In: IEEE, International Symposium on a World of Wireless, Mobile and Multimedia Networks, pp. 1–6. IEEE (2014). https://doi.org/10.1109/wowmom.2014.6918985

8. Nascimento, M.R., Rothenberg, C.E., Salvador, M.R., et al.: Virtual routers as a service: the RouteFlow approach leveraging software-defined networks. In: International Conference on Future Internet Technologies, pp. 34–37. ACM (2011)

9. Hydrogen [EB/OL]. http://www.opendaylight.org/. Accessed 1 Oct 2018

10. Yin, H., Xie, H., Tsou, T., et al.: SDNi: a message exchange protocol for software defined networks (SDNS) across multiple domains. IETF Draft, work in progress (2012)

11. Phemius, K., Bouet, M., Leguay, J.: DISCO: distributed multi-domain SDN controllers. In: Network Operations and Management Symposium, pp. 1–4. IEEE (2014). https://doi.org/10.1109/noms.2014.6838330

12. Gupta, A., Vanbever, L., Shahbaz, M., et al.: SDX: a software defined internet exchange. ACM SIGCOMM Comput. Commun. Rev. 44(4), 579–580 (2014). https://doi.org/10.1145/2740070.2631473

13. Restrepo, J.C.C., Stanojevic, R.: A history of an internet exchange point. ACM SIGCOMM Comput. Commun. Rev. 42(2), 58–64 (2012). https://doi.org/10.1145/2185376.2185384

14. Castro, M., Liskov, B.: Practical Byzantine fault tolerance. Oper. Syst. Des. Implementation 99, 173–186 (1999)

15. Kokoris-Kogias, E., et al.: OmniLedger: a secure, scale-out, decentralized ledger via sharding. In: 2018 IEEE Symposium on Security and Privacy (SP) (2018). https://doi.org/10.1109/sp.2018.000-5

Integrated Multi-featured Android Malicious Code Detection

Qing Yu and Hui Zhao$^{(\boxtimes)}$

Tianjin Key Laboratory of Intelligence Computing and Network Security,
Tianjin University of Technology, Tianjin 300384, China
1477361293@qq.com

Abstract. To solve the problem that using a single feature cannot play the role of multiple features of Android application in malicious code detection, an Android malicious code detection mechanism is proposed based on integrated learning on the basis of dynamic and static detection. Considering three types of Android behavior characteristics, a three-layer hybrid algorithm was proposed. And it combined the malicious code detection based on digital signature to improve the detection efficiency. The digital signature of the known malicious code was extracted to form a malicious sample library. The authority that can reflect Android malicious behavior, API call and the running system call features were also extracted. An expandable hybrid discriminant algorithm was designed for the above three types of features. The algorithm was tested with machine learning method by constructing the optimal classifier suitable for the above features. Finally, the Android malicious code detection system was designed and implemented based on the multi-layer hybrid algorithm. The experimental results show that the system performs Android malicious code detection based on the combination of signature and dynamic and static features. Compared with other related work, the system has better performance in execution efficiency and detection rate.

Keywords: Malicious code · Feature · Optimal algorithm

1 Introduction

In response to malicious code, Android malicious code detection come into being. According to the characteristics of detection schemes, it can be divided into dynamic detection and static detection. Dynamic detection [1] is classified according to the behavior of malicious code. Malicious behavior is judged as malicious code. However, dynamic detection is highly dependent on virtual machine operation, and some malicious code can prevent itself from running in the virtual machine, running in the virtual machine will automatically crash, affecting the effect of dynamic detection. Static detection [2] uses decompilation technology to detect malicious code by analyzing control flow and data flow on smali intermediate code, analyzing the characteristics of the code, and determining malicious code based on matching results. It has high efficiency and code coverage.

© Springer Nature Singapore Pte Ltd. 2019
X. Cheng et al. (Eds.): ICPCSEE 2019, CCIS 1058, pp. 207–216, 2019.
https://doi.org/10.1007/978-981-15-0118-0_16

Because machine learning technology can transform disordered data into useful information, some researchers try to use machine learning technology to analyze the characteristics of malicious applications and detect malicious code. However, the current research methods have four limitations. One is that the single-feature algorithm can not give full play to the advantages of Android multi-features in malicious code discrimination, and the other is different machine learning methods have different detection effects on the same kind of features, so it is impossible to predict which algorithm is the best. Thirdly, the same kind of algorithm does not necessarily have the best detection effect on different types of features. Fourthly, digital signature is not included in the category of static detection.

In view of the above limitations, this paper proposes a new classification algorithm based on three-layer hybrid multi-feature for the first time, and implements a prototype system to automatically detect malicious code on Android platform. The system extracts the multi-class behavior characteristics of Android applications, and builds a prototype system through four-layer hybrid algorithm to detect malicious applications of Android. The specific methods are as follows:

First, create a normal application library and a malicious application library, the normal application is established by crawling the normal application of the 360 application market with python crawler. The malicious application library is downloaded through the virus classification website virushare. A malicious signature library is created by extracting the digital signature of the application in the malicious application library for later use.

Secondly, in the initial screening stage, the digital signature of the application to be detected is extracted and matched with the malicious signature database, and the matching is successfully classified. If it is not successful, the static feature is extracted for training. In the feature extraction phase, the apktool is used to decompile the detection application. Automatically analyze AndroidManifest.xml in the obtained file to extract the permission and system API.

Then, in the feature description stage, the extracted features are formatted to construct a set of various features. Finally, three-level hybrid multi-feature algorithm is used to build the detection model. The first layer uses multiple basic classifiers to train three different types of features and select the classifier with the best classification effect. The second layer optimizes the classifier selected by the upper layer to further improve the classification accuracy. The third layer uses the established model to classify the application to be tested and uses the model fusion algorithm to give the detection result.

2   Key Technologies and Algorithms

2.1   Static Detection Technology

Extraction of Digital Signature. The extraction of the digital signature is based on Google's apktool. By decompressing each Android malicious application, the digital signature file is extracted to form a malicious signature database.

Extraction of Sensitive Permissions. Through the automatic processing of malicious applications and normal applications, the right amount of authority with the highest degree of discrimination is extracted as a feature. In the feature extraction stage, the AndroidManifest.xml configuration file obtained after decompiling the APK file is focused, as shown in the figure, the configuration file information of an Android application is given. The first four lines of the configuration file give the version information, screen settings and other information of the Android application. Starting from the fifth line of the configuration file, the Android application permission is given. From the configuration file, the application requests the permissions WAKE_LOCK, GET_-TASKS, ACCESS_NETWORK_STATE, INTERNET, READ_PHONE_STATE, SET_WALLPAPER, WRITE_EXTERNAL_STORAGE, RECEIVE_BOOT_COMPLETED, INSTALL_SHORTCUT. For this Android app, the permissions of the nine applications are the permissions characteristics of this application (Fig. 1).

```xml
<?xml version="1.0" encoding="utf-8"?>
<manifest android:versionCode="125" android:versionName="1.25" package="cc.romhpjj.shjjtk.tprvv"
    xmlns:android="http://schemas.android.com/apk/res/android">
    <supports-screens android:anyDensity="true" android:smallScreens="true" android:normalScreen
    <uses-permission android:name="android.permission.WAKE_LOCK" />
    <uses-permission android:name="android.permission.GET_TASKS" />
    <uses-permission android:name="android.permission.ACCESS_NETWORK_STATE" />
    <uses-permission android:name="android.permission.INTERNET" />
    <uses-permission android:name="android.permission.READ_PHONE_STATE" />
    <uses-permission android:name="android.permission.SET_WALLPAPER" />
    <uses-permission android:name="android.permission.WRITE_EXTERNAL_STORAGE" />
    <uses-permission android:name="android.permission.RECEIVE_BOOT_COMPLETED" />
    <uses-permission android:name="com.android.launcher.permission.INSTALL_SHORTCUT" />
    <application android:label="@string/app_name" android:icon="@drawable/icon" android:name=".a
        <activity android:theme="@style/Theme0" android:label="@string/app_name" android:icon="@
            <intent-filter>
                <action android:name="android.intent.action.MAIN" />
                <category android:name="android.intent.category.LAUNCHER" />
            </intent-filter>
        </activity>
        <receiver android:name=".b" android:permission="android.permission.RECEIVE_BOOT_COMPLETE
            <intent-filter>
                <action android:name="android.intent.action.BOOT_COMPLETED" />
                <category android:name="android.intent.category.DEFAULT" />
            </intent-filter>
        </receiver>
        <service android:name=".c" android:exported="false" />
    </application>
</manifest>
```

Fig. 1. Configuration file

Extraction of Function Calls. Using Android NDK [5] to extract the sequence of function calls of ELF files generated by compiling and linking native codes. The executable link format ELF object file has three types: relocatable file, executable file and shared object file, and the shared object file is .so file under lib folder, which is the file involved in the Android native code. It is Used to create an ELF object or executable file together with other shared targets or relocatable files to create a process image. This article extracts .so files using readefl tool to extract the dynamic symbol table saved in the .dynsym section and selects the FUNC type.

Extraction of System Calls. System API calls for Android applications can be found in every .smali file. It is worth noting that the instruction opcode used in the Dalvik assembly language for the call to the system API is the invoke directive, so the retrieval of the API call to the Android application is implemented by retrieving the invoke directive in the .smali file.

2.2 Three-Layer Hybrid Algorithm

The following figure is a block diagram of the three-layer hybrid algorithm designed in this paper. The specific idea is to firstly perform preliminary screening according to the digital signature, and hand over the application to the second layer basic classifier for the application that has not given the malicious code judgment, by comparing and selecting the classifier with the best current effect, it is submitted to the next layer for classifier optimization, so as to improve the classification accuracy. Through the processing and delivery of layers, the final delivery to the fourth-level evaluation algorithm gives the classification results (Fig. 2).

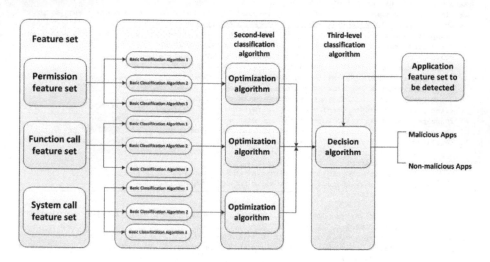

Fig. 2. Three-layer hybrid algorithm block diagram

The following describes the flow of the algorithm in detail:

The first layer uses three basic classifiers $M_m (m = 1, 2, 3)$ for three different types of features $D_i (i = 1, 2, 3)$, and then uses the K-fold cross-validation method [6] to perform the patrol, k select 10. In a feature set, each basic classification algorithm performs a 10-fold cross-validation, and the generated classifier can be described by a matrix as:

$$A = \begin{bmatrix} c_{11} & c_{12} & \cdots & c_{1m} \\ \vdots & \vdots & \ddots & \vdots \\ c_{k1} & c_{k2} & \cdots & c_{km} \end{bmatrix} \tag{1}$$

The average value of k times of detection in each column is used as the final classifier, and the $M_m(m = 1, 2, 3,)$ basic classification algorithm is sequentially iterated for three different feature sets $D_i(i = 1, 2, 3)$, and the generated classification is performed. The matrix is described as:

$$B = \begin{bmatrix} c_{11} & c_{12} & \cdots & c_{1i} \\ \vdots & \vdots & \ddots & \vdots \\ c_{m1} & c_{m2} & \cdots & c_{mi} \end{bmatrix} \tag{2}$$

The optimal classifier is selected in each column, and a total of i classifiers are selected. Finally, an optimal classifier $C_i(i = 1, 2, 3)$ is output for each type of feature set. The basic classification algorithm selected in this paper is KNN, decision tree, SVM, because the system is extensible, you can also add more basic classification algorithms.

The second layer uses the RotationForest algorithm [7] to improve the classification effect of the three optimal classifiers of the first layer output. The RotationForest algorithm randomly divides the sample attribute set into K sub-attribute sets, and uses the principal component analysis method to perform feature transformation on each sub-attribute set for each sub-attribute set to construct new sub-sample data. Principal component analysis [8] reduces the correlation between data, plays a certain role in data preprocessing, lays a foundation for improving classification accuracy, and retains all principal components without changing all the information of the original data. The training of each base classifier is performed on different sub-sample data, so that the difference between the base classifiers is greatly increased, which contributes to the improvement of the prediction accuracy after integration.

The third layer of the application to be detected extracts three types of features, and uses the three optimal classifiers outputted by the second layer to classify. Finally, the Voting function is used to obtain the weighted average to give the decision result. In this paper, 0 is selected as the non-malicious application. 1 represents a malicious application, and the result. Result given by each classifier is a value between 0 and 1 (for example, 0.8, indicating that the probability of the application being determined to be a malicious application is 80%, and the probability of being determined as a non-malicious application is 20%), the value reflects the weight of the malicious feature determination result. The value closer to 1, the weight of the unknown application to be detected is determined to be malicious according to the feature is more, and then the weight of each feature is averaged to obtain the final judgment result. The Voting function formula is defined as follows:

$$\text{Result} = \frac{x*\text{Result1} + y*\text{Result2} + z*\text{Results3}}{x + y + z} \tag{3}$$

x, y ∈ 0, 1, 0 ≤ Result1, Result2 ≤ 1, Where x, y indicates whether the application contains the feature, 1 represents yes, 0 represents no, and Result represents the determination result of various features.

The algorithm flow chart is described as follows (Fig. 3):

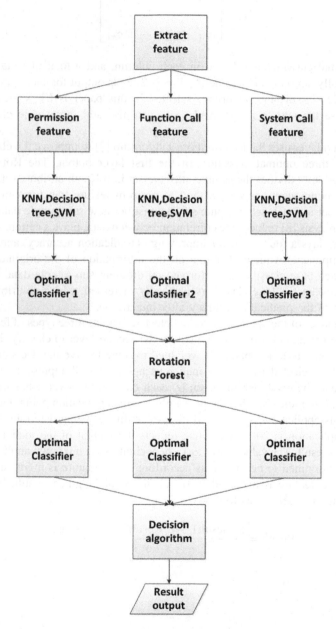

Fig. 3. Algorithm flowchart

3 Experimental Design and Results Analysis

This paper evaluates the validity of the model through experiments, compared with the previous related work, and compared with the famous Android malicious code detection tools.

3.1 Experimental Environment

In this paper, the automatic detection tool AndroTest is implemented in Python and Java language. The feature extraction is implemented in Python. The four-layer hybrid algorithm is implemented in Java. Use AndroTest to detect 1126 malicious apps and 2000 non-malicious apps in reality.

3.2 Experimental Results

First define the evaluation parameters of the classifier performance.

Definition 1 Confusion matrix. Confusion matrix is a useful tool for analyzing classifiers to identify different tuples. As shown in Table 1, this article defines non-malicious applications as positive tuple, malicious applications as negative tuple, and True positives (t_{pos}) means that classifier correctly identifies non-malicious application as non-malicious application tuple, True negatives (t_{neg}) means classifier correctly identifies malicious application as malicious application tuple. False positives (f_{pos}) means that the classifier misidentifies malicious applications as tuples of non-malicious applications; False negatives (f_{neg}) means that the classifier misidentifies non-malicious applications as tuples of malicious applications.

Table 1. Confusion matrix

Prediction	Non-malicious application	Malicious application
Non-malicious application	t_{pos}	f_{pos}
Malicious application	f_{neg}	t_{neg}

The formula for calculating the corresponding classifier evaluation parameters from the confusion matrix is:

$$Accuracy = \frac{t_{pos} + t_{neg}}{t_{pos} + t_{neg} + f_{pos} + f_{neg}} \tag{4}$$

$$TPRate = Recall = \frac{t_{pos}}{t_{pos} + f_{neg}} \tag{5}$$

$$FPRate = \frac{f_{pos}}{f_{pos} + t_{neg}} \tag{6}$$

$$\Pr ecision = \frac{t_{pos}}{t_{pos} + f_{pos}} \tag{7}$$

$$Fmeasure = \frac{2 * \mathrm{Re}call * \Pr ecision}{\mathrm{Re}call + \Pr ecision} \tag{8}$$

The first layer classification uses the selected basic classifier KNN [9], the decision tree, and SVM [10] to use the 10-fold cross-validation test for the two features. The results are shown in Tables 2, 3 and 4, which can be seen from the test results. For each feature, the KNN algorithm performs best. For the function call feature, the SVM algorithm has the highest TP Rate value. Due to the large number of function call features, this paper uses the feature selection algorithm to filter the original features, in order to select 255 features which have great influence on the classifier and meanwhile reducing the dimension to reduce the calculation time, specifically using the InfoGainAttributeEval evaluation strategy method and the Ranker search strategy method; for the system service feature, the KNN algorithm has the highest TP Rate value, Since the Accuracy indicator takes into account the four components of the confusion matrix, this paper uses Accuracy to evaluate the performance of the classifier. Table 5 shows the results of the Accuracy values of the basic classifiers of various feature sets. It is concluded that the SVM algorithm is the optimal classifier for the permission feature, and the KNN algorithm is the optimal classifier for the function call feature, and the decision tree is the optimal classifier for the system call feature.

Table 2. Permission feature classification result

	TP rate	FP rate	Precision	F-measure
KNN	0.974	0.025	0.974	0.974
Decision tree	0.90	0.091	0.90	0.90
SVM	0.971	0.027	0.971	0.971

Table 3. Function call feature classification result

	TP rate	FP rate	Precision	F-measure
KNN	0.871	0.135	0.873	0.872
Decision tree	0.75	0.185	0.805	0.785
SVM	0.905	0.092	0.905	0.905

Table 4. System call feature classification result

	TP rate	FP rate	Precision	F-measure
KNN	0.815	0.23	0.805	0.805
Decision tree	0.645	0.485	0.625	0.625
SVM	0.752	0.283	0.752	0.752

Table 5. Accuracy value of the base classifier

	Permission feature	Function call feature	System call feature
KNN	87.52	97.55	82.32
Decision tree	77.66	91.55	82.58
SVM	90.55	95.22	77.25
Optimal algorithm	SVM	KNN	Decision tree

The second layer classification uses the RotationForest algorithm to retrain the three optimal classifiers selected in the first layer using 10-fold cross-validation, and obtains Table 6. It can be seen that after the RotationForest algorithm, the classifier effect is significantly improved.

Table 6. Second layer algorithm classification result

	TP rate	FP rate	Precision	F-measure
Permission feature	0.915	0.08	0.913	0.912
Function call feature	0.972	0.025	0.975	0.974
System call feature	0.873	0.154	0.872	0.972

The third layer classification uses the three optimal classifier models outputted by the second layer and combines the decision algorithm to detect the new unknown application. It still tests the 3162 applications with 10-fold cross-validation, and the obtained confusion matrix is the Table 7, the accuracy rate is 94.31%, the test results fully prove the validity of the model.

Table 7. Confusion matrix

Prediction	Non-malicious application	Malicious application
Non-malicious application	1940	120
Malicious application	60	1006

3.3 Comparison of Results

The model proposed in this paper combines the advantages of the above research methods and improves their shortcomings. The main advantages are as follows:

(1) Most of the previous research methods use a single feature. This paper proposes to extract three different types of features, which fully reflect the code behavior, and this system is scalable, and more types of features can be added for detection.
(2) Feature extraction using digital signature and machine learning to improve system execution efficiency.

(3) A three-layer hybrid algorithm is designed, which can fully consider the different roles of different types of features in Android special code detection. The optimal basic classifiers are selected for each feature and the comprehensive decision results are given. The experiment proves that the effectiveness of the algorithm.

4   Conclusion

In this paper, machine learning technology is used to extract the multi-class behavior features of Android applications, three-layer hybrid algorithm is designed to build detection model, and the detection tool Android Test is implemented to detect malicious code. Considering the different roles of different types of features in malicious behavior detection of Android, the optimal classifier for each type of features is selected and the comprehensive decision results are given. The experimental results show that AndroidTest performs well in accuracy and execution efficiency, which is superior to other malicious code detection tools. The next step will be to study the basic classifier with more behavioral features and higher precision, simulate more accurate triggering user behavior in the dynamic phase, and implement faster processing by the three-layer hybrid algorithm combined with cloud computing to further improve the system.

References

1. Ham, H.-S., Choi, M.-J.: Analysis of android malware detection performance using machine learning classifiers. Int. Conf. ICT Convergence (ICTC) **8**(3), 156–174 (2013)
2. Seo, S.H., Gupta, A., Mohamed, S.A.: Detecting mobile malware threats to homeland security through static analysis. J. Netw. Comput. **38**(22), 43–53 (2014)
3. Hsu, Y., Lee, D.: Machine learning for implanted malicious code detection with incompletely specified system implementations. In: IEEE International Conference on Network Protocols. IEEE (2011)
4. Elovici, Y., Shabtai, A., Moskovitch, R., Tahan, G., Glezer, C.: Applying machine learning techniques for detection of malicious code in network traffic. In: Hertzberg, J., Beetz, M., Englert, R. (eds.) KI 2007. LNCS (LNAI), vol. 4667, pp. 44–50. Springer, Heidelberg (2007). https://doi.org/10.1007/978-3-540-74565-5_5
5. Zhao, H.: Implementation and application of android NDK development environment. Comput. Knowl. Technol. **06**(35) (2010)
6. Zhouxin, H., Gongjie, Z.: Selective integrated classification algorithm based on k-fold cross validation. Bull. Sci. Technol. **12**, 115–117 (2013)
7. Rodríguez, J.J., Kuncheva, L.I., Alonso, C.J.: Rotation forest: a new classifier ensemble method. IEEE Trans. Pattern Anal. Mach. Intell. **28**(10), 1619–1630 (2006)
8. Yajing, X., Yuanzheng, W.: Improvement of principal component analysis applied method. Pract. Recogn. Math. **36**(6), 68–75 (2006)
9. Larose, D.T.: K-Nearest Neighbor Algorithm. Discovering Knowledge in Data: An Introduction to Data Mining. Wiley, Hoboken (2004)
10. Huaping, S., Zheng, Y., Chengyu, Y.: Research on SVM algorithm and its application. J. Lanzhou Jiaotong Univ. **25**(1), 104–106 (2006)

Computation Offloading Algorithms in Mobile Edge Computing System: A Survey

Zhenyue Chen and Siyao Cheng$^{(\boxtimes)}$

School of Computer Science and Technology, Harbin Institute of Technology,
Harbin, China
chenzyhit@126.com, csy@hit.edu.cn

Abstract. With the rapid development of the internet of things (IoT), the number of devices that can connect to the network has exploded. More computation intensive task appear on mobile terminals, and mobile edge computing has emerged. Computation offloading technology is a key technology in mobile edge computing. This survey reviews the state of the art of computation offloading algorithms. It was classified into three categories: computation offloading algorithms in MEC system with single user, computation offloading algorithms in MEC system with multiple users, computation offloading algorithms in MEC system with enhanced MEC server. For each category of algorithms, the advantages and disadvantages were elaborated, some challenges and unsolved problems were pointed out, and the research prospects were forecasted.

Keywords: Internet of Things · Computation offloading ·
Mobile edge computing

1 Introduction

In recent years, with the development of mobile Internet, more and more mobile devices (such as mobile phones, tablets) have gradually replaced the role of personal computers in people's life, study, work, entertainment and so on. Mobile communication traffic in the network is also growing rapidly. And because of the emergence of the Internet of Things (IoT) [1], a large number of IoT devices have emerged, such as various sensors, smart water meters, cameras and so on. According to CISCO, 50 billion devices will be connected to the Internet by 2020, and 500 billion devices are expected to be connected to the Internet by 2030 [2]. Whether it's for smart manufacturing in the industry, or for smart transportation in the city, or for smart homes, these IoT devices are everywhere in life, and collecting or generating large amounts of data all the time. Although these IoT devices can directly access the cloud computing center to facilitate people's lives and change people's lifestyles, only the single computing model of cloud computing can not meet the needs of the future IoT world.

© Springer Nature Singapore Pte Ltd. 2019
X. Cheng et al. (Eds.): ICPCSEE 2019, CCIS 1058, pp. 217–225, 2019.
https://doi.org/10.1007/978-981-15-0118-0_17

Although cloud computing [3] can provide excellent computational capabilities for some devices with weak processing capabilities, there are still some shortcomings in this computing model. Since the geographical distance of large-scale cloud computing center is generally far away from the crowd, there are inevitable propagation delays in the data transmission process, which can not meet some computing tasks with higher real-time requirements. And if all the terminal devices upload data to the cloud computing center, it will put a huge load on the network [4]. On the other hand, linearly developed cloud computing capabilities will not be able to match explosive growth in IoT data. In addition, centralized storage of data also has hidden dangers of data security. There are also complex requirements for building a cloud computing center, it need take into account geographic location, machine cooling, energy consumption, power costs, and more.

Therefore, driven by a variety of reasons, Mobile Edge Computing (MEC) technology has emerged. The first standardization of this emerging technology is at the European Telecommunications Standards Institute [5], which is defined as providing information technology services and cloud computing capabilities in the nearest radio access network to terminal equipment. The main purpose is to reduce latency, ensure efficient network operations and service delivery, and improve the quality of service. Mobile edge computing provides a new viable computing model for the future IoT world, and a key technology in MEC is computation offloading technology. So in this survey, we summarizes the state of the art of the existing computation offloading algorithms in MEC system, divide them into four categories, elaborates on their strengths and weaknesses, and points out some challenges and unresolved problems. Finally, we also forecast the research prospects in the future.

2 Computation Offloading Algorithms in MEC System with Single User

In the MEC system, since each mobile edge has some data processing capability, it is not necessary to offload all tasks to the remote cloud for processing. Based on this motivation, a series of computation offloading algorithms have been proposed in the MEC system. Early research investigated the problem of individual user offload the computational tasks in the MEC system. This section focuses on a simple single-user MEC system and reviews recent research work on computation offloading algorithms in this system.

2.1 Binary Offloading Algorithms

In [6], the author proposed a theoretical framework for optimizing energy consumption in mobile cloud computing under stochastic wireless channels. The goal is to save energy for mobile devices by optimal performing mobile applications in mobile devices (i.e., mobile execution) or offloading to the cloud (i.e., mobile execution). For mobile execution, they minimize the computational energy by dynamically configuring the chip's clock frequency. For cloud execution, they minimize transmission energy by optimizing scheduled data transmission on a random wireless channel. Finally, the author formulate the scheduling problem as a constrained optimization problem and propose a closed-form solution for the optimal scheduling strategy.

The energy-saving framework has been further developed in [7], the author seamlessly integrates mobile cloud computing and microwave power transfer (MPT) technologies, enabling mobile devices to perform local computing or offloading tasks by using the harvested energy. The purpose of the article is to maximize the probability of successfully calculating given data under the constraints of energy harvesting and deadlines. By translating this optimization problem into an equivalent problem of minimizing the mobile energy consumption for local computing and maximizing the mobile energy savings for offloading, and using convex optimization theory to solve this problem. This article assumes that devices can harvest energy from microwaves for calculation and transmission, but the efficiency of harvesting energy from microwaves is still relatively low, and that energy harvesting from microwaves is not always reliable.

In [8], a convex problem of minimizing transmission energy under computational deadline constraints is proposed. It is pointed out that when the channel power gain is greater than the threshold and the server CPU is fast enough, the task is expected to be offloaded, and it reveal the impact of the wireless channel on the offload decision. However, those papers assumes that the wireless channel remains static throughout the execution of the task, which may be unreasonable.

2.2   Partial Offloading Algorithms

In [9], the application is modeled by task call graphs, which specify dependencies between different subtasks, a heuristic program partitioning algorithm was developed to minimize execution delay by exploiting the concept of load balancing between mobile devices and servers. This paper assumes that the rate of wireless communication is the same, but as far as we know, the upload and download rates of mobile devices vary greatly.

Based on the task call graphs model of [9], a code partitioning scheme is designed to dynamically generate the optimal set of tasks for the offload task in [10]. In this paper, the author analyzes each time slot. If the time slots are divided by different time spans, it will bring different complexity. So how to divide the time slot reasonably is a problem to be considered.

Inspired by parallel computing, partial unloading schemes were proposed in [11], a variable-substitution technique was provided to jointly optimize the offloading ratio, transmission power and CPU-cycle frequency. To further optimize MEC performance, a partial offloading problem were studied in [12], and an approximation algorithm with $(1 + \epsilon)$ ratio bound was given to solve it. In this two work, the data that the task needs to process is arbitrarily separable, but this situation is relatively rare and impractical.

2.3 Summary

All the discussed algorithms, including Binary Offloading algorithms and Partial Offloading algorithms, have achieved high performance for solving the single-user computing offloading problem in the MEC system.

For the Binary Offloading algorithms, if the user has a good channel condition or the local computing power is small, it is more preferred to save energy by offloading the task. If the user has high-speed bandwidth and the MEC server has a huge computation capacity, then computation offloading will also reduce latency.

For the Partial Offloading algorithms, they often offload time-consuming and energy-intensive tasks to the MEC server by splitting tasks, achieving more energy savings and less latency than Binary Offloading.

However, the situations considered by both of them seems a little simple, and more issues are not solved well, where some of them are listed as follows.

1. Firstly, the computation offloading problem for multiple users should be considered in MEC system since a MEC system always serves for many users.
2. Secondly, the fairness of the resource allocation should be considered, because different computation offloading requests may compete the same resources, such as computational resource or transmission resource, in a MEC system.
3. Thirdly, the data distribution in a MEC system should also be taken into account during computation offloading as the data are necessary input for each computation task, and it will save much resource and time when such distribution is considered.

3 Computation Offloading Algorithms in MEC System with Multiple Users

To overcome some shortcomings of the algorithms in Sect. 2, a series of computation offloading algorithms serving for multiple users have been proposed in MEC system. Unlike the system discussed in previous section, a MEC system with multiple users has a single MEC server and a number of mobile devices. Such architecture brings many new challenges during computation offloading, such as how to optimize the wireless and computation resources jointly during task allocation, how to provide a reasonable schedule among tasks to meet users' requirements *etc.*

3.1 Algorithms Based on Game Theory

Many researchers have used game theory to solve resource allocation problems in multi-user MEC system.

The paper [13] has formulated the computation offloading problem of multiple users via a single wireless access point as a computation offloading game, and proposed a decentralized mechanism that can achieve the Nash equilibrium to solve it.

In [14], the authors also formulated the distributed computation offloading decision making problem among mobile device users in multi-channel wireless networks as a multi-user computation offloading game, and designed a distributed computation offloading algorithm that can achieve a Nash equilibrium. Although the latter has been improved on the basis of the former in the two articles, they are not compared with the existing work, and can not highlight the performance of the algorithm.

Based on the system model in [14], the joint optimization of the mobile server's mobile transmission power and CPU cycle allocation is studied in [15]. And the antuors in [16] consider a computing access point (CAP) in system where each mobile user has multiple dependent tasks to be processed using a round by round schedule. Although the algorithms proposed in these articles have reached the Nash equilibrium, the biggest disadvantage of game theory is that many assumptions are not necessarily consistent with our actual life.

3.2   Other Algorithms

In [17], in order to cope with the arrival of a sudden task, the author uses queuing theory to jointly optimize server scheduling and uplink downlink transmission scheduling to minimize the average delay. In [18], a decomposition algorithm was given to control the computation offloading selection, clock frequency control and transmission power allocation iteratively. These work take into account the various situations of multiple users, but do not consider the distribution of data.

Besides, in the recent study, the author in [19] proposed a computation offloading framework for a green mobile edge cloud computing system, and introduced the centralized and distributed Greedy Maximal Scheduling algorithms to solve the computation offloading problem for multiple users with multiple tasks. But the idea of this green mobile edge cloud computing is not very convincing and unreasonable.

3.3   Summary

In these existing work, radio and computation resources in a multi-user MEC system are considered to be limited, and it is desirable to minimize the energy consumption of all mobile devices in the system under such constraints.

If the channel condition of the mobile device is good, offloading the task with high energy consumption locally can save a lot of energy. And if too many offloading tasks, it will cause users to compete for communication and computation resources.

Most of the work uses game theory to make the whole system reach Nash equilibrium. The above algorithm is effective for processing tasks caused by multiple users, however, there are still some problems are not solved well, where some of them are listed as follows.

1. First, the computation offloading problem for multiple servers should be considered in MEC system since a MEC system may have several servers.

2. Second, the cooperation between different MEC servers should be considered, this may increase the efficiency of task processing in the MEC system.
3. Third, the distribution of data in the MEC system should be considered, especially in multi-user MEC system, as tasks may be better scheduled according to data distribution.

4   Computation Offloading Algorithms in MEC System with Enhanced MEC Server

In the existing work mentioned in Sects. 2 and 3, the MEC system has only one server that can handle compute-intensive tasks. However, the computing resources of a single MEC server are limited. When there are too many users who need to offload task, a single MEC server will certainly not meet the needs of all users. If multiple MEC servers can contract and share computing resources, then the workload can be shared for some busy servers. On the one hand, the computation offloading of tasks is more efficient, and on the other hand, the user experience can be better improved. Therefore, this section mainly introduces the computation offloading algorithm in a system with cooperation scheme among multiple MEC servers.

4.1   Cooperation Scheme Among Multi-MEC Server

Resource sharing via server cooperation can not only improve the resource utilization and increase the revenue of computing service providers, but also provide more resources for mobile users to enhance their user experience. This framework was originally proposed in [20].

This study was further extended in [21], which considers not only local resource sharing but also remote resource sharing. The author formulated the problem of resource sharing and cooperation between different servers into a cooperative game, and solves it with a game theory algorithm with stability and convergence guarantee. However, the assumption that competitors are rational in game theory does not apply to the actual situation.

To enhance the cooperations among different base stations, a new cooperation scheme among base stations was given in [22], the author tried to reduce the response latency through caching the results of the popular computation tasks. The paper [23] provided a payment-based incentive mechanism. Such mechanism supports sharing computation resources among base stations and the authors also established a social trust network to manage the security risks among them.

4.2   Three-Level MEC System

More recently, the architecture of a MEC system which involves the three levels and serves for multiple users are sufficiently investigated by [24] and [25]. In [24], a heuristic greedy offloading scheme was given for ultra dense networks. In [25], a distributed computation offloading algorithm that can achieve the Nash equilibrium was provided. In these three-level architecture MEC system, only APs that are lower than the MEC server are considered, and the remote cloud is not considered in the system.

4.3 Summary

Cooperation and resource sharing between servers can improve computing efficiency and resource utilization in the entire MEC system. Recent works can be roughly divided into two types:

The first type of works is to offload tasks from one server to another with more computing power or no tasks to perform the calculations.

The second type of works is to actively cache the results of the calculation and share it between servers.

This can not only meet the needs of users, but also reduce the task delay. However, there are still several aspects of these works that have not been considered.

1. First, if the distribution of data can be considered, the server caches the intermediate results of the calculation according to the distribution, which may improve the efficiency of the computation offloading.
2. Second, cooperation with remote clouds should be considered. The remote cloud can provide huge computation capabilities, although the latency is a little high.

5 Future Work

As we mentioned in the introduction, computation offloading algorithms are very important in MEC system. The existing work has considered various situations and proposed many algorithms with good performance. But there are still some unresolved issues that need to be studied in the future.

First, need to consider the data sharing in the MEC system. In the existing work, the task model considered, whether it is a task oriented to data partitioning or a task oriented to code partitioning, the data that these tasks need to process are all existing in the device itself. When performing a computation offloading, only need to transfer the data already owned by the device itself. However, in reality, some tasks require not only the data of the device itself but also the data of other devices in the process of execution. Such tasks require other devices to share data. Therefore, for the MEC system, the data sharing should also be considered. It is necessary to design a computation offloading algorithm with tasks for data sharing.

Second, the cooperation between the MEC server and the remote cloud should be considered. In most of the existing studies, only the computing power of mobile devices and MEC servers is considered. But in the future IoT world, the number of devices that can connect to the network will explosive growth. Even though computing resources can be shared between servers, the computing resources of the MEC server are always limited. Therefore, in the process of computation offloading, the computing power of the remote cloud needs to be considered, and the MEC system consisting of mobile devices, edge servers, and remote large-scale cloud centers is constructed to make the task offloading calculation more efficient.

Third, the online computation offloading algorithms should be studied. Most of the existing computation offloading algorithms are offline, and it is assumed that the amount of data and complexity that the task needs to process is known, and the situation of the wireless channel is constant. However, in real life, all devices do not send requests for computation offloading at the same time. There are also many real-time tasks that cannot know in advance the amount of data that needs to be processed. The speed of data transmission is also changing at any time. So it is necessary to design some online computation offloading algorithms to take these uncertain factors into account.

Fourth, MEC systems for heterogeneous communications should be considered. Recently, 5G technology has begun to emerge, and the means of communication for devices has increased. How to choose the appropriate communication method when offloading the computation, and how to transfer data in different communication protocols. These are all issues worth considering in the future MEC system.

6   Conclusions

Since computation offloading is one of the key technologies of MEC, many efficient and effective computation offloading algorithms have been proposed. This paper summarizes the existing computation offloading algorithms and classifies them into three categories, introduces their main ideas, analyzes their strengths and points out their shortcomings. Finally, we look forward to future research questions on these topics and provide three unresolved questions that are worthy of careful study.

References

1. Ashton, K., et al.: That "internet of things" thing. RFID J. **22**(7), 97–114 (2009)
2. Alaba, F.A., Othman, M., Hashem, I.A.T., Alotaibi, F.: Internet of things security: a survey. J. Netw. Comput. Appl. **88**, 10–28 (2017)
3. Mell, P., Grance, T., et al.: The NIST definition of cloud computing (2011)
4. Godwin-Jones, R.: Mobile-computing trends: lighter, faster, smarter. Lang. Learn. Technol. **12**(3), 3–9 (2008)
5. Hu, Y.C., Patel, M., Sabella, D., Sprecher, N., Young, V.: Mobile edge computing-a key technology towards 5G. ETSI White Pap. **11**(11), 1–16 (2015)
6. Zhang, W., Wen, Y., Guan, K., Kilper, D., Luo, H., Wu, D.O.: Energy-optimal mobile cloud computing under stochastic wireless channel. IEEE Trans. Wirel. Commun. **12**(9), 4569–4581 (2013)
7. You, C., Huang, K., Chae, H.: Energy efficient mobile cloud computing powered by wireless energy transfer. IEEE J. Sel. Areas Commun. **34**(5), 1757–1771 (2016)

8. Barbarossa, S., Sardellitti, S., Di Lorenzo, P.: Communicating while computing: distributed mobile cloud computing over 5G heterogeneous networks. IEEE Sig. Process. Mag. **31**(6), 45–55 (2014)
9. Jia, M., Cao, J., Yang, L.: Heuristic offloading of concurrent tasks for computation-intensive applications in mobile cloud computing. In: 2014 IEEE Conference on Computer Communications Workshops (INFOCOM WKSHPS), pp. 352–357. IEEE (2014)
10. Mahmoodi, S.E., Uma, R., Subbalakshmi, K.: Optimal joint scheduling and cloud offloading for mobile applications. IEEE Trans. Cloud Comput. (2016)
11. Wang, Y., Sheng, M., Wang, X., Wang, L., Li, J.: Mobile-edge computing: partial computation offloading using dynamic voltage scaling. IEEE Trans. Commun. **64**(10), 4268–4282 (2016)
12. Kao, Y.-H., Krishnamachari, B., Ra, M.-R., Bai, F.: Hermes: latency optimal task assignment for resource-constrained mobile computing. IEEE Trans. Mob. Comput. **16**(11), 3056–3069 (2017)
13. Chen, X.: Decentralized computation offloading game for mobile cloud computing. IEEE Trans. Parallel Distrib. Syst. **26**(4), 974–983 (2015)
14. Chen, X., Jiao, L., Li, W., Fu, X.: Efficient multi-user computation offloading for mobile-edge cloud computing. IEEE/ACM Trans. Netw. **24**(5), 2795–2808 (2016)
15. Lyu, X., Tian, H., Sengul, C., Zhang, P.: Multiuser joint task offloading and resource optimization in proximate clouds. IEEE Trans. Veh. Technol. **66**(4), 3435–3447 (2017)
16. Chen, M.-H., Dong, M., Liang, B.: Multi-user mobile cloud offloading game with computing access point. In: 2016 5th IEEE International Conference on Cloud Networking (CloudNet). IEEE, pp. 64–69 (2016)
17. Molina, M., MuÃoz, O., Pascual-Iserte, A., Vidal, J.: Joint scheduling of communication and computation resources in multiuser wireless application offloading. In: IEEE International Symposium on Personal, Indoor, and Mobile Radio Communication, pp. 1093–1098 (2015)
18. Guo, S., Xiao, B., Yang, Y., Yang, Y.: Energy-efficient dynamic offloading and resource scheduling in mobile cloud computing. In: IEEE INFOCOM, pp. 1–9 (2016)
19. Chen, W., Wang, D., Li, K.: Multi-user multi-task computation offloading ingreen mobile edge cloud computing. IEEE Trans. Serv. Comput. (2018)
20. Kaewpuang, R., Niyato, D., Wang, P., Hossain, E.: A framework for cooperative resource management in mobile cloud computing. IEEE J. Sel. Areas Commun. **31**(12), 2685–2700 (2013)
21. Yu, R., Ding, J., Maharjan, S., Gjessing, S., Zhang, Y., Tsang, D.H.K.: Decentralized and optimal resource cooperation in geo-distributed mobile cloud computing. IEEE Trans. Emerg. Top. Comput. **PP**(99), 1–1 (2018)
22. Elbamby, M.S., Bennis, M., Saad, W.: Proactive edge computing in latency-constrained fog networks. In: European Conference on Networks and Communications, pp. 1–6 (2017)
23. Chen, L., Xu, J.: Socially trusted collaborative edge computing in ultra dense networks. In: Proceedings of the Second ACM/IEEE Symposium on Edge Computing, p. 9. ACM (2017)
24. Guo, H., Liu, J., Zhang, J.: Computation offloading for multi-access mobile edge computing in ultra-dense networks. IEEE Commun. Mag. **56**(8), 14–19 (2018)
25. Tang, L., He, S.: Multi-user computation offloading in mobile edge computing: a behavioral perspective. IEEE Netw. **32**(1), 48–53 (2018)

Wireless Communication Signal Strength Prediction Method Based on the K-nearest Neighbor Algorithm

Zhao Chen[1], Ning Xiong[2], Yujue Wang[2(✉)], Yong Ding[2], Hengkui Xiang[2], Chenjun Tang[2], Lingang Liu[2], Xiuqing Zou[2], and Decun Luo[2]

[1] Key Lab on Cognitive Radio and Information Processing,
School of Information and Communication,
Guilin University of Electronic Technology, Guilin 541004, China
[2] Guangxi Key Laboratory of Cryptography and Information Security,
School of Computer Science and Information Security,
Guilin University of Electronic Technology, Guilin 541004, China
yjwang@guet.edu.cn

Abstract. Existing interference protection systems lack automatic evaluation methods to provide scientific, objective and accurate assessment results. To address this issue, this paper develops a layout scheme by geometrically modeling the actual scene, so that the hand-held full-band spectrum analyzer would be able to collect signal field strength values for indoor complex scenes. An improved prediction algorithm based on the K-nearest neighbor non-parametric kernel regression was proposed to predict the signal field strengths for the whole plane before and after being shield. Then the highest accuracy set of data could be picked out by comparison. The experimental results show that the improved prediction algorithm based on the K-nearest neighbor non-parametric kernel regression can scientifically and objectively predict the indoor complex scenes' signal strength and evaluate the interference protection with high accuracy.

Keywords: Interference protection · K-nearest neighbor algorithm · Non-parametric kernel regression · Signal field strength

This article is supported in part by the National Natural Science Foundation of China under projects 61772150 and 61862012, the Guangxi Key R&D Program under project AB17195025, the Guangxi Natural Science Foundation under grants 2018GXNSFDA281054 and 2018GXNSFAA281232, the National Cryptography Development Fund of China under project MMJJ20170217, the Guangxi Science and Technology Base and Special Talents Program AD18281044, the Innovation Project of GUET Graduate Education under project 2017YJCX46, the Guangxi Young Teachers' Basic Ability Improvement Program under Grant 2018KY0194, and the open program of Guangxi Key Laboratory of Cryptography and Information Security under projects GCIS201621 and GCIS201702.

© Springer Nature Singapore Pte Ltd. 2019
X. Cheng et al. (Eds.): ICPCSEE 2019, CCIS 1058, pp. 226–240, 2019.
https://doi.org/10.1007/978-981-15-0118-0_18

1 Introduction

With the rapid development of mobile communication, signal interference in special scenes is especially important [1], such as the military, government, examination rooms, etc. Mobile communication disturber is a portable device which can shield mobile communication. It use interference signal to effectively block the useful signal network, so that the mobile communication can't obtain the link connection with the base station which have the strongest signal and the closest signal, thereby achieving the purpose of destroying the base station and the mobile phone network, so that the mobile phone can't link to the base station and can't communication with the outside world. Mobile communication disturbers are suitable for all places where mobile phones are prohibited.

There are two main ways of interference in mobile communication interference technology: deceptive interference and repressive interference [14,15]. The existing mobile communication interference technology test method mainly tests whether the mobile phone can talk in the field to manually evaluate the protection effect [13]. However, due to the different technical parameters of the transmitting and receiving signals of various types of mobile phones, some mobile phones in the same location lose communication performance and some still work normally, which may easily cause interference signals in the classified places to shield the dead angle and form a security risk. Therefore, it is difficult to scientifically and objectively evaluate the shielding effect by using manual methods [7]. How to accurately quantify and measure mobile interference signals is a key issue in interference protection assessment and network optimization.

1.1 Research Progress

According to the predicted scenarios, the propagation modal can be divided into indoor model and outdoor model. At present, most of the research on propagation models are outdoor, while the research on indoor propagation models is relatively backward. Because the structure of the interior is more complicated than that of the outdoor, the table and chair will cut the space more fragmented, and there are more dead ends, which brings many difficulties to the study of the indoor propagation model. Many scholars at home and abroad have studied the electromagnetic signal propagation model. According to the nature of electromagnetic radiation prediction, it can be divided into statistical model (also known as "empirical model") and deterministic model (also known as "theoretical model").

In the early days, statistical prediction methods such as Okumura-Hata model [9], COST 231 model [10] and Egli model [3] dominated the electromagnetic prediction algorithm. Based on field measurement data of different propagation environments, this kind of algorithm obtains a graph or formula that can reflect the relation ship between electromagnetic radiation and the height of the transmitting and receiving antenna through statistical analysis. However as the division of the radio wave propagation area becomes smaller and smaller, each cell no longer has statistical similarity. The Okumura-Hata model, which is based on measured data, is based on the field strength median path loss in a quasi-flat terrain metropolitan area, and the correction factors are corrected for different propagation environments and terrain conditions [9]. The COST 231-Hata model extends the Hata model to 2000 MHz for macro cellular systems with a cell radius greater than 1 Km; The COST 231-WI model is based on the Walfisch-Bertoni model and the IKegami model and is mainly used in micro cellular systems. The model considers free space loss, loss from roof to street and street direction, and calculates path loss in both line-of-sight (LOS) and non-line-of-sight (NLOS) cases [10]. Egli concludes from a large number of measured results that he field strength in the uneven region is equal to the field strength calculated by the plane earth reflection formula plus a correction value of $20 * \lg 40/f$, where f is the operating frequency [3].

After the terrain is properly idealized, the electromagnetic radiation is calculated according to the propagation theory of electromagnetic waves, etc., called the deterministic prediction algorithm. Commonly used deterministic prediction methods include Ray-Tracing algorithm (RT) and Finite Difference Time Domain algorithm (FDTD) [16]. The RT algorithm uses optical propagation theory to establish a predictive model, and calculates the influence of topographic features on radio wave propagation. It has been widely used in different propagation environments such as micro cellular. The FDTD algorithm alternately samples the electric field component and the magnetic field component, and uses the update in the time domain to simulate the change of the electromagnetic field [12]. The calculation process should consider the influence of geometric parameters, material parameters and stability of the object in the scene.

Azpilicueta et al. proposed an indoor radio wave propagation model based on ray tracing method based on the physical model of electromagnetic field propagation, geometric optics and diffusion equation. The model uses a combination of ray tracing model and diffusion equation to predict the propagation of radio waves in complex indoor environments, improving computational efficiency and computational accuracy [2]. Salski et al. propose a model of a finite-difference time-domain unit with a thin dispersion layer, which has the advantage of reducing the additional computational effort required to solve a scene containing a non-uniform electrical/magnetically dispersed FDTD unit [11].

However, most of these models are carried out in outdoor scenes, and there are few studies in more complex indoor environments. Zhao et al. [18] studied the classical path loss models such as the dual-ray model, Hata model, and WIM model, and concluded that the signal intensity will be greatly reduced after the radio waves are reflected and refracted in a complex in door environment. The traditional path loss model is not applicable and needs to propose an improved model. However, due to the complex structure of modern buildings and the strong shielding effect of building materials on electromagnetic waves, indoor wireless propagation models are more difficult to study than outdoor models. In the field strength prediction of indoor environment, how to improve the computational efficiency and accuracy of the model has always been the focus and difficulty of the indoor wireless communication model [18].

1.2   Contribution of This Article

This article contributes as follows:

(1) This paper provides a signal field strength prediction algorithm which is applicable to the prediction of wireless communication signals in the whole frequency band and can be measured by professional instruments after reasonable indoor placement.
(2) This algorithm uses the improved prediction algorithm based on K-nearest neighbor non-parametric kernel regression to measure the field strength. By comparing with the measured data, the data corresponding to the K value with the highest accuracy is selected as the prediction result, and the corresponding shielding difference is calculated.
(3) Refer to the effect of signal interference in reality to form an interference protection indicator system, and then achieve the effect of determining interference protection functions.
(4) After using the algorithm proposed in this paper to predict various communication systems, the prediction accuracy of has been greatly improved, and the error range has been further reduced (Compared with the literature [7]).

2   System Model

An effective wireless communication signal detection and interference effect protection evaluation system should be able to solve the problem of objectively quantizing the internal wireless communication signal strength in different scenarios, and should have the following functions:

(1) It can predict full-frequency, full-standard wireless communication signal;
(2) For different security scenarios, it can model the scene and visually display the signal strength values of all points on the virtual scene;
(3) It can differential building, and visually displaying prediction result;
(4) For the prediction effect of the system, it is required that the error of the test points randomly selected from 80% or more is less than 2 dB.

An conventional wireless communication signal detection and interference protection effectiveness evaluation system is shown in Fig. 1. It includes logic modules such as signal collector, scene management, wireless data transmission unit, database, data processing and prediction, interference level evaluation and result display. The scene management module geometrically models the target area, and the user designs a corresponding test point arrangement scheme for the target area, and sends the scheme to the signal collector and the interference level evaluation module. The signal field strength data of each test point is obtained by using a signal collector, and the data is wirelessly transmitted into the background software. The background software stores the data locally and preprocesses the data. The data processing and prediction module uses the algorithm proposed in this paper to predict the field strength of the full-plane signal and calculate the mask difference. The interference level evaluation module

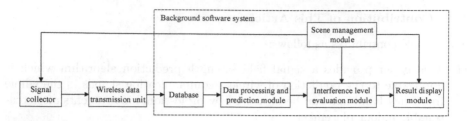

Fig. 1. Wireless communication signal detection and interference protection effectiveness evaluation system

analyzes and evaluates the target area by using the mask difference calculated by the data processing and prediction module.

The scene manager is used to model the real target scene, generate a corresponding virtual scene, and mesh the target scene and set test points in the virtual scene. The scene manager sends the test point scheme to the signal collector, so that the signal collector performs the test point signal field strength value acquisition according to the scheme.

The signal collector is used to measure the signal field strength value before and after the communication signal jammer is turned on of each measured point under different communication systems. The signal collector of the system adopts a hand-held full-band spectrum analyzer, which can simultaneously collect signal field strength values under 19 kind of communication systems. The 19 communication formats include: 2G GSM900 and DCS1800, 3G CDMA, 4G FDD LTE and TDD LTE. The specific communication systems are listed in the below experiments. The signal acquisition time of the signal collector is about 10 s, and 19 line-point signal field strength trajectory lines with a duration of 200 ms and 300 points are generated at the test point, corresponding to 19 kind of communication systems.

The signal collector transmits the collected signal field strength value before and after shielding to the database. The wireless data transmission unit of the system adopts a WiFi communication module to realize communication between the signal collector and the background evaluation software.

For the signal field strength data before and after shielding of the test points collected by the signal collector, the database selects and stores the signal field strength data of each sampling point according to the requirements. In this paper, the signal field strength value before shielding takes the peak value of the trajectory line, and the shielded signal field strength value takes the minimum value of the trajectory line.

The data processing and prediction module uses the signal field strength data before and after shielding of the sample points stored in the database, and uses the prediction algorithm of this paper to predict the full-plane signal strength, and obtains the signal field strength value of each predicted point.

The signal field strength after shielding of all test points obtained by the data processing and prediction module, subtracting the signal field strength before shielding to obtain the shielding difference. Then compare the shielding difference with the preset interference effectiveness difference grading standard, and determine the interference level of each test point in the scene by looking up the table. In the system, the interference level evaluation module determines the interference level of each test point in the scene by querying its built-in interference effectiveness difference grading standard table which showed in Table 1.

Table 1. Interference protection effectiveness difference level grading standard

Shielding difference of the evaluation point	Graded assessment
$Z_{difference\_predict\_i} > 6\,\text{dB}$	Security
$2 \leq Z_{difference\_predict\_i} \leq 6\,\text{dB}$	At risk
$Z_{difference\_predict\_i} < 2\,\text{dB}$	Filed

3 Prediction Algorithm

3.1 Prediction Algorithm Based on Non-parametric Kernel Regression

In the interference protection evaluation system proposed by Li et al., a non-parametric kernel regression algorithm is used to predict the signal field strength value [7]. Nonparametric statistic is to make no assumptions about the specific form of the overall distribution, try to obtain the required information from the sample itself, and gradually establish mathematical descriptions of things and inferential statistics of statistical models [6,8]. Non-parametric statistics have the following characteristics: (1) It has relatively few assumptions about the population and results in better stability; (2) It can handle all types of data and has strong applicability; (3) Non-parametric ideas are easy to understand and calculate [6].

The N-W estimate is a non-parametric kernel regression (kernel estimation) algorithm proposed by Nadaraya and Watson in 1964. There are many non-parametric kernel regression algorithms, and their ideas are basically the same. The predicted value is the conditional expectation $\hat{m}(x_i, y_i)$ given by $x = x_i$, $y = y_i$ is estimated by the following formula [4]

$$\hat{m}(x_i, y_i) = \frac{\frac{\sqrt{2}}{2nh} \sum_{i=1}^{n} \hat{K}(\frac{x-x_i}{h}, \frac{y-y_i}{h}) z_i}{\frac{\sqrt{2}}{2nh} \sum_{i=1}^{n} \hat{K}(\frac{x-x_i}{h}, \frac{y-y_i}{h})} \tag{1}$$

where h is called window width, $\hat{K}(\cdot)$ is the kernel function, and z is the sample value. The window width h determines which samples participate in the operation, and the kernel function $\hat{K}(\cdot)$ determines how these samples are calculated.

In this algorithm, the sample point data is used to estimate the information of the predicted point data. Therefore, the closer the sample point is to the predicted point, the stronger the correlation between the sample point data and the predicted point data, and the more useful information is carried. However, in this algorithm, the window width is a fixed value given at the beginning and cannot be changed dynamically during the calculation process. Therefore, if the window width is too large, some sample points with weak information correlation will be included in the window to increase the error. Conversely, if the window width is too small, the number of sample points in the window will be so small that increase the error.

Secondly, the algorithm only predicts the sampling point data, and does not predict the field strength of the full-plane signal, which makes the interference assessment lack effective basis.

3.2   Improved Prediction Algorithm Based on K-nearest Neighbor Non-parametric Kernel Regression

Aiming at the above two problems of the original algorithm, this paper proposes a new improved algorithm based on the original algorithm, which is an improved prediction algorithm based on K-nearest neighbor nonparametric kernel regression. The signal strength before shielded $P_{predict\_before}$ and the signal strength after shielded $P_{predict\_after}$ of the predicted point are predicted by an improved prediction algorithm based on K-nearest neighbor non-parametric kernel regression.

K-nearest neighbor (KNN) classification algorithm [5,17] is one of the simplest machine learning algorithms. According to the algorithm, given a sample x to be classified, first find the K training sets that are most similar to x in the feature space (the closest in the feature space), and then according to the category labels of the K training samples to determine the category of sample x.

In this algorithm, the K-nearest neighbor algorithm is used to set the distance between the sample point and the predicted point dynamically. The window width of the function is dynamically set, and sample points that are sufficiently correlated with the information of the predicted point data are placed in the window. Since the data measurement requires a lot of time and effort so that can't perform full-plane coverage measurement on the test area, the number of sample points is limited, and the remaining area needs to be predicted. The algorithm expands the sample sequence by increasing the sample points, that is, using the K-nearest neighbor non-parametric kernel regression algorithm to predict the predicted point data from the coordinates $(0, 0)$, and filling the had predicted data into the sample sequence, then used as a sample point to predict the next predicted point data. Skip if it encounters the measured point, and so on to get the full-plane predicted point data. The predicted full-plane data and measured point data are then used to predict each measured point data. The K value can be taken as 2, 3, 4, 5, 6, 7, etc. The system compares the data obtained from different K value predictions with the measured data and calculates the accuracy rate, and selects the data corresponding to the K value which have the highest accuracy as the output. Based on the measured data, an improved prediction algorithm based on K-nearest neighbor non-parametric kernel regression is used to predict the full-plane signal field strength before and after the jammer is turned on, whose flow is shown in Fig. 2.

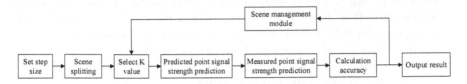

Fig. 2. Flow chart of improved prediction algorithm based on the K-nearest neighbor non-parametric kernel regression

Prediction Point Signal Strength Calculation. The prediction algorithm of this paper is used to predict the signal strength of the full-plane signal. Proceed as follows:

(1) The scene is divided into several points according to a preset step size, and the points except the measured points are all predicted points.
(2) Using the Pythagorean theorem to calculate the distance between each sample point and the predicted point:

$$d_i = \sqrt{(X_i - x)^2 + (Y_i - y)^2} \tag{2}$$

where (X_i, Y_i) is a two-dimensional random variable with independent and identical distribution in a given sample space, (x, y) is the coordinates of the predicted point, and d_i is the distance from (X_i, Y_i) to (x, y).
(3) Sort the distance d between each sample point and the predicted point from small to large.
(4) Select the Kth distance as the window width, ie $R = d(k)$.
(5) The signal strength Z to be predicted can be expressed as a conditional expectation $\hat{m}(x_i, y_i)$ for a given $x = x_i$ and $y = y_i$, and its value is estimated by:

$$\hat{m}(x_i, y_i) = \frac{\sum\limits_{i=1}^{n} \hat{K}(\frac{X_i - x}{R}, \frac{Y_i - y}{R}) \times Z_i}{\sum\limits_{i=1}^{n} \hat{K}(\frac{X_i - x}{R}, \frac{Y_i - y}{R})} \tag{3}$$

where (X_i, Y_i) is the sample point coordinate, and Z_i is the signal strength corresponding to the sample point (X_i, Y_i), which is actually measured according to the layout scheme; (x, y) is the predicted point coordinate, which is directly given by the predicted point; R is the window width and is calculated by the above steps (2) to (4); K is a kernel function and the expression is as follows:

$$\hat{K}(x,y) = \frac{1}{\sqrt{2\pi}} e^{-\frac{1}{2}(x^2+y^2)} I(x,y) \tag{4}$$

(6) The final window width is determined by selecting the window width corresponding to the highest accuracy result from all the results, that is, the optimal K value. The specific determination method is that the window width value $R = d(K)(K = 2,3,4,5,6,7)$ of the candidate window width is sequentially substituted, and then the measured points belonging to the window are selected as the sample points for prediction. The specific judgment method is:

a. When the window type is a square area:

$$I(x,y) = \begin{cases} 1 & \text{if } |x| \leq R, |y| \leq R \\ 0 & \text{otherwise} \end{cases} \tag{5}$$

b. When the window type is a circular area:

$$I(x,y) = \begin{cases} 1 & \text{if } x^2 + y^2 \leq R \\ 0 & \text{otherwise} \end{cases} \tag{6}$$

The known measured point coordinate (X_i, Y_i) in the window and its corresponding measured signal field strength value Z_i, the candidate window width R value and the point (x_i, y_i) to be predicted are substituted into the above-mentioned non-parametric kernel regression model to obtain $\hat{m}(x_i, y_i)$, that is, the signal field strength of point (x_i, y_i).

According to the above method, the signal strengths of all the points to be measured using different K values are predicted, and the predicted point data and the measured point data are used to predict the data of the measured points. The specific method is as follows:

(7) Give predicted data and measured point data to (t_x, t_y, t_z).

(8) Empty the current predicted measured points t_x, t_y, t_z.

(9) Predict the signal strength $Z_{predict\_i}$ of (X_i, Y_i) using the method from step (2) to (5).

(10) The predicted signal strength $Z_{predict\_i}$ is subtracted from the measured signal strength Z_i of (X_i, Y_i) and its absolute values is taken. If the error is within $2\,\text{dBmV}$, it is judged as qualified, that is, $|Z_{predict\_i} - Z_i| \leq 2$. Calculate the accuracy rate corresponding to each K value, and use the set of data corresponding to the K value with the highest accuracy as the output.

4   Experiment and Analysis

4.1   Experiment Scenarios

Select a rectangular classroom with a rule as the test environment. The classroom is $10\,\text{m}$ long and $5\,\text{m}$ width with two aisles in the middle. In the actual measurement process, since the measurement requires a lot of time and effort, the number of measurement points cannot be infinite, and a representative measurement point needs to be selected. When the indoor scene is relatively empty and the indoor objects are evenly distributed, the geometric method can be used to uniformly distribute the points according to certain rules. When the indoor scene is irregular and the irregular and the distribution of indoor objects is complicated, the selection of measurement points must take into account the room structure, building materials, precision and other factors, and arrange at a representative location. This test uses the cross-distribution method to design two layout schemes, which are 50 test points and 150 test points.

The test signal collector uses the W2030A hand-held full-band spectrum analyzer, which can collect the signal field strength values of 19 kind of communication systems at one times. The details are shown in the experiment below. The MA-2006 type secret space mobile communication jammer is used to block the downlink signal of the mobile communication, so as to block the normal link between the mobile phone and the base station within a certain range. The jammer is fixed on the tripod, the bottom of the device is 85 cm from the ground, and the top of the device is 123 cm away from the ground.

4.2   Actual Measurement

K Value Selection Experiment. K represents the number of points to be taken near the predicted point. Using different K values to predict the signal field strength before and after full-plane shielding for various communication systems, and combine with the measured points into a sample set. Using the sample set, predicting the measured points one by one and calculating the prediction error. Then, the ratio less than the error threshold is taken as the accuracy of the corresponding communication system and K value. Table 2 summarizes the accuracy of each K value in different systems.

According to the above table and the K-nearest neighbor principle, the larger the K, the lower the accuracy. For different scenarios and different communication systems, the optimal K value is not completely consistent, so K takes $2 \sim 7$ to meet the demand.

Interference Protection Assessment Experiment. This paper takes China Unicom 2G GSM900 as an example to predict the wireless communication signal and interference signal strength. After the jammer is turned off, the hand-held full-band spectrum analyzer is used to collect the signal field strength value of the point at the test point according to the layout scheme. After completing the before shielding data acquisition for all test points, turn on the jammer and collect the signal field strength values of each test point again. The data collected by the signal collector is transmitted to the background software through the wireless data transmission module for storage. The prediction algorithm of this

Table 2. Accuracy of each K values in different systems (The error threshold is 2 dB)

Communication system	Shielding	Accuracy of K = 2	Accuracy of K = 3	Accuracy of K = 4	Accuracy of K = 5	Accuracy of K = 6	Accuracy of K = 7
China Telecom 2G CDMA 1X	Before	96.0%	90.7%	88.7%	86.7%	86.7%	87.3%
	After	84.0%	79.3%	73.3%	73.3%	72.0%	71.3%
China Telecom 3G EV DO	Before	96.0%	92.0%	91.3%	90.0%	89.3%	88.7%
	After	86.0%	84.0%	80.7%	78.7%	77.3%	74.7%
China Telecom 4G FDDLTE1	Before	92.7%	90.7%	88.0%	86.0%	84.7%	84.7%
	After	84.7%	79.3%	77.3%	78.0%	76.7%	75.3%
China Telecom 4G FDDLTE2	Before	92.7%	92.0%	91.3%	90.0%	90.7%	90.7%
	After	82.0%	79.3%	77.3%	78.7%	76.0%	75.3%
China Mobile 2G DCS1800	Before	86.0%	82.0%	79.3%	78.0%	77.3%	76.7%
	After	82.7%	78.0%	76.7%	73.3%	73.3%	74.7%
China Mobile 2G GSM900	Before	83.3%	79.3%	79.3%	76.7%	79.3%	78.0%
	After	84.7%	80.0%	78.0%	76.7%	76.0%	74.7%
China Mobile 4G TDDLTE1	Before	89.3%	82.7%	81.3%	81.3%	82.0%	83.3%
	After	84.7%	78.0%	75.3%	75.3%	78.7%	77.3%
China Mobile 4G TDDLTE3	Before	92.0%	84.7%	83.3%	82.0%	82.0%	80.0%
	After	85.3%	81.3%	82.7%	80.0%	78.7%	78.0%
China Unicom 2G DCS1800	Before	92.7%	86.7%	84.7%	83.3%	82.7%	84.0%
	After	80.0%	74.0%	76.0%	72.0%	74.7%	73.3%
China Unicom 2G GSM900	Before	96.7%	89.3%	87.3%	86.7%	86.0%	84.0%
	After	82.0%	78.0%	80.0%	76.7%	77.3%	75.3%
China Unicom 3G WCDMA	Before	94.7%	87.3%	87.3%	84.7%	85.3%	85.3%
	After	84.7%	79.3%	80.7%	80.0%	77.3%	76.7%
China Unicom 4G FDDLTE1	Before	94.7%	89.3%	88.0%	85.3%	84.0%	82.7%
	After	81.3%	75.3%	76.0%	75.3%	76.0%	76.0%
China Unicom 4G FDDLTE2	Before	92.7%	87.3%	86.7%	86.7%	86.0%	86.0%
	After	92.7%	88.7%	89.3%	86.7%	88.0%	86.0%
China Unicom 4G TDDLTE2	Before	94.7%	91.3%	88.7%	88.7%	88.7%	88.0%
	After	81.3%	78.7%	79.3%	80.7%	81.3%	80.0%

paper is used to predict the signal field strength before and after shielding of the whole plane, and the predicted signal strength before and after shielding of the full-plane subtracted each other to obtain the shielding difference. By comparing the built-in Interference protection effectiveness difference level grading standard table, analyze and show the full-plane interference protection effect.

The interference protection effect is shown in Fig. 3. A white area is formed on the lower right of the figure, indicating that the area is unqualified, but the mobile phone continues to be out of service in this area. After checking the trajectory diagram, it is judged that the shielding effect of the jammer is unstable during the test period, and an abnormal signal appears in the signal sent by the base station, but the risk possibility is not excluded, and the trajectory is as shown in Fig. 4.

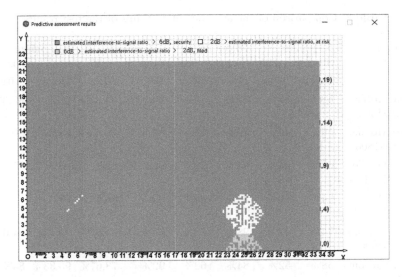

Fig. 3. Interference protection effect display (China Unicom 2G GSM900)

4.3 Prediction Accuracy Evaluation

The prediction algorithm proposed in this paper first predicts the field strength before and after the full-plane shielding, and then combines the predicted prediction point data with the measured point data to form a new sample sequence. Reverse prediction of signal field strength values at each measured point using a new sample sequence. The measured field strength value is subtracted from

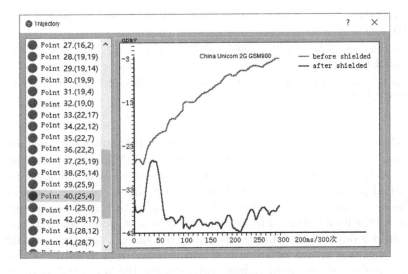

Fig. 4. Trajectory of a test point

Table 3. Comparison of the accuracy of 4 groups of the test results

Test program	50 points Step size 0.25		50 points Step size 0.05		150 points Step size 0.25		150 points Step size 0.05	
	Before	After	Before	After	Before	After	Before	After
China Mobile 2G GSM900	58%	52%	82%	72%	84.67%	84.33%	90%	85.33%
China Mobile 2G DCS1800	56%	44%	80%	74%	84.67%	79.33%	88.67%	80%
China Mobile 4G TDDLTE1	68%	38%	84%	72%	90.67%	82%	93.33%	85.33%
China Mobile 4G TDDLTE3	64%	70%	84%	88%	92%	88.67%	93.33%	88%
China Unicom 2G GSM900	74%	62%	98%	78%	97.33%	84.67%	99.33%	84.67%
China Unicom 2G DCS1800	68%	52%	84%	68%	92%	78.67%	95.33%	84%
China Unicom 3G WCDMA	82%	58%	94%	74%	94.67%	82%	98%	90%
China Unicom 4G TDDLTE1	92%	58%	96%	76%	94.67%	84.67%	97.33%	85.33%
China Unicom 4G FDDLTE1	70%	56%	90%	66%	96%	80%	98%	84%
China Unicom 4G FDDLTE2	80%	100%	92%	100%	95.33%	99.33%	98.67%	100%
China Telecom 2G CDMA 1X	98%	98%	100%	100%	96%	98.67%	99.33%	99.33%
China Telecom 3G EV DO	98%	100%	100%	100%	96.67%	98.67%	99.33%	98.67%
China Telecom 4G FDDLTE1	86%	68%	90%	76%	94%	87.33%	98%	85.33%
China Telecom 4G FDDLTE2	84%	36%	92%	68%	95.33%	82%	98%	84.67%

the field strength value predicted by the corresponding position of the measured point, and the point where the absolute value of the difference is less than 2 dB is taken as the qualified point. The percentage obtained by dividing the number of qualified points by the number of measured points is taken as the accuracy, and the set of data corresponding to the K value with the highest accuracy is selected as the final output. This test uses a total of 50 points and 150 points to carry out 4 groups of test, and the accuracy of different step size and different layout schemes is shown in Table 4. It can be seen from Table 3 that increasing the number of measured points and reducing the predicted step size can improve the prediction accuracy, but the measurement time, workload an the

Table 4. Comparison of accuracy between existing algorithm and the algorithm of this paper (CDMA communication system)

Algorithm	Accuracy
Nonparametric kernel regression algorithm [7]	75.86% (the error within ±3.5 dB)
	82.76% (the error within ±4 dB)
FDTD algorithm [7]	84.48% (the error within ±3.5 dB)
Algorithm of this paper	99.33% (the error within ±2 dB)

background softwares computational complexity will also increase, reducing the work efficiency.

According to the method proposed by Li et al. And the data given [7], the prediction accuracy of the non-parametric kernel regression algorithm is 75.86%, the error is within ±3.5dB, and 82.57% whose error is within ±4dB; The prediction accuracy of FDTD algorithm is 84.48%, and the error is within ±3.5dB. Compared with CDMA communication system which used the same in the literature [7] test, after using the prediction algorithm proposed in this paper for prediction, the accuracy rate is as high as 99.33%, and the error is within ±2dB. In the CDMA communication system, the comparison of accuracy between the existing algorithm and the algorithm proposed in this paper is shown in Table 4.

5   Conclusion

For indoor secret places, professional instrument is used to collect the signal field strength value of the sampling point. The algorithm of this paper is used to predict the full-plane signal field strength value, analyze the data error and improve the algorithm to improve the data prediction accuracy. The experimental results of the prediction algorithm show that the prediction accuracy of signal strength before and after shielding of all communication systems exceeds 80%, and the error is within ±2 dB. The prediction accuracy of signal strength before shielding is more than 90%, and the error is within ±2 dB. For the field of interference assessment, which has low prediction accuracy and lack of effective evaluation criteria, a new method for predicting full-space signal strength and evaluating the interference effect of mobile communication wireless signals is provided, which has great guiding significance in network optimization and interference implementation.

References

1. Ansah, A.K.K.: Design and implementation of a GSM mobile detector and jammer. Proc. World Congr. Eng. Comput. Sci. **1**, 95–101 (2018)
2. Azpilicueta, L., Falcone, F., Janaswamy, R.: A hybrid ray launching-diffusion equation approach for propagation prediction in complex indoor environments. IEEE Antennas Wirel. Propag. Lett. **16**, 214–217 (2017). https://doi.org/10.1109/LAWP.2016.2570126

3. Egli, J.J.: Radio propagation above 40 MC over irregular terrain. Proc. IRE **45**(10), 1383–1391 (1957). https://doi.org/10.1109/JRPROC.1957.278224
4. Hazelton, M.: Nonparametric regression an introduction (2014). https://doi.org/10.1002/9781118445112.stat05768
5. Jiang, S., Pang, G., Wu, M., Kuang, L.: An improved k-nearest-neighbor algorithm for text categorization. Expert Syst. Appl. **39**(1), 1503–1509 (2012). https://doi.org/10.1016/j.eswa.2011.08.040. http://www.sciencedirect.com/science/article/pii/S0957417411011511
6. Khakshooy, A.M., Chiappelli, F.: Nonparametric statistics. In: Practical Biostatistics in Translational Healthcare, pp. 123–137. Springer (2018). https://doi.org/10.1007/978-3-662-57437-9
7. Li, S., Li, Y., Wang, J., Ding, Y.: Study on prediction model of interference signal of mobile communication indoors based on finite-difference time-domain method and nonparametric kernel regression method. Appl. Res. Comput. **4**, 1213–1216 (2017). (In Chinese)
8. Lloyd, J.R., Duvenaud, D., Grosse, R., Tenenbaum, J., Ghahramani, Z.: Automatic construction and natural-language description of nonparametric regression models. In: Twenty-Eighth AAAI Conference on Artificial Intelligence (2014)
9. Okumura, Y.: Field strength and its variability in VHF and UHF land-mobile radio service. Rev. Electr. Commun. Lab. **16**, 825–873 (1968). https://ci.nii.ac.jp/naid/10010001461/en/
10. Pedersen, G.F.: Cost 231 - digital mobile radio towards future generation systems. Cost 231 - Digital Mobile Radio Towards Future Generation Systems, pp. 92–96 (1999)
11. Salski, B.: An FDTD model of a thin dispersive layer. IEEE Trans. Microw. Theory Tech. **62**(9), 1912–1919 (2014). https://doi.org/10.1109/TMTT.2014.2337286
12. Santos, M., et al.: Maxwell's equations based 3D model of light scattering in the retina. In: 2015 IEEE 4th Portuguese Meeting on Bioengineering (ENBENG), pp. 1–5 (February 2015). https://doi.org/10.1109/ENBENG.2015.7088869
13. Scarfone, K., Souppaya, M., Cody, A., Orebaugh, A.: Technical guide to information security testing and assessment. NIST Spec. Publ. **800**(115), 2–25 (2008)
14. Shah, S.W., et al.: Cell phone jammer. In: 2008 IEEE International Multitopic Conference, pp. 579–580 (December 2008). https://doi.org/10.1109/INMIC.2008.4777805
15. Stenumgaard, P., Fors, K., Wiklundh, K., Linder, S.: Electromagnetic interference on tactical radio systems from collocated medical equipment on military camps. IEEE Commun. Mag. **50**(10), 64–69 (2012). https://doi.org/10.1109/MCOM.2012.6316777
16. Sullivan, D.M.: Electromagnetic Simulation Using the FDTD Method. Wiley, Hoboken (2013)
17. Yu, X., Pu, K.Q., Koudas, N.: Monitoring k-nearest neighbor queries over moving objects. In: 21st International Conference on Data Engineering (ICDE 2005), pp. 631–642 (April 2005). https://doi.org/10.1109/ICDE.2005.92
18. Zhao, Y., Li, M., Shi, F.: Indoor radio propagation model based on dominant path. Int. J. Commun. Netw. Syst. Sci. **3**(03), 330 (2010)

Prediction Model for Non-topological Event Propagation in Social Networks

Zitu Liu, Rui Wang, and Yong Liu$^{(\boxtimes)}$

Heilongjiang University, Harbin, China
liuyong123456@hlju.edu.cn

Abstract. The spread of events happens all the time in social networks. The prediction of event propagation has received extensive attention in data mining community. In prior studies, topologies in social networks are usually exploited to predict the scope of event propagation. User's action logs can be obtained in reality, but it is difficult to get topologies in social networks. In this paper, NT-GP, a prediction model for non-topological event propagation, is proposed. Firstly a time decay sampling method was used to extract the walk paths from user's action log, and then deep learning method was applied to learn the sampling paths and predict the future propagation range of the target event. Extensive experiments demonstrate effectiveness of NT-GP.

Keywords: Social network · Non-topological · Action log · Time decay sampling

1 Introduction

With the wide application of social systems and the surge of users and data, the research on how to extract hidden information from massive data becomes more and more important. In social networks, there are large social networks with huge amounts of data, and they have a large number of users. Most users in the social network log in to the social network to view the information as the main purpose, and a large number of action logs are generated while browsing. At the same time, it also brings a lot of research topics that can be studied in depth. For example, many companies conduct information on product promotion and market analysis by analyzing microblogs, blogs, user forwarding, comments, and sharing. How to correctly analyze and predict the information dissemination process in social networks, and the subsequent trend of communication become more important.

Information diffusion prediction in social networks and communication structure analysis rely on information prediction models. The independent cascade model (IC) model [1] and the linear threshold (LT) model proposed in 2003 are based on the graph structure and are continuously studied in depth. From the initial stage of exploring the structure, to the later research on the structure of the propagation network, as well as content analysis, to the recent network representation learning. Researchers are constantly exploring the process of information dissemination and the value of its applications. Among them, the research on predicting social network communication [2] and information path propagation [3] according to the graph

© Springer Nature Singapore Pte Ltd. 2019
X. Cheng et al. (Eds.): ICPCSEE 2019, CCIS 1058, pp. 241–252, 2019.
https://doi.org/10.1007/978-981-15-0118-0_19

structure is gradually becoming more and more mature. There is relatively little research on the social network model based on no topology. What this paper does is to construct a non-topological event propagation prediction model.

With the advent of the information age, various data on the Internet have accumulated rapidly. Combining the functions of existing social networks and the way event propagation is transmitted, most of the models at this stage are not suitable for the analysis of existing social networks. In social networks, most users will post an event by forwarding, and we want to get the source of the event to obtain the original source of an event after multiple searches. If it is not the original creation of Weibo, we usually do not know where the source of the event is, and more is to obtain a triple of action sequence similar to a user, Weibo, and time. Even if we can get the source of the information, we can only observe the partial connections of several users, and some of the observed information can be used for analysis. But this information does not represent the true diffusion [4] of information in a social network. In real social networks, users get information from different channels. Intuitively, we don't know who is influencing a particular user, but users participate in the dissemination of information. In practice, we usually only capture incomplete propagation processes. And there is no personal privacy involved in the local message propagation process.

To solve the above problems, the research data we used is similar to the data set in the form of (users, events, time), and it is assumed that there is no relationship between users in the social network in the given data set. A random walk strategy is used to describe the process of message propagation and a vector containing the characteristics of user message propagation is obtained by learning the process of event propagation. The influence range of the event is predicted by learning the obtained vectors and the path of the event propagation. Combining the methods of this part, we construct a topological structure free social information diffusion model.

This model compares the functions of other topless social network event propagation models. We consider the randomness in the process of message diffusion, and avoid the artificially determined feature representation diffusion or the overall network structure, so as to predict the scale of information flow transmission. We propose the NT-GP model to solve the above problems. We use the idea of expressive learning and the deep learning framework to extract general characteristics, and finally adopt an end-to-end prediction method. By analyzing the sampled data, we can obtain the feature vectors of social network users and predict the scale of information diffusion through the neural network. The contributions of this paper are as follows:

(1) We propose a non-topology propagation model NT-GP. It applies to large social networks by directly using the time series of event propagation without the assumption of potential influence relationships that may exist in the unknown graph structure.

(2) In the existing propagation model, the prediction propagation process uses the transmitted feature vector, but there are some new propagation terms in the propagation items in the test set. In order to improve the accuracy of the test, a new propagation item learning method is also proposed.

(3) The experimental results show that compared with the previous social network propagation model, our propagation model has a better effect on the prediction of the propagation range.

2   Related Work

In this part, we mainly introduce social network information. The work related to the dissemination model can be divided into two main categories: (a) Prediction of propagation influence range without topology; (b) Representation learning of information dissemination in a social network without topology.

Specific models can achieve a certain degree of prediction by learning a specific data set, such as message dissemination in tweets or microblogs [5, 6], and prediction of cooperative relationships in academic papers [6]. The existing literature mainly carries out two types of methods to extend. One type of method is to analyze through the existing topology and some other factors involved in the propagation process, to get a model to make many strong assumptions [7], and to simplify the actual propagation process [10]. They tend to underperform in actual forecasts. Or some obvious factors can be analyzed manually, extracted from data, weighted by machine learning algorithm, and then classified or returned. In general, some extracted features are largely dependent on the topology of social network, and some accurate models are also dependent on specific social network structure. The other method comes from the calculation of graph structure. According to the similar structure in the graph structure or the expression vector of nodes [8, 9], the obtained vector is used to make connection prediction and influence prediction [12, 13].

We not only use the past effective experience and means, but also summarize the shortcomings. For example, the algorithm is not suitable for a large amount of data, some experiments are only valid for a specific data set, and some only consider the user's influence on the user, ignore Issues such as the important role of the communication item itself. Our model advantages over previous models are: (a) for larger data sets; (b) higher accuracy for real-world topology-free data sets; (c) building social network propagation models In the process, if the reason for forwarding a certain propagation item is not clear, it is impossible to know who forwards the message. We propose a time series based random walk strategy to solve the impact.

3   Method

3.1   Problem Definition

We select the social network structure at time t and define it as $G = <V, E>$. V represents a social network user node. $E \in V * V$ is the edge in the social network. This is a topological free model, and we don't care about the connections in the network.

Definition $g_c^t = (V_c^t, E_c^t)$ A represents an information diffusion network that is finally formed by a stream of information $c \in C$ during the t period. In this paper, we mainly focus on the situation that the number of nodes affected by the above network changes with time. Define $\Delta S_C = \left| V_C^{t+\Delta t} \right| - \left| V_C^t \right|$ to represent how a message affects the number of users over a time period of t. We will propose a mapping function f, which represents the predicted value of the propagation term affecting the user in the time period t. The smaller the difference between the predicted number and the actual number, the smaller the objective function is. The objective function is shown in formula (1), and the mapping function f here is the prediction model we proposed, namely NT-GP model for short.

$$\theta = \frac{1}{|C|} \sum_C (f(g_C^t) - \Delta S_C)^2 \tag{1}$$

When Δt is small, we can predict the early propagation range of information, that is, the short-term propagation of information flow can be predicted. When Δt A is large, the information we predict is close to the later propagation prediction, that is, the propagation of the given information flow after Δt A considerable period of time. Finally, it should be noted that the overall network structure does not change at any time.

3.2   Sampling of a Propagation Sequence

This section introduces how to select multiple propagation sequences for the same propagation item. We process the initial data through two methods, namely window movement selection method and time decreasing selection method. The same propagation item can be divided into multiple propagation sequences according to the length of the propagation sequence. We define an array B to store all the propagation sequences of each propagation item.

The input form of this model is <user, propagation item, time> triplet, which is a record of the message propagation process without topology, representing the user's participation in the propagation of time at a certain moment. Our choice of propagation sequence is for the same propagation item, we can easily extract the user time collection belonging to the same propagation item from the original data. For example $E = \{[u_1, t_1], [u_2, t_2], \ldots, [u_L, t_L]\}$. t indicates the time when the user participates in the time propagation. The length of the window defined in this chapter is T, which means that each segment of data is intercepted from the data set E to form the length of the propagation sequence. Specifically, the length of the window is T T, indicating that the data in the E is divided into T data. A group, each group represents a propagation sequence, if the last user's time data is less than T, then the last few users in the previous propagation sequence complement the propagation sequence. Table 1 details this process.

Table 1. Window movement selection:

```
Form move method selects propagation path
```

Input: number of users of the same propagation item L, data set
E, window length T

Output: propagation path set B

1: *for i=0 to $\frac{L}{T} - 1$*

1:      $B_i = \{E_{i*T}, E_{i*T+1}, ..., E_{i*T+T}\}$

2: *end for*

3: *if $(L\%T)$ == 0 then*

4:      *output $B = \{B_1, B_2, ..., B_{\frac{L}{T}-1}\}$*

5: *else*

6:      $B_{\frac{L}{T}+1} = \{E_{L-T}, E_{L-T+1}, ..., E_L\}$

7:      *output $B = \{B_1, B_2, ..., B_{\frac{L}{T}-1}, B_{\frac{L}{T}}\}$*

8: *end if*

Time Decrement Selection Method: First, the explicit relationship is to spread two user nodes of the same propagation item. If the time interval is longer, the influence of the previous user node on the latter user node is smaller. Conversely, the greater the influence. This section selects function $w(t_i, t_j) = e^{-\mu(t_i - t_j)}$ to illustrate this influence relationship. This function reflects the influence of the distance between two nodes on the propagation information. The specific selection method of the propagation sequence is as shown in Fig. 1. When we get $E = \{[u_1, t_1], [u_2, t_2], ..., [u_L, t_L]\}$, we can construct an initial propagation graph. Since the data is chronological, the user who is in front of the user may have information about the user propagation. The effect, as shown by the directed edges in the figure, but not all edges are present, so a weight is given to each edge by a time function. The size of the reference weight is used to select a reasonable fixed length propagation sequence.

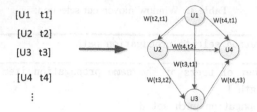

Fig. 1. Construct a propagation sequence using a time function

Random Selection Method: In a complete message propagation sequence, randomly select an event record as the starting node u_i. According to the time series of event sequence, $E = \{[u_1, t_1], [u_2, t_2], \ldots, [u_i, t_i], [u_{i+1}, t_{i+1}], \ldots, [u_L, t_L]\}$ node u_i is randomly selected in E as an initial node of message propagation, and a node is randomly selected from $\{[u_{i+1}, t_{i+1}], \ldots, [u_L, t_L]\}$ as the next node selected. Repeat multiple times until the path length, or the last node in the time series, is reached.

3.3 User Vector Representation Learning

Spread by introduction last, be able to get each item than the fixed length of spreading sequences, but haven't got user node corresponding vectors on each of the sequence, which is the input vector of the neural network needs, this section also each user in the sequence is obtained by two methods to the corresponding vector, word vector training method respectively.

Word Vector Training Method: Skip-gram is a form of word2vec, and word2vec [9, 10, 14] can be understood as "word vector", which is used to transform words in natural language into dense vectors that can be understood by computers. Although it is more applied to natural language, in social networks, each propagation sequence can also be regarded as a sentence, and each user node is equivalent to the word in the sentence. The algorithm of this model is used to process and train each user vector. The specific calculation method is as follows.

The first step is to represent the user nodes in each sequence as a one-hot vector. In particular, when a large data set is used, that is, when the length of each propagation sequence node T is selected to be large, the vector is very sparse, and considerable computing resources will be consumed in the calculation. In this case, dimensionality reduction of the vector is also required. We will each vector F the same embedding matrix multiplication, the size of the matrix H * Nnone, including Nnone said user node, the number of H representative for processing after the vector dimension reduction. After user nodes are processed by one-hot coding, they are put into skip-gram model for training. The purpose of the skip-gram model is to find the relationship between a user in the same sequence and other users, that is, the probability of the user's image in which the user propagates information. We introduce a technique for finding the probability of the Softmax method, such as the formula (2). As shown in (4), the current user node w is known to need to predict the node con(w) affected by it, thus constructing a conditional function

$$p(con(w)|w) = \prod_{u\in con(w)} p(u|w) \tag{2}$$

p(u|w) is solved using the Hierarchical Softmax technique, which is a key technique for improving vector learning performance, specifically expressed as

$$p(u|w) = \prod_{j=2}^{l^u} p(d_j^u|v(w), \theta_{j-1}^u) \tag{3}$$

Here each user node is a node in the Huffman tree, where p_1^u represents the root node of the tree, l^u represents the number of nodes in the path p^u, $d_j^u \in \{0,1\}$, and the Huffman code representation path of the node u The code corresponding to the j-th node in p^u, θ_j^u represents the vector corresponding to the jth non-leaf subnode in the path p^u. According to Softmax's definition of $p(d_j^u|v(w), \theta_{j-1}^u)$, we want the probability we want. v(w) represents the vector of user w.

$$p(d_j^u|v(w), \theta_{j-1}^u = [\sigma(v(w)^T\theta_{j-1}^u)]^{1-d_j^u} \cdot [1 - \sigma(v(w)^T\theta_{j-1}^u)]^{d_j^u} \tag{4}$$

The output of the model is still the user node vector, where each dimension represents the probability of occurrence of the node behind the node, that is, the probability obtained by the above formula.

3.4 GRU Neural Network Training

When we successfully obtain the user node vector in the propagation sequence, that is, (x1, x2, x3, x4) in Fig. 2, we need to put it into the GRU (Gated Recurrent Unit) neural network for training. The event propagation sequence can be Read from left to right, or read from right to left. In this way, the earlier user nodes in the sequence can be informed which nodes are affected by the propagation items they pass, and then through the two-way GRU unit, the splicing results in an implicit vector representation representing the propagation trend of the sequence, as shown in Fig. 2 (h1, h2, h3, h4).

Specifically, the user vector needs to be trained by the formula (5), where i represents the i-th node in a propagation path, and the model uses the input user node vector xi together with the previous hidden state hi-1 as an input, and passes The GRU formula calculates the updated hidden state hi, where σ(*) is the sigmoid activation function, * represents the concrete form of the formula, W and U are the GRU parameters learned during training, and H is the dimension size representing the vector.

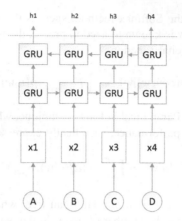

Fig. 2 Neural network training process

$$r_i = \sigma(W^r x_i + U^r h_{i-1} + b^r)$$

$$u_i = \sigma(W^u x_i + U^u h_{i-1} + b^u)$$

$$\tilde{h} = \tanh(W^h x_i + r_i U^h h_{i-1} + b^h) \tag{5}$$

$$\tilde{h} = u_i \tilde{h}_{i-1} + (1 - u_i) h_{i-1}$$

$$y_i = \sigma(W^o h_i)$$

3.5 NT-GP Model

We make predictions about the extent of the spread by trying to indicate the current scale of the event. The attention mechanism is used to learn the relationship between the propagation sequence and the propagation term to predict the range of event propagation. For the propagation of an event, the propagation sequence is extracted, and the K propagation sequences are extracted and the length of each propagation sequence is T. The user vector obtained previously is assembled, for example for a propagation sequence representing only one event in Fig. 2. Through the integration of the propagation structure of an event and representation by vectorization, we obtain the vector and use it as the input of the function to obtain the propagation range of the elapsed time event. Which is expressed as:

$$h_g(E) = \alpha I_{S_E} + \sum_{k=1}^{K} \sum_{i=1}^{T} ((1 - a_c)^{\lfloor k/B \rfloor} a_c) \lambda_i h_i \tag{6}$$

In formula (6), g represents the social network, B represents the number of all sequences of a propagation item, each B represents a small batch of sets, T is the length of each propagation sequence, and the kth sequence will fall. In the (k/B + 1) small batch set, in practice, it can be assumed that the polynomial distribution of T nodes is $\lambda_1, \lambda_2, \ldots, \lambda_T$, and makes $p_i(\lambda_i) = 1$, and the value of λ is in the nerve The hidden layer in the network will be trained, it represents the same propagation item, each propagation.

The weight of the sequence, $\left((1 - a_c)^{\left[\frac{k}{B}\right]} a_c\right)$ in the formula is also a weight parameter, which represents the parameter corresponding to each propagation item. It represents the parameter corresponding to each propagation item, and the initial value we give it

$$a_c = p^{\left[\log_2(sz(g_c) + 1)\right]} \tag{7}$$

After obtaining h_g, the final impact range f_g can be obtained through the MLP multilayer perceptron, which represents the number of users affected in a period of time, that is, the range of influence predicted by the NT-GP model.

$$f_g = MLP(h_g) \tag{8}$$

4   Experiments

4.1   Datesets

For the experiments in this article, we used to download 2 sets of data sets online.

Digg: Diggers is a user-driven news site where a large number of users write, comment and submit news stories from the web, and users can also bookmark and forward them. We used the propagation of 500 propagation events for 4,916 users in 159 timestamps.

Flixster [11]: is a movie social networking site that allows users to share movie ratings and discuss new movies. We used 109,816 users to spread the 1000 events in 155 timestamps.

4.2   Evaluation Method

This chapter only uses the MSE mean square error to evaluate the effect of the model. The calculation method is the average of the squared sum of the measured value errors to calculate the overall error. The smaller the value, the more accurate it is. Assume that the actual value of the number of users affected by the fixed time period of the model output is Pi, and the predicted value is defined as follows:

$$MSE = \frac{1}{N} \sum_{i=1}^{n} (p_i - \overline{p_i})^2 \tag{9}$$

Since the value of Pi is an integer representing the number of people, in order to avoid the value of the error being too large, the overall value of the MSE is too large, or the calculation is difficult. We calculate the mean square error after taking the logarithm of Pi. $Pi = log_2(\Delta Si + 1)$. Since the value of Pi is an integer representing the number of people,

4.3 Baseline Methods

EmbeddingIC: An embedded version of the independent cascade model, taking into account the interaction between users, speculating the propagation probability, embedding the user in the hidden projection space, and finding the sender and receiver in the form of the Q function in the EM algorithm. User parameters to further build the model.

Deepcas: A method of predicting the extent of event propagation. By adopting the path of obtaining event diffusion. The path is converted to the expression vector of the path through the attention mechanism and the GRU network. A propagation expression vector of an event is obtained by integrating the propagation path expression vector of an event. Finally, the attention mechanism is used to predict the scope of the event.

NT-GP-R: Using a variant of NT-GP, use a random walk to extract the path. Convert randomly sampled paths to vectors and make predictions.

5 Experimental Result

In Table 2, the number of nodes relative to the digg data set is small. Using the three path selection methods of NT-GP, the path that can better express the node order can be obtained. The node vector obtained by skip-gram can better measure the relationship of other nodes appearing after the node. A better result can be obtained by learning the vector GRU obtained. In Table 3, using Flixster's dataset, each event has a larger impact than the digg's dataset Flixster's dataset. The Flixster dataset contains more users. When you choose the walk strategy, the time decay walk strategy will get a better structural expression. After the skip-gram, the vector expression will be better. The effect after the neural network is somewhat different from the effect of the digg data set, but for the other two models, the effect is significantly improved. According to the size of the experimental data, there are also differences in the parameter selection of word2vec. The count parameter indicates that the spatial representation vector of the node is calculated when the frequency of the node is calculated more than a certain frequency. We may not need all node vectors for the impact range prediction of events when performing calculations in the GRU. For example, many users with a large number of fans in Weibo are used as the source of information, so some nodes may be more important as information sources when information is transmitted. Through the results obtained by EmbeddingIC, the probability of one person participating in information dissemination, we obtain the impact range of the event through probability summation, and the obtained result is affected by the action log, the number of users is less, the action log is more, and the predicted result Will be more accurate.

Table 2. MSE values of different models under the digg data set

Count	0	200	400	600
NT-GP	1.131	1.124	1.125	1.117
NT-GP-R	1.399	1.400	1.401	1.400
NT-GP-W	1.885	1.886	1.878	1.894
Deepcas	1.452	1.448	1.568	1.623
EmdeddingIC	1.464			

Table 3. MSE values for different models under the Flixste data set

Count	0	200	400	600
NT-GP	0.864	0.839	0.856	0.845
NT-GP-R	1.408	1.400	1.400	1.415
NT-GP-W	1.384	1.395	1.405	1.465
Deepcas	1.452	1.523	1.586	1.598
EmdeddingIC	2.757			

6 Conclusion

An event prediction model based on deep learning, the purpose of which is to predict the range of propagation over a certain period of time. In general, this model can determine the number of users that a certain propagation item will affect in a fixed period of time. In the process of implementation, the topological structure of the social network, that is, the process of selecting the propagation path, is first learned. The user vector is then constructed by constructing the word vector, and the training mechanism of the GRU neural network is used to further train the user vector to obtain a vector that can represent the trend of event propagation. After the vector is integrated, the MLP mechanism can successfully predict the spread of event.

Acknowledgment. This work was supported by the National Natural Science Foundation of China (No. 61602159), the Natural Science Foundation of Heilongjiang Province (No. F201430), the Innovation Talents Project of Science and Technology Bureau of Harbin (No. 2017RAQXJ094, No. 2017RAQXJ131), and the fundamental research funds of universities in Heilongjiang Province, special fund of Heilongjiang University (No. HDJCCX-201608).

References

1. Chen, W., Wang, Y., Yang, S.: Efficient influence maximization in social networks. In: ACM SIGKDD International Conference on Knowledge Discovery and Data Mining, pp. 199–208. ACM (2009)
2. Liu, L., Li, X., Cheung, W.K., et al.: A structural representation learning for multi-relational networks. In: Twenty-Sixth International Joint Conference on Artificial Intelligence, pp. 4047–4053 (2017)

3. Ribeiro, L.F.R., Saverese, P.H.P., Figueiredo, D.R.: struc2vec: learning node representations from structural identity, pp. 385–394 (2017)
4. Bourigault, S., Lamprier, S., Gallinari, P.: Representation learning for information diffusion through social networks: an embedded cascade model. In: ACM International Conference on Web Search and Data Mining, pp. 573–582. ACM (2016)
5. Jenders, M., Kasneci, G., Naumann, F.: Analyzing and predicting viral tweets, pp. 657–664. ACM (2013)
6. Weng, L., Menczer, F., Ahn, Y.Y.: Predicting successful memes using network and community structure. Eprint Arxiv (2014)
7. Sitnikov, K.: Learning diffusion probability based on node attributes in social networks. In: Proceedings of the Foundations of Intelligent Systems - International Symposium, ISMIS 2011, Warsaw, Poland, 28–30 June 2011, pp. 153–162. DBLP (2011)
8. Zhang, J., Liu, B., Tang, J., et al.: Social influence locality for modeling retweeting behaviors. In: International Joint Conference on Artificial Intelligence, pp. 2761–2767. AAAI Press (2013)
9. Mikolov, T., Corrado, G., Chen, K., et al.: Efficient estimation of word representations in vector space. In: International Conference on Learning Representations, pp. 1–12 (2013)
10. Mikolov, T., Sutskever, I., Chen, K., et al.: Distributed representations of words and phrases and their compositionality. In: International Conference on Neural Information Processing Systems, pp. 3111–3119. Curran Associates Inc. (2013)
11. Li, D., Luo, Z., Ding, Y., et al.: User-level microblogging recommendation incorporating social influence. J. Assoc. Inf. Sci. Technol. 68(3), 553–568 (2017)
12. Xie, X., Li, Y., Zhang, Z., et al.: A joint link prediction method for social network. IFIP Adv. Inf. Commun. Technol. 503, 56–64 (2015)
13. Haghani, S., Keyvanpour, M.R.: A systemic analysis of link prediction in social network. Artif. Intell. Rev. 3, 1–35 (2017)
14. Li, C., Ma, J., Guo, X., et al.: DeepCas: an end-to-end predictor of information cascades. In: International Conference on World Wide Web. International World Wide Web Conferences Steering Committee, pp. 577–586 (2017)

Experimental Research on Internet Ecosystem and AS Hierarchy

Lv Ting, Donghong Qin[(✉)], and Lina Ge

School of Information Science and Engineering,
Guangxi University for Nationalities, Nanning, China
donghong_qin@163.com

Abstract. The network architecture has undergone great changes. For example, the network topology of autonomous system level tends to be flattened. In this paper, actual network topology map is constructed through actual network routing data, to study and analyze the network structure changes of the Internet recently and the dependence of the Internet on the core AS. In order to achieve the above goals, the experimental analysis framework and some related algorithms were designed. Experimental results show that the Internet architectures gradually become flattening, and the dependence of the Internet on core AS is gradually reduced but the importance of the core ASes is still indispensable. Most traffic between local ASes can be transited through the local infrastructure (such as IXP) rather than the upper ISPs. The observations may guide the design of inter-domain routing protocols and Internet exchange architectures to achieve higher performance and reliability.

Keywords: Internet ecosystem · Experimental framework ·
Autonomous system · AS hierarchy

1 Introduction

With the development of the Internet, the characteristics of a multi-layered structure have been formed. The core of the Internet consists of 10 to 20 network providers' networks. The core autonomous systems (ASes) establish peer-to-peer interconnections and provide network services to the lower regional autonomous systems. Other network organizations form multiple regional secondary networks by connecting to these core ASes. These secondary network providers provide network services to smaller three-tier network providers, and so on, and then provide services downwards layer by layer until the last stub AS, which eventually forms a hierarchical Internet.

In the traditional autonomous system interconnection structure, the overall structure of the Internet is hierarchical. Hierarchy is a basic feature of the existence of the Internet. In the early stage, based on the intuitive understanding of "hierarchy", the network was divided into Stub domain or Transit domain, and a static topology model of Internet with strict hierarchy structure was proposed, transit-stub [1]. In [4], hierarchical modularity is proposed after studying hierarchical organization in complex networks. In the study of network core structure, Gaertler and Patrignani [7] proposed the kernel and k-core decomposition to quantify the centrality of nodes. In [8], the

© Springer Nature Singapore Pte Ltd. 2019
X. Cheng et al. (Eds.): ICPCSEE 2019, CCIS 1058, pp. 253–263, 2019.
https://doi.org/10.1007/978-981-15-0118-0_20

relationship between node degree and core is analyzed, and a hierarchical model is established.

With the rapid development of the Internet, the level of the network is becoming more and more blurred, and the Internet is flattening gradually. In [2], the flattening phenomenon of the Internet is analyzed through graph theory research. Because some AS in the Internet are transmitted through IXP (Internet exchange point) to establish a peer-to-peer connection, the hierarchical structure of the network is gradually blurred. The influence of IXP on network topology is analyzed in [9, 10].

Inter-domain traffic was measured in [14, 15], and the results showed that most inter-domain communication flows directly between large providers and data centers/CDNs. Commercial interests are an important driving force for the development of autonomous systems. With the addition of large content providers such AS Google, Amazon and Microsoft, a large number of AS directly establish connections with these large content providers, the cross-domain traffic of the Internet gradually becomes an integration trend, and the structure of the Internet gradually flattens.

In this paper we aim to study and analyze changes in the structure of the Internet, and the extent to which the Internet is dependent on core ASes. According to the characteristics of the Internet, we designed a network layering algorithm to conduct hierarchical research on the constructed AS topology graph. At the same time, in order to more accurately analyze the network structure changes, we use the relevant parameter indicators in the complex network analysis method to help the experimental analysis.

The rest of the structure in this paper is as follows: In the second section, we introduce the Internet analysis model and give relevant experimental indicators. In the third section, the data set is introduced. In the fourth section, the experimental framework and some important algorithms are introduced. In the fifth section, we use the experimental design method to analyze the actual network data collected. In the sixth section, we summarize this article.

2   Internet Analysis Model

2.1   Internet Structure Characteristics and Its Measurement

Clustering Coefficient
The clustering coefficient is used to measure the parameters of consistency in the graph, and it mainly investigates the community characteristics of one node and its neighbor in the network. In the experiment, we use the global average clustering coefficient to understand the aggregation degree of AS network topology. The global average clustering coefficient is defined as the average value of local clustering coefficient [3].

Centralization Coefficient
Typical centralization indexes include degree centralization, closeness centralization, and betweenness centrality. These parameters are often used together to form multiple centrality [6].

Degree centrality is used to study the basic parameters of scale-free network topology, to describe the direct influence of a node on the surrounding network nodes in a static network, and at the same time, the degree value of a node reflects the ability of the node to establish a direct connection with the surrounding nodes.

Closeness centrality. This metric is used to describe how easy it is for a node in a network to pass through its host node to other nodes. The calculation of closeness centrality is shown in (1), where N represents the number of nodes in the network (the same below), and d_{iy} represents the shortest path from node i to node y. The compactness centrality reflects the ability of nodes to exert influence on other nodes through the network, so the compactness centrality can better reflect the overall structure of the network.

$$C_C(i) = (N - 1)\left(\sum_{y=1}^{N} d_{iy}\right)^{-1} \tag{1}$$

Betweenness centrality. This indicator reflects the influence of nodes in the network on the flow of information, in other word, a measure of the shortest path formed by the node. In the network, if the value of the betweenness centrality of a node is larger, it means that the shortest path formed by any two nodes in the network passes through the number of nodes. Therefore, it is possible to determine the nodes with heavy information load in the network through the median centrality. The specific calculation method is as shown in the (2), where g_{mn} represents the shortest path number between the node m and the node n, and $g_{mn}(i)$ represents the shortest path number between the node m and the node n passing through the node i.

$$C_B(i) = \left(\sum_{m<n} g_{mn}(i)/g_{mn}\right)/[(N - 1)(N - 2)/2] \tag{2}$$

2.2   Connectivity and Connected Subgraph Analysis

The AS network topology graph is a large one connected by AS nodes and paths. We define the AS graph as G(N, E), where N represents all nodes in the graph, and E represents the edges of all paths in the graph. After the AS map is layered, we define the graph formed by each layer as $G_i(n, e)(i \in N^*)$, where n represents all nodes in the G_i, and e represents the edges of all paths in the G_i. In the process of layering, the connected edges will be disconnected between each layer. We define these edges as set U, where $U = \{U_1, \cdots, U_n\}$, and U_n represents the disconnected edges between the n layer and the n + 1 layer. In each layer, graph G_i is not necessarily a connected graph, but there may be many connected subgraphs, so $G_i(n, e) = \sum_{x=1}^{n} g_x(n_x, e_x)$.

According to the above definition, after layering, the smaller the number of $g_x(n_x, e_x)$ in $G_i(n, e)$, the stronger the ability to form a complete connected graph in this layer and the smaller the dependence on the upper layer. At the same time, the smaller value of U, the lower degree of contact between the layer and the upper layer.

3 Data Sets

The Routeview is a project created by the center for advanced network technology at the University of Oregon that allows Internet users to view global BGP routing information from other locations around the Internet [11].

We build the AS topology by using the data on the Routeview. The Routeview project provides a snapshot of the routing table. These route table snapshots are data from the actual network environment, through which we conducted experimental analysis.

4 Experimental Design and Analysis Framework

4.1 Experimental Analysis Framework

To analyze the change of Internet structure, the AS graph needs to be constructed first. The AS graph need use the data collected by the actual network BGP dump. We build the AS graph using the AS path [13]. In the construction of AS graph with BGP routing table, we need to preprocess the data of routing table and extract the AS path. The experimental analysis process is shown in Fig. 1.

By constructing the AS topology graph, we calculate and analyze the corresponding experimental index coefficients of the AS nodes in the graph. Then, we performed multiple layered experiments on the AS topology by using a network layering algorithm. The results of the stratification can be used to obtain multiple subgraphs. We observe the subgraphs and analyze the experimental parameters of the subgraphs.

Finally, the structure of the AS network topology graph is studied and analyzed by the recorded experimental index coefficients. In order to observe the impact of changes in Internet structure on BGP routing paths, we perform statistical analysis of the length of AS paths.

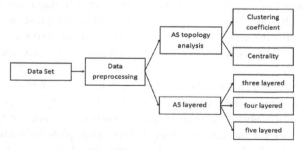

Fig. 1. Experimental analysis framework

4.2 Design and Implementation of Graph Layering Algorithm

The hierarchical structure of the Internet is more and more fuzzy, and the network structure tends to be flat. In order to help experimental research, we designed a network layering algorithm to layer the AS topology graph we built. The topology diagram of AS network constructed by actual network data is a huge connected graph G, which is segmented by hierarchical algorithm. The connected edges established between each layer are disconnected, and graph G is segmented into several independent sub-graphs according to the hierarchy (see in Fig. 2).

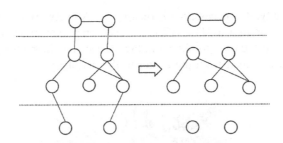

Fig. 2. Layering example

The layering algorithm consists of three main steps. In the first step, we need to calculate the degree of each node in the AS topology graph, and distinguish the AS size by the degree of the AS node. Because the greater the degree of the AS node, the more connections it establishes with, the higher the level it represents, and the AS tends to choose the AS that is equivalent to its own degree to establish a peer-to-peer connection [5]. We believe that when two AS have the same degree, they are at the same network level. AS is stratified by the degree of the AS node. Since the AS in the core of the Internet is known [12], these AS are located at the highest level, so we only need to stratify the remaining AS.

In the second step, according to the divided AS level, the connected edges between each level are disconnected, and graph G is divided into sub-graphs of each level. To facilitate the construction of sub-graphs at each level, we delete all AS nodes and edges that are not in the same layer.

In the third step, according to the subgraphs of each level obtained by the segmentation, the statistical calculation of the relevant parameter indicators is performed. Because the graph G is a huge connected graph, but the subgraphs obtained by the segmentation do not necessarily have connectivity, the subgraph needs to be judged first. If the subgraph is connected, the statistical calculation is directly performed. On the contrary, separate statistical calculations are needed for the subgraphs.

5 Experimental Results and Analysis

We selected data from October 31 of each year from 2014, 2015, 2016, 2017, and 2018 as experimental data. We preprocessed the experiment data, because the data volume of the routing table data is relatively large. In order to provide the execution efficiency of the algorithm, we extract the AS path data from the routing table data as the initial experimental data.

5.1 Building Internet AS Topology Graph

Before layering study, it is necessary to construct an AS network topology graph. This graph is generated through reading AS paths. The graph constructed through the AS paths is a huge connected one. As shown in Fig. 3, each red dot represents an AS. From the figure, the AS is spread from the center to the periphery.

Fig. 3. The AS graph constructed by Oct. 31 2014 data (Color figure online)

In order to facilitate the layering research, we conducted basic data statistics for each set of experimental data. As shown in Table 1, the links between AS and the average degree of AS are in direct proportion to the number of AS, the path of AS is on the whole in an increasing trend, the global average clustering coefficient is basically stable, and the average path degree of AS is relatively short. From the above analysis, we can see that the Internet as a whole is becoming more and more connected.

Table 1. AS-RELATED characteristics

Year	2014	2015	2016	2017	2018
Number of AS nodes	48851	52391	55828	59567	63026
Number of AS paths	18863822	22322803	28781757	33551661	30390186
AS edge number	119441	132660	153042	163252	166202
Global average clustering coefficient	0.20	0.21	0.21	0.21	0.21
Average degree	4.89	5.06	5.48	5.48	5.27
Average path length	4.92	4.34	4.41	4.43	4.47

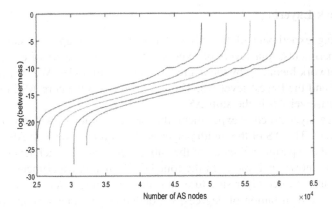

Fig. 4. Betweenness centrality

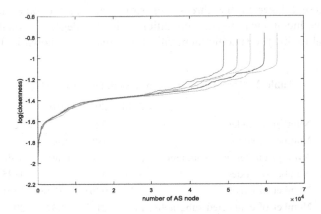

Fig. 5. Closeness distribution

Figure 4 shows the distribution diagram of the betweenness centrality of nodes for five groups of data. In Fig. 4, the abscissa represents the number of AS nodes, and the ordinate represents the logarithm of the betweenness value of number of AS nodes. It can be seen that the distribution of betweenness centrality does not change much each year, and only a few nodes have relatively large betweenness centrality values. This shows that the overall transmission capacity of the Internet has not become complicated with the increase of AS. Figure 5 shows the distribution diagram of the closeness centrality of the five groups of data. In Fig. 5, the abscissa represents the number of AS nodes, and the ordinate represents the logarithm the closeness value of number of AS nodes. Although the nodes are growing every year, the closeness centrality of most AS still maintains a small change, and some AS have a large change, but the trend of function is roughly the same. On the whole, the connection ability between AS is relatively stable compared with the Internet.

5.2   Network Layering

In the layering experiment of the AS network topology graph, we conducted three layers of experiments, four layers of experiments, and five layers of experiments. We divide the network hierarchy by the degree distribution of nodes. We first identified the top-level AS and the lowest-level AS. The top-level AS is the core AS in the network, and the lowest-level AS is the stub AS.

In the three-layer structure experiment, the topology graph of AS network is divided into three layers. The AS of the middle layer is all AS minus the core AS and stub AS. We calculate the experimental index of the sub-graph formed at each level. The specific stratification is shown in Table 2. In the four-layer structure experiment, we divide the second layer in the three-layer structure, and the remaining layers remain unchanged. According to the distribution of degrees, we divide the original second layer into two (based on the degree equal to 5): nodes with the degree less than 5 belong to the third layer, and nodes with the degree greater than 5 belong to the second layer. The specific stratification is shown in Table 3. In the five-layer structure experiment, we further divide the second layer in the three-layer structure, and the other levels remain unchanged. The hierarchical division condition is that the degree is equal to 5 and the degree is equal to 50. The specific hierarchical information is shown in Table 4.

Table 2. Three-layer structure experiment result table

Year		2014	2015	2016	2017	2018
First layer	Number of nodes	22	22	22	22	22
	Number of edges	199	195	189	188	185
	Average clustering coefficient	0.87	0.88	0.89	0.89	0.85
Second layer	Number of nodes	31744	34068	36509	38855	40862
	Number of edges	82856	93613	112188	120140	121368
	Number of connected subgraphs	2133	2108	2037	2073	1995
Third layer	Number of nodes	17085	18301	19297	20690	22142

Table 3. Four-layer structure experiment result table

Year		2014	2015	2016	2017	2018
First layer	Number of nodes	22	22	22	22	22
	Number of edges	199	195	189	188	185
	Average clustering coefficient	0.87	0.88	0.89	0.89	0.85
Second layer	Number of nodes	8606	9473	10704	11434	11969
	Number of edges	37448	44792	60133	64492	62214
	Number of connected subgraphs	160	166	161	178	176
Third layer	Number of nodes	23138	24595	25805	27421	28893
	Number of edges	8747	9288	9787	10396	11102
	Number of connected subgraphs	21718	23152	24462	25881	27329
Fourth floor	Number of nodes	17085	18301	19297	20690	22142

Table 4. Five-layer structure experiment result table

Year		2014	2015	2016	2017	2018
First layer	Number of nodes	22	22	22	22	22
	Number of edges	199	195	189	188	185
	Average clustering coefficient	0.87	0.88	0.89	0.89	0.85
Second layer	Number of nodes	334	375		471	477
	Number of edges	2730	3813	6232	6616	5575
	Number of connected subgraphs	12	13	14	15	18
Third layer	Number of nodes	8272	9098	10257	10963	11492
	Number of edges	13117	14138	15810	16569	17512
	Number of connected subgraphs	2634	3033	3673	3837	3998
Fourth floor	Number of nodes	23138	24595	25805	27421	28893
	Number of edges	8747	9288	9787	10396	11102
	Number of connected subgraphs	21718	23152	24462	25881	27329
Fifth floor	Number of nodes	17085	18301	19297	20690	22142

5.3 Analysis of Experimental Results

It can be seen from the experimental data in Sect. 5.B that the average clustering coefficient of the top-level AS is higher than 0.8, and the high-level AS maintains a close relationship. In the five groups of data, the number of links between the top layer AS and the bottom layer is 19301, 20551, 21368, 22234 and 22507, accounting for 16.18%, 15.49%, 13.96%, 13.61% and 13.54% of the total number of links respectively. Through calculation, it is found that although the number of links between the top AS and the bottom AS is increasing year by year, it is still decreasing year by year for the Internet AS a whole.

We use a layer algorithm to get a three-tier structure. The second layer has many sub-graphs. Most ASes can form a large connected sub-graph. The number of nodes contained in the maximal connected sub-graph of each group of data is 29553, 31902, 34424, 36727, 38823. The proportion of the largest sub-graph nodes in the number of the second layer nodes is 93.1%, 93.64%, 94.29%, 94.52% and 95.01%.

At the same time, the number of edges formed by the largest sub-graph is 82000, 92714, 111304, 119220, and 120502 respectively, accounting for more than 90% of all links in the second layer structure. In the maximum sub-graph, the global average clustering coefficients are 0.20, 0.22, 0.24, 0.25, 0.24, respectively, while the global average clustering coefficients of the second layer are 0.19, 0.21, 0.23, 0.24, 0.23, respectively. This shows that most of the ASes can form a large scale connected graph in the case of the top layer ASes. In the independent sub-graph, ASes can complete the corresponding transport service.

From the above analysis, we can conclude that the dependence of the Internet on the top AS is gradually decreasing, but the top AS is also indispensable to the Internet. Although some transit services between ASes do not require the assistance of top-level AS, the role of top-level AS in increasingly complex networks is enormous.

In the hierarchical experiment, we divide the network into three levels. Through experiments, we find that the more levels the network divides, the more sub-graphs the network forms. In the three-layer structure, the number of sub-graphs in the second layer is 2133, 2108, 2037, 2073, 1995. In the four-layer structure, the total number of sub-graphs is 21878, 23318, 24623, 26059 and 27505. In the five-layer structure, the total number of sub-graphs is 24364, 26198, 28149, 29733, 31345. As the level is divided, the number of sub-graphs formed between ASes is more, and most of the sub-graphs only consist of a few ASes, but part of ASes in each layer can also form a larger sub-graph. At the same time, we have made statistics on the edge loss of hierarchical partitioning between non-stub and non-top-level AS, shown in Table 5. After AS layering, the more edges are lost between AS, indicating that the connection between ASes is less and less.

Each layering does not form a sub-graph with relatively uniform size and scale, and most sub-graphs only consist of a few ASes. For example, IXP can help transfer between AS. The hierarchy of networks is becoming blurred. Some transmissions between AS do not need to go through the upper AS. This proves that the overall network structure flattening trend.

Table 5. Loss link statistics table

Level change	Level	2014	2015	2016	2017	2018
Three-layer structure to four-layer structure	Amount	36621	39533	42268	45252	48052
	Percentage	44.20%	42.23%	37.67%	37.67%	39.59%
Three-layer structure to five-layer structure	Amount	58562	66374	80359	86559	87179
	Percentage	70.68%	70.90%	71.63%	72.05%	71.83%

6   Conclusion

In this paper, AS network topology, Internet structure changes and the importance of core ASes in the network are studied and analyzed. At the same time, we help to analyze and capture the characteristics of AS network topology by using the method of complex network. In addition, a simple layering algorithm is designed to study and analyze the changes of the Internet hierarchy. Our experimental results show that the hierarchical structure of the Internet is gradually blurred, and most traffic between local ASes can be transited through the local infrastructure (such as IXP) rather than the upper ISPs. The overall structure of the Internet is gradually flattening, and the Internet is becoming less dependent on core AS. However, core AS still plays an indispensable role in the Internet.

Acknowledgments. This work was supported by the National Natural Science Foundation of China (No. 61462009, 61862007), the Natural Science Foundation of Guangxi (No. 2018GXNSFAA281269, No. 2018GXNSFAA138147), the Innovation Project of Guangxi Graduate Education (No. Gxun-chxjg201707).

References

1. Zegura, E.W., Calvert, K.L., Donahoo, M.J.: A quantitative comparison of graph-based models for Internet topology. IEEE/ACM Trans. Netw. **5**(6), 770–783 (1997)
2. Masoud, M.Z., Hei, X., Cheng, W.: A graph-theoretic study of the flattening internet AS topology (2013)
3. Kemper, A.: Valuation of Network Effects in Software Markets: A Complex Networks Approach. Springer, Heidelberg (2009). https://doi.org/10.1007/978-3-7908-2367-7
4. Ravasz, E., Barabási, A.L.: Hierarchical organization in complex networks. Phys. Rev. E **67** (2), 12–20 (2003)
5. Gao, L.: On inferring autonomous system relationships in the Internet. IEEE/ACM Trans. Netw. **9**(6), 733–745 (2001)
6. Porta, S., Crucitti, P., Latora, V.: The network analysis of urban streets: a dual approach. Phys. A **369**(2), 853–866 (2006)
7. Gaertler, M., Patrignani, M.: Dynamic analysis of the autonomous system graph. In: Proceedings of IPS 2004, Budapest, Hungary (2004)
8. Zhang, J., Zhao, H., Zhou, Y.: Relationship between degree and core number of internet nodes at router level. J. Northeast. Univ. Nat. Sci. **29**(5), 653–656 (2008)
9. Ahmad, M.Z., Guha, R.: [ACM Press the ACM CoNEXT Student Workshop - Philadelphia, Pennsylvania (2010.11.30–2010.11.30)]. In: Proceedings of the ACM CoNEXT Student Workshop on - CoNEXT\"10 Student Workshop - Understanding the Impact of Internet Exchange Points on Internet Topology and Routing Performance (2010)
10. Ahmad, M.Z., Guha, R.: Impact of Internet exchange points on Internet topology evolution. In: Local Computer Networks. IEEE (2010)
11. http://www.routeviews.org/routeviews/
12. https://en.wikipedia.org/wiki/Tier_1_network
13. Chang, H., Jamin, S., Wang, W.: Live streaming performance of the Zattoo network. In: ACM IMC, pp. 417–429 (2009)
14. Qin, D., Yang, J., Wang, H.: Experimental study on diversity and novelty of interdomain paths. Chin. J. Electron. **22**, 160–166 (2013)
15. Labovitz, C., Iekel-Johnson, S., et al.: Internet inter-domain traffic. SIGCOMM Comput. Commun. Rev. **40**(4), 75–86 (2010)

Speed-Grading Mobile Charging Policy in Large-Scale Wireless Rechargeable Sensor Networks

Xianhao Shen[✉], Hangyu Xu, and Kangyong Liu

Guangxi Key Laboratory of Embedded Technology and Intelligent System,
Guilin University of Technology, Guilin 541006, China
25337698@qq.com

Abstract. As the technological breakthrough is made in wireless charging, the wireless rechargeable sensor networks (WRSNs) are finally proposed. In order to reduce the charging completion time, most existing works use the "mobile-then-charge" model—the Wireless charging vehicles (WCV) moves to the charging spot first and then charges nodes nearby. These works often aim to reduce the node's movement delay or charging delay. However, the charging opportunities during the movement are overlooked in this model because WCV can charge nodes when it goes from one spot to the next. In order to use the charging opportunities, a speed grading method is proposed under the circumstance of variable WCV speed, which transformed the problem of final charging delay into a traveling salesman problem with speed grading. The problem was further solved by linear programming method. The simulation experiments show that, compared with the existing charging methods, the proposed method has a significant improvement in charging delay.

Keywords: Wireless rechargeable sensor networks · Wireless charging · Speed grading · Charging delay

1 Introduction

The energy issue is one of the key impediment in wireless sensor large-scale applications and deployment today, most existing sensor networks are powered by batteries with limited capacity [1–3], which results in their limited life cycle and the inability to complete data collection and communication tasks successfully. With the breakthrough of wireless charging technology by Zeng et al. [4, 5], it provides a promising alternative to the traditional power supply for sensor nodes. We expect that new WRSNs will be widely used in our daily lives in the near future, such as smart furniture, internet of things, artificial intelligence.

This research work was supported by Natural Science Foundation under Grant NO. 61662018; Natural Science Foundation of Guangxi province under Grant NO 2015GXNSFBA139254 and Natural Science Foundation of Guangxi province under Grant NO 2018GXNSFAA294061.

X. Cheng et al. (Eds.): ICPCSEE 2019, CCIS 1058, pp. 264–280, 2019.
https://doi.org/10.1007/978-981-15-0118-0_21

Recently, in order to improve the charging efficiency of WRSNs, most works have focused on the internal design of microelectronics [6–8], while improving the underlying microelectronics technology is important for wireless transmission systems, in practice, we have found that the charging time of a single wireless rechargeable node is not negligible, and it plays an important role in the performance of the entire wireless charging system. For typical wireless rechargeable sensor nodes, such as the Wireless Identification and Sensing Platform (WISP) developed by Intel [9], the sensor node requires that the wireless charging energy must exceed a certain threshold for proper operation of various sensing, computing, and communication component tasks. Due to the limited wireless charging speed, the charging process is usually time consuming. For example, it requires about 155 s to fully charge a WISP node [10] in 10 m distance, which greatly affects the performance of WRSNs, especially in large-scale sensor networks. The mentioned problem of low charging efficiency mainly emphasizes the design improvement inside the microelectronics and neglects the effective optimization of the delay in the charging process. In fact, the charging completion time can be shortened by reasonable and effective optimization in the charging process. This plays an important role in the overall performance of WRSNs. In recent years, the optimization on charging completion time for WRSNs has been raised with more and more attentions for us to research on it. For example, Fu proposed to use the method of planning the optimal path to minimize the charging delay and solve the delay by linear programming method [11]. Most of the previous research work is based on the "move-then-charge" model: the WCV moves to each charging spot and then charges the sensor nodes near the spot [12], which goes until all networks nodes are fully charged. However, the charging opportunities during the movement are overlooked in this model, because the WCV can charge nodes when it goes from one spot to the next. This kind of charging scheme can better reduce the charging completion time. In order to further explore the charging opportunities during the movement, the speed of WCV becomes an important factor. There is a paradox about the speed of WCV. On the one hand, in order to charge more nodes in the network, in principle, the speed is slower, because the nodes in the network can get more energy, but on the other hand, in order to reduce the moving time, the speed is required to be as fast as possible. Therefore, the speed of WCV has a great influence on charging delay.

This paper discusses the problem about the WCV speed grading in WRSNs, and the goal is to reduce the charging completion time. In order to achieve the goal, we propose a speed grading scheme based on joint consideration of movement and charging delay. First, the nodes in the sensor network are clustered, and the nodes are divided into clusters of specific values (the values set according to the situation). After the combination, the charging spots are obtained, and then the distance of two neighbor spots in the running path are calculated. The WCV traverses all the spots and charges the rechargeable nodes around the spot, this problem can be transformed into a Traveling Salesman Problem with speed grading. We use the K-nearest Neighbor algorithm to solve this TSP problem and get the proper charging path. The nodes are charged at different speeds in different operating intervals, which ultimately reduces the total charging delay, improves the charging efficiency, and prolongs the life cycle of the sensor network.

2   Network Model

Wireless Identification and Sensing Platform (WISP): Intel's WISP platform is the most representative platform in WRSNs. It has traditional RFID reading capabilities and also supports information acquisition and computing capabilities. The WISP nodes inherit the functionality of traditional RFID tags and also support perceptual and computational functions. When approaching an RFID reader, the WISP node can collect energy from the reader signal. The energy obtained by charging this node is stored in a capacitor and can be used for future data sensing, recording, calculation and transmission [13].

Energy Charging Model: In this paper, we propose a wireless charging model as an energy charging model, which is expressed by the following formula:

$$P_r = \frac{G_s G_r \eta}{L_p} \left(\frac{\lambda}{4\pi(d+\beta)} \right)^2 P_0$$

Where d is the distance between the WISP reader and the sensor node, P_0 is the transmit power of WISP, G_s is the transmit antenna power, G_r is the accept antenna power, L_p is the polarization loss, λ is the charging wavelength, η is the rectifier efficiency, β is the parameter to adjust the Friis' free space equation for short distance transmission. Where d is the only variable in the formula, and the rest are constants based on the environment and device settings. The above model is based on the Friis' free space equation and it is proved by experiments that this is a great approximation of the charging energy [14].

In order to simplify the description of the design part that follows, we simplify the formula 1, and the simplified formula is as follows:

$$P_r = \frac{\alpha}{(d+\beta)^2}$$

Where d is the distance between the WCV and the sensor node, α represents the constants affected by the environment, Including P_0, G_s, G_r, L_p, λ, η in formula 1.

The Network Charging Model: In existing research, some special charging spots are selected for charging network nodes nearby. The spots can be obtained by clustering algorithms [15]. The WCV first moves to the charging spots and then charges the nodes in the surrounding charging range. When all of rechargeable nodes around the spot are charged completely, the WCV will move along the planned path to the next charging spot and charge the neighbor nodes, and this process will continue till all nodes in the network are fully charged.

Charging Radius: Since the WCV relies on its own transmitting antenna to charge the node wirelessly, the charging radius of WCV is limited. In this paper, the charging radius of WCV is assumed as 15 m.

It is assumed that N fixed rechargeable sensor nodes are deployed in the sensor network. The position of a single node i in the network can be obtained by the positioning technique [16]. The node coordinates are expressed as $\left(W_x^i, W_y^i\right)$. At the same time, it is assumed that the wireless charger can move at a certain speed by WCV in the deployment area. When the sensor node is in the charging radius of the charger, it charges the node wirelessly by using the internal storage energy carried by itself, when the energy obtained by the node is above a certain threshold δ the deployed sensor nodes can complete works such as signal sensing, communication, and calculation. Therefore, in order to minimize the total charge cycle of all nodes in the network, we need to find an optimal charging strategy to minimize the charging delay. Especially, in this paper, the main goal is to find a suitable charging spot and the staying time at the spot by using the different ranges of the speed to achieve the purpose of minimizing the charging delay.

In order to express this charging problem explicitly by formula, it is assumed that the charger stays at several different spots, for example, the coordinate at the spot j near the node i is $\left(R_x^j, R_y^j\right)$, t_j is the staying time, the distance between node i and spot j is $d_{ij} = \sqrt{(W_x^i, R_x^j)^2 + (W_y^i, R_y^j)^2}$, according to formula (2), the corresponding charging power is $P_{ij} = \frac{\alpha}{(d_{ij} + \beta)^2}$, the charging energy E obtained by the node i at the spot j is $P_{ij}t_j$. Given the interest area of N nodes, the node charging energy threshold is δ, the charger can stay at any spots in this area, we can use mathematical formulas to represent this minimization of the charging node delay problem. It is computed as follows:

$$\min T = \sum \frac{p_i}{s_i} + \sum t_i$$

$$s.t. \begin{cases} \forall n_i \in N, \sum (\frac{p_i}{s_i}.e_i + t_j.e_j \gg \delta) \\ \forall i, \ 0 \le s_i \le s_T \\ \forall i, \ t_i \ge 0 \end{cases}$$

$$P_{ij} = \frac{\alpha}{(d_{ij} + \beta)^2}, \ (i \in N, j \in (1.2.....\infty))$$

$$d_{ij} = \sqrt{(W_x^i, R_x^j)^2 + (W_y^i, R_y^j)^2}$$

The goal is to find the minimum of the formula (3) under the constraint conditions, and using the linear programming method to solve the final charge completion time.

3 WRSNs Charging Tour Planning Optimization Algorithm Under Speed Grading

3.1 Determining the Charging Spot

The charging power is first discretized into several different charging levels. After this processing, we reduce the search space of the Linear Programming (LP) planning problem in formula (3), and the potential spot position changes from infinite to finite. The optimal charging delay may consist of a large number of charging spots, but in the actual charging scenario, it is not practical for the charger to move in a large number of spots, so the number of spots should be reduced relative to the number of nodes deployed. We introduce a charging spot combination design, which can reduce the number of spots effectively within a certain threshold.

In order to combine the charging spot position and reduce the number of spots, we use the famous K-means clustering algorithm, Lloyd's algorithm [17], combining chargers to multiple clusters based on geographic location. Lloyd's clustering algorithm divides n observation points into k clusters, each of which belongs to the nearest cluster center. For the charger spot combination problem, the first n observation points are the total number of charging spots obtained by the charging power discrete processing. We set the k value according to the following rules. It starts with a smaller value and increments by one in each round of recalculation until the average distance between the clusters is less than a given threshold. After combining the charging spots, each sensor node is assigned to the nearest cluster center node, and the corresponding charging level is available through formula (5).

3.2   Generating Movement Path

The path selection problem can be transformed into a Traveling Salesman Problem with speed grading. We propose a greedy algorithm to solve this problem. The pseudocode of the algorithm will be given below. First, we add all the minimum closed spot sets to the path P, and then we add the nodes other than the path P in turn according to the algorithm.

In each round of node selection, we choose to traverse a node V_i that can achieve the minimizing delay, then add the node to path P, and update the path to get $V_i + P$. For no spot, we calculate its expected delay at each insertion spot position P_i. By listing all possible nodes and insertion spots, we can pick the appropriate spot and the corresponding insertion position to achieve a minimum total charge completion time. This process continues until all nodes in the network are added to path P. The initial charging location can be selected as any charging spot and the total charging delay remains the same.

Obviously, calculating the total charging completion time involves the speed of WCV. We will describe the WCV speed grading in detail in the next section.

3.3   Speed Grading Process

For two neighbor spots, we refer to the line segment formed by their connection as edge, which is divided into several segments objects according to different charging levels, and the charger takes different speeds on different segments of the edge. Before introducing the speed grading scheme, we first introduce a theorem related to the charging energy of the sensor.

Theorem 1 : When the speed of WCV for different strategies in a segment changes, if the movement time at a given charging level is the same on each sub-segment, then the charging completion time is the same in this segment.

Using theorem 1, we can discretize each edge according to the charging level into different segments, they adopt different speeds s_i and corresponding charging levels. The speed selection factor is related to the amount of energy received by the node being charged. When the energy of the sensor node is low, the energy received unit time during charging is high, and the corresponding charger speed should be slower, so that the node receives more energy during the movement. Correspondingly, when the energy of the node is sufficient, and the charging efficiency unit time is low, the charger should move at a faster speed to the spot to charge the neighbor nodes, because there are not any charging opportunities during the movement, and swift movement will be conducive to reduce exercise time. The definition of the total charge S in the unit time and the correlation with the speed is given below.

The total charge total S in the unit time is defined as: $S = \sum_{i=1}^{n} P_i$ (n is the rechargeable nodes around the spot). The relationship between S and v is as follows

$$v = \frac{K}{S}(K = 1)$$

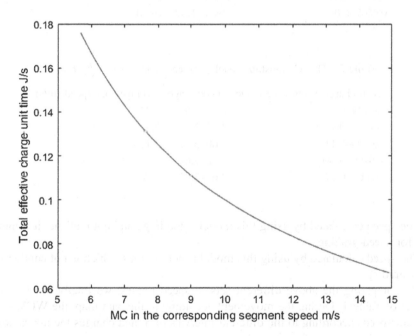

Fig. 1. The relationship between S and v

According to the obtained S-v curve, the speed is divided into the following five gradings (Fig. 1).

Table 1. The relationship model between S and v is $v = \frac{K}{S}(K = 1)$

Total charge S unit time range	Speed range	Average split speed (m/s)
0.1429–0.2	5–7	6
0.1111–0.1429	7–9	8
0.0909–0.1111	9–11	10
0.0769–0.0909	11–13	12
0.0667–0.0769	13–15	14

Table 2. The relationship model between S and v is $v = \frac{K}{S^2}(K = 1)$

Total charge S unit time range	Speed range	Average split speed (m/s)
0.1429–0.2	25–49.0	37.0
0.1111–0.1429	49.0–81	65.0
0.0909–0.1111	81–121	101
0.0769–0.0909	121.0–169.1	145.1
0.0667–0.0769	169.1–224.8	197.0

Table 3. The relationship model between S and v is $v = \frac{K}{\sqrt{S}}(K = 1)$

Total charge S unit time range	Speed range	Average split speed (m/s)
0.1429–0.2	2.24–2.65	2.45
0.1111–0.1429	2.65–3.00	2.83
0.0909–0.1111	3.00–3.32	3.16
0.0769–0.0909	3.32–3.61	3.47
0.0667–0.0769	3.61–3.87	3.74

The speed calculated by using this model is too large and not realistic, it cannot be used for speed grading.

The speed calculated by using this model is not obvious, which is not conducive to speed grading.

By comparing the above different sets of speed ranges, we finally use the speed scheme of Table 1. During the movement according to the path map, the WCV selects different speeds according to the different intervals of S, and charges the nodes around the spot.

Obviously, different speeds have a significant impact on the charging efficiency and the final charging delay. The WCV matches different speeds according to different charging efficiencies during the movement from the previous spot to the next. For example, when the number of rechargeable nodes in the neighbor charging range is large, the WCV should select a low speed and move slowly, so that more sensor nodes can be charged, and using the charging opportunities as much as possible. When the neighbor nodes are too sparse, passing through at a high rate, which involves the balance between speed and charging efficiency. The charging efficiency can be approximated by the total amount of effective charging unit time in formula 7 (Tables 2 and 3).

The Balance Between Optimal Solution and Complexity: In a dense deployed network, the edge formed by two charging spots may contain multiple different charging levels. If all segments have different charging levels and assumed different speeds, then the computational complexity will be quite large. To solve this problem, we have designed a segment combination scheme that combines multiple segments into a new segment. The charging speed s_{np} of the new segment is calculated as follows:

$$s_{np} = \sum E(d_n, c_{P_i}), i = 1, 2, 3$$

$E(d_n, c_{P_i})$ represents the charging speed from segment P_i to node n, it can be calculated by formula 1.

We set a threshold for multiple segments of each edge, and the above combination scheme adopts when the number of segments exceeds this threshold.

The pseudocode is as follows:

Algorithm 1 speed grading processing algorithm

Input: The total effective charge unit time S, The path P, Subsection Set $N=\{ n_i \}$
Output: The speed set V
1 Begin
2 Construct travel path P by TSP solution algorithm
3 For all $n_i \in P$ do
4 Calculate the value of S according to $S= \sum_{i=1}^{m} P_i$
5 If $S_i != 0$ then
6 Get the speed set V according to formula(7)
7 End if
8 End for
9 Return: V

---

4   Simulation and Performance Evaluation

We conduct simulation experiments to evaluate the performance of speed grading scheme, and compared it with the scheme proposed in the literature [11] in terms of the charging completion time. The delay performance of the two schemes under different network topologies and parameter settings is further compared.

This paper evaluates the excellent performance of the two schemes by the following two indexes.

(i) Completion time: the completion time refers to the time when the charger starts to move to the initial charging spot after the charging is completed.

(ii) Charging energy variance: Obviously, when all nodes are charged but not over-charged, the minimizing completion time can be achieved. In addition, no energy is wasted, and when the variance becomes larger, more energy and charging time are wasted.

Simulation parameter setting: We set four charging levels for each node according to the charging model, and the network topology is generated randomly. We compare our scheme and the minimizing charging delay (MCD) scheme [11] under the same topology. The deployment area is within a two-dimensional square area of 100 m * 100 m, the maximum speed of WCV is $v = 15$ m/s, the maximum charging radius of WCV is $r = 15$ m. In the formula (2), $\alpha = 36$, $\beta = 30$, d is the distance between the WCV and the sensor node. When using the speed grading method to control the movement of WCV, the step size of dividing edge into segments is taken as 2, the node energy threshold is 2 J (the charging energy more than 2 J is qualified), and the number of nodes initially deployed is 100.

The path map generation process in Fig. 2 is as follows. Firstly, the appropriate number of charging spots are obtained according to the clustering algorithm, and then the heuristic algorithm is used to obtain the path map for solving the TSP problem. The path map obtained by 100 random nodes, the generation path changes with the change in the qualification of K-means clustering algorithm, and it is also related to the initial deployment.

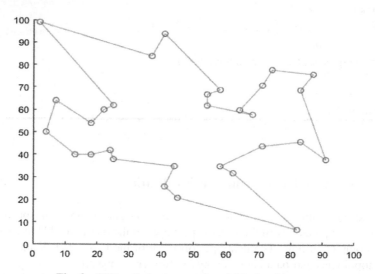

Fig. 2. TSP optimization results of 30 charging spots

The above path optimization is only obtained by simple clustering, the final charging path and charging delay are not optimal solutions. We consider the clustering optimization scheme in the charging strategy of large-scale WRSNs, we cluster the nodes in the network by relying on the energy limitation of each node in WRSNs and other constraints.

The calculation of locating the charging spot obtained by the K-means clustering algorithm did not involve whether the number of the spots finally obtained has a large improvement on the charging efficiency of WCV. Therefore, we decided to further optimize the delay by adopting the method of splitting the charging cluster to improve the charging efficiency. The wireless charging transmission power depends on the distance between the node and the WCV, and the shorter distance can reduce the charging time of the sensor node. So, the charging cluster with a large number of nodes into multiple sub-charge clusters is tried to be split based on the previous K-means clustering to obtain the charging spot. If the total completion time of the multiple subtasks is less than the original charging task, confirming the splitting operation and updating the charging spot position list. The separation basis and method are as follows:

It is necessary to measure the travel time of the subtask aggregation when confirming whether the task or the calculation can be split within the task verification. The calculation process is as shown in formula 9:

$$\theta = \sum_{I=1}^{\epsilon} w_i' + \sum_{i=1}^{\epsilon-1} \frac{D_{\tau_{i},\tau_{i+1}}}{V}$$

When $\theta < w_i$, it indicates that splitting charging cluster can reduce the charging time, then we need to decide to split the number of new task subsets, the method can be given by the following pseudocode:

Algorithm 2 Determining $\in$

Input: $\tau 1$
Output: $\in$
1 Begin
2 for all $i \leftarrow 2$ to $\lceil \tau \cdot c_i \rceil$ do
3 Calculate θ according to formula (9)
4 if $\theta < w_i$ then
5 $\in \leftarrow$ i
6 End if
7 End for
8 Return: $\in$

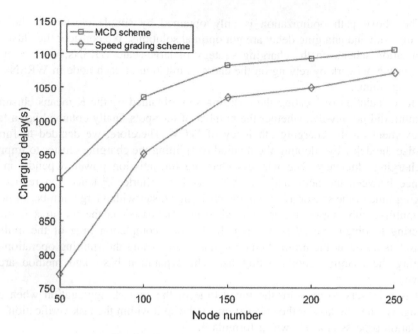

Fig. 3. Charging delay and node number

As shown in Fig. 3, it compares the charging delay of speed grading and MCD scheme [11] with different node numbers. We can get the following conclusions: (i) As the number of deployed nodes increases, the charging delay also begins to increase. However, the time delay of speed grading scheme is effectively reduced compared with the MCD scheme. The charging completion time is shortened, and the early delay reduction effect is obvious. But the improvement between comparisons is reduced as the number of deployment nodes increases. For possible reasons, the number of charging spots also increases when the number of deployment nodes increases, and the energy obtained by WCV during the movement is limited. Therefore, the later improvement is not obvious. (ii) The subsequent charging delays of two schemes are not increased greatly when the deployment node is larger than a certain number. For example, the subsequent charging delays of the two schemes are not increased much, and the delay growth tends to be slow when the number of nodes is 150. The possible reason is that there is overlap in the charging radius of all nodes and the number of charging spots does not have a linear relationship with the number of nodes when the deployment density increases. It may be a better choice for WCV to charge at the late stop.

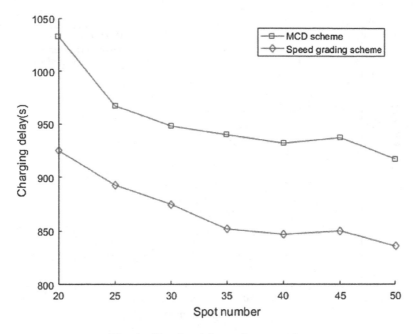

Fig. 4. Charging delay and spot number

As shown in Fig. 4, it compares the charging delay of speed grading and MCD scheme [11] with different spot numbers. We can get the following three conclusions: (i) As the number of charging spots increases, the delay of both scheme begins to decrease. The reason is that when the number of charging spots increases, the path segment generated by it is also more, and the WCV has more charging opportunities during the movement, which is beneficial to reduce the delay. But the MCD scheme is when the number of charging spots increases, the charging time at the spot becomes longer, and the node obtains more charging energy than a small number of spots. The delay tends to be flat and limited. (ii) When the number of spots is same, the delay of the speed grading scheme is reduced compared with the MCD scheme, and the effect of the previous period reduction is obvious. The reason is that the speed grading scheme utilizes the charging opportunities during the movement, and the exquisite speed grading is beneficial to the charging of WCV, so that the charging opportunities are more and the charging is more reasonable. (iii) When the number of charging spots continues to increase in the later period, there is a possibility that a delay of a high number of charging spots is slightly larger or has the same delay than a low number of delays. The reason is that the nodes are deployed randomly, the charging spots obtained by the K-means clustering algorithm will have outliers. The position also changes when the number of charging spots changes. Some outliers and charging spots will combine and appear in a more appropriate position. This allows different numbers of charging spots to have the same delay or a higher number of it with a slightly longer delay than a lower number one.

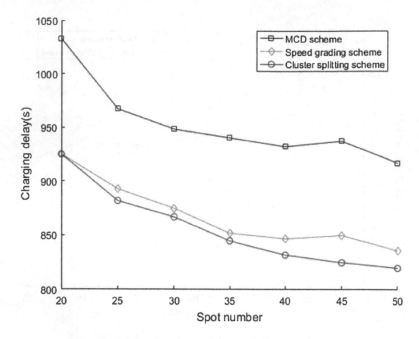

Fig. 5. Charging delay comparison with cluster splitting scheme

After adopting the cluster splitting scheme, we recalculated the delay and made a comparison curve with the previous two schemes as shown in the Fig. 5. The red one is the delay curve after splitting the charging clusters. It can be found that it is similar to the previous law. As the number of charging spots increases, the charging delay also gradually decreases, and the downward trend in the early period is obvious. As the number of charging spots increases to a certain amount about 40, the decrease in the charging delay tends to be slow. The reason is that when the number of spots increase to a certain amount, the optimization scheme of dividing the charging cluster does not improve the charging efficiency significantly. It indicates that it is not suitable for splitting, and it is difficult to reduce the charging delay by continuing to increase the spots by cluster splitting scheme. Moreover, it can be seen from the Fig. 5 that compared with the scheme of speed grading processing, the cluster splitting scheme has no obvious fluctuation in the later period, and the delay has been reduced with the increase of the charging spots until it is slow. The reason is that the number of outliers is reduced by the cluster splitting, and the number of nodes gathered around the charging spot is more reasonable. There is no large charging loss when the WCV charges the neighbor nodes at the charging spot, and enhancing the efficiency of WCV.

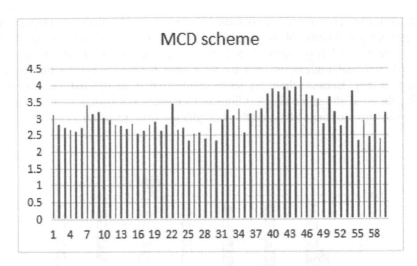

Fig. 6. Node energy statistics for MCD scheme

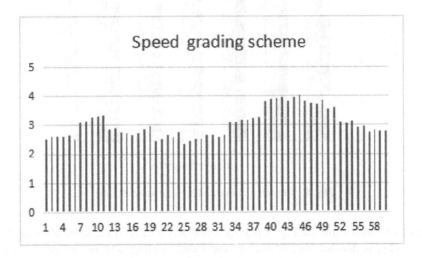

Fig. 7. Node energy statistics for Speed grading scheme

As shown in Figs. 6 and 7, they summarize the charging energy obtained by 60 nodes in two different charging schemes under the same deployment conditions. The following data is obtained through analysis and calculation. The MCD scheme [11] compare with the speed grading scheme, the average value of the charging energy of 60 nodes is reduced from 3.0455 J to 2.7136 J, and the variance is reduced from 0.7463 to 0.4357. This means that the proposed speed grading scheme can reduce the energy waste in the speed optimization process. The reason is that with a more exquisite division of the speed, the charging time of the charger at each charging spot can be more accurately scheduled, which further reduces the energy waste.

The following conclusions are obtained by comparing the above two figures: with the speed grading scheme, 60 sensor nodes deployed randomly meet the threshold of energy greater than 2 J. But there are still a few nodes that are overcharged—searching its coordinates and finding out that the overshoot nodes are neighboring in dense deployment area. Because there are some segments that are divided too much when dividing the edge, the speed changes little and the speed is too slow in these segments causing overshoot of the nodes, and these neighbor nodes are also in dense deployment areas.

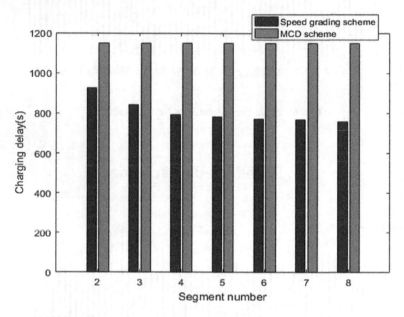

Fig. 8. Charging delay and segmentation number

It can be seen from the Fig. 8 that the delay has remained the same after the MCD scheme [11] is divided into edges. As the speed grading scheme increases with the division of the edge, the charging delay decreases, and the improvement over MCD scheme is obvious. However, when the number of divisions increases to a certain amount, the charging delay remains unchanged or increases slightly. This is because when the number of divided segments is increased to a certain amount, the difference in the number of different segments becomes smaller. As a result, the improvement of time delay is not obvious compared with the MCD scheme.

5 Conclusion

In this paper, the wireless charging sensor problem is identified as a charging planning problem under speed grading. In order to overcome the contradiction between the charging delay and the movement delay of the WCV, we propose to use the charging opportunities during the movement to achieve a great balance between them. This wireless sensor charging problem can be transformed into a Traveling Salesman Problem with speed grading. The charging delay is solved by the linear programming method, and through comparison of several simulation experiments, it can be found that the speed grading scheme which proposed by us can greatly improve the charging efficiency and reduce the total charging delay compared with the MCD scheme [11]. We will consider solving the problems of time delay optimization and efficiency improvement under different topology deployment when using multiple WCVs to charge sensor nodes.

References

1. Xu, C., Cheng, R.H., Wu, T.K.: Wireless rechargeable sensor networks with separable charger array. Int. J. Distrib. Sens. Netw. **14**(4), 1550147718768990 (2018)
2. Chen, C.C.: A novel data collection method with recharge plan for rechargeable wireless sensor networks. Wirel. Commun. Mob. Comput. **2018**, 1–19 (2018)
3. Shih, K.P., Yang, C.M.: A coverage-aware energy replenishment scheme for wireless rechargeable sensor networks. EURASIP J. Wirel. Commun. Netw. **2017**(1), 217 (2017)
4. Zeng, Y., Clerckx, B., Zhang, R.: Communications and signals design for wireless power transmission. IEEE Trans. Commun. **65**(5), 2264–2290 (2017)
5. Jiang, X., Xiang, L.: Improved decomposition-based global EDF scheduling of DAGs. J. Circuits Syst. Comput. **27**(7), 1–23 (2018)
6. Fu, L., Cheng, P., Gu, Y., et al.: Optimal charging in wireless rechargeable sensor networks. ACM Trans. Embed. Comput. Syst. (TECS) **65**(1), 278–291 (2016)
7. Li, S., Fu, L., He, S., et al.: Near-optimal co-deployment of chargers and sink stations in rechargeable sensor networks. ACM Trans. Embed. Comput. Syst. (TECS) **17**(1), 10 (2018)
8. Chabalko, M.J., Shahmohammadi, M., Sample, A.P.: Quasistatic cavity resonance for ubiquitous wireless power transfer. PLoS ONE **12**(2), e0169045 (2017)
9. Deng, R., He, S., Cheng, P., et al.: Towards balanced energy charging and transmission collision in wireless rechargeable sensor networks. J. Commun. Netw. **19**(4), 341–350 (2017)
10. Sample, A.P., Yeager, D.J., Powledge, P.S., et al.: Design of an RFID-based battery-free programmable sensing platform. IEEE Trans. Instrum. Meas. **57**(11), 2608–2615 (2008)
11. Fu, L., Cheng, P., Gu, Y., et al.: Minimizing charging delay in wireless rechargeable sensor networks. In: 2013 Proceedings IEEE INFOCOM, pp. 2922–2930. IEEE (2013)
12. Zhang, S., Qian, Z., Wu, J., et al.: Optimizing itinerary selection and charging association for mobile chargers. IEEE Trans. Mob. Comput. **16**(10), 2833–2846 (2017)
13. Farris, I., Militano, L., Iera, A., et al.: Tag-based cooperative data gathering and energy recharging in wide area RFID sensor networks. Ad Hoc Netw. **36**, 214–228 (2016)
14. He, S., Chen, J., Jiang, F., et al.: Energy provisioning in wireless rechargeable sensor networks. IEEE Trans. Mob. Comput. **12**(10), 1931–1942 (2013)

15. Hu, C., Wang, Y.: Minimizing energy consumption of the mobile charger in a wireless rechargeable sensor network. J. Internet Technol. **16**(6), 1111–1119 (2015)
16. Chen, Y., Lu, S., Chen, J., et al.: Node localization algorithm of wireless sensor networks with mobile beacon node. Peer-To-Peer Netw. Appl. **10**(3), 795–807 (2017)
17. Aoyama, K., Saito, K., Ikeda, T.: Accelerating a lloyd-type k-means clustering algorithm with summable lower bounds in a lower-dimensional space. IEICE Trans. Inf. Syst. **101**(11), 2773–2783 (2018)

Security

Solar Radio Burst Automatic Detection Method for Decimetric and Metric Data of YNAO

Guowu Yuan[1,2(✉)], Menglin Jin[1], Zexiao Cui[1], Gaifang Luo[1], Guoliang Li[1], Hongbing Dai[1], and Liang Dong[3]

[1] School of Information Science and Engineering, Yunnan University, Kunming 650091, China
yuanguowu@sina.com
[2] CAS Key Laboratory of Solar Activity, National Astronomical Observatories of Chinese Academy of Sciences, Beijing 100012, China
[3] Yunnan Observatories, Chinese Academy of Sciences, Kunming 650011, China

Abstract. With the development of solar radio spectrometer, it is difficult to process a large number of observed data quickly by manual detection method. Yunnan astronomical observatories (YNAO) have two solar radio spectrometers with high time and frequency resolution. An automatic detection method of solar radio burst for decimetric and metric data of YNAO is proposed in this paper. The duration of solar radio burst was counted and analyzed. Channel normalization was used to denoise the original solar radio image. Through experimental comparison, Otsu method was selected as a binary method of solar radio spectrum, and open and close operations were used to smooth the binary image. Experiments show that the proposed method for automatic detection of solar radio bursts is effective.

Keywords: Solar radio spectrum · Solar radio burst · Image denoise · Otsu

1 Introduction

The Sun is the only star that can be finely observed. At the same time, the solar system is the only star system known to humans that has life-proliferating behavior. Therefore, solar physics research is not only the most active and important field in astronomy, but also extremely important for studying the origin and evolution of life in the universe [1]. The violent burst of the Sun, mainly referred to flare and coronal mass ejection (CME), is the most intense energy release process in the solar system. In particular, CME throws a large amount of plasma materials into the interplanetary space at speeds of more than one thousand kilometers per second, causes dramatic disturbances to the space and geomagnetic fields of the Sun and earth, and directly affects spacecraft, communications and electricity. Therefore, CME has many adverse effects on human production and life [2]. Taking a solar radio burst which occurred on December 6, 2006

© Springer Nature Singapore Pte Ltd. 2019
X. Cheng et al. (Eds.): ICPCSEE 2019, CCIS 1058, pp. 283–294, 2019.
https://doi.org/10.1007/978-981-15-0118-0_22

as an example, this burst disturbed the GPS signal and caused the serious situation of the GPS ground station's navigation losing lock [3].

The energy release process of the solar burst event is manifested in the entire electromagnetic spectrum, and the observation of the radio band is an indispensable means. By studying solar radio bursts, space weather can be forecasted effectively and unnecessary risks can be avoided in time [4].

According to changes in radio intensity, solar radiation can be divided into three types: quiet solar radio, slowly varying solar radio and solar radio bursts [5].

The quiet solar radio is a component that does not change with time, and it also exists in the absence of obvious solar activity. Its source is bremsstrahlung radiation from the atmosphere above the photosphere, also known as the "B component".

Solar slowly varying radio is a kind of radiation produced when small disturbances occur on the sun. The flux density of solar slowly varying radio shows well in statistical. Its intensity increases with the increase of wavelength, which is also called "S component".

The phenomena of intense flux enhancement observed in the radio band are called solar radio bursts. The duration of solar radio bursts varies from less than a second to more than an hour.

At present, the discrimination of solar radio burst mainly depends on the manual observation of solar radio spectrum. Solar radio bursts are small probability events, usually there are dozens to hundreds of solar radio bursts in one year, but the total time is only a few dozen hours, while the annual solar radio spectrum is more than 3000 h, which is inefficient and boring by manual detection. Moreover, manual detection often takes a long time to determine the type and parameters of a burst. The manual method cannot meet the real-time requirements, but also has many drawbacks, such as omission, randomness and so on.

With the rapid development of digital image processing technology in recent years, considering that solar radio spectrum is a digital image, digital image processing technology is applied to automatically detection solar radio spectrum, or at least to find possible solar radio bursts and then submit them to manual detection. It can greatly reduce the amount of manual detection. This is of great significance to the early warning of space weather and the radio observation of the sun.

The automatic detection of solar radio bursts has been studied by many scholars in recent years. Lobzin et al. proposed an automatic detection method for solar radio bursts [6]. This method can detect single and group bursts in type III bursts with 84% success rate, but it cannot detect intermittent bursts in type II bursts and type III bursts. Zhang et al. at the University of Chinese Academy of Sciences have achieved the automatic detection and analysis of type III radio bursts by using Hough transform to detect straight lines [7]. The disadvantage is that they cannot detect type II radio bursts and recognize the fine structure of bursts. Lin Ma et al. applied the deep learning method to solar radio burst detection [8], but only achieved the burst detection of 2.2–3.8 GHz band and stored data, and could not distinguish between type III burst and type II burst, and could not identify the fine structure of the burst.

In this paper, we proposed an automatic detection method of solar radio burst, which is suitable for high temporal and frequency resolution data collected by the solar radio spectrometers from Yunnan Astronomical Observatories of the Chinese Academy of Sciences. In this method, channel normalization is used to denoise, Otsu method is used to achieve binarization of the solar radio spectrum image, and Open and Close operations are used to smooth the binary image. Our experimental results are compared with those of Nobeyama Solar Radio Observatories.

2 Acquisition and Format of Solar Radio Spectrum

2.1 Data Acquisition Device

Yunnan Astronomical Observatories (YANO), located in Fenghuang Mountain at the eastern suburb of Kunming, has two digital solar radio spectrometers. The working frequency band of the metric solar radio spectrometer located at the Fuxianhu Solar Observatory base of Yunnan Astronomical Observatories is 70–700 MHz, and the working frequency band of the decimetric solar radio spectrometer located at the headquarters of Yunnan Astronomical Observatories is 625–1500 MHz. Because of the strong mobile communication interference in the 800 MHz–975 MHz band, the decimetric solar radio spectrometer located at the Yunnan Astronomical Observatories only recorded signals in 625–800 MHz and 975–1500 MHz (Fig. 1).

    (a) Decimetric spectrometer   (b) Metric spectrometer

Fig. 1. Two solar radio spectrometers of Yunnan astronomical observatories of the Chinese academy of sciences [5]

The maximum time resolution and frequency resolution of the two solar radio spectrometers are 2 ms and 200 kHz respectively. The time and frequency resolution of the two spectrometers are set to 80 ms and 200 kHz respectively in daily work, and the working time is from 8:00 to 18:00 when converted to Beijing time (Table 1).

Table 1. Parameters of 10 m and 11 m solar radio telescopes of Yunnan astronomical observatories of the Chinese academy of sciences [5]

Parameter	Metric spectrometer	Decimetric spectrometer
Frequency range	70–700 MHz	625–1500 MHz
Recording time	00:00–10:00 UT	00:00–10:00 UT
Highest time resolution	2 ms	2 ms
Highest spectral resolution	200 kHz	200 kHz

2.2   Data Format

Solar radio spectrum is recorded in a special format and stored in a file with the extension.dat. Each file is 100M Bytes, be named by Beijing time when recording the data, and recorded solar radio data of 512 s and 3500 frequency channels. The daily data volume of each solar radio spectrometer is about 8G Bytes. After deleting the redundant channels and rearranging the matrix, the radio spectrum image can be obtained. A radio spectrum image is shown in Fig. 2.

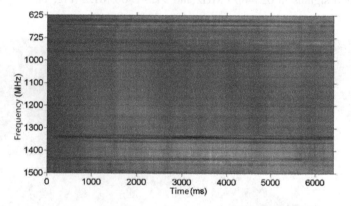

Fig. 2. Solar radio spectrum obtained by decimetric solar radio spectrometer at Yunnan astronomical observatories

3   Statistics and Analysis of the Duration of Solar Radio Burst

In order to facilitate the determination of the algorithm, we have statistics on the solar radio burst data of the last 10 years. Since the Nobeyama Solar Radio Observatory of National Astronomical Observatory of Japan (NAOJ) published detailed data and some of the data from the Yunnan Observatory was incomplete, we counted the data of Nobeyama Solar Radio Observatory of NAOJ.

Statistical results of all solar bursts from January 1, 2001 to March 31, 2015 show that there were 581 solar radio bursts during the 15 years and 3 months [9]. Most of them have a short duration. The longest burst duration is 10 h, 31 min and 21 s, and the shortest duration is only 3 s. The average burst duration is 27 min and 33 s. As can be seen from Fig. 3, in most cases, the burst duration is less than half an hour.

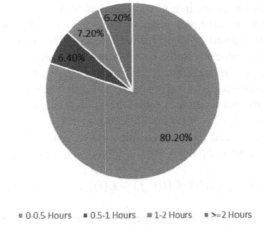

■ 0-0.5 Hours   ■ 0.5-1 Hours   ■ 1-2 Hours   ■ >=2 Hours

Fig. 3. Statistics of the duration of the solar radio burst

According to the statistical results, the solar radio burst only accounts for a very small part of the total observation duration. The amount of radiation is generally stable as the sun is in a calm period. Therefore, we can consider the process of finding a burst as the detection of the outlier of the radiation intensity in the same frequency band.

According to data statistics, the longer bursts have the greater values of observations. The dynamic method is used to detect the value of the burst interval as an outlier. Since the duration of a data file is 8 min, a burst more than 8 min cannot be identified as an outlier. So the detection time should be extended as long as possible under ideal conditions. As the proportion of the state of quiet sun increases, the burst as an outlier becomes more obvious. For example, we use the observation data in the previous month as the foreground. Even lower bursts are easier to identify in long periods of quiet sun. However, due to the huge amount of data, the time cost will increase significantly if we use this method.

4   Automatic Detection Method of Solar Radio Bursts

The solar radio spectrum exists in the form of two-dimensional matrix $I(t, f)$. t represents time, f represents frequency, and the value of matrix element $I(t, f)$ represents the value of solar radio flux at time t on frequency f. $I(t, f)$ is small when solar is quiet. $I(t, f)$ increases when the solar radio bursts occurred. If we can find a suitable threshold, find out the part whose radio flux value is greater than this value, and then determine the area where the possible solar radio burst occurs, we can find the possible solar radio burst, which greatly reduces the amount of manual detection data.

4.1   Denoising of Solar Radio Spectrum Image

The solar radio signal received by the solar radio spectrometer will inevitably be disturbed by the environment, and there will be a variety of noise in the spectrum. There are space electromagnetic interference, gradual background noise, interference of instrument, quantization error, rounding error and so on. The interference of the instrument is mainly caused by the channel effect of the instrument. The intrinsic factors of the instrument can hardly be eliminated or weakened in hardware. The interference reflected in the spectrum is long horizontal stripes.

Channel normalization can greatly reduce the difference between channels. The method is to divide the net amount of solar current each channel by the average of the channel [10]. If we record the value of radio spectrum as $I(t, f)$, the value after channel normalization as $g(t, f)$, and the average value of f-th row in $I(t, f)$ as $\bar{I}_f(t)$, then the net solar radio flux is $\Delta I(t, f) = I(t, f) - \bar{I}_f(t)$ and $g(x, y)$ is calculated as

$$g(t, f) = \Delta I(t, f)/\bar{I}_f(t) = (I(t, f) - \bar{I}_f(t))/\bar{I}_f(t) = I(t, f)/\bar{I}_f(t) - 1 \qquad (1)$$

It should be noted that the pixel value after channel normalization does not represent the solar radiation flux value, but it does not affect the image display effect and subsequent processing steps.

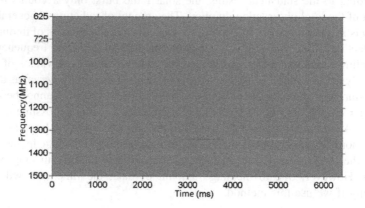

Fig. 4. Grayscale image after channel normalization of Fig. 2

4.2   Binarization of Solar Radio Spectrum Image

After channel normalization, the noise on the solar radio spectrum is reduced. According to statistics, the pixel value of the solar radio spectrum corresponded to solar radio burst area is higher, and the pixel value corresponded to the quiet solar radio area is lower. So it is feasible to find out the solar radio burst area by choosing the appropriate threshold to binarize the solar radio spectrum image.

When binarizing the solar radio spectrum image, Lobzin proposed the maxima method [6], and Zhang P proposed a second-order maxima method based on this [7].

The basic idea of the maximum value method is that if the pixel value of a point is not less than the two points on the left and right, the pixel value of the point is set to 1, otherwise the pixel value of the point is set to 0. The formula is expressed as follows.

$$BI = \begin{cases} 1, & if\ g(t-1, f) \le g(t, f) \le g(t+1, f) \\ 0, & otherwise \end{cases} \tag{2}$$

The spectrum images collected by the decimetric solar radio spectrometer at Yunnan Astronomical Observatory are processed and the result of Fig. 4 is obtained (Fig. 5).

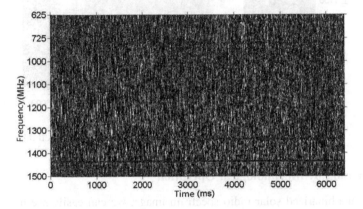

Fig. 5. Result of binarization using local maximum method

After using the local maximum method to binarize gray scale image, binary image are full of irregular noise points. It is difficult to distinguish the burst part from the quiet part. The reason is that the time resolution of YNAO solar spectrum is much higher than that of RSTN solar spectrum used by Lobzin [6]. The time resolution of YNAO is 80 ms, but that of RSTN is 3 s. A small change in solar radio will produce a considerable number of extreme points, which will result in so many irregular noises.

The maximum inter-class variance is commonly used in binarization, and this method was proposed by Japanese scholar Otsu in 1979, so this method is also called Otsu Method [11]. For image G, the threshold of binarization is T, ω_0 is the proportion of foreground points to the whole image, and the average gray level of foreground is u_0; ω_1 is the proportion of background points to the whole image, and the average gray level of background is u_1. The total average gray level of the whole image is $u_T = \omega_0 \times u_0 + \omega_1 \times u_1$. Traversing from the minimum gray value, when T makes the variance $\sigma^2 = \omega_0 \times \omega_1 \times (u_0 - u_1)$ maximum, then T is the best threshold for image segmentation. Variance is a measure of the uniformity of gray distribution. The larger the variance value, the larger the difference between the two parts of the image. When part of the foreground is misclassified into the background or part of the background is divided into the foreground, the difference between the two parts will become smaller. Therefore, to minimize the probability of misclassification, it is necessary to ensure that the variance between classes is the largest.

We use the maximum inter-class variance method to binarize Fig. 4, and the results are shown in Fig. 6. Through experimental comparison, the maximum inter-class variance method is better than the maximum value method for the solar radio spectrum with high time and frequency resolution from Yunnan Astronomical Observatories.

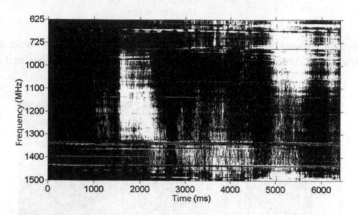

Fig. 6. Binarization result of spectrum image using Otsu method

4.3 Denoising by Morphological Operation

Observing the binarized solar radio spectrum image, we can easily see that the white bright burst area is obvious, but there are still irregular noises in the quiet part of the spectrum image. Some of these noises exist naturally, and others are increased in the process of gray image converting to binarized image. These noises are very disadvantageous to detect the burst boundary. An effective method is to smooth the binary image. Therefore, open and closed operations are needed.

The operation of first eroding and then expanding is called Open operation. A is the original image, B is the structural element, and the morphological opening operation of A with B can be recorded as A ∘ B. Open operation can eliminate small objects, separate objects at fine points and smooth the boundary of larger regions.

The operation of first expanding and then eroding is called Close operation. A is the original image, B is the structural element, the morphological closing operation of A with B can be recorded as A • B. Close operation has the function of filling small holes in the object and connecting adjacent objects.

By using different sizes structural elements to Open and Close the binary image of solar radio spectrum, it is found that the structural elements of square 3×3 size have the best effect, which can effectively reduce the noise interference of the quiet part without affecting the shape of the explosion area (Fig. 7).

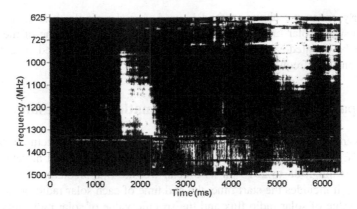

Fig. 7. Open and close operations using structural elements of 3×3 size

4.4   Detection of Burst Boundary

By observing the spectrum image smoothed by morphological operation, we can see that the noise of the quiet part has been greatly reduced, compared with the burst area, it is more obvious. Because the binary image has only two discrete values of 0 and 1, we can use the method of threshold setting, by calculating the sum of column and row pixel values, and judging the burst when the value is larger than a certain set value. The output column and row numbers plus displacement can be used to obtain the burst parameters.

Figure 8 is a histogram with time as the horizontal axis, column pixels as vertical axis. Observing the histogram, we can clearly see that there are three peaks of different shapes in the graph. The last peak is not an outbreak because of the short time. We can add conditions to remove this interference in the process of setting threshold method. The first two peaks indicate that two solar radio bursts occurred within the time range recorded in this data segment. Therefore, we can set a threshold to determine the start and end time of the burst.

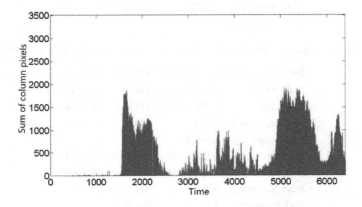

Fig. 8. Statistical histogram of time versus the sum of column pixels

Similarly, when the above method is used to find a single solar radio burst, the image of a single solar radio burst can be separated. Then, each line of the separated image can be counted, and a suitable threshold can be determined to get the start frequency and end frequency of the solar radio burst.

5　Comparison and Analysis of Automatic Detect Results

The Nobeyama Solar Radio Observatories is located in Nagano Prefecture (35°56′30″N 138°28′33″E), at an altitude of 1350 m. The Observatories recorded in detail the solar radio bursts from 1988 to 2015 on its official website (http://solar.nro.nao.ac.jp/norp/html/event/). It includes the start time, the end time of each solar radio burst event, the maximum value of solar radio flux and the specific value of solar radio flux (SFU) in each frequency range. At the same time, the Yunnan Astronomical Observatories has recorded the records of meter wave bursts after March 2012 and decimeter wave bursts after September 2009 on its official website (http://www.ynao.ac.cn/solar/) [5], which can be used for comparison and analysis.

We randomly selected three data segments and detected and identified four solar radio bursts. The following is a comparison of these bursts with observations from Yunnan Astronomical Observatories and Japan Nobeyama Solar Radio Observatories (Table 2).

Table 2.　Automatic detection results of solar radio bursts on February 9, 2011.

Burst parameter	Our method	YNAO	NSRO
Start time	012626128	0126	012626
End time	012711888	0127	013250
Start frequency	643	625	/
End frequency	1500	1500	/
Start time	012720001	0128	012626
End time	012921761	0129	013250
Start frequency	625	625	/
End frequency	1500	1500	/

Table 3.　Automatic detection results of solar radio bursts on March 5, 2012.

Burst parameter	Our method	YNAO	NSRO
Start time	034533360	0346	022144
End time	034622880	0347	084708
Start frequency	775.6	625	/
End frequency	1346.4	900	/
Start time	035007360	Not have	022144
End time	035153680	Not have	084708
Start frequency	625.4	Not have	/
End frequency	1020.8	Not have	/

Because the recorded data of the Nobeyama Solar Radio Observatories include all the solar radio spectrum from 1 GHz to 80 GHz, the corresponding bursts always start earlier and end later than those recorded by Yunnan Astronomical Observatories only for 625 MHz–1500 MHz, so the burst parameters recorded by the Nobeyama Solar Radio Observatories can only be used as reference. We focus on the burst records of Yunnan Astronomical Observatories (Table 3).

From the above table, it is shown that our proposed automatic detection method has better coincidence with the manual detection results of Yunnan Astronomical Observatories in the start time and the end time, and the coincidence effect of the two bursts on March 5, 2012 in the frequency range is better. However, the first burst on February 9, 2011 has errors, but it is still within the acceptable error range. It should be noted that a burst that was not recorded by Yunnan Astronomical Observatory was identified in the data of February 9, 2011 by automatic detection method. Because Yunnan Astronomical Observatory has no corresponding record, it can only be judged as a false detection. This false detection results can be eliminated in the subsequent manual review. The automatic detection method greatly reduces the scope of manual detection and improves the efficiency, and the time resolution and frequency resolution are better than manual detection. In a word, the solar radio automatic detection method proposed in this paper is feasible.

6   Summary and Prospect

6.1   Summary

An automatic detection method of solar radio burst based on Otsu binarization is proposed in this paper. In this method, channel normalization is used to denoise the original solar radio image, and good results are obtained. Otsu method is used to achieve binarization of the solar radio spectrum image, and Open and Close operations are used to smooth the binary image.

6.2   Prospect

Although it is simple and quick to set a threshold to detect solar radio bursts, there is a disadvantage that the threshold is difficult to select when the bursts overlap in frequency. At the same time, there are also the drawbacks of the same burst across data segments being identified as two different bursts. Another major shortcoming of this paper is that there is no fine structure detection for solar radio bursts. In recent years, with the continuous development of artificial intelligence technology, some scholars have begun to apply convolutional neural networks to solar radio burst detection [12], which is also a future development direction.

Acknowledgments. This work is supported by the Natural Science Foundation of China (Grant No. 11663007, 61802337, U1831210), Chinese Academy of Sciences "Western Light" Talent Development Program, the Youth Top Talents Project of Yunnan Provincial "Ten Thousands Plan", and the Action Plan of Yunnan University Serving for Yunnan.

References

1. Zhao, R., Jin, S., Qijun, F.: Solar Radio Microwave Burst. Science Press, Beijing (1997)
2. Zhao, R.: Theory of Solar Radio Radiation. Science Press, Beijing (1999)
3. Dong, L.: The Assessment and Pre-alarm Methods Research for the Solar Radio Burst Events Interfering the Communication Systems. Doctor, Yunnan University (2016)
4. Klein, K.L., Matamoros, C.S., Zucca, P.: Solar radio bursts as a tool for space weather forecasting. Comptes Rendus Phys. **19**(1–2), 36–42 (2018)
5. Gao, G.: The Solar Radio Bursts and Fine Structuresin Metric and Decimetric Bands. Doctor, University of Chinese Academy of Sciences (2015)
6. Lobzin, V.V., Cairns, I.H., Robinson, P.A., Steward, G., Patterson, G.: Automatic recognition of type III solar radio bursts: automated radio burst identification system method and first observations. Space Weather **7**(4), 1–12 (2009)
7. Zhang, P.J., Wang, C.B., Ye, L.: A type III radio burst automatic analysis system and statistic results for a half solar cycle with Nançay Decameter Array data. Astron. Astrophys. **618**, A165 (2018)
8. Lin, M., Zhuo, C., Long, X., et al.: Multimodal deep learning for solar radio burst classification. Pattern Recognit. **61**, 573–582 (2017)
9. Nobeyama Radio Observatory: Nobeyama Radio Polarimeters Event List (2015). http://solar.nro.nao.ac.jp/norp/html/event/. Accessed 26 Jan 2019
10. Long, X.: Wavelet Analysis and Its Applications to the Processing of Solar Radio Data Observed. Xidian University, Master (2002)
11. Tang Z., Wu Y.: One image segmentation method based on Otsu and fuzzy theory seeking image segment threshold. In: International Conference on Electronics, pp. 2170–2173. IEEE (2011)
12. Chen, S., Xu, L., Ma, L., Zhang, W., Chen, Z., Yan, Y.: Convolutional neural network for classification of solar radio spectrum. In: IEEE International Conference on Multimedia & Expo Workshops, pp. 198–201. IEEE (2017)

Internet Web Trust System Based on Smart Contract

Shaozhuo Li, Na Wang[✉], Xuehui Du, and Aodi Liu

He'nan Province Key Laboratory of Information Security,
Zhengzhou 450001, Henan, China
twftina_w@126.com

Abstract. The current Internet web trust system is based on the traditional PKI system, to achieve the purpose of secure communication through the trusted third party. However, with the increase of network nodes, various problems appear in the centralization system of public key infrastructure (PKI). In recent years, in addition to cryptographic problems, attacks against PKI have focused on the single point of failure of certificate authority (CA). Although there are many reasons for a single point of failure, the purpose of the attack is to invalidate the CA. Thus a distributed authentication system is explored to provide a feasible solution to develop distributed PKI with the rise of the blockchain. Due to the automation and economic penalties of smart contracts, a PKI system is proposed based on smart contracts. The certificate chain was constructed in the blockchain, and a mechanism was adopted for auditing access to CA nodes in the blockchain. Experimental results show that security requirements of CA are met in this system.

Keywords: Public key infrastructure · Blockchain · Smart contract

1 Introduction

The current Internet web trust system is based on the traditional PKI system. Public Key Infrastructure (PKI) is a key management platform that follows established standards. It provides cryptographic services such as encryption and digital signatures and the key and certificate management systems necessary for all web applications. PKI technology is the core of information security technology and the key and basic technology of e-commerce.

This third-party-based trust mechanism is now facing serious security challenges, resulting in frequent security incidents. For example, DigiNotar was invaded in 2011 [1]. DigiNotar is a Dutch company whose main business is to issue certificates to the public and is the first CA to be completely invaded. The forged certificate caused a very serious man-in-the-middle attack in Iran, collecting a large number of Gmail passwords, and it's root certificate was revoked.In December 2013, TurkTrust CA issued a false certificate [2]. TURKTRUST Inc. incorrectly created two CA branches (*.ego.-gov.tr and e-islem.kktcmerkezbankasi.org). The CA of the.EGO.GOV.TR branch was subsequently used to issue a false digital certificate to *.google.com. This deceptive certificate may be used to perform phishing attacks or man-in-the-middle attacks on

© Springer Nature Singapore Pte Ltd. 2019
X. Cheng et al. (Eds.): ICPCSEE 2019, CCIS 1058, pp. 295–311, 2019.
https://doi.org/10.1007/978-981-15-0118-0_23

several Google Webs. With the increase of network nodes, the problem that single-point CA is not credible has surfaced, and the PKI system has a crisis. There are several key problems that need to be solved:

Certificates Can Be Issued Without the Authorization of the Domain Name Owner [3]

In principle, the biggest problem we have now is that any CA can issue certificates to any domain name without the authorization of the domain name owner. The key problem here is that there is no effective technical means to ensure that CA does not produce omissions and security errors. In those days when there were only a few CAs, this may not be a problem, but now there are hundreds of CAs, which has become a big problem.

Lack of Trust Flexibility

Another theoretical problem is the lack of trust flexibility. The relying party maintains a root certificate store containing a certain number of CA certificates, so that the CA is either completely trusted or completely untrustworthy, with no intermediate conditions. In theory, the relying party can remove the CA from the root certificate store. But in fact, this can only happen if there is a serious incompetence or security breach, or if the CA is very small. Once a CA issues a large number of certificates, it will be a big deal.

Invalidity of Revocation

There are two main reasons, the first is that it takes time to transfer revocation information between systems, and the second is the current soft failure strategy implemented by all browsers.

Therefore, we urgently need to construct a distributed decentralized identity management system. The emergence of blockchain gives us an opportunity.

Blockchain technology is a new distributed infrastructure and computing method, which uses blockchain data structure to verify and store data, uses distributed node consensus algorithm to generate and update data, uses cryptography to ensure data transmission and access security, and uses smart contract composed of automated script code to program and operate data. Blockchain has the following characteristics: distributed, open consensus, autonomy, information not tampering, anonymity, traceability, and programmability.

However, the current research on distributed PKI based on blockchain is just in its infancy, and there are many shortcomings, such as using only the blockchain as a data storage tool or ignoring the reality to adopt a fully distributed structure.

Based on this, this paper proposes a PKI system construction based on smart contract. Its main principle is to use the characteristics of blockchain to solve many problems of existing PKI. Firstly, its open consensus and information not tampering guarantee the openness and transparency of the whole authentication process and the unforgeability of certificates. At the same time, traceability also ensures the realization of the accountability system. The open consensus ensures that the certificate revocation can take effect in time. We have adopted smart contracts in this article, which has the characteristics of automated execution and programmability. We solve the problem that any CA can issue certificates to any domain without consent by setting the conditions required for the smart contract automation execution in the program. At the same time, we use its automated execution to realize the automatic reward and punishment mechanism for erroneous CA, and solve the problem of trust flexibility. Moreover, the existing PKI system does not have an effective audit access mechanism for the CA node, and cannot guarantee the security and reliability of the CA in the blockchain.

2 Research Status

2.1 Improvement Scheme of PKI

At present, the main problem with PKI is that it does not have the consent of the domain name owner to issue a certificate. At present, there are mainly the following solutions to this problem:

Certificate Transparency CT:
Certificate transparency is an open framework for auditing and monitoring certificates. The CA submits each certificate they issued to the public certificate log and obtains a certificate of encryption submitted this time. Anyone can monitor every new certificate issued by the CA. For example, a domain owner can monitor each certificate that has their domain name issued. If this idea is implemented, then the fake certificate can be quickly discovered. The proof of submission can be delivered to the client in a different way so that it can be used to confirm that the certificate has been made public. Through this transparent and open way to supervise CA, the problem of issuing certificates without the domain name owner's consent is solved.

Public Key Pinning:
Public key pinning generates a fixed hash value for the certificate used by the website, and then passes the hash value to the browser when the user visits it. The browser will check whether the certificate hash value is the same as the previous hash value when the user visits it next time. If it is different, the connection will be disconnected. With pinning, network owners can choose one or more CAs they trust to create their own trusted ecosystems that are much smaller than the global ecosystem. Public key pinning can now be implemented through Chrome's proprietary mechanism, and HTTP public key pinning standards are being developed.

Although these two methods can solve the untrustworthy problem of CA to a certain extent, they all have certain defects. For example, CT lacks a feedback mechanism for error certificates. The public key pinning does not solve the existing PKI system problem, but adopts a way to narrow the credibility range to reduce the probability that CA is not credible.

2.2 Research Status of Blockchain

Blockchain is a decentralized, non-trusted distributed ledger technology, which is composed of distributed data storage, point-to-point transmission, consensus mechanism, encryption algorithm and other technologies. Blockchain is a decentralized and trustless Distributed Accounting technology, which is composed of distributed data storage, point-to-point transmission, consensus mechanism, encryption algorithm and other technologies. Blockchain was first proposed in 2008, which is the underlying technology of Bitcoin proposed by Zhongbencong [4]. Since its emergence, blockchain has been applied more and more.

In 2014, Ethereum founders Vitalik Buterin, Gavin Wood and Jeffrey Wilcke began researching a new generation of blockchain, trying to achieve a smart contract platform that does not require a trust base in general [5]. In the narrow sense, Ethereum refers to a series of protocols that define the decentralized application platform. Its core is the Ethereum Virtual Machine (EVM), which can perform the encoding of any complex algorithm. Smart contracts rely on blockchain on EVM bytecode to run. The default execution environment of Ethereum is no process, nothing happens, and the status of each account remains the same. However, each user can trigger an action by sending a transaction from an external account to launch Ethereum. If the destination of the transaction is another foreign account, then the transaction may transfer some Ethereum, otherwise nothing will be done. But if the destination is a contract, the contract will be activated and the code will run automatically.

Contracts usually serve four purposes:

Saving as a database represents something useful to other contracts or the outside world;

As a foreign account with a more complex access protocol, it is called a "forward contract";

Manage an ongoing contract or relationship between multiple users;

Provide functionality to other contracts, essentially as a software library.

2.3 Distributed PKI Based on BlockChain

The characteristics of blockchain technology make it an ideal technology for a variety of applications. In particular, these characteristics demonstrate the applicability of blockchain to PKI. Since blockchain-based PKI solutions are distributed, they do not have centralized failure points. More importantly, blockchain technology has several open source implementations that help build cost-effective and efficient solutions.

According to different trust systems, the main distributed PKI research based on blockchain can be divided into two categories:

The first category is to improve the existing PKI trust system. Its main idea is to make use of the information of blockchain that can not be tampered with to modify and open consensus, so as to achieve transparent and open certificate storage and authentication process, such as:

Instant Karma PKI (IKP) [6]

The Instant Karma PKI (IKP) framework extends the traditional CA approach by recording CA behavior into blockchain. In this way, the network can detect a misbehaving or attacked CA and must react. The blockchain's event logging feature helps blockchain users track and monitor CAs and helps detect misbehaving CAs. This method can reduce trust problems in traditional CA-based algorithms, because abnormal CA can be detected eventually.

Pemcor [7]

Pemcor uses the blockchain database as a distributed secure data store [7]. The idea is to have the CA issue an unsigned certificate, the hash of the certificate is stored in the blockchain, and the blockchain is controlled by an authority such as a bank or government. These organizations share two blockchain databases, one for the generated certificate and the other for the revoked certificate. At the time of verification, these agencies check the blockchain data storage they maintain. If the hash value of the certificate exists in the generated certificate blockchain and is not in the revoked certificate blockchain, the certificate is valid; otherwise it is invalid. The idea is simple, with verification with low latency guarantees.

Certcoin [8]

Certcoin is a completely decentralized PKI that relies on Namecoin [9] to build its platform. The core idea is to record the user certificate through the public general ledger, and associate the user identity with the certificate public key in an open manner to realize the decentralized PKI construction. Any user can query the certificate issuance process to solve the problem of certificate transparency and CA single point of failure faced by traditional PKI systems. The registration, updating and revocation of certificates are realized by issuing users and their public keys in the form of blockchain transactions, and the normal operation of PKI is guaranteed by the unalterable attributes of blockchain. Merkle root only records the hash value of the transaction. Users can complete the certificate verification without downloading all blockchain transaction data.

At present, typical applications in this mechanism are too limited for the use of blockchain characteristics, and can only be used for storage and public authentication, ignoring other characteristics of the blockchain. And there is no complete PKI system construction, and the problem it solves is very limited.

The second type of mechanism adopts a fully distributed trust system. Its main idea is to use the distributed nature of the blockchain to better implement existing distributed trust systems, such as pgp networks. Its typical application is SCPKI.

SCPKI [10]

SCPKI is a fully distributed PKI system that uses a network trust model (PGP) and a smart contract on the Ethereum blockchain. Using the design of the network trust model, an entity in the system can verify (or guarantee) the identity of another entity as a replacement for the CA.

This type of mechanism is an ideally perfect distributed trust system, but it faces great challenges in its practical application. The PGP trust model largely reflects the relationship between people in real life. When used in a wide range, it is difficult to trust other nodes unconditionally and trust any node through trust transmission. The lack of trust management is a fatal defect of PGP. It is easy to cause security risks [11]. In the PGP trust system, each node is very important and is an indispensable part of the trust network. Once a certain point or even more points are manipulated or bought, the transfer chain will be interrupted, thus losing trust to many nodes. This causes irreversible damage to the trust network.

3   Construction of PKI System Based on Smart Contract

The whole system is divided into five parts: admission mechanism for CA nodes, certificate chain structure, conventional certificate management function, error certificate feedback mechanism and automatic economic reward and punishment mechanism. These five functions are realized through contracts, and make full use of the characteristics of automatic contract execution to improve the existing mechanisms. Before introducing these five functions, we first make a description of the nodes in the system.

Node settings:
We classify the nodes into CA nodes, end-user nodes, and ordinary nodes. The CA nodes participate in the blockchain transaction for certificate authentication. The end-user node has no similar permissions and can only be used as an entity for certificate application and certificate query. Ordinary nodes, they can only view certificates and participate in the maintenance of the blockchain. The joining of a CA node with a certificate issuing function requires an admission mechanism.

3.1   Admission Mechanism for CA Nodes

Moreover, the current distributed PKI system based on blockchain is purely public, and any node can join the blockchain, which is a big security risk for CA. The system in this paper introduces an admission mechanism for the joining of CA nodes, which distinguishes the joining of CA nodes from other nodes. Other nodes, they can only have certificates or view certificates, and cannot issue certificates. The joining of a CA node with a certificate issuing function requires an admission mechanism.

The addition of a CA node requires auditing to ensure the security of the CA in the system, similar to the consortium blockchains structure. At the same time, other nodes join without auditing, maintain the characteristics of the public blockchains, and ensure large-scale application on the Internet.

The CA node wants to join the blockchain. First, you need to obtain the certificate issued by the advanced CA offline. After obtaining the certificate, the node joins the blockchain to upload its own certificate to the contract, and the advanced CA that issues the certificate to it also uploads the same certificate to the contract. At this point, the contract will automatically detect whether the two certificate information is consistent, and detect whether the two nodes meet the information in the certificate. After the contract detection information is correct, the node can successfully become a CA node. After the initial CA chain is constructed, the root CA is normally in the offline state. When there is a demand, the root CA goes online again. Thereby ensuring the security of the root CA.

The certificate uploaded by both parties includes two parts. The first part is the data required by the X.509 format certificate, including Sequence number, signature algorithm identifier, signer name, validity period, principal name, principal public key. The second part is the verification node information. The required two-party address, Nodeaddress and Signeraddress (Fig. 1).

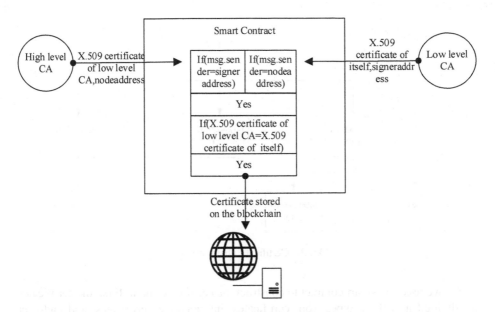

Fig. 1. Admission mechanism for CA nodes.

3.2 Construction of Certificate Chain

At present, the PKI system is centralized and hierarchical, and there are different levels among CAs. Except for root CA, every level of CA needs higher CA to issue certificates to it. Except for the root certificate, all the other certificates are intermediate certificates. We use intermediate certificates as proxies, because we must put the root certificate behind many security layers to ensure that its key is absolutely inaccessible. At the same time, in the real internet, the root CA is limited. If all authentication needs root CA to implement, its speed and efficiency can not be guaranteed. So we need to decentralize its operation and function, and finally establish trust relationship through certificate chain.

However, the current blockchain-based distributed PKI does not build such a certificate chain. Just use the blockchain for certificate storage, there is no logical relationship between certificates, they are independent of each other. As a result, the root CA can be freely accessed in the blockchain, without the protection of the security layer, and is easily attacked by malicious nodes. At the same time, the certificate stored in the blockchain cannot form a certificate chain. When the user views the certificate, the root CA trusted by the certificate cannot be determined, and the user cannot judge the trust level of the certificate. And it is also impossible to clearly define the level of CA in the blockchain, so it is impossible to determine who to apply for (Fig. 2).

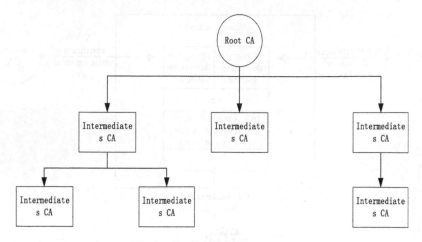

Fig. 2. Certificate chain structure

So we used the smart contract to construct the certificate chain. First, the certificates are divided into three types: root certificates, intermediate certificates, and end-user certificates. The organizational structure of CA is a tree structure. A root CAs contains multiple intermediates CAs, and intermediates can contain multiple intermediates. Both the root CAs and the intermediates CAs can issue certificates to users. The certificate that the end user uses to authenticate the public key is called end-user Certificates. The composition method of the certificate chain are shown in the Fig. 3 below.

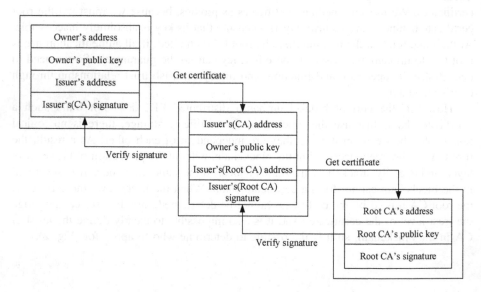

Fig. 3. The composition method of the certificate chain

The main method of forming a certificate chain in a contract is to map the CA's address to its certificate in the contract. Through its address, it can query the certificate it holds. The certificate records the public key of the signed CA as the signature. By analogy, we can form the entire certificate chain.

3.3   Conventional Certificate Management Function

As a PKI system, conventional certificate management is necessary, including the functions of issuing certificates, revoking certificates and updating certificates. The function of issuing certificates is limited to issuing end certificates. To achieve the functions of conventional certificate management, multiple functions in the contract need to work together. Take the complete step of adding a certificate to the blockchain as an example. First, the user submits an application through the Request function. After CA certification, it calls Add Certificate function to upload certificates. From the submission of applications by users to the final completion, two functions in the contract are invoked, namely Request function and Add Certificate function.

The process of revoking and updating the certificate function is slightly different from the above process. This process does not need the Request function, but only needs to verify whether the user of the function meets the requirements. The specific content will be described later.

Add Certificate Function

Adding certificate function is one of the main functions of PKI system. After a node applies for service, CA will authenticate and sign the certificate to the node, and then call adding certificate function to publish the certificate to the contract.

The parameters needed to add certificates are:

The parameters required here are divided into two parts. The first part is the data required for X.509 certificates.

Sequence number, signature algorithm identifier, signer name, validity period, principal name, principal public key.

The second part is the data needed for the contract.

Rank: The level of the requesting authentication entity in the blockchain is determined by the signature CA. And the level must be lower than the signature CA, because only the high-level node can sign the low-level node.

Nodeaddress: The address of the requesting authentication node. This address is used by CA for traceability authentication of the node. It is also used for transfer after certificate addition or for penalty after error.

Signeraddress: The address of the CA that signs the certificate. It is used to identify signature entity, transfer after certificate addition and punishment after error.

Addcertificates event records every call to the certificate addition function, and records the attributes of the certificate issued in the log. By indexing Nodeaddress, Signeraddress, principal name and principal public key, we can query the log with certificate function through these indexes, so that we can view the log in some cases. At the same time, we map Nodeaddress of each certificate, principal public key and node certificate. When we need to look at the relevant information of a node certificate, we can query it as long as we know its address or public key.

Revoke Certificate Function

Certificate revocation is used to revoke expired certificates and certificates with security problems. We stipulate here that only the node issuing the certificate can revoke the certificate. Other nodes have no authority to revoke the certificate. If the certificate revocation is successful, it will be put into the revocation certificate list.

The list of revoked certificates here stores key information for revoking certificates. Because of the inextricable modification of the blockchain, we can not delete the revoked certificate. Can only tell the user that the certificate has been revoked by recording its key information.

Certificate revocation function algorithm:

```
Function revokecertificate(address anodeaddress) public{
    Node revokenode=list[anodeaddress];
    Require(msg.sender=revokenode.signeraddress);
    RevokeCertificates.push(Certificate({nodeaddress:anodeaddress,signeraddr
ess:msg.sender}));
    REVOKECERTIFICATES(anodeaddress,msg.sender);
}
```

Update Certificate Function

Update certificate function is used to update certificate information. We stipulate here that only the node issuing the certificate can update the certificate. Other nodes do not have permission to update, and the certificate update will be put into the blockchain after success. Its algorithm is similar to certificate revocation function.

3.4   Error Certificate Feedback Mechanism

The feedback mechanism for discovering error certificates in existing PKI systems is limited. And there is no automated execution process. So we implement a reliable and secure error certificate feedback mechanism through two functions. As shown in Fig. 4, it is a process of error certificate feedback. The first function called is the Question function. The Question function is to upload the relevant information of the certificate to the contract, which is equivalent to proposing a proposal to vote. After uploading, each CA will see the certificate, and then the CA that meets the voting requirements will vote for the validity of the certificate according to the specific situation. After the voting time limit is reached, the contract will automatically terminate the Vote function and review the voting results. Check that the voting result meets the requirements and the contract will automatically add the certificate to the list of revoked certificates.

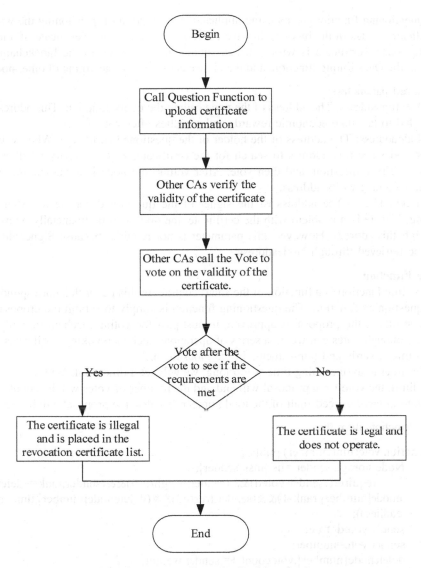

Fig. 4. Error certificate feedback mechanism

Question Function

The questioning function is based on the automation execution of smart contracts and effective economic rewards and punishments. The questioning function is not an essential part of the existing PKI system, but it is an important function in the PKI system based on smart contracts in this paper. The questioning function compensates for the lack of supervisors in the traditional PKI system, and also increases the economic penalties for the erroneous CA and the wrong certificate holder, so as to ensure the correctness of the nodes and certificates on the blockchain as much as possible.

Questioning function, as its name implies, is the function of questioning the wrong certificate. Nodes in the blockchain have the right to question any certificate. If a node thinks that a certificate is wrong, it can submit the certificate to the blockchain by calling the Questioning function, and then it needs to wait for the voting of other nodes.

Required parameters:

Questionaddress: The address of the node calling the query function. This address is recorded to facilitate economic rewards and penalties after voting.

Nodeaddress: The address of the holder of the questioned certificate. When voting, other nodes use this address to search for the certificate, so as to verify whether the certificate is in question, and then vote. After voting, the node is rewarded and punished according to its address.

Signeraddress: The address of the signer of the questioned certificate. After the voting, if there is a problem with the certificate, the node will be financially punished through this address. However, this parameter is not required because Signeraddress can be retrieved through Nodeaddress.

Vote Function

The voting function is a function of the smart contract in this paper that corresponds to the questioning function. The questioning function is simply to submit the proposal to the contract. If the proposal is approved, it must pass the voting mechanism. Only by getting enough votes can we do a series of operations such as revoking certificates and economic rewards and punishments in the blockchain.

In order to avoid voting time is too long, we will set a time limit. Within a certain time limit, the vote for a proposal will end and the number of votes will be counted. If the vote in favor exceeds half of the total number of votes, the proposal will be passed.

Vote function algorithm:

```
function vote(uint number) public{
    Node storage sender =list[msg.sender];
        require((!sender.voted)&&(sender.weight!=0)&&(sender.rank==delet
    enode[number].rank-1)&&(sender.trust>0)&&(deletenode[number].time<d
    eadline));
    sender.voted=true;
    sender.vote=number;
    deletenode[number].votecount += sender.weight;
}
```

3.5 Automatic Economic Reward and Punishment Mechanism

In order to maintain the high efficiency and stability of the whole system and ensure that each node can consciously maintain the entire blockchain, we use the smart contract to set up an automatic economic reward and punishment mechanism. Because it is executed automatically, it will not be disturbed by the outside world. As long as the node achieves the goal, it can harvest the set economic reward. Correspondingly, if the node is evil, it will automatically be punished economically, avoiding the trust flexibility problem in the traditional PKI.

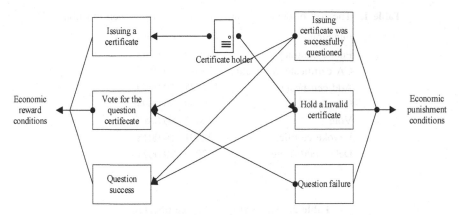

Fig. 5. The framework of the entire economic reward and punishment mechanism

As shown in Fig. 5, it shows the framework of the entire economic reward and punishment mechanism, including the conditions for economic rewards and punishments, as well as the flow of funds for penalties and rewards. The conditions required for economic rewards include: certificate certification, participation in certificate question voting, and questioning certificate success. The reward for certification comes from the node that obtains the certificate; the reward for participating in certificate questioning voting comes from the economic penalty for CA that issues and holds the wrong certificate or the economic penalty for CA that has invalid questioning. The reward for questioning the success of certificates comes from the economic penalties for CAs that issue and hold incorrect certificates. The conditions for economic punishment include: successful issuance of certificates, holding of invalid certificates and questioning invalidity. When successful issuance of certificates is questioned, not only economic punishment is imposed on the CA issuing certificates, but also all the economy of the holder of the certificates is liquidated. When the CA is questioned invalidity, the penalty amount is equally divided by the CA participating in the certificate questioning voting.

4 Feasibility and Safety Analysis

4.1 Feasibility Analysis

Lab environment: We wrote smart contracts by using remix and used a JavaScript VM to test the smart contract-based PKI system in this article. All testing processes were tested in the VM.

For the Ethereum system, the economic cost of contract construction and operation is the most important part of the feasibility. Therefore, we regard the post-deployment operating cost as the main content of the feasibility analysis.

After the deployment is complete, we tested each function in the contract, and then got the cost of each function call, we mainly analyze these costs. As of March 2019, 1 gas = 0.000000018ether [12], and 1 ether = $4.755 [13] (Table 1).

Table 1. The cost of deploying the contract and using each function.

Function	Gas	Gas in USD
Public contract	893002	$0.0763
CA certificate verification	225611	$0.0193
Add certificate	174828	$0.0149
Question	110138	$0.0094
Vote	22497	$0.0019
Revoke certificate	22001	$0.0018
Delete certificate	27099	$0.0021

Table 2. The cost of adding certificates.

Date(bytes)	Gas	Gas in USD
500	350081	$0.0299
600	415278	$0.0355
700	480231	$0.0411
800	575130	$0.0492
900	650001	$0.0556
1000	720526	$0.0616

This table shows the cost of deploying the contract and using each function, and the cost of deploying the contract requires −¥. In addition to the deployment contract, the most expensive cost are the CA certificate verification function and Add certificate function, because the two functions are the main functions of the contract, the most parameters are required to be passed in, and the node data required in the contract is passed through this function. The second is the challenge function. Because the node is to be challenged, it is necessary to upload the challenge certificate information, which also requires a relatively large overhead. Other functions are basically to operate on the data already in the blockchain, so the amount of money required is relatively small (Table 2).

The price of the certificates in the existing PKI is not uniform, but the cheapest dv certificate also needs about 100 yuan, and the general price of other certificates is about 10,000 yuan. We can see that all the costs of the contract are small compared to the fees for certification services in the PKI system, so the PKI system on smart contracts is realistic and feasible.

4.2 Safety Analysis

Previously, we mentioned several existing problems of PKI. Next, we make a security analysis of the system in this paper to solve these problems.

Certificates Can Be Issued Without the Authorization of the Domain Name Owner

We all know that smart contracts are only executed when certain conditions are met. Therefore, by setting conditions, we can realize that only when a node applies for authentication service, the corresponding certificate adding function in the contract can be invoked. And we add the process of authentication to ensure that the information of the application node and the certificate-holding node is the same. Through this series of settings, the problem of CA issuing arbitrary certificates directly without domain name owner's authorization is avoided.

Lack of Trust Flexibility

The error certificate feedback mechanism and economic reward and punishment mechanism in this paper are to solve this problem. When a CA is not credible, we use the error certificate feedback mechanism, and the certificate of an untrustworthy CA will be revoked. When a CA issues an error certificate, we use the economic reward and punishment mechanism to punish it economically, and the amount of money in the blockchain also represents its credibility. Through the cooperation of these two mechanisms, we can solve the problem of lack of trust flexibility.

Invalidity of Revocation

To solve this problem, we add a list of revoked certificates to the blockchain, which is realized by revoking certificates function in the contract. As long as the node participates in the blockchain, it needs to update the block. After updating the block, you can know the revoked certificate in the list of revoked certificates in the block. The main reason why the existing revocation does not take effect is that the revocation information can not be updated in time [14]. However, it only takes 10 s to update the block in Ethereum, so that the revocation certificate list can be updated quickly, thus solving the problem of revocation does not take effect.

4.3 Comparison with Current Research

In the current research situation, we analyze the advantages and disadvantages of each research. In view of the above research shortcomings, we propose a new smart contract-based PKI system in this paper.

Compared to IKP, Pemcor and Certcoin, these three improvements. We effectively use the automatic response and economic reward and punishment mechanisms of smart contracts to punish the issuing of wrong certificates and nodes that have experienced wrong behavior, and even revoke node certificates in the blockchain. It fills the gap of the existing distributed PKI for node management, and it is no longer simply a storage certificate.

Compared with SCPKI, we have not chosen a PGP network that cannot be used in a large number of nodes, but based on the traditional PKI to avoid the problem of trust chain breakage that occurs during large-scale application. And by setting the question function and vote function, it provides a error certificate feedback mechanism, not just offline feedback.

At the same time, we also set up a revocation certificate list, using the advantage of short blockchain update time to achieve faster update of the revocation certificate, avoiding the problem of revocation not effective.

5 Conclusions and Future Work

This paper describes a Internet web trust system based on smart contract. In the article, the specific content and implementation methods of each function are described, and the architecture is also elaborated. Finally, through experiments, the feasibility of this distributed PKI system is proved. At present, the smart contract-based PKI system of this paper has many advantages for the existing distributed PKI system, and it is very feasible, but it also has its own shortcomings. These shortcomings are mainly due to the disadvantages of the blockchain itself. The blockchain itself is difficult to complete the storage of big data, so the storage of a large number of certificates becomes the bottleneck of the system. At the same time, key recovery is also a problem that needs to be solved. In the identity management based on blockchain, we must first guarantee the security of the certificate. Secondly, the number of nodes needs to be guaranteed, and everyone can join the blockchain to view the certificate. So our consensus mechanism must ensure scalability and security. In blockchain-based identity management, it is inevitable to store some private identity information on the blockchain, but the blockchain itself is publicly accessible. Therefore, we must ensure the accessibility of the identity information on the blockchain and the privacy protection of the information [15]. The above questions are the next step for us.

Acknowledgements. This work is supported by the National Natural Science Foundations of China (grant No. 61802436 and No. 61702550) and the National Key Research and Development Plan (grant No. 2018YFB0803603 and No. 2016YFB0501901).

References

1. Is This MITM Attack to Gmail's SSL?(5). https://productforums.google.com/forum/#!msg/gmail/3J3r2JqFNTw/oHHZLJeed-HMJ. Accessed 20 Mar 2019
2. http://www.cnbeta.com/articles/tech/220690.htm. Accessed 20 Mar 2019
3. Ellison, C., Schneier, B.: Ten risks of PKI: What you're not being told about public key infrastructure. Comput. Secur. J. **16**(1), 1–7 (2000)
4. Nakamoto, S.: Bitcoin: a peer-to-peer electronic cash system. Consulted (2008)
5. A next-generation smart contract and decentralizedapplication platform (5) (2016). https://github.com/ethereum/wiki/wiki/WhitePaper/784a271b596e7fe4e047a2a585b733d631fcf1d4. Accessed 20 Mar 2019
6. Matsumoto, S., Reischuk, R.M.: IKP: turning a PKI around with decentralized automated incentives. In: 2017 IEEE Symposium on Security and Privacy (SP), pp. 410–426. IEEE (2017)
7. Corella, F.: Implementing a PKI on a Blockchain. Pomcor Research inMobile and Web Technology (5). https://pomcor.com/2016/10/25/implementing-a-pki-on-a-blockchain/. Accessed 20 Mar 2019

8. Fromknecht, C., Velicanu, D., Yakoubov, S.: A decentralized public key infrastructure with identity retention. IACR Cryptology ePrint Archive 2014/803 (2014)
9. Wikipedia: Namecoin (5). https://en.wikipedia.org/wiki/Namecoin. Accessed 20 Mar 2019
10. Al-Bassam, M.: SCPKI: a smart contract-based PKI and identity system. In: Proceedings of the ACM Workshop on Blockchain, Cryptocurrencies and Contracts, pp. 35–40. ACM (2017)
11. Garfinkel, S.: PGP: Pretty Good Privacy. O'Reilly & Associates, Newton (1995)
12. https://ethstats.net/. Accessed 20 Mar 2019
13. https://coinmarketcap.co. 20 Mar 2019
14. Orman, H.: Blockchain: the emperors new PKI? IEEE Internet Comput. **22**(2), 23–28 (2018)
15. Jiang, W., Li, H., Xu, G., et al.: PTAS: Privacy-preserving Thin-client Authentication Scheme in Blockchain-Based PKI, Future Generation Computer Systems (2019). https://doi.org/10.1016/j.future.2019.01.026

Destructive Method with High Quality and Speed to Counter Information Hiding

Feng Liu, Xuehu Yan, and Yuliang Lu[✉]

National University of Defense Technology, Hefei 230037, China
publicLuYL@126.com

Abstract. Hidden information behavior is becoming more difficult to be detected, causing illegal information spreading. To counter information hiding, more complex and high-dimensional steganalysis methods are applied, and have good detection for some specific information hiding algorithms. However, huge time consumption is needed and only suitable for offline usage with this method. In this paper, a destructive method with high quality and speed is proposed to counter information hiding. The experimental result presents that the extracting error rate of hiding information is around 50%, which means the method can destroy the possible covert communication completely. At the same time, the quality of covers after the destructive operation is good and the average time of operating an image is millisecond.

Keywords: Information hiding · Steganalysis · High quality and speed · Destructive

1 Introduction

With the continuous development of information technologies, a wide variety of digital medium are rapidly emerging on the Internet. And information hiding technologies develop rapidly with the rise of digital media. Information hiding is an efficient technique that hides secret information into the medium. Information hiding uses the normal digital medium, such as images, audios and videos as the covers, to embed the secret information into the covers. At the same time, the behavior of embedding data will not change the visual and auditory effect of the covers, nor does it change the size and format of the covers. So it's hard to find the hiding information.

As the rapid development of information hiding, a lot of steganographic algorithms are proposed. Steganographic algorithms can be generally divided into two categories: the spatial domain and the frequency domain. The spatial do- main algorithms directly embed the secret data into the covers, such as the classic Least Significant Bit (LSB) algorithm. The frequency domain algorithms embed the secret data into the spectrum space, such as Discrete Cosine Transform (DCT) and Discrete Wavelet Transformation (DWT). The early information hiding algorithms are just for the purpose of hiding data. While this can easily be detected using statistics techniques. So some researchers proposed lots of more secure algorithms that have the ability of adaptive emdedding, such as WOW [1] and HUGO [2] algorithms for uncompressed

© Springer Nature Singapore Pte Ltd. 2019
X. Cheng et al. (Eds.): ICPCSEE 2019, CCIS 1058, pp. 312–322, 2019.
https://doi.org/10.1007/978-981-15-0118-0_24

images, JUNIWARD [3] and nsF5 [4] algorithms for JPEG images. At the same time, lots of unknown information hiding algorithms are more subtle [5].

When information hiding is widely used, it also provides a new way for illegal organizations and intelligence personnel to covert communication. Even it can do huge harm to society and country. For example, the September 11 attacks. The terrorists just used the technology to convey information. Therefore, we need to be able to effectively detect information hiding. So steganalysis algorithms against information hiding rise in response to the proper time and conditions.

Steganalysis, as Fig. 1 is the opposite of information hiding. It is mainly based on the statistical analysis to detect whether the cover contains secret information. However, with the development of information hiding, current steganographic algorithms can achieve hiding information while maintaining the statistical characteristics, as we mentioned above. So some reaearchers consider more complex and high-dimensional statistical characteristics in order to counter the advanced adaptive information hiding algorithms [6]. And this is becoming a trend in steganalysis field. Hence, the cost of time is more and more expensive while improving detection accuracy.

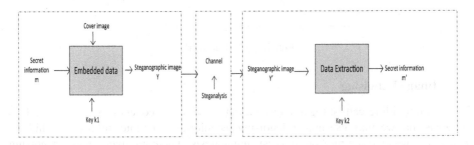

Fig. 1. Steganalysis framwork

However, the current steganalysis algorithms are almost offline, so that the illegal information may have been successfully transmitted. Therefore, it is necessary to propose a method that can deal with the suspicious information in time to assist steganalysis. In this paper, we propose a destructive method with high quality and speed to assist steganalysis destroying information hiding.

The rest of the paper is organized as follows. The preliminaries are briefly introduced in Sect. 2. Our method is proposed in Sect. 3. Section 4 provides the experimental results and the evaluations. And Sect. 5 gives the concludes of this paper.

2 Preliminaries

In this section, some preliminaries are given as the foundation for our method in order to conveniently understand our work.

2.1   Syndrome-Trellis Codes Framwork

STC framwork, as Fig. 2, is proposed by Filler and Fridrich in 2010. It is an adaptive information hiding framwok which defines a general form of distortion function. Users can minimize distortion by embedding secret information. In this paper, we use the framwork to embed and extract the secret information.

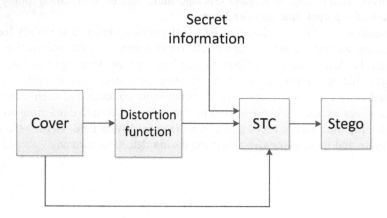

Fig. 2. SCT framwork

2.2   Image Denoising

Information hiding can be regarded as adding noise to the covers to some extent. This suggests that we may use image denoising techniques to counter information hiding. Therefore, we take the method to assist steganalysis. There are many image denoising methods, such as neighborhood average method, median filtering, Gaussian filtering, etc.

Neighborhood Average Method. Neighborhood average method is using the average value of surrounding pixel values instead of the original pixel value in the image. First, we will set a parameter which determines the size of a template composed of the neighbor pixels. Then we use the template to deal with each pixel $p(x, y)$, and calculate the average value. Finally, we assign the average value to the current pixel $p(x, y)$. The Eq. (1) is shown as follow:

$$p(x,y) = \frac{1}{m}\sum f(x,y) \tag{1}$$

where m is the total number of pixels in current template.

Median Filtering. Median filter uses the median value to replace the current pixel. Similarly, first we will set a parameter which determines the size of a filter window composed of the neighbor pixels. Then we arrange the pixel values in the window from smallest to largest. Finally, we assign the median value to the point $p(x, y)$. Let's assume the filter window A. So it can be defined as Eq. (2).

$$p(x, y) = MED\{p(x, y)\} \quad p(x, y) \in A \tag{2}$$

Gaussian Filtering. The Gaussian Filter will assign different weights to the surrounding pixels and then calculate the average pixel values for each point in the image, as Eq. (3). Similarly, firs we will set an odd-sized parameter which determines the size of the template, such as $(3 \times 3; 5 \times 5; 7 \times 7;...)$. Then we will assign different weights to the surrounding pixels, which the closer the pixel is, the greater the weight is. Finally, we calculate the weighted average value in the template and assign the value to the template center.

$$H(u, v) = e^{\frac{-D^2(u,v)}{2D_0^2}} \tag{3}$$

2.3   Image Quality Evaluation Index

When we process the image using our method, we need to evaluate the quality of the destructive image. So we need to introduce some common image evaluation indexes.

Peak Signal to Noise Ratio. *PSNR* [7], as Eq. (4), is used to assess the images similarity between the two images I and I^t, where *MSE* as Eq. (5) proves the mean square error.

$$PSNR = 10\log_{10}\left(\frac{255^2}{MSE}\right)dB \tag{4}$$

$$MSE = \frac{1}{W \times H}\sum_{i=1}^{W}\sum_{j=1}^{H}[I'(i,j) - I(i,j)]^2 \tag{5}$$

Structural Similarity. SSIM [8], is used to evaluate the visual impact of the three characteristics in an image, contrast, luminance, and structure, in Eq. (6), which gets a multiple combination of the three terms. SSIM value is in -1 and 1. The larger SSIM value is, the higher image similarity is.

$$SSIM(x, y) = [l(x, y)]^{\alpha} \cdot [c(x, y)]^{\beta} \cdot [s(x, y)]^{\gamma} \tag{6}$$

where

$$l(x, y) = \frac{2u_x u_y + C_1}{u_x^2 + u_y^2 + C_1}$$

$$c(x,y) = \frac{2\sigma_x\sigma_y + C_2}{\sigma_x^2 + \sigma_y^2 + C_2}$$

$$s(x,y) = \frac{2\sigma_{xy} + C_3}{\sigma_x + \sigma_y + C_3}$$

$\mu_x, \mu_y, \sigma_x, \sigma_y,$ and σ_{xy} are the local means, standard deviations, and cross-covariance for images x, y. In this paper, we set $C3 = \frac{C_2}{2}$, $\alpha = \beta = \gamma = 1$. Where D_0^2 represents the radius of passband.

3   Proposed Method

The details of our method framwork is shown in Fig. 3. For the suspicious area and people, we can not confirm whether the data stream on the internet from them contains secret information. Meanwhile, we only have a very short time to deal with the data stream in case arousing suspicion. However, the current steganalysis will spend a lot of time in detecting suspicious targets. Therefore, when we capture the data stream on the internet, we have no enough time to distinguish between legal or illegal information. In order to prevent illegal organizations from transmitting secret information, we propose our method to destroy the possible secret information, meanwhile costs milliseconds and maintains the image quality.

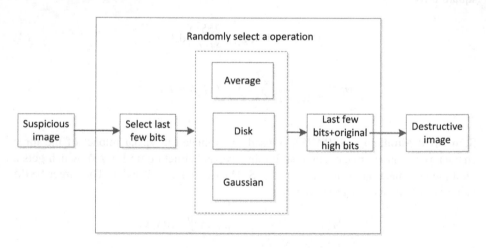

Fig. 3. Framework of proposed method

Secret information is usually embedded in the whole cover using steganographic methods, and we cannot locate the embedding position using existing technologies. Especially, for the unknown steganographic methods, we may even not find them using existing methods. So we propose the simple method to deal with these suspicious images whether they contain secret information or not. The algorithmic steps are described in Algorithm 1. First, we will select the last few bits of each pixel in the image in order to keep the image quality and the number of last few bits is set by the user. Second, we will calculate the last bits value by the surrounding pixel's least few bits values. Third, the new pixel value is obtained by combining the high pixel bits of the original image with the obtained low pixel bits. Finally, we will replace the original value with the new pixel value. In our method, we randomly choose three different type operations of *gaussian*, *disk* and *average*. And we also select the smallest parameters of each operation to further ensure the image quality.

Algorithm 1: the destructive method
Input: A $M \times N$ gray suspicious image   **Output**: A destructive image
Step 1: Select the last few bits of each position $(i,j) \in \{(i,j)\|1 \le i \le M, 1 \le j \le N\}$, repeat Steps 2-5.   **Step 2:** Randomly select a operation of *gaussian*, *disk* and *average*   **Step 3:** Set the corresponding parameters of the selected operation .   **Step 4:** For each last few bits of the position (i,j), do the selected operation.   **Step 5:** Combine the high pixel bits of the original image with the obtained low pixel bits.   **Step 6:** Output the destructive image.

4   Experiments and Evaluations

In this section, we will present our experiments and show the effectiveness.

4.1   Experiments

In our experiments, we use the four gray bmp images, as shown in Fig. 4, Lena, Man, Lake, and Baboon, with the size of 512×512, as covers. First, we utilize STC framework to embed the secret information with the payload of 0.05 bpp. Second, we select the last four or three (set by yourself) bits of the pixels in the image for experiment.

(a) Lena   (b) Man   (c) Lake   (d) Baboon

Fig. 4. Cover images (1) Lena. (2) Man. (3) Lake. (4) Baboon

Third, we randomly select a operation above and set the parameter for each selected last bits to destroy the possible secret information. Meanwhile we will record the total time of the second and third parts as the operation time. Finally, we try to extract the secret information and compare it with the original information to get the extraction error rate. And we will repeat 20 times for each image to avoid the randomness in our experiments.

We can see the relationship between data extracting error and repeat times in Figs. 5 and 6 for the four covers, Lena, Man, Lake, and Baboon.

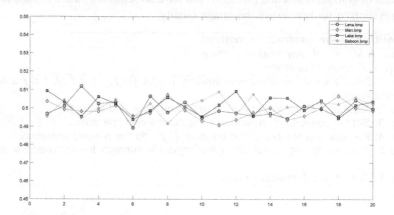

Fig. 5. Relationship between data extracting error and repeat times of last three bits with payload 0.05 bpp

From Figs. 5 and 6, we can see the values of extracting error are all around 50%, which means that the extracted bits are approximately random for the embedded information. So the possible secret information can not be obtained. The results of our experiments guarantee that the possible existing secret data is destroyed completely. The rest of this section is introduced the evaluations of quality and time with the experiments using the last three bits of each pixel.

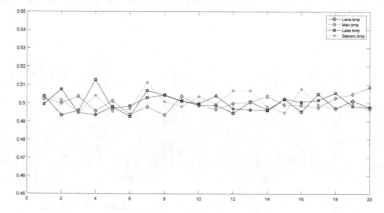

Fig. 6. Relationship between data extracting error and repeat times of last four bits with payload 0.05 bpp

4.2   Quality Evaluations

In this paper, we use PSNR and SSIM as the criterions to evaluate the quality of images. The qualities of destructive covers are shown in Tables 1 and 2. For each cover we experiment twenty trials and record the time of everytime. And we calculate the average quality of the twenty times as the last quality of the corresponding cover. According to the Tables 1 and 2, we can get that the destructive covers have high qualities.

Table 1. PSNR of the images

Repeat times	Lena	Man	Lake	Baboon
1	43.7	43.6	42.2	43.5
2	42	42	40.8	41.9
3	41.1	41.1	40.6	41.6
4	40.8	40.80	40.8	41.2
5	40.7	40.9	40.9	40.9
6	40.7	40.7	40.7	40.8
7	40.7	40.6	40.7	40.7
8	40.8	40.7	40.7	40.7
9	40.8	40.6	40.7	40.7
10	40.8	40.7	40.7	40.7
11	40.8	40.7	40.7	40.7
12	40.7	40.6	40.7	40.7
13	40.8	40.6	40.7	40.7
14	40.7	40.6	40.7	40.7
15	40.7	40.5	40.7	40.7
16	40.7	40.5	40.7	40.7
17	40.7	40.5	40.7	40.7
18	40.7	40.5	40.7	40.6
19	40.7	40.5	40.7	40.6
20	40.7	40.5	40.7	40.6
Average	40.965	40.86	40.79	40.97

Table 2. SSIM of the images

Repeat times	Lena	Man	Lake	Baboon
1	0.9408	0.9610	0.9491	0.9912
2	0.9178	0.9458	0.9345	0.9875
3	0.9032	0.9354	0.9311	0.9868
4	0.8972	0.9303	0.9333	0.9853
5	0.8950	0.9315	0.9345	0.9845
6	0.8943	0.9289	0.9329	0.9841
7	0.8953	0.9274	0.9321	0.9840
8	0.8961	0.9280	0.9326	0.9840
9	0.8967	0.9271	0.9322	0.9841
10	0.8972	0.9275	0.9325	0.9841
11	0.8964	0.9278	0.9323	0.9841
12	0.8961	0.9271	0.9325	0.9842
13	0.8963	0.9267	0.9324	0.9842
14	0.8962	0.9265	0.9326	0.9842
15	0.8964	0.9263	0.9325	0.9843
16	0.8963	0.9263	0.9326	0.9843
17	0.8964	0.9262	0.9326	0.9843
18	0.8964	0.9263	0.9327	0.9843
19	0.8965	0.9263	0.9328	0.9843
20	0.8966	0.9264	0.9328	0.9843
Average	0.8999	0.9304	0.9335	0.9849

4.3   Time Evaluations

Table 3 records the time taken for each experiment and the unit of record is seconds. We can see from Table 3 that each trial takes a very short time. Also we calculate the average time of the twenty times as the last time of the corresponding cover. According to the Table 3, we can know that our destructive method takes milliseconds.

Through the experiment, it show that our method can destroy the possible secret information completely [9] so as to realize the purpose of assisting steganalysis. And the quality of the destructive covers is high, meanwhile it takes only milliseconds.

Table 3. Times of the images(s)

Repeat times	Lena	Man	Lake	Baboon
1	0.0701	0.0052	0.0047	0.0039
2	0.0079	0.0080	0.0042	0.0040
3	0.0082	0.0046	0.0043	0.0042
4	0.0056	0.0049	0.0040	0.0041
5	0.0150	0.0046	0.0045	0.0042
6	0.0090	0.0044	0.0045	0.0043
7	0.0082	0.0040	0.0045	0.0040
8	0.0066	0.0042	0.0042	0.0044
9	0.0052	0.0040	0.0044	0.0042
10	0.0107	0.0043	0.0044	0.0043
11	0.0052	0.0042	0.0044	0.0043
12	0.0051	0.0039	0.0043	0.0043
13	0.0047	0.0046	0.0041	0.0045
14	0.0048	0.0045	0.0046	0.0043
15	0.0042	0.0046	0.0044	0.0043
16	0.0065	0.0041	0.0043	0.0045
17	0.0040	0.0045	0.0040	0.0042
18	0.0041	0.0040	0.0046	0.0041
19	0.0041	0.0042	0.0041	0.0042
20	0.0045	0.0044	0.0044	0.0044
Average	0.0062	0.0046	0.0043	0.0042

5 Conclusion

In this paper, we propose a destructive method with high quality and speed to counter information hiding on the suspicious images to assist steganalysis. And experimental result presents that the extracting error rate of hiding information is around 50%, which means our method can destroy the possible covert communication completely. At the same time, the quality of the covers after using our destructive operation is high and the average time of each cover is milliseconds.

Acknowledgement. The authors would like to thank the anonymous reviewers for their valuable comments. This work is supported by the National Natural Science Foundation of China (Grant Number: 61602491).

References

1. Holub, V., Fridrich, J.: Designing steganographic distortion using directional filters. In: 2012 IEEE International Workshop on Information Forensics and Security (WIFS), pp. 234–239. IEEE (2012)
2. Filler, T., Fridrich, J.: Gibbs construction in steganography. IEEE Trans. Inf. Forensics Secur. **5**(4), 705–720 (2010)

3. Holub, V., Fridrich, J., Denemark, T.: Universal distortion function for steganography in an arbitrary domain. EURASIP J. Inf. Secur. **2014**(1), 1 (2014)
4. Fridrich, J., Pevný, T., Kodovský, J.: Statistically undetectable jpeg steganography: dead ends challenges, and opportunities. In: Proceedings of the 9th Workshop on Multimedia & Security, pp. 3–14. ACM (2007)
5. Yan, X., Liu, X., Yang, C.N.: An enhanced threshold visual secret sharing based on random grids. J. Real-Time Image Process. **14**(1), 61–73 (2018)
6. Denemark, T., Sedighi, V., Holub, V., Cogranne, R., Fridrich, J.: Selection-channel - aware rich model for steganalysis of digital images. In: 2014 IEEE International Workshop on Information Forensics and Security (WIFS), pp. 48–53. IEEE (2014)
7. Huynh-Thu, Q., Ghanbari, M.: Scope of validity of PSNR in image/video quality assessment. Electron. Lett. **44**(13), 800–801 (2008)
8. Wang, Z., Bovik, A.C., Sheikh, H.R., Simoncelli, E.P.: Image quality assessment: from error visibility to structural similarity. IEEE Trans. Image Process. **13**(4), 600–612 (2004)
9. Qian, Z., Wang, Z., Zhang, X., Feng, G.: Breaking steganography: slight modification with distortion minimization. Int. J. Digit. Crime Forensics (IJDCF) **11**(1), 114–125 (2019)

Defence Against Adversarial Attacks Using Clustering Algorithm

Yanbin Zheng[1,2,3], Hongxu Yun[1], Fu Wang[1], Yong Ding[1], Yongzhong Huang[1], and Wenfen Liu[1(✉)]

[1] Guangxi Key Laboratory of Cryptography and Information Security,
Guilin University of Electronic Technology, Guilin, China
{zhengyanbin,liuwenfen}@guet.edu.cn
[2] School of Computer Science and Engineering,
South China University of Technology, Guangzhou, China
[3] School of Computer Science and Network Security,
Dongguan University of Technology, Dongguan, China

Abstract. Deep learning model is vulnerable to adversarial examples in the task of image classification. In this paper, a cluster-based method for defending against adversarial examples is proposed. Each adversarial example before attacking a classifier is reconstructed by a clustering algorithm according to the pixel values. The MNIST database of handwritten digits was used to assess the defence performance of the method under the fast gradient sign method (FGSM) and the DeepFool algorithm. The defence model proposed is simple and the trained classifier does not need to be retrained.

Keywords: Deep learning · Adversarial example ·
Adversarial attack · Clustering algorithm · Defence method

1 Introduction

Deep learning has been widely used in many fields, such as image classification, speech recognition and natural language processing. However, deep learning models are vulnerable to malicious attacks. In 2015, Goodfellow et al. [4] added some perturbations into an image of a panda, and then the perturbed "panda" was recognized as a "gibbon" with a probability of 99.3% by an advanced neural network classifier. The carefully-perturbed input samples are called adversarial examples, which aim to mislead detection at test time. Recent years, many algorithms (such as FGSM [4] and DeepFool [13]) are proposed to compute adversarial examples, and the perturbations generated by these algorithms are usually very small. But they can induce classifier to make false predictions with high-confidence. This phenomenon poses a potential threat to the applications that require high recognition accuracies, such as identity authentication [15] and autonomous driving [2]. Hence devising new methods for defending against adversarial examples is critical to the development of deep learning.

© Springer Nature Singapore Pte Ltd. 2019
X. Cheng et al. (Eds.): ICPCSEE 2019, CCIS 1058, pp. 323–333, 2019.
https://doi.org/10.1007/978-981-15-0118-0_25

To detect adversarial examples, Xu et al. [16] proposed a feature squeezing method. They coalesced samples that corresponded to many different feature vectors in the original space into a single sample, and so the search space available to an adversary was reduced. In this way, adversarial examples can be detected and then rejected. Ma et al. [10] characterized the dimensional properties of adversarial regions by using the "local intrinsic dimensionality", and showed how these could be used as features in the process of detecting adversarial examples, then rejected the adversarial examples. The disadvantage of defence mechanisms in [10,16] is that a lot of useful information will be lost.

Recently, the widely used defence mechanism is adversarial processing, which includes three methods: adversarial training, classifier hiding, and sample preprocessing. In the adversarial training method, the classifier is trained by adding adversarial examples to the training set, which can improve the robustness of the classifier, see for example [3,14]. Its disadvantage is that classifiers can still be deceived by other different adversarial examples generated algorithms. The classifier hiding method achieves the defensive effect by hiding the information of the classifier, such as weight matrix, activation function, and loss function. Its shortcoming is that adversarial examples could remain effective even for the models other than the one used to generate it. The attacker can train a shadow classifier by simulating prediction of the target classifier, and then use the shadow classifier to generate new adversarial examples aimed to attack the target classifier, see for example [1]. The third method, sample preprocessing, aims at reducing the impact of perturbations on the classifier. In [8], the perturbations in adversarial examples were treated as a kind of noise and introduced two classic image processing techniques, scalar quantization and smoothing spatial filter, to reduce its effect.

In [9], a high-level representation guided denoiser was proposed as a defence for image classification, which was trained by using a loss function defined as the difference between the target model's outputs activated by the original image and denoised image. McCoyd et al. [12] presented a defence of expanding the training set with a single, large, and diverse class of background images, striving to fill around the borders of the classification boundary. Background images can help detect and neutralize adversarial examples. Guo et al. [5] investigated strategies that defend against adversarial-example attacks on image-classification systems by transforming the inputs before feeding them to the system, such as bit-depth compression, JPEG compression, total variance minimization and image quilting. In summary, we can use Fig. 1 to demonstrate the main idea of the sample preprocessing defence method.

Fig. 1. The flowchart of the sample preprocessing mechanism

In this paper, in order to improve the effectiveness of defence strategies, we propose a cluster-based method for defending the attacks of adversarial examples, which belongs to the sample preprocessing mechanism. In our method, the adversarial examples, before attacking the classifier, are reconstructed by a clustering algorithm, and the classifier needs not to be retrained. Different from the artificially designed denoising algorithm, our approach does not require prior knowledge about adversarial examples and can retain the outline information.

The rest of the paper is organized as follows. Section 2 introduces the CNN, adversarial attack models and the K-means algorithm. Section 3 describes our defence method based on the K-means algorithm. Section 4 is the experimental part, and the performance of our defence method is verified on the MNIST dataset of handwritten digits. Section 5 summarizes the work of this paper.

2   Background

The algorithms used in our defence method will be introduced in this section.

2.1   Convolutional Neural Network

Convolutional Neural Network (CNN) is composed of convolutional layers and pooled layers. Each convolution layer consists of several convolution kernels of the same size. The convolution kernels can extract the local features of the sample during the backpropagation process, and they are independent of each other. Different convolution kernels of the same layer can extract different features. In the field of image recognition, the pooling layer compresses the features by usually taking the maximum value of the extracted features, which can help CNN to achieve higher accuracy in the task.

At present, almost all classical CNN models have high accuracy in image recognition. LeNet model is one of the classical CNN models, which was designed by LeCun [6] in 1998. LeNet model is simple and requires low training cost. The MINIST dataset used in this experiment is a set of grayscale images, and the size of the dataset is small. Using LeNet classifier can also achieve high accuracy. Therefore, LeNet model is selected in this paper. The architecture of the classic LeNet model is shown in the following Fig. 2:

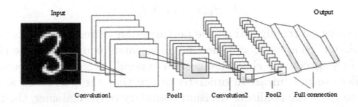

Fig. 2. Classic LeNet model architecture

2.2   Adversarial Attack Model

The classifier based on CNN can achieve high accuracy in the classifying task for the MNIST dataset. However, adding some perturbations to the sample could cause the classifier to makes mistakes. In the task of digital handwriting recognition, the classifier is defined as $f : R^n \to R^c$, where n (resp. c) is the dimension of the input (resp. output) vector. Assume that

$$\widehat{k}(x) = \underset{k}{\mathrm{argmax}}\, f_k(x), \tag{1}$$

where $f_k(x)$ is the output of the k-th class, the predicted class $\widehat{k}(x)$ is the one with the largest probability. We next briefly describe two algorithms for generating adversarial examples: FGSM and DeepFool.

FGSM (Fast Gradient Sign Method), proposed by GoodFellow et al. in [4], can quickly generate adversarial examples. This algorithm is a white box algorithm, and the attacker needs to know the loss function of the target classifier and the gradient matrix generated when the model is backpropagated. Given an image x, the FGSM sets

$$x' = x + \varepsilon \cdot \mathrm{sign}(\nabla_x \mathrm{loss}_f(x, y)), \tag{2}$$

where ε is the strength of perturbation and y is the correct label of x. When $\widehat{k}(x) = y$ and $\widehat{k}(x') \neq y$, the attack is successful. Intuitively, for each pixel, the FGSM uses the gradient of the loss function to determine in which direction the intensity of the pixel should be changed (whether it should be increased or decreased) to minimize the loss function; then it moves all pixels simultaneously. It is worth noting that FGSM was designed to be fast rather than optimal. The adversarial perturbation it produces is not necessarily minimal.

The DeepFool algorithm [13] used a L_2 minimization-based formulation to search for adversarial examples. The algorithm assumes that the decision boundary of the target classifier is linear, and the different classes are separated by some hyperplanes. Then DeepFool searches within these hyperplanes for the minimal perturbation that can change the decision of target classifier. First, one can compute an optimal perturbation

$$\begin{aligned} \widehat{r} = &\underset{r}{\mathrm{argmin}} \|r\|_2 \\ &\mathrm{s.t.}\ \widehat{k}(x) = y \\ &\exists k : f_k(x + r) \geq f_{\widehat{k}}(x + r), \end{aligned} \tag{3}$$

At each iteration of the algorithm, the perturbation vector that reaches the decision boundary is computed, and the current estimate is updated. The algorithm stops until the predicted class changes. In general, the decision boundaries of the classifier are not linear. In order to ensure that the perturbed adversarial example can cross the nonlinear decision boundary of the classifier, the final perturbation vector $\widehat{r}$ is multiplied by $1 + \eta$, where η is a control parameter. Then one generates an adversarial example:

$$x' = x + (1 + \eta)\,\widehat{r}. \tag{4}$$

DeepFool and FGSM are both white-box algorithms. However, compared with FGSM, the adversarial examples generated by DeepFool are very close to the original samples.

2.3   K-means Clustering Algorithm

The K-means algorithm [11] is a clustering algorithm based on the partitioning method. It divides N data objects into k clusters, and the data objects in the same cluster have higher similarity, the similarity between different clusters is lower. The similarity is calculated by the centroid of the cluster, that is, the average of all objects in a cluster. The K-means algorithm aims to minimize the value of an objective function:

$$E = \sum_{1 \le i \le k} \sum_{x \in C_i} \|x - \mu_i\|^2, \tag{5}$$

where x is the input sample, C_i is the partition of the k clusters of x,

$$\mu_i = \frac{1}{|C_i|} \sum_{x \in C_i} x \tag{6}$$

denotes the centroid of a cluster.

3   Defence Method Based on K-means Algorithms

In order to defend against adversarial attacks, we propose a defence method based on K-means algorithms. Our approach can be divided into four steps: (1) training the LeNet based classifier by using the MNIST database of handwritten digits, (2) generating adversarial examples by FGSM and DeepFool, respectively, (3) clustering each adversarial example by employing the K-means algorithm, reconstructing the adversarial example and then inputting it into the classifier, (4) computing the classification accuracy of the model. We use Fig. 3 to summarize our defence process.

The clustering algorithm can divide all pixel values in one sample into different clusters according to the similarity. The added perturbation makes tiny changes to the local pixel values without affecting the global features of the original sample. Therefore, in the clustering process, the perturbation points are assigned to the corresponding clusters according to the similarity as other pixels, which is the reason why we chose a clustering algorithm to process the adversarial examples. In fact, the value of the perturbation point will be averaged into the cluster which it belongs to in the next pixel reconstruction step. Therefore, the reconstructed samples do not contain adversarial perturbations. In this way, the impact of adversarial attacks is eliminated.

The process of clustering an adversarial example by using the K-means algorithm can be divided into four steps: (1) selecting k points from the pixels of the adversarial examples as the initial centroid. The samples in the MNIST database

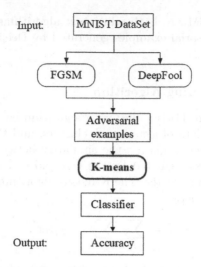

Fig. 3. The defence model based on the K-means algorithm

of handwritten digits are grayscale, so we use the K-means algorithm (with the hyperparameter $k = 2$) to divide all the pixels in the sample into two categories. (2) dividing them into clusters with the largest similarity by calculating the similarity between the remaining pixels and the centroid. (3) recalculating the average of each cluster as the new centroid. (4) repeating (2) and (3) until the objective function converges. From now on, all pixel values in the same cluster are replaced with the value of centroid of this cluster. Hence all the pixels in the adversarial example are pixel-reconstructed according to the category. In this process, the information carried by the pixel values is greatly simplified, and the position information of the pixels (i.e., the boundary of the digital image) is well preserved.

Our defence approach has three advantages over other defence methods appeared recently. First, our model is simple and does not require human prior knowledge when clustering the pixels in the sample. Second, our defence needs not to retrain the classifier. Third, even if the attacker knows our model structure and parameters, our defence method is effective under the condition that the classifier is misjudged by adding small perturbations. However, our method exhibits a weak defence effect when the perturbation is very large.

4   Experiment and Result Analysis

The experiment was programmed in the PyTorch framework using the Python language. All models were trained using the GeForce GTX 1060 graphics card under the Linux operating system. The experimental evaluation index is

$$\text{Accuracy} = \frac{TP + TN}{TP + TN + FP + FN}, \tag{7}$$

where TP, TN, FP and FN are defined in Table 1.

Table 1. The confusion matrix of classification results

Hypothesis	True class	
output	True	False
Positive	TP	FP
Negative	TN	FN

4.1 Dataset

The MNIST database of handwritten digits [7] has a training set of 60,000 examples and a test set of 10,000 examples. It is a subset of a larger set available from NIST. Each image in this subset contains 28×28 pixels and has a corresponding label, namely an integer in $[0, 9]$ shown in the image.

4.2 Experimental Process

First, we trained a classifier using the MNIST database of handwritten digits, and the classification accuracy on the test set was 99.28%. The structure and parameters of the classifier are shown in Tables 2 and 3.

Table 2. The structure of classifier

Layer type	Classifier
Convolution + ReLU	$5 \times 5 \times 20$
Max pooling	2×2
Convolution + ReLU	$5 \times 5 \times 50$
Max pooling	2×2
Fully connected + ReLU	500
Softmax	10

Table 3. The parameters of classifier

Parameter	Value
Learning rate	0.01
SGD momentum	0.5
Dropout	0.5
Batch size	64
Epochs	30

After the classifier is trained, the perturbations are added to the MNIST test set using the FGSM and DeepFool, respectively. Then the generated adversarial examples are directly inputted into the classifier and then count the output of the classifier. If the corresponding label is different between the output result and the input, the attack is successful. It should be noted that FGSM and DeepFool generate adversarial examples based on the gradient information of the target classifier. It means that for the same sample, the algorithm will generate different adversarial examples if the classifier is different.

After recording the success rate of the attack algorithm, we began to test our proposed defence method. In our defence method, the adversarial examples are clustered before its input the classifier. That is, all the pixels in the sample are clustered, and then the sample pixel values are reset according to the category of each pixel. Finally, we count the classification results, and then compare the accuracy of classification after the attack and defence.

4.3   Analysis of Experimental Results

As shown in Fig. 4(a), with the increase of the strength of perturbation ε, the classification accuracy curve (blue) of the FGSM attack classifier shows a rapid downward trend on the whole. The classifier accuracy curve (orange) tends to be gentle when $0 \leq \varepsilon < 0.15$. In this case, the classifier can accurately classify the adversarial examples after clustering, and the defence effect is obvious and stable. However, when $\varepsilon \geq 0.15$, the accuracy of the classifier begins to show a significant downward trend, and the classification accuracy decreases rapidly as the attack strength increases. Since FGSM adds perturbations to the opposite direction of the gradient descent and using FGSM to generate adversarial examples will change the value of all pixels in the original sample. It can be seen from Fig. 5 that as the perturbation coefficient increases, the perturbation in the adversarial example gradually becomes more visible, and the clustering effect becomes worse. When $\varepsilon \geq 0.25$, the perturbation added by FGSM in the adversarial example has made the clustering processing unable to work normally. In this case, the adversarial example loses the almost original features after it is clustered, and the classifier cannot recognize the image. Therefore, the defensive accuracy is lower than the attack accuracy if $\varepsilon > 0.25$.

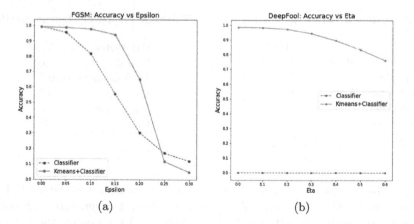

(a)   (b)

Fig. 4. The experimental results under the attack methods of FGSM and DeepFool. The horizontal axis represents the strength of perturbation and the vertical axis represents the classification accuracy. The blue line indicates the accuracy curve of the adversarial example directly into the classifier, and the orange line indicates the accuracy curve of the defence method based on the K-means clustering. (Color figure online)

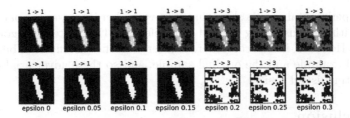

Fig. 5. The first row is the adversarial examples generated by the FGSM algorithm for the same sample at different attack strengths, and the second row is the result of clustering the adversarial examples in the first row. The clustering classifier can correctly classify 1 when $\varepsilon \leq 0.15$. When $\varepsilon \geq 0.2$, the clustered 1 is classified as 3.

It can be seen from Fig. 4(b) that the classification accuracy (blue line) of the DeepFool attack classifier is 0. The main reason is that the DeepFool algorithm does not stop until the attack is successful. More precisely, the DeepFool algorithm finds the location of pixels that have a greater impact on the classifier decision, and then add perturbation at these locations until the classifier is confused.

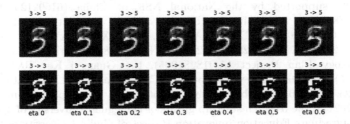

Fig. 6. The first row is the adversarial examples generated by the DeepFool algorithm for the same sample at different attack strengths, and the second row is the result of clustering the adversarial examples in the first row. The clustered 3 is classified as 3 when $\eta \leq 0.2$, and is classified as 5 when $\eta > 0.3$.

As the attack strength increases, the classifier accuracy curve (orange) in Fig. 4(b) tends to decrease slowly. Most of the adversarial examples generated by the DeepFool algorithm can be correctly identified by the classifier after being clustered. As can be seen from Fig. 6, as the degree of perturbation increases, the clustering effect gradually becomes worse. In this experiment, the parameter η cannot change the addition position of the pixel, and only change the value of the added perturbations. The pixel position of the perturbations is the same when $\eta = 0$ or 0.6, but the perturbed pixel values are different. When DeepFool adds a large perturbation to the key pixel position, it affects the local features after the sample is clustered, which causes the classifier to make errors. This is the reason why the classifier accuracy curve will slowly decline with the increase of attack intensity.

The experimental results show that if the attacker increases the perturbation strength without considering the cost, the defence of the clustering algorithm can be broken. However, in this case, there is a very significant difference between the adversarial example and the original sample, and so these adversarial examples can be distinguished easily.

5   Conclusion

Aiming at the problem that deep learning is vulnerable to adversarial examples, this paper designs a method based on clustering algorithm to defend against adversarial attacks. Experiments show that when the attacker adds less perturbation, our defence method can effectively average the perturbation, and can improve the classification accuracy of the image.

In the future, we will apply this defence method to RGB images. Due to the uncertain number of RGB image colour categories, the k-value selection problem of the K-means algorithm becomes more difficult. How to select the clustering algorithm and set the hyperparameter is the main research content in the future.

Acknowledgment. We would like to thank Zhichao Xia, Zhi Guo, Xiaodong Mu, Bixia Liu and Professor Yimin Wen for their helpful suggestions. The work was partially supported by the National NSF of China (61602125, 61772150, 61862011, 61862012), the China Postdoctoral Science Foundation (2018M633041), the NSF of Guangxi (2016GXNSFBA380153, 2017GXNSFAA198192, 2018GXNS-FAA138116, 2018-GXNSFAA281232, 2018GXNSFDA281054), the Guangxi Science and Technology Plan Project (AD18281065), the Guangxi Key R&D Program (AB17195025), the Guangxi Key Laboratory of Cryptography and Information Security (GCIS201625, GCIS201704), the National Cryptography Development Fund of China (MMJJ20170217), the research start-up grants of Dongguan University of Technology, and the Postgraduate Education Innovation Project of Guilin University of Electronic Technology (2018YJCX51, 2019YCXS052).

References

1. Bose, A.J., Aarabi, P.: Adversarial attacks on face detectors using neural net based constrained optimization. In: IEEE International Workshop on Multimedia Signal Processing, Vancouver, BC, Canada, 29–31 August 2018. https://doi.org/10.1109/MMSP.2018.8547128
2. Eykholt, K., Evtimov, I., Fernandes, E., et. al.: Robust physical-world attacks on deep learning visual classification. In: IEEE Conference on Computer Vision and Pattern Recognition, Salt Lake City, UT, USA, 18–22 June 2018, pp. 1625–1634 (2018)
3. Goodfellow, I.J., Pouget-Abadie, J., Mirza, M., et. al.: Generative adversarial nets. In: Advances in Neural Information Processing Systems, Montreal, Quebec, Canada, 8–13 December 2014, pp. 2672–2680 (2014). http://papers.nips.cc/paper/5423-generative-adversarial-nets
4. Goodfellow, I.J., Shlens, J., Szegedy, C.: Explaining and harnessing adversarial examples. In: International Conference on Learning Representations (2015). http://arxiv.org/abs/1412.6572

5. Guo, C., Rana, M., Cissé, M., et. al.: Countering adversarial images using input transformations. In: International Conference on Learning Representations (2018). https://openreview.net/forum?id=SyJ7ClWCb

6. Lecun, Y., Bottou, L., Bengio, Y., et. al.: Gradient-based learning applied to document recognition. Proc. IEEE **86**(11), 2278–2324 (1998)

7. LeCun, Y., Cortes, C., Burges, C.J.C.: The MNIST database of handwritten digits. http://yann.lecun.com/exdb/mnist/

8. Liang, B., Li, H., Su, M., et. al.: Detecting adversarial image examples in deep neural networks with adaptive noise reduction. IEEE Trans. Dependable Secure Comput. (2018). https://doi.org/10.1109/TDSC.2018.2874243

9. Liao, F., Liang, M., Dong, Y., et. al.: Defense against adversarial attacks using high-level representation guided denoiser. In: IEEE Conference on Computer Vision and Pattern Recognition, Salt Lake City, UT, USA, 18–22 June 2018, pp. 1778–1787 (2018). https://doi.org/10.1109/CVPR.2018.00191

10. Ma, X., Li, B., Wang, Y., et. al.: Characterizing adversarial subspaces using local intrinsic dimensionality. In: International Conference on Learning Representations (2018). https://openreview.net/forum?id=B1gJ1L2aW

11. MacQueen, J.B.: Some methods for classification and analysis of multivariate observations. In: Proceedings of the Fifth Berkeley Symposium on Mathematical Statistics and Probability, vol. 1, pp. 281–297. University of California Press (1967)

12. McCoyd, M., Wagner, D.A.: Background class defense against adversarial examples. In: IEEE Security and Privacy Workshops, San Francisco, CA, USA, 24 May 2018, pp. 96–102 (2018). https://doi.org/10.1109/SPW.2018.00023

13. Moosavi-Dezfooli, S., Fawzi, A., Frossard, P.: DeepFool: a simple and accurate method to fool deep neural networks. In: IEEE Conference on Computer Vision and Pattern Recognition, CVPR 2016, Las Vegas, NV, USA, 27–30 June 2016, pp. 2574–2582 (2016). https://doi.org/10.1109/CVPR.2016.282

14. Samangouei, P., Kabkab, M., Chellappa, R.: Defense-GAN: protecting classifiers against adversarial attacks using generative models. In: International Conference on Learning Representations (2018). https://openreview.net/forum?id=BkJ3ibb0-

15. Sharif, M., Bhagavatula, S., Bauer, L., Reiter, M.K.: Accessorize to a crime: real and stealthy attacks on state-of-the-art face recognition. In: Proceedings of ACM SIGSAC Conference on Computer and Communications Security, Vienna, Austria, 24–28 October, 2016, pp. 1528–1540 (2016). https://doi.org/10.1145/2976749.2978392

16. Xu, W., Evans, D., Qi, Y.: Feature squeezing: detecting adversarial examples in deep neural networks. In: 25th Annual Network and Distributed System Security Symposium, San Diego, California, USA, 18–21 February 2018 (2018). http://wp.internetsociety.org/ndss/wp-content/uploads/sites/25/2018/02/ndss2018_03A-4_Xu_paper.pdf

A Privacy-Preserving TPA-aided Remote Data Integrity Auditing Scheme in Clouds

Meng Zhao[1], Yong Ding[2(✉)], Yujue Wang[2], Huiyong Wang[3], Bingyao Wang[3], and Lingang Liu[2]

[1] School of Mechanical and Electrical Engineering,
Guilin University of Electronic Technology, Guilin 541004, China
[2] Guangxi Key Laboratory of Cryptography and Information Security,
School of Computer Science and Information Security,
Guilin University of Electronic Technology, Guilin 541004, China
stone_dingy@126.com
[3] School of Mathematics and Computing Science,
Guilin University of Electronic Technology, Guilin 541004, China

Abstract. The remote data integrity auditing technology can guarantee the integrity of outsourced data in clouds. Users can periodically run an integrity auditing protocol by interacting with cloud server, to verify the latest status of outsourced data. Integrity auditing requires user to take massive time-consuming computations, which would not be affordable by weak devices. In this paper, we propose a privacy-preserving TPA-aided remote data integrity auditing scheme based on Li et al.'s data integrity auditing scheme without bilinear pairings, where a third party auditor (TPA) is employed to perform integrity auditing on outsourced data for users. The privacy of outsourced data can be guaranteed against TPA in the sense that TPA could not infer its contents from the returned proofs in the integrity auditing phase. Our construction is as efficient as Li et al.'s scheme, that is, each procedure takes the same time-consuming operations in both schemes, and our solution does not increase the sizes of processed data, challenge and proof.

Keywords: Cloud storage · Integrity auditing ·
Provable Data Possession · Proofs of Storage · Proofs of Retrievability

1 Introduction

Cloud computing provides powerful storage and computing services for end users, which has been widely adopted in peoples' daily work and life. With the cloud computing technology, users are able to engage cloud servers to store large-scale data [6]. In this way, the local storage costs of users could be significantly saved and the outsourced data could be accessed freely at any time and place via Internet. This is extremely useful for the resource-restricted end devices of users, where such devices do not have enough storage spaces for massive data.

© Springer Nature Singapore Pte Ltd. 2019
X. Cheng et al. (Eds.): ICPCSEE 2019, CCIS 1058, pp. 334–345, 2019.
https://doi.org/10.1007/978-981-15-0118-0_26

However, data outsourcing also brings security and privacy issues to user data [10,15,25]. In such cloud-aided storage scenario, the cloud server totally controls the outsourced data of users, and usually does not fully trustworthy by end users. To save storage costs, a malicious cloud server may delete some rarely accessed data. Also, for some economic interests, it may change outsourced data in a way without being detected by data owner. These behaviors of the cloud server pose the integrity issue to outsourced data.

To address the above mentioned integrity issue in cloud storage scenario, several cryptographic primitives such as *Provable Data Possession* (PDP) [2], *Proofs of Retrievability* (PoR) [11], and *Proofs of Storage* (PoS) [4,12] have been introduced. Publicly verifiable PDP/PoR/PoS schemes allow any user to audit the integrity of outsourced data, whereas privately verifiable ones only support integrity auditing by someone holding the private key of data owner. Besides, the *remote data integrity checking/auditing* technology has also been extensively studied to solve the same security problem as by PDP/PoR/PoS.

1.1 Our Contributions

In this paper, we consider TPA-aided data integrity auditing in clouds, where only the specified TPA is able to perform integrity auditing for users. Also, the integrity auditing procedure would not compromise the privacy of outsourced data, that is, in the integrity auditing phase, TPA cannot learn the real contents of outsourced data. To address these issues, we introduce a privacy-preserving TPA-aided remote data integrity auditing system (P^2RIA) and present a concrete P^2RIA construction based on Li et al.'s data integrity auditing scheme without bilinear pairings (WiBPA). Particularly, in the data processing procedure, two secret elements are randomly chosen by user to generate metadata for data blocks. These two elements are shared with TPA to perform integrity auditing. Thorough analysis and comparison demonstrate that our P^2RIA construction is as efficient as Li et al.'s original scheme in each procedure.

1.2 Related Works

To address the integrity issue of outsourced data in clouds, Ateniese et al. [3] introduced the notion of *Provable Data Possession*. With their PDP technology, the integrity of outsourced data can be verified without retrieving the entire data from the server. To achieve this goal, user data should be processed before being uploaded to the cloud server. A generic way in processing user data is to split it into blocks and then produce a verifiable metadata for each block. Since Ateniese et al.'s scheme [3], a large number of PDP-related schemes have been proposed by following the same data processing framework.

Yu et al. [27] presented an identity-based integrity checking scheme for secure cloud data, which provides privacy guarantee against the third party verifier. To support applications in a multiuser setting, an identity-based PDP scheme is presented from pre-homomorphic signatures in [26]. In [19], Wang developed a secure scheme to address the integrity issue of outsourced data in multi-cloud storage setting. He et al. [8] investigated the integrity protection issue for smart grid data in clouds and presented an efficient certificateless PDP solution. A certificateless PDP scheme with privacy guarantee of outsourced data against auditor is proposed in [9]. Wang et al. [21] identified the integrity issue for outsourced electronic medical/health records of critical patients in public clouds, where the proposed solution allows some authorized entities and hospital to share some key medical/health data of patient in emergency case.

In multi-user setting, delegated data outsourcing brings additional security issue to user data. For example, the proxy may abuse the delegation to process and outsource data, which has been studied and solved in [20,23]. Wang et al. [24] observed that the data processing procedure in public verifiable PDP schemes takes too many time-consuming exponentiations, which may not affordable by resource-restricted user devices. They introduced a semi-generic online/offline framework to convert such schemes satisfying some conditions into their online/offline version, so that the time-consuming operations can be precomputed in the offline phase before seeing the real data.

Juels and Kaliski [11] introduced *Proofs of Retrievability*, which is slightly stronger than PDP in that PoR can ensure the retrievability of outsourced data in clouds. In their original proposal, some special sentinels are inserted into the processed data for auditing the behavior of cloud server. Thus, their proposal only supports a fixed number rounds of integrity auditing. Shacham and Waters [16] further presented both privately and publicly verifiable PoR schemes, and for the first time proved their security in strong model. Fan et al. [7] designed a secure remote storage scheme to provide data privacy guarantee against auditor. Cui, Mu, and Au [5] considered related-key attacks to PoR scheme, and presented a publicly verifiable PoR scheme proved secure against such attacks. Armknecht et al. [1] introduced outsourced PoR to delegate the integrity auditing task to a TPA, where cloud user is able to verify the auditing work of TPA.

The privacy of outsource data against cloud server is also an important issue in clouds. In [13], Kononchuk et al. noticed that user-data oriented services only enjoy privacy-preserving property in static settings, and further provided a privacy-preserving cryptographic solution for dynamic settings. Tupakula and Varadharajan [18] presented a mechanism that allows the cloud service provider to monitor and identify malicious transactions. Wang et al. [22] considered the privacy of outsourced road condition data in clouds, and presented a cryptographic solution with one-wayness privacy guarantee. The cloud server can be authorized to conduct many works for users on outsourced data, e.g., processing k nearest neighbors [17], without compromising data privacy.

1.3 Paper Organization

The remainder of this paper is organized as follows. Section 2 defines the system model of secure cloud storage with TPA-aided integrity auditing, and summarizes the security requirements. Section 3 reviews the details of Li et al.'s WiBPA scheme. Section 4 provides a concrete P^2RIA construction followed with security analysis and efficiency comparison. Finally, Sect. 5 concludes the paper.

2   System Model and Requirements

2.1   System Model

As shown in Fig. 1, a privacy-preserving TPA-aided remote data integrity auditing system (P^2RIA) consists of three types of entities, that is, users, cloud server, and TPA. To save local storage costs, user data are outsourced to a cloud server. The cloud server has powerful storage and computing resources. It is not fully trustworthy in that it may remove some rarely accessed data or forge a valid proof to respond the user's integrity challenge, e.g., for saving storage resources. TPA can perform integrity auditing on outsourced data on behalf of users, by interacting with the cloud server. However, TPA is curious about the contents of outsourced data.

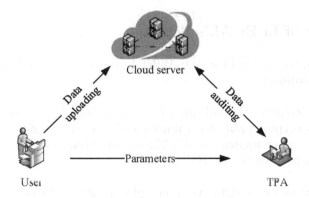

Fig. 1. P^2RIA system model.

Before outsourcing data to the cloud server, user performs data processing procedure using her private key, in this way to generate the corresponding verifiable tags for the data. In this procedure, the data is split into many blocks such that the data tags are separately generated for every data block. These tags and the original data are uploaded to the cloud server, while the parameters employed in the data processing procedure are sent to TPA for performing integrity auditing. Thus, only TPA is able to audit the integrity of outsourced data in the P^2RIA system, except the user herself.

To further protect the privacy of outsourced data, TPA should not be able to learn the real data contents in the integrity auditing phase. In fact, in standard PDP-related schemes, the auditor is able to obtain a number of aggregated data blocks, which would let the auditor deduce the real values of data blocks. As required in Li et al.'s WiBPA system, in responding an integrity challenge from TPA, the integrity proof generated by cloud server must be masked using random values. Thus, TPA would not obtain the real value of aggregated data blocks.

2.2 System Requirements

A privacy-preserving integrity auditing system for cloud data should satisfy the following security requirements.

Efficient Auditing : To auditing the integrity of outsourced data, TPA does not need to request the whole data in a challenge, and the cloud server does not need to respond the entire data.

Data Security : For outsourced data, the cloud server is unable to launch forgery attacks and replace attacks without being caught by TPA.

Data Privacy : During the integrity auditing process, TPA is unable to recover the real contents of outsourced data.

3 Review of Li Et Al.'s Scheme

In this section, we review Li et al.'s WiBPA scheme [14], which consists of the following procedures.

KeyGen : User constructs a multiplicative cyclic group $G = \langle g \rangle$ with prime order p. She picks a cryptographic hash function $H : \{0,1\}^* \rightarrow Z_p$. User randomly picks $x \in_R Z_p^*$ and computes $y \leftarrow g^x$. Thus, user obtains the private key $\mathsf{sk} = x$ and public key $\mathsf{pk} = (g, G, p, y, H)$.

TagGen : For an input data M, user splits it into n blocks such as $M = (m_1, m_2, \cdots, m_n)$, where $m_i \in Z_p$ for each $1 \leq i \leq n$. User picks a unique file name $\mathcal{N}$ for data M. For each data block m_i $(1 \leq i \leq n)$, user sets the version number $V_i \leftarrow 1$ and the timestamp T_i, and computes metadata as follows

$$\sigma_i \leftarrow \left(g^{m_i + H(W_i)} \right)^x \in G$$

where $W_i \leftarrow \mathcal{N} \| B_i \| V_i \| T_i$ and $\|$ denotes concatenation of strings.

At last, user uploads the processed data $M' = \{M, \mathcal{N}, \Phi = (\sigma_1, \sigma_2, \cdots, \sigma_n)\}$ to the cloud server, and sends the data parameters $\{\mathcal{N}, n, V = (V_1, V_2, \cdots, V_n), T = (T_1, T_2, \cdots, T_n)\}$ to TPA.

ChalGen : TPA randomly picks a subset $I \subset \{1, 2, \cdots, n\}$. Then, he randomly picks an element $v_i \in_R Z_p^*$ for each $i \in I$. TPA sends the challenge $chal = \{\mathcal{N}, (i, v_i) : i \in I\}$ to the cloud server.

PrfGen : Cloud server randomly picks a value $r \in_R Z_p^*$, and computes as follows using the outsourced data M' with name $\mathcal{N}$:

$$\theta \leftarrow y^r \in G$$
$$\nu' \leftarrow \sum_{i \in I} v_i \cdot m_i$$
$$\nu \leftarrow \nu' + r \cdot H(\theta) \mod p$$
$$\sigma \leftarrow \left(\prod_{i \in I} \sigma_i^{v_i}\right)^r \in G$$
$$\mu \leftarrow \theta^r$$

Then cloud server returns the proof $Proof = \{\theta, \nu, \sigma, \mu\}$.

PrfVrf : TPA checks the following equality:

$$\mu^{H(\theta)} \cdot \sigma \stackrel{?}{=} \theta^{\nu + \sum_{i \in I} H(W_i) v_i} \tag{1}$$

If it holds, then TPA outputs "1", which means the outsourced data M' is kept intact; otherwise, TPA outputs "0", which implies the outsourced data M' has been tampered.

In [14], Li et al. also presented the procedures to support data update, which do not affect the security analysis in this paper and are thus omitted here. In [14, Section 4], Li et al. discussed that outsourced data can be protected from forgery and substitution attacks. However, the cloud server is able to tamper every data block in the outsourced data M' and forge valid metadata. For example, for data block m_i $(1 \leq i \leq n)$, the cloud server is able to pick a random element $\tilde{m} \in_R Z_p^*$, and make changes to (m_i, σ_i) as $(m_i' \leftarrow m_i + \tilde{m} \mod p, \sigma_i' \leftarrow \sigma_i \cdot y^{\tilde{m}})$.

4 Our P²RIA Construction

In this section, we present an efficient P²RIA construction based on Li et al.'s WiBPA scheme [14] (Fig. 2).

4.1 Concrete Construction

KeyGen : User constructs a multiplicative cyclic group $G = \langle g \rangle$ with prime order p. She picks two cryptographic hash functions $H_1 : \{0,1\}^* \rightarrow Z_p$ and $H_2 : G \rightarrow Z_p$. User randomly picks $x \in_R Z_p^*$ and computes $y \leftarrow g^x$. Thus, user obtains the private key $sk = x$ and public key $pk = (g, G, p, y, H_1, H_2)$.

TagGen : For an input data M, user splits it into n blocks such as $M = (m_1, m_2, \cdots, m_n)$, where $m_i \in Z_p$ for each $1 \leq i \leq n$. User picks a unique file name $\mathcal{N}$ for data M, and selects two random values $a, b \in_R Z_p^*$. For each data block m_i $(1 \leq i \leq n)$, user sets the version number $V_i \leftarrow 1$ and the timestamp T_i, and computes metadata as follows

$$\sigma_i \leftarrow \left(g^{a \cdot m_i + b \cdot H_1(W_i)} \right)^x \in G \tag{2}$$

where $W_i \leftarrow \mathcal{N} \| n \| B_i \| V_i \| T_i$.

At last, user uploads the processed data $M' = \{M, \mathcal{N}, \Phi = (\sigma_1, \sigma_2, \cdots, \sigma_n)\}$ to the cloud server, and sends the data parameters $\{\mathcal{N}, n, V = (V_1, V_2, \cdots, V_n), T = (T_1, T_2, \cdots, T_n)\}$ to TPA. She also sends a and b to TPA via secure channel.

ChalGen : TPA randomly picks a subset $I \subset \{1, 2, \cdots, n\}$. Then, he randomly picks an element $v_i \in_R Z_p^*$ for each $i \in I$. TPA sends the challenge $chal = \{\mathcal{N}, (i, v_i) : i \in I\}$ to the cloud server.

PrfGen : Cloud server randomly picks a value $r \in_R Z_p^*$, and computes as follows using the outsourced data M' with name $\mathcal{N}$:

$$\theta \leftarrow y^r \in G$$
$$\nu' \leftarrow \sum_{i \in I} v_i \cdot m_i$$
$$\nu \leftarrow \nu' + r \cdot H_2(\theta) \mod p$$
$$\sigma \leftarrow \left(\prod_{i \in I} \sigma_i^{v_i} \right)^r \in G$$
$$\mu \leftarrow \theta^r$$

Then cloud server returns the proof $Proof = \{\theta, \nu, \sigma, \mu\}$.

PrfVrf : TPA checks the following equality:

$$\mu^{a H_2(\theta)} \cdot \sigma \stackrel{?}{=} \theta^{a\nu + b \sum_{i \in I} H_1(W_i) v_i} \tag{3}$$

If it holds, then TPA outputs "1", which means the outsourced data M' is kept intact; otherwise, TPA outputs "0", which implies the outsourced data M' has been tampered.

Theorem 1. *The proposed P^2RIA construction is sound, that is, if the outsourced is kept intact, then Eq. (3) must hold in each round of integrity auditing.*

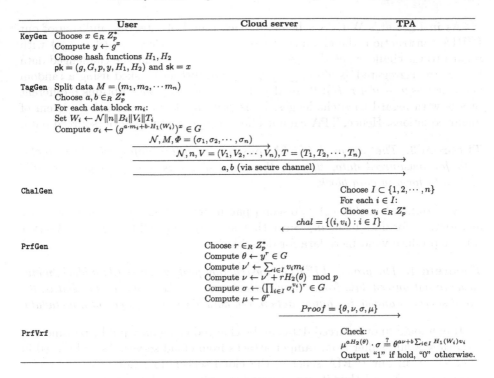

	User	Cloud server	TPA
KeyGen	Choose $x \in_R Z_p^*$		
	Compute $y \leftarrow g^x$		
	Choose hash functions H_1, H_2		
	pk $= (g, G, p, y, H_1, H_2)$ and sk $= x$		

TagGen — Split data $M = (m_1, m_2, \cdots m_n)$
Choose $a, b \in_R Z_p^*$
For each data block m_i:
Set $W_i \leftarrow \mathcal{N}\|n\|B_i\|V_i\|T_i$
Compute $\sigma_i \leftarrow (g^{a \cdot m_i + b \cdot H_1(W_i)})^x \in G$

$$\mathcal{N}, M, \Phi = (\sigma_1, \sigma_2, \cdots, \sigma_n) \longrightarrow$$
$$\mathcal{N}, n, V = (V_1, V_2, \cdots, V_n), T = (T_1, T_2, \cdots, T_n) \longrightarrow$$
$$a, b \text{ (via secure channel)} \longrightarrow$$

ChalGen — Choose $I \subset \{1, 2, \cdots, n\}$
For each $i \in I$:
Choose $v_i \in_R Z_p^*$
$$\xleftarrow{\quad chal = \{(i, v_i) : i \in I\} \quad}$$

PrfGen — Choose $r \in_R Z_p^*$
Compute $\theta \leftarrow y^r \in G$
Compute $\nu' \leftarrow \sum_{i \in I} v_i m_i$
Compute $\nu \leftarrow \nu' + r H_2(\theta) \mod p$
Compute $\sigma \leftarrow (\prod_{i \in I} \sigma_i^{v_i})^r \in G$
Compute $\mu \leftarrow \theta^r$
$$\xrightarrow{\quad Proof = \{\theta, \nu, \sigma, \mu\} \quad}$$

PrfVrf — Check:
$$\mu^{a H_2(\theta)} \cdot \sigma \stackrel{?}{=} \theta^{a\nu + b \sum_{i \in I} H_1(W_i) v_i}$$
Output "1" if hold, "0" otherwise.

Fig. 2. The proposed P^{2}RIA construction

Proof. We only need to show Eq. (3) holds in each round of integrity auditing.

$$\mu^{a H_2(\theta)} \cdot \sigma = (\theta^r)^{a \cdot H_2(\theta)} \cdot \left(\prod_{i \in I} \sigma_i^{v_i} \right)^r$$

$$= \theta^{a \cdot r \cdot H_2(\theta)} \cdot \prod_{i \in I} \left(g^{a \cdot v_i \cdot m_i + b \cdot v_i \cdot H_1(W_i)} \right)^{x \cdot r}$$

$$= \theta^{a \cdot r \cdot H_2(\theta)} \cdot \prod_{i \in I} \theta^{a \cdot v_i \cdot m_i + b \cdot v_i \cdot H_1(W_i)}$$

$$= \theta^{a \cdot r \cdot H_2(\theta)} \cdot \theta^{\sum_{i \in I} a \cdot v_i \cdot m_i} \cdot \theta^{\sum_{i \in I} b \cdot v_i \cdot H_1(W_i)}$$

$$= \theta^{a(r \cdot H_2(\theta) + \sum_{i \in I} v_i \cdot m_i)} \cdot \theta^{b \sum_{i \in I} v_i \cdot H_1(W_i)}$$

$$= \theta^{a\nu + b \sum_{i \in I} H_1(W_i) v_i}$$

Thus, Eq. (3) holds. □

4.2 Security Analysis

Theorem 2. *The proposed P^{2}RIA construction can protect the privacy of outsourced data against TPA, that is, TPA is unable to learn the real contents of outsourced data in performing integrity auditing.*

As in Li et al.'s WiBPA scheme, in each round of integrity auditing of our P^2RIA construction, the cloud server returns a proof $Proof = \{\theta, \nu, \sigma, \mu\}$ with regard to the challenge $chal = \{\mathcal{N}, (i, v_i) : i \in I\}$. Note that the challenged data blocks are aggregated as $\nu' \leftarrow \sum_{i \in I} v_i \cdot m_i$ and further masked using a random element r as $\nu \leftarrow \nu' + r \cdot H_2(\theta) \mod p$. Thus, even when TPA obtains n different proofs with regard to n challenges, he is still unable to construct a system of linear equations. Hence, TPA cannot infer the real contents of outsourced data.

Theorem 3. *The proposed P^2RIA construction offers unforgeability of meta-data for outsourced data, that is, the cloud server is unable to forge a valid metadata for any data block.*

As shown in Equality (2), two secret parameters a, b are needed in generating metadata σ_i for data block m_i. Since they are only sent to TPA, the cloud server cannot produce valid metadata for data blocks.

Theorem 4. *The proposed P^2RIA construction protects pairs (data block, meta-data) in outsourced data from tamper attacks against the cloud server, that is, the cloud server is unable to change outsourced data blocks and forge valid metadata.*

If metadata in outsourced data can be changed using only public parameters, then they would suffer from tamper attacks from cloud server. As elaborated in Theorem 3, in our P^2RIA scheme, the cloud server does not hold two secret parameters a, b, and thus it cannot produce valid metadata for tampered data blocks.

4.3   Comparison

In this section, we analyze the computational complexity of the procedures in our P^2RIA construction and compare with Li et al.'s WiBPA scheme [14]. The computational complexity analysis only focuses on the most time-consuming exponentiations in multiplicative cyclic group G, while modular additions, mod-ular multiplications and hash evaluations are light-weight and omitted. Let γ_G denote the evaluation time of an exponentiation in G. Table 1 summarizes the computation costs of each procedure in two schemes. As shown in Sect. 4.1, the challenge generation procedure ChalGen does not require any time-consuming operation. In the table, the computation costs of procedures PrfGen and PrfVrf are analyzed for one round of integrity auditing, where $|I|$ denotes the cardinal-ity of subset $I \subset [1, n]$. Thus, each procedure in our P^2RIA scheme enjoys the same efficiency as in Li et al.'s scheme [14].

With the TagGen procedure, the processed data M' contains n blocks and n metadata. Thus, its size is exactly the same as in Li et al.'s WiBPA scheme in processing the same data M. The elements in the challenge $chal$ and proof $Proof$ are not changed, which means they also have the same sizes as in Li et al.'s original scheme. Thus, our P^2RIA construction does not incur additional storage costs and communication costs compared with Li et al.'s WiBPA scheme.

Table 1. Comparison of computation cost for each procedure

	WiBPA scheme [14]	P^2RIA scheme
KeyGen	$1\gamma_G$	$1\gamma_G$
TagGen	$n\gamma_G$	$n\gamma_G$
ChalGen	–	–
PrfGen	$(|I|+3)\gamma_G$	$(|I|+3)\gamma_G$
PrfVrf	$2\gamma_G$	$2\gamma_G$

5   Conclusion

In this paper, we studied the security issues in a TPA-aided integrity auditing scenario for cloud data. To address the integrity issue against the cloud server and the privacy issue against TPA, we proposed an efficient P^2RIA construction based on Li et al.'s WiBPA scheme without bilinear pairings [14]. In the data processing procedure, we introduced two secret parameters in generating metadata for data blocks. Security analysis showed that our P^2RIA construction can resist tampering attacks from the cloud server. Through theoretical analysis, our P^2RIA construction is as efficient as Li et al.'s scheme in every procedure.

Acknowledgment. This article is supported in part by the National Natural Science Foundation of China under projects 61772150 and 61862012, the Guangxi Key R&D Program under project AB17195025, the Guangxi Natural Science Foundation under grants 2018GXNSFDA281054 and 2018GXNSFAA281232, the National Cryptography Development Fund of China under project MMJJ20170217, the Guangxi Young Teachers' Basic Ability Improvement Program under Grant 2018KY0194, and the open program of Guangxi Key Laboratory of Cryptography and Information Security under projects GCIS201621 and GCIS201702.

References

1. Armknecht, F., Bohli, J.M., Karame, G.O., Liu, Z., Reuter, C.A.: Outsourced proofs of retrievability. In: Proceedings of the 2014 ACM SIGSAC Conference on Computer and Communications Security, CCS 2014, pp. 831–843. ACM, New York (2014). https://doi.org/10.1145/2660267.2660310
2. Ateniese, G., et al.: Remote data checking using provable data possession. ACM Trans. Inf. Syst. Secur. **14**(1), 12:1–12:34 (2011). https://doi.org/10.1145/1952982.1952994
3. Ateniese, G., et al.: Provable data possession at untrusted stores. In: Proceedings of the 14th ACM Conference on Computer and Communications Security, CCS 2007, pp. 598–609. ACM, New York (2007). https://doi.org/10.1145/1315245.1315318
4. Ateniese, G., Kamara, S., Katz, J.: Proofs of storage from homomorphic identification protocols. In: Matsui, M. (ed.) ASIACRYPT 2009. LNCS, vol. 5912, pp. 319–333. Springer, Heidelberg (2009). https://doi.org/10.1007/978-3-642-10366-7_19

344   M. Zhao et al.

5. Cui, H., Mu, Y., Au, M.H.: Proof of retrievability with public verifiability resilient against related-key attacks. IET Inf. Secur. **9**(1), 43–49 (2015). https://doi.org/10.1049/iet-ifs.2013.0322
6. Date, S.: Should you upload or ship big data to the cloud? Commun. ACM **59**(7), 44–51 (2016). https://doi.org/10.1145/2909493
7. Fan, X., Yang, G., Mu, Y., Yu, Y.: On indistinguishability in remote data integrity checking. Comput. J. **58**(4), 823–830 (2015). https://doi.org/10.1093/comjnl/bxt137
8. He, D., Kumar, N., Zeadally, S., Wang, H.: Certificateless provable data possession scheme for cloud-based smart grid data management systems. IEEE Trans. Ind. Inf. **14**(3), 1232–1241 (2018). https://doi.org/10.1109/TII.2017.2761806
9. He, D., Kumar, N., Wang, H., Wang, L., Choo, K.K.R.: Privacy-preserving certificateless provable data possession scheme for big data storage on cloud. Appl. Math. Comput. **314**, 31–43 (2017). https://doi.org/10.1016/j.amc.2017.07.008
10. Islam, S., Ouedraogo, M., Kalloniatis, C., Mouratidis, H., Gritzalis, S.: Assurance of security and privacy requirements for cloud deployment models. IEEE Trans. Cloud Comput. **6**(2), 387–400 (2018). https://doi.org/10.1109/TCC.2015.2511719
11. Juels, A., Kaliski Jr., B.S.: PoRs: Proofs of retrievability for large files. In: Proceedings of the 14th ACM Conference on Computer and Communications Security, CCS 2007, pp. 584–597. ACM, New York (2007). https://doi.org/10.1145/1315245.1315317
12. Kamara, S.: Proofs of storage: theory, constructions and applications. In: Muntean, T., Poulakis, D., Rolland, R. (eds.) CAI 2013. LNCS, vol. 8080, pp. 7–8. Springer, Heidelberg (2013). https://doi.org/10.1007/978-3-642-40663-8_4
13. Kononchuk, D., Erkin, Z., van der Lubbe, J.C.A., Lagendijk, R.L.: Privacy-preserving user data oriented services for groups with dynamic participation. In: Crampton, J., Jajodia, S., Mayes, K. (eds.) ESORICS 2013. LNCS, vol. 8134, pp. 418–442. Springer, Heidelberg (2013). https://doi.org/10.1007/978-3-642-40203-6_24
14. Li, C., Wang, P., Sun, C., Zhou, K., Huang, P.: WiBPA: an efficient data integrity auditing scheme without bilinear pairings. Comput. Materi. Cont. **58**(2), 319–333 (2019). https://doi.org/10.32604/cmc.2019.03856
15. Li, X., Kumari, S., Shen, J., Wu, F., Chen, C., Islam, S.H.: Secure data access and sharing scheme for cloud storage. Wirel. Pers. Commun. **96**(4), 5295–5314 (2017). https://doi.org/10.1007/s11277-016-3742-6
16. Shacham, H., Waters, B.: Compact proofs of retrievability. J. Cryptol. **26**(3), 442–483 (2013). https://doi.org/10.1007/s00145-012-9129-2
17. Singh, G., Kaul, A., Mehta, S.: Secure k-NN as a service over encrypted data in multi-user setting. In: 2018 IEEE 11th International Conference on Cloud Computing (CLOUD), pp. 154–161, July 2018. https://doi.org/10.1109/CLOUD.2018.00027
18. Tupakula, U., Varadharajan, V.: Trust enhanced security for tenant transactions in the cloud environment. Comput. J. **58**(10), 2388–2403 (2014). https://doi.org/10.1093/comjnl/bxu048
19. Wang, H.: Identity-based distributed provable data possession in multicloud storage. IEEE Trans. Serv. Comput. **8**(2), 328–340 (2015). https://doi.org/10.1109/TSC.2014.1
20. Wang, H., He, D., Tang, S.: Identity-based proxy-oriented data uploading and remote data integrity checking in public cloud. IEEE Trans. Inf. Forensics Secur. **11**(6), 1165–1176 (2016). https://doi.org/10.1109/TIFS.2016.2520886

21. Wang, H., Li, K., Ota, K., Shen, J.: Remote data integrity checking and sharing in cloud-based health Internet of Things. IEICE Trans. Inf. Syst. **99**(8), 1966–1973 (2016). https://doi.org/10.1587/transinf.2015INI0001
22. Wang, Y., Ding, Y., Wu, Q., Wei, Y., Qin, B., Wang, H.: Privacy-preserving cloud-based road condition monitoring with source authentication in vanets. IEEE Trans. Inf. Forensics Secur. **14**(7), 1779–1790 (2019). https://doi.org/10.1109/TIFS.2018.2885277
23. Wang, Y., Wu, Q., Qin, B., Shi, W., Deng, R.H., Hu, J.: Identity-based data outsourcing with comprehensive auditing in clouds. IEEE Trans. Inf. Forensics Secur. **12**(4), 940–952 (2017). https://doi.org/10.1109/TIFS.2016.2646913
24. Wang, Y., Wu, Q., Qin, B., Tang, S., Susilo, W.: Online/offline provable data possession. IEEE Trans. Inf. Forensics Secur. **12**(5), 1182–1194 (2017). https://doi.org/10.1109/TIFS.2017.2656461
25. Wang, Y., Pang, H., Deng, R.H., Ding, Y., Wu, Q., Qin, B.: Securing messaging services through efficient signcryption with designated equality test. Inf. Sci. **490**, 146–165 (2019). https://doi.org/10.1016/j.ins.2019.03.039
26. Wang, Y., Wu, Q., Qin, B., Chen, X., Huang, X., Lou, J.: Ownership-hidden group-oriented proofs of storage from pre-homomorphic signatures. Peer-to-Peer Networking and Applications, pp. 1–17 (2016).https://doi.org/10.1007/s12083-016-0530-8
27. Yu, Y., et al.: Identity-based remote data integrity checking with perfect data privacy preserving for cloud storage. IEEE Trans. Inf. Forensics Secur. (2016). https://doi.org/10.1109/TIFS.2016.2615853

Relation Extraction Based on Dual Attention Mechanism

Xue Li[✉], Yuan Rao, Long Sun, and Yi Lu

Lab of Social Intelligence and Complex Data Processing, School of Software,
Xi'an Jiaotong University, Xi'an 710049, China
1255651833@qq.com

Abstract. The traditional deep learning model has problems that the long-distance dependent information cannot be learned, and the correlation between the input and output of the model is not considered. And the information processing on the sentence set is still insufficient. Aiming at the above problems, a relation extraction method combining bidirectional GRU network and multi-attention mechanism is proposed. The word-level attention mechanism was used to extract the word-level features from the sentence, and the sentence-level attention mechanism was used to focus on the characteristics of sentence sets. The experimental verification in the NYT dataset was conducted. The experimental results show that the proposed method can effectively improve the F1 value of the relationship extraction.

Keywords: Bidirectional GRU · Multi-attention · Relation extraction

1   Introduction

The explosive growth of information has caused people to be submerged in a disorderly database. It is difficult to quickly and accurately search for valid information, so that the use of computers to extract structured information from unstructured texts, that is, information extraction technology, is particularly important. Information extraction is an important basic task in natural language processing, including: entity extraction, relationship extraction, and event extraction. Relation extraction [1] is an important branch of information extraction. Its important role is to identify the semantic relationship defined between entities in text, and to solve the problem of classification between target entities in the original text. It is also an important processing step for constructing complex systems. It helps to construct knowledge maps. At the same time, entity relationship extraction technology is of great significance in the fields of massive information processing, automatic knowledge base construction and search engine. Through the extraction of relationships, text data in uniform format is extracted from unstructured text data. Digging deeper into the semantic relationships in the text, providing a more comprehensive and accurate search for user search, helping to process large amounts of text data and improving efficiency. Therefore, entity relationship extraction has good research significance and a wide range of application scenarios.

© Springer Nature Singapore Pte Ltd. 2019
X. Cheng et al. (Eds.): ICPCSEE 2019, CCIS 1058, pp. 346–356, 2019.
https://doi.org/10.1007/978-981-15-0118-0_27

In this research task, the e_1 and e_2 is an entity, and the relationship between them is defined as $<e_1, r, e_2>$, for example, in the sentence "Steve Jobs is the founder of Apple." containing the entity and relationship, the entity pairs "Steve Jobs" and "Apple" The relationship between them is "founder".

2 Relevant Works

In the relation extraction task, the deep learning method can obtain more effective features than the traditional method, and its accuracy and recall rate are relatively high. The Recurrent neural network model can encode variable-length sentences into fixed lengths, taking into account the contextual dependencies of entity pairs, and retaining sentence global features.

The deep neural network can automatically learn the underlying feature information and is applied to the relationship extraction including Convolutional neural network [2, 3]. Recurrent neural network and its variants in relation extraction. Zhang et al. [4] Using the Recurrent neural network based on word position information to complete the relation extraction task, the context information of the entity is better utilized. Sundermeyer et al. [5] proposed Recurrent neural network improvement Model—Long short-term Memory network, by storing a dedicated memory unit to store historical information, each time state saves the previous input information, the word sequence affects the output behind, effectively solving the long-distance dependence between two entities. Liu [6] propose to use SDP-LSTM (short dependence paths) model implements open domain entity relationship extraction, model division deal with multiple dependent paths that might appear in a sentence. Zhang [7] proposes to model the entire sentence using a Bidirectional long short-term Memory network for relation classification. Hu [8] added five combined feature as the input of BiLSTM including relative position feature, part of spcech tagging feature, entity tagging feature, adverbial feature, dependent feature (package including the dependency feature and the relative dependency feature). Relative max pooling is also used for getting sentence-level feature, this method improves the relation classification performance. Christopoulou et al. [9] proposed a graph-based method that uses BiLSTM and other models to spread information through iterations, so that the edges in the graph get more text information to help identify. Phi et al. [10] creative conversion of automatic selection and data noise reduction tasks in relation extraction into sorting problems. Although these methods have achieved good results, it requires a lot of data to be labeled when training the model, which consumes a lot of manpower and material resources.

The attention mechanism is a model proposed by Treisam and Gelade [11] to simulate the mechanism of human brain attention. It calculates the influence distribution of attention and highlights the influence of a key output, which is very good for the traditional model. Based on this advantage, some algorithms use the attention mechanism to extract the semantic information and reduce the noise problem in the encoding process. Among them, Zhang began [4] to try to use RNN to process text in order to get more feature information. Zhou [12] integrated the attention mechanism into BiLSTM based on Zhang. Lin [13] et al. proposed a sentence-level attention model to reduce the noise caused by false labels in the remote supervisory relationship extraction model. Wang [14] uses CNN and attention mechanism and proposes to use Attention Pooling instead of Max Pooling to reduce noise interference and strengthen the weight of related words. However, there are still insufficient representations of local and global information of sentences. Feng [15] proposes to use the Memory Network idea to extract relationships and use Attention to introduce correlations between relationships. The above results have certain implications for this study.

Previous studies have shown that sentence-level semantics are determined by the semantics of the words they contain and the way words are combined, while sentences are mainly composed of words. This paper solves the shortcomings in characterizing sentence local and global information by proposing a BiGRU relation extraction model based on word level and sentence level attention. GRU [16] maintains the effects of LSTM while making the structure simpler, reducing training parameters and increasing the rate of model training. First, character embedding is performed on each Chinese character in the sentence, and then input into the attention GRU model at the word level. The characteristics of each sentence output are re-entered into the Bi-GRU, and the sentence-level attention is added to solve the problem that the global information cannot be represented.

3   The Dual Attention Mechanism Model

In order to make better use of the sentence information, it can capture the important parts of the sentence and reduce the noise problem caused by the wrong label. This paper combines the attention mechanism at the word level and sentence level and proposes a dual attention relationship extraction model. There are five steps:

- Word-level feature representation: the position vector between word and entity pairs is input as the neural network model.
- Bi - GRU layer.
- Attention at the word level.
- Introduce sentence-level attention.
- Pooling, feature fusion and classification.

3.1   Input Layer

In this paper, the statement is determined as positive sample input through entity pair co-occurrence, and some negative samples are randomly input. In order to capture the syntactic and semantic information of words, words in the input sentence need to be mapped into word vectors. For a sentence containing W words $M = \{W_1, W_2, \ldots, W_n\}$, where each word is represented as the real value vector W_i (Fig. 1).

$$w_i = W^{wrd}v^i \tag{1}$$

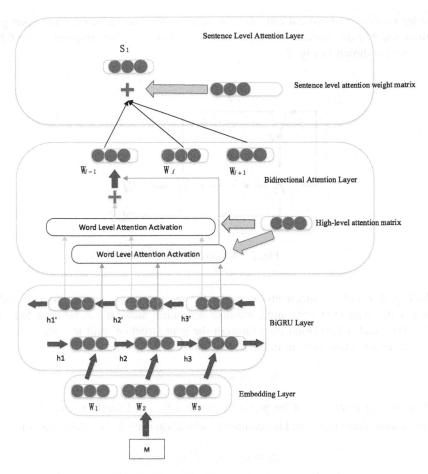

Fig. 1. Dual attention mechanism model

$W^{wrd} \subseteq R^{d_w \times |V|}$ is the vector matrix trained by word2vec. d_w is the dimension of word vector, $|V|$ is the size of dictionary, and v_i is the word bag representation of the input word. Thus, a vector sequence $M = \{W_1, W_2, \ldots, W_n\}$ is obtained.

3.2 The BiGRU Layer

In general, the main problems of RNN structure are gradient disappearance and gradient explosion during training, which make it impossible to apply to large machine learning tasks. GRU is an improved structural type of hidden layer node in RNN to deal with the above two problems. Like the LSTM unit, the GRU controls the memory storage through update and reset gates. Where, update gate controls the proportion of preorder memory and current candidate memory in the current state, while reset gate controls whether the current content is remembered or not. The structure of the GRU component is shown in Fig. 2:

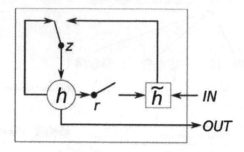

Fig. 2. A GRU Network

In Fig. 2, r and z respectively represent the reset door and update door, h and $\tilde{h}$ represent the activation state and candidate activation state. It can be seen that this model deals with information flow through the gate structure built into the structural unit, T-time activation state h_t in GRU structure:

$$h_t = (1 - z_t)h_{t-1} + z_t h \tag{2}$$

The activation state h_t of the previous moment h_{t-1} is linearly related. In formula (2), the update gate state z_t and the candidate activation state $\tilde{h}$ are calculated as follow:

$$z_t = \sigma(w_z x_t + U_z h_{t-1}) \tag{3}$$

$$\tilde{h} = \tan h(Wx_t + U(r_t \otimes h_{t-1}) \tag{4}$$

$$r_t = \sigma(W_r x_t + Urh_{t-1}) \tag{5}$$

Equation (5) represents the reset door. GRU has fewer parameters, simpler structure and easier convergence than LSTM.

In this article, in order to get the context information, BiGRU helped network can do so within a specific time more effective use of the characteristics of the past (state) via the former and future features (state) via the after. Each sentence downstream (from the first word to the last word recursion) and reverse (from the last word to the first word recursion) input, it is two different hidden layer, said then will forward, reverse said in every moment of the hidden layer vector pieced together. As shown in Fig. 3.

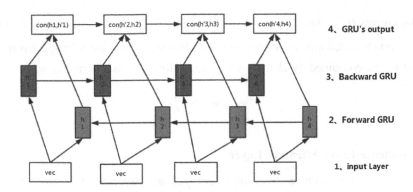

Fig. 3. BiGRU network structure

3.3   Word-Level Attention Layer

In the model of this paper, a mechanism of attention force is used to extract features from a single sentence, which has been proved by Zhou et al. As an effective method of relationship classification.

For relational extraction tasks, the set of relationships used for classification is different for the importance of words in a sentence. Therefore, this paper uses the word-level two-way attention weighting mechanism matrix to obtain the information of the relationship between the sentence and the target. In order to get more effective cor-relation characteristics, this paper uses the pooling operation of the attention mechanism. The forward hidden layer and the backward hidden layer of the bidirectional GRU are multiplied by the attention weight matrix, and then the most efficient feature representation is obtained by using the maximum pooling operation.

The output W_j is calculated from the following formula:

$$\beta = \text{soft max}(VYH) \tag{6}$$

$$w_j = \text{max pool}(H\beta) \tag{7}$$

is the input matrix composed of vectors $[h_1, h_1, h_1, \ldots, h_T]$ generated by the Bi-GRU layer, where d_w is the dimension of the vector, and T is the length of the sentence. Y is a set of all relationships, and V is the parameter matrix (updated during the training process). The dimension of the vector β are T. The sentence vectors $\left[\overrightarrow{h_1}, \overrightarrow{h_2}, \overrightarrow{h_3}, \ldots, \overrightarrow{h_T}\right]$ and $\left[\overleftarrow{h_1}, \overleftarrow{h_2}, \overleftarrow{h_3}, \ldots, \overleftarrow{h_T}\right]$ represented by the BiGRU and respectively multi-plied by the weight matrix to highlight the important features of the word level, and then the maximization strategy is used to find the most significant features. The output of this layer is the eigenvector w_j representing a single sentence j.

The outputs $H^f = \left[\overrightarrow{h_1}, \overrightarrow{h_2}, \overrightarrow{h_3}, \ldots, \overrightarrow{h_T}\right]$ and $H^b = \left[\overleftarrow{h_1}, \overleftarrow{h_2}, \overleftarrow{h_3}, \ldots, \overleftarrow{h_T}\right]$ of the GRU network into bidirectional attention layer, and obtained the w_j^f and w_j^b respectively. and w_j^f and w_j^b are combined by element-wise sum to form the output matrix w_j.

$$w_j = w_j^f \oplus w_j^b \tag{8}$$

3.4 Sentence-Level Attention Layer

Due to contain the same entity (entity1, entity2) and expressed the same relation r sentences may carry a lot of noise or useless information, so sentence layer attention mechanism concentrated extract features from a sentence, which contains relation r. As with the previous level of word-level attention, the representation of the sentence set s_j is calculated by the following equation.

$$r_j = soft \max\left(s_j A r\right) \tag{9}$$

$$s_j^* = \sum_j r_j s_j \tag{10}$$

$$s_j = soft \max\left(s_j^*\right) \tag{11}$$

A is a weighted diagonal matrix, r is a vector representation of the corresponding prediction relationship r of the entity pair in the knowledge base. Since the relationship vector is unknown during the test, it corresponds to the entity pair in the knowledge base during the training process. The prediction relationship is obtained by random initialization during the test.

The last *soft max* activation is used for output s_j prediction. The cost function of the model is the cross entropy of the real category label y, as defined below:

$$C = -\frac{1}{n_r}\sum\left[y ln s_j + (1 - y) ln\left(1 - s_j\right)\right] + \frac{\lambda}{2n_r}\sum_w w^2 \tag{12}$$

The L2 regularization is added to the cost function, where $\lambda >$ is called regularization parameter, $\sum_w w^2$ is the sum of squares of weights in w softmax activation function. To optimize the function, use Adam to minimize the target function.

4 Experiments

4.1 Datasets

The purpose of our experiment is to evaluate the performance of the dual attention mechanism in the model and the overall performance of the model in relation extraction. For this reason, data set and experimental setup are introduced. Then, assess the effectiveness of dual attention mechanisms. Finally, the performance of our model is analyzed and compared with other methods.

Our experiment was based on data developed by Riedel and others called NYT. The New York times corpus extraction relationship has 12 predefined relationship types, including a special relationship "None" that indicates that the relationship between two entities does not belong to any predefined relationship.

The mirroring mechanism and regularization are combined to avoid over-fitting, and loss is adopted in the output of sentence-level attention layer. In the experiments, There are 3 position indicators marking the position of the entity "<e1>", "</e1>", "<e2>" and "</e2>" to indicate that the entity right. Last season at Carthon, brown scored the fewest points in the league. The entities are Carthon and Browns. "E2> brown scored the fewest points in the league last season with carthon." The four position indicators are treated as a single word and will be converted to a word vector.

In this paper, precision, recall and F-score are adopted as the performance evaluation indicators of the model. The indicators are calculated as follows:

$$precision = \frac{out\_cor}{out\_all}$$

$$recall = \frac{out\_cor}{this\_all}$$

$$F_{score} = \frac{2 \times precision \times recall}{precision + recall}$$

Among them: *out_cor* represents the number of correct relations judged by the output *out_all* represents the number of all relations exported, and *this_all* represents the number of all the relationships in the test set.

4.2 The Role of Dual Attention Mechanisms

In order to prove the effect of dual attention mechanism, an experiment was designed to compare the performance of different parts of the model and evaluate them by relative comparison. In this paper, the results of CNN network are taken as the basis. At the same time, two different types of performance with attention mechanism are compared:

(1) BiLSTM + attention at the word level;
(2) BiLSTM + attention based on sentence level (Table. 1).

Table 1. The F1_score value of the above model

Model	Feature Set	F1
CNN	WV(Turian et al.,) (dim = 50)	69.7
(Zeng et al. 2014)	+PF+WordNet	82.7
RNN	WV(Turian et al.,) (dim = 50)+PI	80.0
(Zhang 2015)	WV(Mikolov et al.,) (dim = 300)+PI	82.5
GRU	WV(dim = 300)+PI	82.9
BiLSTM	WV(Turian et al.) (dim = 50)+PI	80.7
Att-BiLSTM	WV(Turian et al.) (dim = 50)+PI	82.5
BiLSTM	WV(Pennington et al.) (dim = 100)+PI	82.7
Att-BiLSTM	WV(Pennington et al.,) (dim = 100)+PI	84.0
DaulAtt-BiGRU	WV(dim = 50)+PI	83.1
DaulAtt-BiGRU	WV(dim = 100)+PI	84.5

WV and PI respectively represent word vector and position identification, while dim represents the number of neurons in the hidden layer

CNN: Zeng et al. treated sentences as sequential data and used the convolution neural network to learn sentence level features, and used a special position vector to represent each word. Finally, the sentence level and the lexical features are connected into a vector and input to softmax classifier for prediction. Got 82.7% F1.

RNN: Zhang and Wang used bidirectional RNN networks with two different latitude vectors for relationship classification. They obtained 82.8% of the F1 value using a 300-dimensional word vector pre-trained by Mikolovet et al. Use the Turian and others' pre-trained 50-dimensional word vector to get an F1 value of 80.0%. Peng Zhou et al. added attention mechanism to bidirectional BiLSTM and achieved an F1 score of 82.5% with the same word vector of 50 dimensions, which was about 2.5% higher than theirs. They also changed the word vector to 100 dimensions, and the F1 value has exceeded the model of RNN and GRU in 300 dimensions of the word vector. Therefore, it can be seen that attention mechanism can improve the F1 value of relationship classification.

Table 2. Compares our model with other excellent models

Test datasets	One-sentence dataset				Two-sentence dataset				All-sentence dataset			
Precision@N	100	200	300	Mean	100	200	300	Mean	100	200	300	Mean
CNN+SATT	76.2	65.2	60.8	67.4	76.2	65.7	61.1	68.0	76.2	68.6	59.8	68.2
BiLSTM +BSATT	81.0	72.2	**69.3**	74.2	82.2	75.1	**72.0**	76.4	83.2	76.8	**72.1**	77.4
BiGRU+DATT	**82.9**	**74.7**	68.9	**75.5**	**83.1**	**75.9**	71.5	**76.8**	**84.5**	**78.3**	71.5	**78.2**

From the Table 2, we can draw the following conclusions: (1) the GRU network with the Daul-attention mechanism achieves the state-of-the-art performance in relation extraction; (2) the daul attention mechanism has an excellent ability to extract features from a single sentence and exhibits better performance in relation extraction; and (3) when the number of sentences in a subdataset increases, the sentence-level attention mechanism plays a more important role and the contribution of the bidirectional attention mechanism is limited.

5 Conclusion

In this paper, we propose a model that uses a multi-dimensional dual attention mechanism. The attention mechanism is used at both the lexical level and the sentence level. Fully used the influence of relations on the words in the sentence, and considered the influence of the same entity on the prediction relationship of sentence collections in the sentence collection. Experiments show that the model presented in this paper is suitable for remote entity relationship extraction tasks. Future work will try to use multiple types of models to characterize sentence vectors; and in the aspect of sentence attention mechanisms, explore different ways to solve the noise problems caused by multiple instances.

References

1. Bohui, Z., Wei, F., Yu, H., et al.: Entity relationship extraction based on multi-channel convolutional neural network. Appl. Res. Comput. **34**(3), 689–692 (2017). (in Chinese)
2. Nguyen, T.H., Grishman, R.: Relation extraction: perspective from convolutional neural networks. In: Proceedings of the 1st Workshop on Vector Space Modeling for Natural Language Processing, vol. 39–48 (2015)
3. Zeng, D., Liu, K., Chen, Y., et al.: Distant supervision for relation extraction via piecewise convolutional neural networks. In: Proceedings of the 2015 Conference on Empirical Methods in Natural Language Processing, pp. 1753–1762 (2015)
4. Zhang, D, Wang, D.: Relation classification via recurrent neural network. arXiv preprint arXiv:1508.01006 (2015)
5. Sundermeyer, M., Schlüter, R., Ney, H.: LSTM neural networks for language modeling. In: Thirteenth Annual Conference of the International Speech Communication Association (2012)
6. Wei, L.: Mining of Chinese entity relationship for "BIGCILIN". Harbin Institute of Technology, 1–z (2016). (in Chinese)
7. Zhang, S., Zheng, D., Hu, X., et al.: Bidirectional long short-term memory networks for relation classification. In: Proceedings of the 29th Pacific Asia Conference on Language, Information and Computation, pp. 73–78 (2015)
8. Hu, X.: Research on semantic relationship classification based on LSTM. Harbin Institute of Technology (2015). (in Chinese)
9. Christopoulou, F., Miwa, M., Ananiadou, S.: A walk-based model on entity graphs for relation extraction. In: Proceedings of the 56th Annual Meeting of the Association for Computational Linguistics (Volume 2: Short Papers), vol. 2, pp. 81–88 (2018)
10. Phi, V.T., Santoso, J., Shimbo, M., et al.: Ranking-based automatic seed selection and noise reduction for weakly supervised relation extraction. In: Proceedings of the 56th Annual Meeting of the Association for Computational Linguistics (Volume 2: Short Papers), vol. 2, pp. 89–95 (2018)

11. Luong, M.T., Pham, H., Manning, C.D.: Effective approaches to attention-based neural machine translation. arXiv preprint arXiv:1508.04025 (2015)
12. Zhou, P., Shi, W., Tian, J., et al.: Attention-based bidirectional long short-term memory networks for relation classification. In: Proceedings of the 54th Annual Meeting of the Association for Computational Linguistics (Volume 2: Short Papers), vol. 2, pp. 207–212 (2016)
13. Lin, Y., Shen, S., Liu, Z., et al.: Neural relation extraction with selective attention over instances. In: Proceedings of the 54th Annual Meeting of the Association for Computational Linguistics (Volume 1: Long Papers), vol. 1, pp. 2124–2133 (2016)
14. Wang, L., Cao, Z., De Melo, G., et al.: Relation classification via multi-level attention CNNs (2016)
15. Feng, X., Guo, J., Qin, B., et al.: Effective deep memory networks for distant supervised relation extraction. In: IJCAI, pp. 4002–4008 (2017)
16. Dey, R., Salemt, F.M.: Gate-variants of gated recurrent unit (GRU) neural networks. In: 2017 IEEE 60th International Midwest Symposium on Circuits and Systems (MWSCAS), pp. 1597–1600. IEEE (2017)

Analysis and Defense of Network Attacking Based on the Linux Server

Dapeng Lang[1,2], Wei Ding[1(✉)], Yuhan Xiang[1], and Xiangyu Liu[1]

[1] College of Computer Science and Technology,
Harbin Engineerning University, Harbin 150001, China
706473838@qq.com
[2] Key Laboratory of Network Assessment Technology,
CAS (Institute of Information Engineering, Chinese Academy of Sciences),
Beijing 100093, China

Abstract. The kernel of the Linux server is analyzed to find out the main cause of the server's denial of service when it is attacked. In the kernel, when the connection request information memory is full, the new connection request is discarded. Therefore, the printk function was used to alert the kernel output log when the memory was full, the processing of discarding the connection request in the kernel was changed, and the function tcp_syn_flood_action was applied to full memory processing. In the function tcp_syn_flood_action, the free function was used to release the memory according to the condition, so that the new connection request has a storage space, thereby offering the server's normal service. Finally, the proposed defense technology is verified to be effective.

Keywords: TCP/IP · DDOS · SYN Flood · System kernel

1 Introduction

In recent years, network technology has developed rapidly, and the network has penetrated into every corner of the world. People use online shopping, find information, chat, etc., and people are increasingly inseparable from the Internet. But at the same time as the rapid development of the network, hackers also set their sights on the network. In a survey in the United States, 60% of the more than 500 organizations surveyed were attacked by cyber attacks, and their combined losses totaled more than $4 million. This shows that cybersecurity issues are related to national and national security. So studying the issue of cybersecurity is extremely important. Network technology continues to advance, attack technology is constantly evolving, and a denial of service attack (DDoS) attack technology has emerged. And as network traffic increases, DDoS attacks against large traffic are more difficult to detect and defend. In a denial of service attack, the SYN Flood attack is the most typical. Commonly known as flood attacks.

© Springer Nature Singapore Pte Ltd. 2019
X. Cheng et al. (Eds.): ICPCSEE 2019, CCIS 1058, pp. 357–372, 2019.
https://doi.org/10.1007/978-981-15-0118-0_28

2 Research Actuality

The current network connection is based on the TCP/IP protocol, but there are also many problems in the TCP/IP protocol, such as: the source IP address can be arbitrarily changed, the unrestricted SYN connection, and the lack of effective authentication of the information source [1]. In the DDoS attack, the TCP-based attack mode accounts for more than 94% of the total number of attacks. The UDP attack and the ICMP attack account for about 4%. The SYN flood attack is based on the TCP attack mode. The most proportioned one, accounting for about 90%.

In order to solve the increasingly serious DDoS attacks, many researchers at home and abroad are studying DDoS attacks and are working to find a way to effectively detect, track and defend against DDoS attacks [2]. Papadopoulos [3] proposed a method to prevent concurrent attacks in the aspect of DDoS detection. The main principle of this method is to use the components of the edge network and other edge networks to defend against DDoS attacks. Savage [4] proposed a method of probabilistic packet marking for DDoS tracking. The main principle of this method is to let the router mark the passing packets by probability. TAO Peng [5] proposed a filtering method to mitigate DDoS attacks against the defense of DDoS. There are also products that are specifically designed to protect against DDos attacks. Among them, Captus' CaptIO G-2 can be configured to monitor and mitigate network traffic, effectively preventing flood attacks like SYN Flood attacks. In China, China United Green League's black hole products use a new algorithm of reverse detection and fingerprint recognition to effectively prevent SYN Flood attacks.

In recent years, domestic and foreign defense strategies against SYN Flood attacks are mainly divided into: defense strategies based on concessions and feedbacks, defense strategies based on gateway interception and defense strategies based on TCP protocol stack defects [6]. Defense strategies based on backoff and negative feedback typically prevent SYN Flood attacks by setting some values on the TCP/IP connection in the operating system. The method of reducing the timeout period, increasing the length of the semi-join queue, and reducing the number of retransmissions of the SYN +ACK packet can alleviate the phenomenon of denial of service caused by the attack on the server side [7]. By reducing the timeout period, the server can disconnect from the attacker as soon as possible. During the first handshake, the server sends a connection request sent by the client to the semi-connected queue after sending the SYN +ACK to the client. After receiving the ACK from the client, the connection request information is added to the connection queue. When the semi-join queue is full, the server discards the new connection request, so by increasing the semi-join queue, the server can handle more connection requests. When the ACK of the client is not received, the server will continuously retransmit the SYN+ACK until timeout, and discard the connection request packet. By reducing the number of retransmissions of SYN+ACK, the server can quickly disconnect disconnected connection requests and save resources. The main principle of the defense policy based on the gateway is to process the IP address that sends the connection request, mainly by judging whether the IP address is legal or not, and judging whether to discard the request packet. SYN Cookies technology is the most commonly used defense strategy based on TCP protocol stack defects. The SYN cookies technology mainly performs the second handshake of TCP. When receiving the connection request packet and returning the SYN+ACK packet, the connection request is not added to the semi-connection queue, but the value of the cookie is calculated according to some information of the connection request. The third handshake is based on the value of this cookie to determine whether it is the third handshake. When the value of the cookie is legal, space is allocated for this connection request and a connection is established.

At present, although many technologies for defending against SYN flood attacks have been developed at home and abroad, no matter which method or product can defend the SYN Flood attack fundamentally, the defense technology of the SYN Flood attack still needs further improvement.

3 Analysis of the SYN Flood

3.1 SYN Flood Attack

A SYN flood attack is a type of denial of service attack. The SYN Flood exploits a defect in the TCP protocol to send a large number of semi-connected requests to the target host, consuming resources such as the server's memory and CPU, making it unable to serve connection requests from legitimate users.

3.1.1 Principle of SYN Flood Attack

The SYN flood attack is that the attacker deliberately does not perform a three-way handshake with the server, causing the server to exhaust resources. First, the handshake process of the TCP protocol is introduced. Based on the handshake process of the TCP protocol, the principle of the SYN Flood attack is introduced.

The TCP protocol provides a connection-oriented, reliable byte stream service that must be established before the client and server exchange data. The establishment of the connection requires the three ends to successfully complete the three-way handshake [8]. The three-way handshake process [9] of the TCP protocol is:

(1) The first handshake is that the client sends a SYN packet to the server port to be connected, and the initial sequence number;

(2) The second handshake is that the server receives the connection request from the client, sends a SYN segment containing the initial serial number of the server to the client as a response, and sets the confirmed sequence number to the initial serial number of the client plus one to the client. The SYN segment is confirmed. Both the server sends the SYN+ACK phase;

(3) The third handshake is the response of the client receiving the server. The initial number of the server is incremented to confirm the SYN segment of the server. The server receives the confirmation sequence number and establishes a connection with the client. At this time, the client and the server can exchange data. Both the client sends the ACK phase. The three-way handshake process of the TCP protocol is shown in Fig. 1.

After the server sends a SYN+ACK packet to the client, it does not receive the ACK confirmation from the client. The server retransmits the SYN+ACK packet to the client multiple times within the timeout period. The connection status of the server is semi-connected. status. If the timeout period and the number of retransmissions are exceeded, the server still does not receive the ACK acknowledgement packet from the client, and the server will release the space for the semi-join request at this time. The SYN Flood attack utilizes the feature of the TCP protocol to send a large number of connection request packets with forged IP addresses to the server. After receiving the data packets, the server sends a SYN+ACk response packet to the forged IP address, but The IP address does not send a packet to the server, so it does not send an ACK to the server. The server does not receive an ACK to confirm the packet. In a certain period of time, it consumes a large amount of resources to save the semi-connected state. The information, when the memory of the server-side semi-connection request is saturated, the server will discard the new connection request packet, and the server will not be able to provide normal services for legitimate users. The process of a SYN Flood attack is shown in Fig. 2.

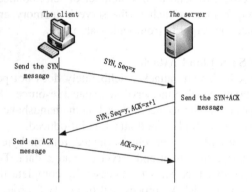

Fig. 1. TCP three handshake process diagram

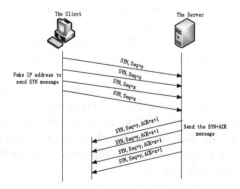

Fig. 2. SYN Flood attack schematic map

Since the packets sent by the attacker are not significantly different from the normal data packets, it is difficult for the server to distinguish and filter the data packets from the attacker, which makes it difficult to fundamentally defend against the SYN Flood attack.

3.2 TCP Connection Process Analysis for Linux Servers

This chapter will analyze the process of establishing a TCP connection for a Linux server from three aspects. The first is to analyze the three-way handshake process of the passive connection TCP protocol in the Linux server kernel, and then analyze the main data structure applied to achieve the defense. Finally, the defense-related function implementation is analyzed.

3.2.1 Analysis of TCP's First Handshake

The server has different processing functions according to different states. In the kernel of the Linux server, the TCP receiving process is completed by the tcp_rcv_state_ process function. This function also implements the state transition of the TCP protocol. When the function is in the listening state, when it is received, When the SYN flag is packaged, it is called by the tcp_v4_conn_ request function. This function is used to process the first handshake of TCP.

When the server receives a new connection request, it first determines whether the address sent by the request is a broadcast or multicast address. For the SYN segment sent to the broadcast or multicast, the server returns without processing. If the address is normal, the function inet_csk_reqsk_queue_is _full is used to determine whether the length of the semi-join queue has reached the upper limit. If the upper limit is reached, the connection request is directly discarded. If the Linux server does not take any defense measures, the effect of the attack will be denied. The new connection request is dropped. However, there is a handling of SYN Flood in the Linux kernel, namely SYN Cookiess technology, which can be used to defend against attacks by enabling SYN Cookiess technology to receive new connection requests. Then the full connection queue is judged. If the full connection queue is full and there is a connection request that has not been retransmitted in the semi-join queue, the server will also discard the connection request. When all the conditions are met, the server will call the function reqsk_alloc() to allocate a connection request block that holds the connection information for this connection request. Call the function tcp_parse_options to resolve the TCP option in the SYN segment. If the Linux server starts the SYN Cookiess technology, the resolved TCP option is cleared. The function tcp_openreq_init is then called to initialize the information in the connection request block with the information in the received SYN segment.

Assign the parsed information to the structure that holds the connection request information, and obtain the IP option and save it by calling the function tcp_v4_save_options. If the cookies technology is enabled, the client's initial serial number is encrypted to obtain the serial number. If not, the sequence number is obtained. The IP address and port calculate the initial serial number of the server and save it, then send SYN+ACK to the client, and save the data storing the connection request information to the hash table of the semi-connected queue. Finally, set the timeout period and start the timing. Device.

Assign the parsed information to the structure that holds the connection request information, and obtain the IP option and save it by calling the function tcp_v4_save_options. If the cookies technology is enabled, the client's initial serial number is encrypted to obtain the serial number. If not, the sequence number is obtained. The IP address and port calculate the initial serial number of the server and save it, then send SYN+ACK to the client, and save the data storing the connection request information to the hash table of the semi-connected queue. Finally, set the timeout period and start the timing.

By analyzing the process of establishing a connection by passively opening in a Linux server, it can be seen that the calling relationship between the first handshake function of the server processing the TCP protocol is as shown in Fig. 3.

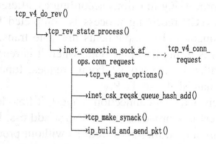

Fig. 3. Function call to complete the first handshake

3.2.2 Analysis of the Second TCP Handshake

The query routing entry is completed by the function inet_csk_route_req. The function is based on the information of the listening transmission control block for processing the server connection process and the connection request block established by the connection, including the output network device, IP address and port number, etc., for sending SYN+The ACK segment queries the route entry and returns null if the query route entry fails. This valid route cache entry is returned if the query is successful.

Constructing the SYN+ACK segment is done by the tcp_make_synack function. The function first allocates the SKB for the SYN+ACK segment. If the allocation length and the upper limit are reached, the function alloc_skb() is forced to allocate the SKB for this purpose. The necessary space is then reserved for each layer header of the datagram, and some information about the SYN+ACK segment is calculated based on the information in the SYN segment. Initialize the TCP header and the items in the SKB, set the values of the fields in the TCP header, and perform some settings on the receiving window of the local end, and return the SKB of the generated SYN+ACK segment after the information in the datagram is set.

After constructing SYN+ACK, the function tcp_v4_send_check is called to complete the calculation of the TCP header checksum. The function ip_build_and_send_pkt is called to add the IP header and send the SYN+ACK segment to the client.

3.2.3   Analysis of the Third TCP Handshake

When the server receives the ACK segment, the last handshake is handled by the function tcp_v4_hnd_req. First, the corresponding connection request information is searched in the hash table of the connection request according to the destination IP address, the source IP address, and the source port number. If the search succeeds, the semi-connection has been established, and then the function tcp_check_req is called to perform the last handshake. confirm.

If no search is successful, it will be looked up in the ehash hash table. If it is still not found in the ehash table, the invalid message will be discarded. If the cookies technology is enabled, the function cookies_v4_check is called for detection. If the detection is successful, the connection is not required to be queried in the semi-join queue, and the last handshake is completed directly.

When the server is in the semi-connected state, the received TCP segment is processed by calling the function tcp_check_req. The function first obtains the value of each flag bit in the received TCP header, and determines whether the length of the received TCP header is greater than the length of the header without the TCP option. If yes, it indicates that there is a TCP option in the message, and the function tcp_parse_options is called. To resolve this TCP option. Check if the serial number in TCP is valid by calling the function tcp_paws_check. If the received message is a SYN segment, the message is processed as a SYN segment and the function rtx_syn_ack is called for processing. If the ACK acknowledgement sequence number or sequence number is invalid, no processing is performed immediately, if the received report is received. If the text is an RST segment or a new SYN segment, the RST is returned to the client for reset. If all the checks pass, a subtransmission control block is created and the connection request is moved from the semi-join queue to the completed full connection queue, waiting for the user function's accept function call.

4   Defense Realization and Test

This chapter introduces the defense techniques of the SYN Flood attack adopted in this paper from three aspects. The first is the detection of the SYN Flood attack. After detecting the attack, it will alert the kernel, and then introduce the implementation of the defense technology. Finally, the advantages of the defense technology described in this paper and the areas for improvement are compared with other defense technologies.

4.1   Detection and Warning of SYN Flood Attack

4.1.1   SYN Flood Attack is Detected

From the analysis of the function of the first handshake in the Linux kernel in Sect. 3, it is known that there is a judgment on whether the semi-join queue reaches the upper limit in the Linux kernel, and the reason why the Linux server will refuse the service after being attacked by the SYN Flood is half. The connection queue has reached the upper limit, so new connection requests will be dropped by the Linux server. Therefore, when the semi-connected queue reaches the upper limit, it can be determined that the server is attacked by the SYN Flood.

The conditional function that determines whether the number of semi-join queues reaches the upper limit in the server kernel is inet_csk_reqsk_queue_is_full. Through the analysis of the data structure, it is known that there is a number of connections in the listener_sock for storing the current semi-join queue and the number of semi-joins for the maximum connection. Therefore, when the current number of semi-joins is equal to the maximum number of connections, the semi-join queue reaches the upper limit at this time.

Table 1. Pringk log level description table

Macro definition	Level	Describe
KERN_EMERG	0	An emergency message indicating that the system is unavailable before crashing
KERN_ALERT	1	Sources in the report said measures must be taken immediately
KERN_CRIT	2	Critical condition, usually involving a serious hardware or software operation failure
KERN_ERR	3	Error conditions, where the driver reports a hardware error
KERN_WARNING	4	Warning conditions, warning of possible problems
KERN_NOTICE	5	A normal but important condition for reminder
KERN_INFO	6	Prompt messages, such as driver startup, print hardware information
KERN_DEBUG	7	Debug level messages

4.1.2   Warning of SYN Flood Attack

In order to distinguish SYN Flood attack from DOS attack, this paper mainly analyzes through log. When a SYN flood attack is detected, an alarm is generated in the kernel. The alarm is generated by outputting the information about the SYN flood attack in the log. The user can find out whether the SYN flood attack is received by querying the log. The Linux kernel outputs information to the log through the printk function, and the information output by the kernel through the pringk function is log level. The level of the log is a macro-defined output level character or a tip before the string output by printk. The integers in parentheses are used to control. A total of eight different log levels are provided in the kernel. The description of the pringk log level is shown in Table 1.

According to the description of Table 1, the level of output and the actual application, when using dmesg to view the log output of the system, only the highest level output of KERN_EMERG can be displayed in the log, so the alarm information in this article is by printk ("KERN_EMERG") Function implementation. The function first gets the port it wants to attack from the connection request from the attacker. The alarm function displays the port number of the SYN Flood attack in the output. The alarm that the server is attacked by the SYN Flood is shown in Fig. 4.

```
743.327427|  syn_flood_action:after unlink sum_part[11]  =1
743.327429|  syn_flood_action:after unlink sum_part[12]  =1
743.327431|  syn_flood_action:after unlink sum_part[13]  =1
743.327433|  syn_flood_action:after unlink sum_part[14]  =1
743.327435|  syn_flood_action:after unlink sum_part[15]  =1
743.327437|  syn_flood_action: after unlink sum2 =15
743.327439|  isfull =0
743.327476|  TCP Possible SYN flooding on port 80
743.327480|  syn_flood_action:before unlink max_sum_part[7]  =2
743.327483|  syn_flood_action: before unlink sum1 =16
743.327485|  th->source= 0x4a0ca8c0, iph->raddr = 20
743.327487|  syn_flood_action: reqs = 0xffff880076466d00, reqs->expires = 0x10
006c7bd
743.327489|  syn_flood_action: reqs->s_timeout = 0x0
743.327491|  syn_flood_action: reqs->rsk_ops = 0xffffffff8196c340, reqs->rsk_op
s->slab = 0xffff88007f807900
743.327493|  syn_flood_action: reqs = 0xffff880076466d00, (prev = &reqs) = 0xf
fff88007c603938
743.327496|  syn_flood_action:after unlink sum_part[0]  =0
743.327498|  syn_flood_action:after unlink sum_part[1]  =1
743.327500|  syn_flood_action:after unlink sum_part[2]  =1
743.327502|  syn_flood_action:after unlink sum_part[3]  =1
743.327504|  syn_flood_action:after unlink sum_part[4]  =1
743.327506|  syn_flood_action:after unlink sum_part[5]  =1
```

Fig. 4. Linux server attack warning diagram

4.2 SYN Flood Defense Realization

In the kernel of the Linux server, when the semi-join queue that holds the connection request information is full, the newly received connection request is processed by DROP. It can be seen that when the server is attacked, the root cause of the denial of service is that the server discards the connection request. Therefore, the defense technology is to let the server be subjected to a SYN Flood attack. The process is not to discard the new connection request, but to release an item in the semi-connected queue according to the condition. At this time, there will be a new space in the semi-connected queue. New connection request.

For the analysis of the connection process of the TCP protocol in the Linux server kernel, the processing function tcp_v4_conn_request of the first handshake is known. In the function tcp_v4_conn_request, after receiving a new connection request, it first determines the state of the connection queue at this time. When the SYN Flood attack is detected in Sect. 4.1.1, it can be known that when the semi-connected queue is full, the new connection request is DROP. So when it is judged to be full, it is processed by the function tcp_syn_flood_action. The function first acquires the transport control block sk that processes the TCP segment, the received TCP segment skb, and the protocol TCP. Convert the transport control block sk to the lopt of the listen_sock structure. In the listen_sock structure, there is an item set to be attacked by the SYN Flood. Set this value to 1 and output attack information. Loop through the entire hash table, the end of the loop is that the number of loops is equal to the maximum value nr_table_entries of the record hash table in the listen_sock structure. Defines the sum_part record to record the number of connection requests stored in each hash table, and defines a max_-sum_part to record the largest number of connection requests stored in the hash table, and output the number of connection requests before releasing the connection request. Go to the log so that the connection is released successfully after the connection is released. Find the largest number of hash table key values, and release the longest connection request saved in this hash table. The analysis of the hash hash table in Sect. 4 shows that the longest stored item in the hash table is the linked list. The tail, so the circular list finds the end of the list and records the previous item at the end. If there is only one connection request in the hash table, the previous item is the same as the tail, then the hash table is set to null, otherwise the previous one will be The dl_next of the item is set to null. Through this algorithm, the connection request to be released is separated from the linked list. At this time, the number of connection requests should be decremented by one. The function reqsk_queue_removed reduces the value of the number of connection requests in the listen_sock structure by one, and finally releases the connection request. The SYN Flood attack defense implementation process is shown in Figs. 5 and 6.

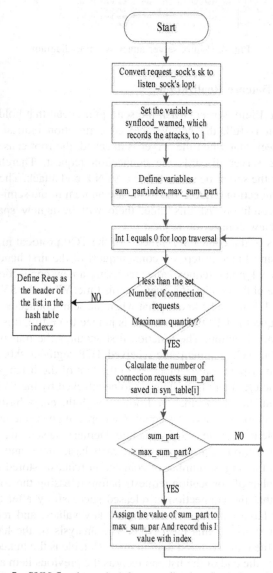

Fig. 5. SYN flood attack defense realization flow chart part I

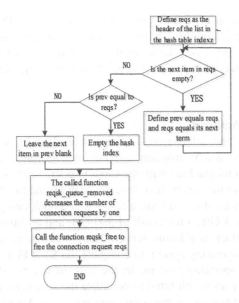

Fig. 6. SYN flood attack defense realization flow chart part II

4.3 Advantages and Disadvantages of Defense Technology

The current defense technology used in the Linux kernel is SYN Cookies. From the introduction of SYN Flood prevention technology, the principle of SYN Cookies technology is that the connection request information is not stored on the server, but the information is encrypted and saved to SYN+ACK. In the segment, the information is stored on the server only after receiving the correct ACK from the client. This principle of SYN Cookies technology can effectively defend against SYN Flood attacks. However, this technology cannot encrypt all the information contained in the SYN segment. Therefore, the use of this technology will result in loss of information and decrease. The quality of service of the server. In the TCP protocol, if the client does not receive the ACK packet within a certain period of time, the server will resend the SYN +ACK packet, but if the server uses the SYN cookies technology, the server does not save any information about the connection request, so the server Retransmission cannot be implemented. At this time, the connection request of the legitimate user may fail to connect successfully. The SYN cookies technology mainly determines whether to establish a connection according to the ACK packet replied by the client, and has nothing to do with the data packet sent in the first handshake and the second hand-shake. If the attacker exploits this defect, the ACK packet is continuously sent to the server. It may cause the server to be attacked by another.

Defense technology in this paper, the principle is random release the TCP stack, a new connection request will not be discarded by the server because TCP stack full, and the technology solved the defects existing in the SYN Cookies, the server will still save the SYN segment connection request information and allocate memory for the con-nection request, this article's defense technology also ensures that a connection queue will not cause the collapse of a Linux server by overflow. This topic of defense technology in the Linux kernel to add the traversal of the list, if the user in the use of the half connection queue is very large, then under attack, the traversal of the list will consume some time, may also consume some resources of the server.

4.4   Test

4.4.1   Verification Platform

The SYN Flood attack and defense verification test needs to be completed under three operating systems. The attack program and the defense function are all completed under the VMware virtual machine with the CentOS operating system version 3.10.90, and the normal access to the server. It is done under the Windows 10 operating system. The hardware platform for the verification test is a notebook computer with a CPU of 1.9 GHz, a memory of 4 GB, a hard disk of 500 GB, and an Ethernet card of 100 M.

The SYN Flood attack is a Linux server. Therefore, you need to set up an httpd server on the CentOS operating system to complete the SYN Flood attack and defense tests. In the CentOS operating system, in the case of a network, the terminal can execute the command yum install httpd to complete the automatic installation of httpd. Yum is a background program for managing rpm packages. After the yum server is set up, you can use the yum command to install and update the rpm packages that exist in all resource libraries. Install is the meaning of the installation in the yum command, httpd is the httpd server.

4.4.2   Compile the Kernel

The defense technology is implemented in the Linux kernel, and the defense function is added in the net/ipv4/tcp-ipv4.c file of the Linux operating system. After changing the Linux kernel, you can run the changed operating system after recompiling the kernel and rebooting. The steps to compile the kernel are:

(1)   Download the source code of the Linux kernel and extract it to the/usr/src directory.
(2)   Enter the kernel source directory and execute the make menuconfig command. This command is used to configure the Linux kernel at compile time. The kernel for configuration compiling can only be executed in the first compiled kernel. When compiling the kernel again, you do not need to execute this command.
(3)   Execute the make command, the make command is actually an abbreviation of make bzImag and make modules, where the make bzImag command is used to create the core file, and the make modules command is used to create the module-related files;
(4)   Execute the command make moudles_install, the make moudles_install command is used to install the module;
(5)   Execute the make install command to complete the final kernel installation;
(6)   Execute the reboot command, restart the kernel, and the recompiled operating system will be displayed on the Linux system startup interface.

4.4.3 Verification Test

The test method is to attack the server with the defense technology and the non-defense technology by the same number of times, continuously increase the value of the attack times, and test the time when the defense technology and the defense technology refuse the service under different attack times. The test is shown, for example, in Table 2, X represents the time of service denial of service with defense technology, and Y represents the time when the server without defense technology refuses service. It can be found from Fig. 7 that the more attacks, the longer the time, and the more stable the state of the increased defense server.

Table 2. Defense technology test cases table

Attack number of times (times)	X (seconds)	Y (seconds)
100	24	73
200	20	78
300	16	80
500	15	79
800	21	85
1000	14	87

Run the attack program to attack the Linux server without defense function. The client accesses this Linux server multiple times and cannot get any response. Run the attacker again, attack the Linux server with defense function, and the client accesses the server, and it will get a normal response. Attack the server with defense function, and use the dmesg command to view the output log of the kernel. You can see that the server is under a lot of attacks. There is a warning in the output log to warn that the port is attacked by SYN Flood on port 80. The effect is shown in Fig. 8. It is known by max_sum_part that the largest number of hash tables obtained by traversing the linked list is that two connection request information is saved in syn_table, and the total number of connection requests is 16.

After releasing the connection request, iterate through all the items in the hash table and calculate the number of connection requests saved in each item to be output to the log. You can see that there is only one connection request in syn_table and re-count the total number of connection requests. It can be found that the total number changes from sixteen to fifteen, and the effect is shown in Fig. 9.

As can be seen from Fig. 9, when the server is attacked, it is processed by the function tcp_syn_flood_action. In the hash table in which the semi-join queue is saved, the original full state is changed to a free state. Prove the feasibility of defense technology.

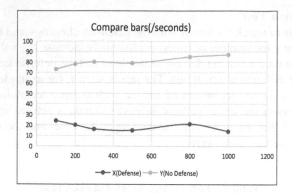

Fig. 7. Defense technology test comparison chart

```
743.327476|  TCP Possible SYN flooding on port 80
743.327480|  syn_flood_action: before unlink max_sum_part[7]  = 2
743.327483|  syn_flood_action: before unlink sum1  = 16
743.327485|  th->source= 0x4a0ca8c0, iph->raddr = 120
743.327487|  syn_flood_action: reqs = 0xffff880076466d00, reqs->expires = 0x10
)06c7bd
743.327489|  syn_flood_action: reqs->s_timeout = 0x0
743.327491|  syn_flood_action: reqs->rsk_ops = 0xffffffff8196c340, reqs->rsk_op
s->slab = 0xffff88007f807900
743.327493|  syn_flood_action: reqs = 0xffff880076466d00, (prev = &reqs) = 0xf
fff88007c603938
743.327496|  syn_flood_action: after unlink sum_part[0]  = 0
743.327498|  syn_flood_action: after unlink sum_part[1]  = 1
743.327500|  syn_flood_action: after unlink sum_part[2]  = 1
743.327502|  syn_flood_action: after unlink sum_part[3]  = 1
743.327504|  syn_flood_action: after unlink sum_part[4]  = 1
743.327506|  syn_flood_action: after unlink sum_part[5]  = 1
```

Fig. 8. SYN Flood attack processing schematic diagram

```
743.327493|  syn_flood_action: reqs = 0xffff880076466d00, (prev = &reqs) = 0xf
fff88007c603938
743.327496|  syn_flood_action: after unlink sum_part[0]  = 0
743.327498|  syn_flood_action: after unlink sum_part[1]  = 1
743.327500|  syn_flood_action: after unlink sum_part[2]  = 1
743.327502|  syn_flood_action: after unlink sum_part[3]  = 1
743.327504|  syn_flood_action: after unlink sum_part[4]  = 1
743.327506|  syn_flood_action: after unlink sum_part[5]  = 1
743.327508|  syn_flood_action: after unlink sum_part[6]  = 1
743.327510|  syn_flood_action: after unlink sum_part[7]  = 1
743.327513|  syn_flood_action: after unlink sum_part[8]  = 1
743.327515|  syn_flood_action: after unlink sum_part[9]  = 1
743.327517|  syn_flood_action: after unlink sum_part[10]  = 1
743.327519|  syn_flood_action: after unlink sum_part[11]  = 1
743.327521|  syn_flood_action: after unlink sum_part[12]  = 1
743.327523|  syn_flood_action: after unlink sum_part[13]  = 1
743.327525|  syn_flood_action: after unlink sum_part[14]  = 1
743.327527|  syn_flood_action: after unlink sum_part[15]  = 1
743.327529|  syn_flood_action: after unlink sum2 = 15
```

Fig. 9. Schematic diagram of hash table after release of connection request

4.5   The Summary of This Chapter

The main content of this chapter is to implement detection, alarm, and defense technologies for SYN Flood attacks in the kernel. The detection of the attack is done by the
kernel function, the alarm of the attack is completed by the printk function, and then the
defense is implemented by adding the function tcp_syn_flood_action in the kernel, and
finally the feasibility of the attack program and the defense function is tested.

5 Conclusion

The Internet is a double-edged sword. It brings convenience to people and brings security risks. With the development of network technology, the technology of cyber attack is also constantly improving. Research on cyber attacks and defenses has become very important. Many enterprises use a Linux operating system with better stability as a server. Therefore, it is very important to study cyber attacks and defenses against Linux servers. One of the most popular cyber attacks is the DDoS attack, which is a distributed denial of service attack. DDoS attacks are popular among many hackers because of their high attack power and are easy to hide. They also bring huge losses to the attackers. The most typical DDoS attack is the SYN Flood attack. Therefore, this paper mainly does the following work for SYN Flood attack and defense:

(1) Analyze the principle and attack mode of the DDoS attack, and carefully analyze the principle of the SYN Flood attack, and study the current defense technology features for the SYN Flood attack.
(2) Analyzing the TCP protocol of the application layer according to the principle of the SYN Flood attack;
(3) Analyze the three-way handshake process of the TCP protocol in the Linux server kernel; analyze the data structure applied in the TCP protocol in the kernel; analyze the implementation of the implementation function related to the defense technology in the kernel;
(4) According to the knowledge of the TCP protocol in the analysis kernel, find out the root cause of the service denial of service when the server is attacked by the SYN Flood, and add a defense function;
(5) Recompile the kernel, start the SYN Flood attack program, and attack the server with defense function and non-defense function respectively to test the feasibility of the defense function.

This paper proposes a SYN flood prevention method based on Linux kernel. The experiment proves that the method proposed in this paper is feasible. Compared with the method of using software for flood prevention, the method proposed in this paper can save the consumption of memory and reduce the inconvenience caused by the use of tools. However, there is still some room for improvement in the method adopted in this paper. This paper only analyzes the feasibility of the method and does not compare it with the functional method of the security software. Therefore, in the next stage, this paper will conduct an experimental analysis of the method of this kind of software and make an experimental comparison with the method in this paper.

References

1. Zhao, K., Li, X.: Principle, detection and defense technology of SYN Flooding network attack. J. South Mech. Electr. Coll. 18(03), 45–46+ 112 (2010)
2. Wei, D.: I have to know the server three attack killers. Comput. Netw. 42(22), 51 (2016)
3. Fapadopoulos, C., Lindell, R.: COSSACK: coordinated suppression of simultaneous attacks. In: Proceedings of DARPA Information Survivability Conference and Exposition, vol. 2, pp. 94–96 (2003)

4. Savage, S., Wetherall, D.: Network support for IP trace back. IEEE/ACM Trans. Netw. **9**(3), 226–237 (2001)
5. Peng, T., Leckie, C.: Protection from distributed denial of service attacks using history-based 1P filtering. In: IEEE International Conference on Communications, Alaska, USA, pp. 186–482. Electronic and Electrical Engineering publishing company (2003)
6. Liu, Y., He, Y.: SYN Flood detection method based on statistical features. Softw. Eng. **20** (04), 4–8 (2017)
7. Han, X.: Design and implementation of DDoS attack defense system. Harbin Engineering University (2015)
8. Huang, H., Hu, X., Ma, M., Li, P.: Research on comprehensive experiment teaching reform of SYN flood network security. Comput. Knowl. Technol. **14**(14), 127–130 (2018)
9. Yang, X.: TCP/IP related protocols and their applications. Commun. World **26**(01), 27–28 (2019)

Anti-quantum Cryptography Scheme Based on the Improvement of Cubic Simple Matrix and LRPC Code

Zhong Wang and Yiliang Han[✉]

College of Cryptographic Engineering,
Engineering University of PAP, Xi'an 710086, China
hanyil@163.com

Abstract. Coding cryptography can resist quantum computing attacks with high efficiency. It is similar to multivariate public key cryptography when constructing core mapping. Data compression is an advantage of coding cryptography. Therefore, combining the coding cryptography with the core mapping of multivariate public key cryptography to enhance the security of multivariate public key cryptography is a good choice. This paper first improved the Cubic Simple Matrix scheme in multivariate cryptography, and then combined the improved version scheme with the low rank parity check (LRPC) code to construct a new scheme. Compared with the Cubic Simple Matrix scheme, the ciphertext expansion rate is reduced by 50%, and the security of the scheme has been improved. The new solution is based on the improved version of the Cubic Simple Matrix, which reduces the dimensional constraints on the code when selecting LRPC codes.

Keywords: Cubic Simple Matrix · LRPC code · Security analysis · Efficiency analysis

1 Introduction

In 1994, mathematician Peter Shor pioneered a polynomial time algorithm that can run on quantum computers, the Shor algorithm [1]. Short algorithm can be used to solve the difficult problems of decomposition of large integers and the dependence of traditional public key cryptography such as discrete logarithm, which makes the attack of traditional public key cryptography possible. With the development of quantum computers, the traditional public key cryptography based on classical number theory is facing more and more threats, such as RSA, elliptic curve cryptography ECC, backpack ciphers, etc. become more and more insecure. Once the practical quantum computer appears, half of the cryptography community will face a huge threat.

At present, there are four main passwords for anti-quantum computing attacks, namely the cryptosystem based on hash function, the cryptosystem based on encoding, the cryptosystem based on lattice and the cryptosystem based on multivariate [2]. These cryptosystems are based on the NP-hard problem. Electronic computers and quantum computers have the same effect in the face of NP-hard problems, and cannot

© Springer Nature Singapore Pte Ltd. 2019
X. Cheng et al. (Eds.): ICPCSEE 2019, CCIS 1058, pp. 373–385, 2019.
https://doi.org/10.1007/978-981-15-0118-0_29

be effectively solved. Based on the multivariate cryptosystem, the research on the above four anti-quantum computing ciphers started earlier because it is more efficient and consumes less resources, making it implementable on lightweight devices [3]. The first multivariate public key cryptosystem is an MI (or C*) encryption system jointly proposed by Matsumoto and Imai [4]. At present, there are many mature multivariate public key cryptosystems such as HFE system, triangle (TS) system [5], oil vinegar (OV) system [6], etc. The Simple Matrix (ABC) scheme is a multivariate public key encryption scheme proposed by Tao et al. at the PQCrypto 2013 conference, which adopts a new multivariate cryptosystem [7]. In 2014, in PQCrypto 2014, Jintai Ding et al. improved the Simple Matrix scheme and proposed the Cubic Simple Matrix (Cubic ABC) scheme [8]. The secret ciphertext ratio of these two schemes is fixed to a double relationship, which provides some clues for the attacker and is not flexible enough to meet a variety of requirements for plaintext and ciphertext. In order to improve this deficiency, this paper proposes an improved version of Cubic Simple Matrix, which provides security, while providing flexible cryptographic ratios, and lays the foundation for combining with LRPC codes.

In 1978, McEliece proposed the first public key encryption scheme based on coding theory design [9], in which the private key is a binary irreducible Goppa code, and the public key is the result of the randomization of the generator matrix. Niederreiter proposed the famous Goppa code based Niederreiter cryptosystem in 1986 [10], compared with the McEliece scheme, the Niederreiter scheme has a reduced amount of public key storage and a higher signaling rate. Inspired by the McEliece scheme and the Niederreiter scheme, many new schemes have also emerged. Up to the current research stage, the focus of public key cryptography based on coding theory is mainly on the choice of code, from the initial Goppa code, Reed-Solomon code, to the current QC-MDPC code [11], LDPC (Low Density Parity Check codes) [12] and QC-LDPC codes [13]. With the development of coding theory, the LRPC codes [14] that appeared later have received extensive attention due to their excellent characteristics.

The cryptography based on coding applies the operations between matrices, which is similar to multivariate ciphers. This similarity provides the idea and direction for the combination of the two. The cryptography based on coding mainly uses the nature of the error-correcting code. For example, in 2012, Wang et al. proposed a class of multivariable public key encryption schemes based on error-correcting coding by using the public key in Niederriter system instead of a linear mapping in multivariable cryptographic schemes, which provided a new platform for constructing the core mapping of multivariable public key cryptographic schemes [15, 16]. This paper combines this idea, first improve the Cubic Simple Matrix, and then combine the LRPC code with the Cubic Simple Matrix improvement scheme, so that the new scheme has the advantage of coding cipher while reducing the ciphertext expansion rate compared to the Cubic Simple Matrix scheme. On the other hand, due to the adoption of the improved version of Cubic Simple Matrix, the size of the LRPC code check matrix is more flexible, and it is not necessary to limit the number of columns to twice the number of rows. The second section of this paper introduces the MQ problem with the Cubic Simple Matrix scheme and the coding cryptosystem. The third section details the improved version of the Cubic Simple Matrix and the new scheme combined with the

LRPC code. The fourth section analyzes the security of the new scheme. Section five for efficiency analysis, the final section is the summary of the full text to found insufficient.

2 Preliminaries

2.1 MQ Problem and IP Problem

The MQ problem is the basis of multivariate public key cryptography. The so-called MQ problem refers to the solution of the equations from $y_1, y_2, \cdots, y_m \in F$, (F is finite field F = GF(q)) and equation transformation. This problem has been proved to be an NP-hard problem. The specific definition is as follows:

Definition 2.1.1 (MQ problem): Given a set of quadratic multivariate polynomial equations with n variables and m polynomials on a finite field F, the structure is as shown in Eq. (1), consisting of a set of known $y_1, y_2, \cdots, y_m \in F$ and equation transformation, solve a set of solutions $x_1, x_2, \cdots, x_n \in F$ to satisfy the equations.

$$
\begin{cases}
y_1 = f_1(x_1, \ldots, x_n) = \sum_{1 \le i,j \le n} a_{1,i,j} x_i x_j + \sum_{i=1}^{n} b_{1,i} x_i + c_1 \\
y_2 = f_2(x_1, \ldots, x_n) = \sum_{1 \le i,j \le n} a_{2,i,j} x_i x_j + \sum_{i=1}^{n} b_{2,i} x_i + c_2 \\
\vdots \\
y_m = f_m(x_1, \ldots, x_n) = \sum_{1 \le i,j \le n} a_{m,i,j} x_i x_j + \sum_{i=1}^{n} b_{m,i} x_i + c_m
\end{cases}
\tag{1}
$$

Definition 2.1.2 (IP problem): Let F and P be a set of stochastic multivariable quadratic polynomial equations with m polynomials and n variables on a given finite field F = GF(q). Find two reversible linear transforms $S: K^m \to K^m$ and $T: K^n \to K^n$ on the finite field so that F and P satisfy the relation: $P = S \circ F \circ T$, then the problem of finding these two reversible linear transforms is called IP problem.

2.2 Cubic Simple Matrix Scheme

In 2014, in PQCrypto 2014, Jintai Ding et al. proposed the Cubic Simple Matrix (Cubic ABC) scheme. The scheme uses the operations between matrices to subtly construct the core map. The main parameters are n, m, s. Where n represents the length of the plaintext and m represents the length of the ciphertext. The relationship between them satisfies $m = 2n$, that is, the length of the ciphertext is twice that of the plaintext, and the plaintext variable is used $X = (x_1, x_2 \cdots, x_n)$, $s \in N$, and $s^2 = n$.

There are three $s \times s$-dimensional matrices A, B, C. The elements in matrix A are quadratic equations for the plaintext variable $X = (x_1, x_2 \cdots, x_n)$, while the elements in matrices B and C are linear combinations of plaintext variables.

$$A = \begin{bmatrix} a_1 & \cdots & a_s \\ \vdots & \ddots & \vdots \\ a_{(s-1)s+1} & \cdots & a_n \end{bmatrix}$$

$$B = \begin{bmatrix} b_1 & \cdots & b_s \\ \vdots & \ddots & \vdots \\ b_{(s-1)s+1} & \cdots & b_n \end{bmatrix}$$

$$C = \begin{bmatrix} c_1 & \cdots & c_s \\ \vdots & \ddots & \vdots \\ c_{(s-1)s+1} & \cdots & c_n \end{bmatrix}$$

Two $s \times s$-dimensional matrices E_1 and E_2, $E_1 = A \cdot B$, $E_2 = A \cdot C$, the elements in E_1 and E_2 are all cubic equations for variables. There are m quadratic polynomials for both E_1 and E_2. These m quadratic polynomials are combined into the center map F of the Cubic Simple Matrix algorithm: $K^n \rightarrow K^m$. There are two reversible linear maps T: $K^n \rightarrow K^n$, $S:K^m \rightarrow K^m$, and they are combined with the center map F to obtain the public key, $P = S \circ F \circ T: K^n \rightarrow K^m$. The private key contains two matrices B, C and reversible linear map S, T.

Specific encryption and decryption process in reference [8].

2.3   LRPC Code and RSD Problem

The LRPC code is modeled after the classic LDPC code. The difference between the two is that the LRPC code is a low-rank parity check code based on the rank matrix, which is measured by the rank distance rather than by the hamming distance as the LDPC code. The definition is as follows: The LRPC code is a rank code with a dimension k, a rank d, and a length n on the domain $GK(q^m)$, having a matrix $L(h_{ij})$, and the $(n - k) \times n$ matrix L is the check matrix of the $[n, k]$ LRPC code. The subvector space dimension on GK (q^m) generated by the coefficient l_{ij} of the matrix L is at most d. The $k \times n$-dimensional matrix G is the matrix for which it is generated.

Rank Syndrome Decoding Problem (RSD). Let L be a $(n - k) \times n$ matrix on the domain $GK(q^m)$, $k \leq n, s \in GK(q^m)^k$, r is a integer. The rank maximum likelihood decoding problem is to find an x that satisfies $rank(x) = r$, and $Lx^t = s$. Although the RSD problem has not been proven to be NP-HARD, the SD problem based on hamming distance has been proved to be NP-HARD, and the RSD problem is similar to the SD problem.

3 Improvement of Cubic Simple Matrix and the New Scheme with LRPC Code

3.1 Improvement of Cubic Simple Matrix

The Cubic Simple Matrix Improvement is different from the Cubic Simple Matrix scheme in that the matrix B is transformed. In the form of the matrix, they are no longer square matrices of $s \times s$, but $s \times u$ dimensional flat matrix, $u \in N$, $u \geq s$, the elements in them are still unchanged linear combination. The element in matrix A is still a quadratic equation about the variable $X = (x_1, x_2 \cdots, x_n)$.

$$A = \begin{bmatrix} a_{11} & \cdots & a_{1s} \\ \vdots & \ddots & \vdots \\ a_{s1} & \cdots & a_{ss} \end{bmatrix}$$

$$a_{ij} = \sum_{r=1}^{n} \sum_{s=r}^{n} \alpha_{rs}^{(ij)} x_r x_s + \sum_{r=1}^{n} \beta_r^{(ij)} x_r + \varepsilon \tag{2}$$

$$B = \begin{bmatrix} b_{11} & b_{12} & \cdots & b_{1s} & \cdots & b_{1u} \\ b_{21} & b_{22} & \cdots & b_{2s} & \cdots & b_{2u} \\ \vdots & \vdots & & \vdots & & \vdots \\ b_{s1} & b_{s2} & \cdots & b_{ss} & \cdots & b_{su} \end{bmatrix}$$

$$b_{ij} = \sum_{r=1}^{n} \alpha_r^{(ij)} x_r + \mu \tag{3}$$

$$C = \begin{bmatrix} c_{11} & \cdots & c_{1s} \\ \vdots & \ddots & \vdots \\ c_{s1} & \cdots & c_{ss} \end{bmatrix}$$

$$c_{ij} = \sum_{m=1}^{n} \alpha_m^{(ij)} x_m + \theta \tag{4}$$

Then, a matrix of $s \times u$ dimensional matrix E_1 and $s \times s$ dimensional matrix E_2 are defined, $E_1 = A \cdot B$, $E_2 = A \cdot C$.

$$E_1 = \begin{bmatrix} e_{11} & e_{12} & \cdots & e_{1s} & \cdots & e_{1u} \\ e_{21} & e_{22} & \cdots & e_{2s} & \cdots & e_{2u} \\ \vdots & \vdots & & \vdots & & \vdots \\ e_{s1} & e_{s2} & \cdots & e_{ss} & \cdots & c_{su} \end{bmatrix}$$

$$e_{ij} = \sum_{r=1}^{n} \sum_{s=r}^{n} \sum_{t=s}^{n} \alpha_{rst}^{(ij)} x_r x_s x_t + \sum_{r=1}^{n} \sum_{s=r}^{n} \beta_{rs}^{(ij)} x_r x_s + \sum_{r=1}^{n} \sigma_r^{(ij)} x_r + \delta \tag{5}$$

$$E_2 = \begin{bmatrix} \bar{e}_{11} & \cdots & \bar{e}_{1s} \\ & \ddots & \vdots \\ \bar{e}_{s1} & \cdots & \bar{e}_{ss} \end{bmatrix}$$

$$\bar{e}_{ij} = \sum_{a=1}^{n}\sum_{b=a}^{n}\sum_{c=b}^{n} \alpha_{abc}^{(ij)} x_a x_b x_c + \sum_{a=1}^{n}\sum_{b=a}^{n} \beta_{ab}^{(ij)} x_a x_b + \sum_{a=1}^{n} \sigma_a^{(ij)} x_a + \gamma \qquad (6)$$

As we can see from the above, there is a total of $m = su + s^2$ polynomials about plaintext variables in E_1 and E_2. No longer the fixed relationship of two times in the original plan. Its core mapping is $F : K^n \rightarrow K^m$. Two reversible linear transformations are also needed $T : K^n \rightarrow K^n$, $S : K^m \rightarrow K^m$. The public key of the scheme is $P = S \circ F \circ T : K^n \rightarrow K^m$. The private key are the matrix B and C, the linear transformations T, S.

(1) Encryption process: the plaintext $X = (x_1, x_2 \cdots, x_n)$ is brought into the formula of the public key P to obtain the ciphertext $Y = (y_1, y_2, \cdots, y_m)$.
(2) Decryption process: when decrypting, first use the linear transformation S to transform the ciphertext $Y = (y_1, y_2, \cdots, y_m)$. And then look at whether matrix E_2 is invertible, if invertible, we can obtain $s \times s$ equations for n variables by solving the equations $C \cdot E_2^{-1} \cdot E_1 = B$. If the matrix E_2 is irreversible, it depends on whether the matrix A is invertible. If the matrix A is invertible, the elements in the matrix A^{-1} can be regarded as new elements and multiplied by the matrix E_1, E_2, respectively, resulting in $A^{-1}E_1 = B, A^{-1}E_2 = C$, there are $2n(2s^2)$ variables, $m = s \times u + s^2$ linear equations, $m \geq 2n$, so the plaintext variable X can be solved finally. Last the linear transformation T is used to manipulate the variables to get the plaintext $X = (x_1, x_2 \cdots, x_n)$.

3.2   New Cryptography Scheme Based on the Improvement of Cubic Simple Matrix Scheme and LRPC Code

Same as the Cubic Simple Matrix Improvement, there are m, n, s and $m \in N, n \in N, s \in N, s^2 = n$. A matrix A of $s \times s$ dimension and a flat matrix B of $s \times u$ dimension and C of $s \times s$ dimension. The elements of A are quadratic polynomials of variables $X = (x_1, x_2 \cdots, x_n)$, and the elements of matrix B and C are linear combinations of variables $X = (x_1, x_2 \cdots, x_n)$. Then, according to the requirements, a LRPC code C_{LRPC} of $[u + s, s]$ is selected on the domain $GK(q^s)$, and the matrix G_{LRPC} of $s \times (u + s)$ dimension is the generator matrix of LRPC code C_{LRPC}, the matrix L_{LRPC} is the check matrix. Let $s \times u$ dimension matrix E_1, $s \times s$ dimension matrix E_2 as shown in Sect. 3.1, $E_1 = A \cdot B, E_2 = A \cdot C$, core mapping F contains parts of E_1, E_2. Now E_1 and E_2 are combined into the following form:

$$F(X) = \begin{bmatrix} e_{11} & e_{12} & \cdots & e_{1s} & \cdots & e_{1u} & \bar{e}_{11} & \bar{e}_{12} & \cdots & \bar{e}_{1s} \\ e_{21} & e_{22} & \cdots & e_{2s} & \cdots & e_{2u} & \bar{e}_{21} & \bar{e}_{22} & \cdots & \bar{e}_{2s} \\ \vdots & \vdots & & \vdots & & \vdots & \vdots & \vdots & & \vdots \\ e_{s1} & e_{s2} & \cdots & e_{ss} & \cdots & e_{su} & \bar{e}_{s1} & \bar{e}_{s2} & \cdots & \bar{e}_{ss} \end{bmatrix}$$

A new core mapping H is constructed by transforming the core mapping of the improved Cubic Simple Matrix version with the generating matrix G_{LRPC}.

$$H = F \circ G_{LRPC}^T + E \tag{7}$$

$e = (e_1, e_2 \ldots, e_s)$ is an error vector with rank r selected on $GK(q^s)$ according to user identity $s_{ID} \in GK(q^{s \times s})$. E is expressed as a circular matrix and called error matrix E. It is then added to the construction of core mapping as a perturbation part. The error matrix E can be defined as follows:

$$E = \begin{bmatrix} e_1 & e_2 & e_3 & \cdots & e_s \\ e_s & e_1 & e_2 & \cdots & e_{s-1} \\ e_{s-1} & e_s & e_1 & \cdots & e_{s-2} \\ \vdots & \vdots & \vdots & \ddots & \vdots \\ e_2 & e_3 & \cdots & e_s & e_1 \end{bmatrix}$$

Then, by using two reversible linear transformations $T : K^n \rightarrow K^n, S : K^n \rightarrow K^n$, we perform a compound operation on the core mapping H and get a new core mapping $P = S \circ H \circ T$, then the public key mapping is $P = S \circ H \circ T : K^n \rightarrow K^n$, and the private key is the matrix B of $s \times u$ dimension, the matrix C of $s \times s$ dimension and the generating matrix G_{LRPC} of $s \times (u+s)$ dimension, as well as two reversible linear transformations, $T : K^n \rightarrow K^n, S : K^n \rightarrow K^n$.

(1) Encryption process: assuming that the given information $X = (x_1, x_2 \cdots, x_n)$ is to be encrypted, it is directly brought into the public key mapping, that is, $P = S \circ H \circ T(X) = Y = (y_1, y_2, \cdots, y_n)$.

(2) Decryption process:
First, the inversion of reversible linear transformation S is used to operate the ciphertext to get the variable $W = (w_1, w_2, \cdots, w_n) = S^{-1}Y = S^{-1}(y_1, y_2, \cdots, y_n)$. By calculating $H^{-1}(W) = U$, a set of variables $U = (u_1, u_2, \cdots, u_n)$ is found to make $H(U) = W$.

Second, calculate the syndrome , $s_{ID} = L_{LRPC} \cdot e^t$, recover the error vector E by LRPC code, and get the error vector matrix E. Look for $\bar{X}$, and satisfy $F(\bar{X}) = (H(U) - E) \cdot G_{LRPC}^{T-1}$.

Third, the variables in $\bar{X}$ can be represented by matrix $\bar{E}_1, \bar{E}_2$, as follows:

$$
\bar{E}_1 =
\begin{bmatrix}
\bar{x}_1 & \bar{x}_2 & \cdots & \bar{x}_s & \cdots & \bar{x}_u \\
\bar{x}_{u+1} & \bar{x}_{u+2} & \cdots & \bar{x}_{u+s} & \cdots & \bar{x}_{2u} \\
\vdots & \vdots & \cdots & \vdots & \cdots & \vdots \\
\bar{x}_{(s-1)u+1} & \bar{x}_{(s-1)u+2} & \cdots & \bar{x}_{(s-1)u+s} & \cdots & \bar{x}_{su}
\end{bmatrix}
$$

$$
\bar{E}_2 =
\begin{bmatrix}
\bar{x}_{su+1} & \bar{x}_{su+2} & \cdots & \bar{x}_{su+s} \\
\bar{x}_{su+s+1} & \bar{x}_{su+s+2} & \cdots & \bar{x}_{su+2s} \\
\vdots & \vdots & \cdots & \vdots \\
\bar{x}_{su+(s-1)s+1} & \bar{x}_{su+(s-1)s+2} & \cdots & \bar{x}_{su+s^2}
\end{bmatrix}
$$

The solution to $F^{-1}(\bar{X})$ is the same as the Sect. 3.1 of the decryption process of the improved version of Cubic Simple Matrix scheme.

Last, through the calculation of $X = T^{-1}(\bar{X})$, we can get the plaintext $X = (x_1, x_2 \ldots, x_n)$.

4   Security Analysis

4.1   Scheme Integrity Analysis

It can be seen from the structural ideas and specific scheme descriptions in Sects. 3.1 and 3.2 that since the elements in matrix A are all quadratic polynomials about plaintext variables, and the choice of these quadratic polynomials is random, even if the same plaintext is encrypted, the result is basically impossible. Reference [8] pointed out that when directly solving the MQ problem, under $O(2^{100})$ security, the time complexity of direct attack is μ: $\mu \geq 3 \cdot \partial \cdot \delta^2 \geq 2^{102}$ (δ is the number of the highest order automorphism mononomial, ∂ is the number of non-zero monomials in each polynomial). In addition, the anti-quantum cryptography scheme of this paper is constructed based on the improved version of Cubic Simple Matrix. In the improved version of Cubic Simple Matrix, the number of equations is additionally increased, the redundancy is increased, and the complexity of MQ problem is further enhanced. It is not feasible to have an attacker O wanting to directly break the MQ problem in polynomial time. Secondly, because the anti-quantum cryptography scheme of this paper combines the coded cryptography with the multivariate public key cryptography in the design process of the core mapping, the attacker must first crack the RSD problem before attacking the MQ problem. The complexity of the RSD problem is about $q^{r \cdot s}$, it has a close relationship with its parameter selection, and its complexity changes exponentially with the choice of parameters. It can be seen from the above analysis that it is not feasible for the attacker to break the scheme in polynomial time after selecting the appropriate parameters.

4.2   Algebraic Attacks

Algebraic attack is a more effective way to attack multivariable public key cryptography directly. Among them, the most famous algorithms are XL algorithm [17] and

Gröbner basis algorithm, which are often used to solve the nonlinear equation [18]. The main idea of algebraic attack is to substitute ciphertext into public key polynomial equations and solve them directly to get plaintext. XL algorithm is similar to F4 algorithm and can be regarded as a redundant variant of F4 algorithm [19]. Their essence is still to calculate Gröbner basis. For solving Gröbner basis, the more effective methods are F4 and F5 algorithm proposed by Faugere.

When using algebraic attacks to attack the new cryptography scheme based on Cubic Simple Matrix scheme and LRPC, attackers first need to face the RSD problem, which will greatly increase the complexity of the solution. Secondly, we will face a system of equations consisting of n variables and m ($m \geq 2n$) cubic multivariate polynomial equations. The number of equations is larger than the number of variables, which is equivalent to increasing external disturbances and improving the redundancy of the system. Apart from the RSD problem of encoding ciphers, the improved version of Cubic Simple Matrix can also resist algebraic attacks. Because the system obtained by computing between matrices is a cubic system, its solution difficulty is not lower than that of random multivariate quadratic system, and because of the participation of two reversible linear transformations, IP problem is introduced. Isomorphism also adds security. It can be seen from the above that algebraic attacks cannot effectively attack the new cryptography scheme based on Cubic Simple Matrix scheme and LRPC codes.

4.3   Rank Attacks

Rank attacks is an effective attack method for multivariable public key cryptography. It can be divided into LowRank attack [20] and HighRank attack [21].

The purpose of LowRank attack is to find a linear combination in the public key, and its rank is the smallest rank r. It is defined as finding a non-trivial linear combination, i.e. for m matrices whose dimension are $n \times n$ on a finite field K, find a non-trivial linear combination, namely: $M = \mu_1 M_1 + \mu_2 M_2 + \cdots + \mu_m M_m$, and the rank of this linear combination is not greater than r (rank(M) $\leq r$). The minimum rank problem is also a NP problem, but when r is small enough, its complexity is acceptable and can be solved. For example, in the HFE scheme, a low rank polynomial corresponds to a polynomial in the center mapping. The attacker can recover the reversible linear transformation T by finding a low rank linear combination, and finally get the plaintext. In HighRank attack, the attacker finds a linear combination of variables with the least number of occurrences in the central polynomial. For example, the oil variable in the last layer of Rainbow scheme [22] is the variable with the least number of occurrences. By repeating this attack on each layer, the attacker can restore the reversible linear transformation S and finally restore the plaintext.

In the improvement of Cubic Simple Matrix version, the elements of matrix A are random selected quadratic polynomials. Random selected quadratic polynomials make the rank of matrix close to n, so low rank cannot be effectively found and then cannot be effectively cracked. Because multivariate quadratic polynomials are composed of plaintext variables, every plaintext variable may appear in every equation, and the number of occurrences is random, the number of occurrences is nearly the same, which makes high rank attack invalid. LRPC code further transforms the core mapping, which increases its complexity. It needs to solve the difficulty of decoding before rank attack can be carried out. Therefore, rank attack is not effective for anti-quantum cryptography schemes based on the improvement Cubic Simple Matrix and LRPC codes.

5 Efficiency Analysis

5.1 Selection of LRPC Codes

Based on the improvement of Cubic Simple Matrix and LRPC codes, the improved version of Cubic Simple Matrix is combined with LRPC codes, which makes the selection of LRPC codes more flexible than the combination of Cubic Simple Matrix and LRPC codes.

In the process of constructing the combination of multivariable public key cryptography and LRPC code, the choice of LRPC code is mainly limited by the dimension of matrix in the core mapping. The core mapping of Cubic Simple Matrix scheme is composed of two matrices E_1 and E_2 with dimension of $s \times s$. When choosing LRPC codes, considering that its generator matrix G_{LRPC} needs to be operated with $[E_1 \mid E_2]$ of $s \times 2s$ dimension, the choice of LRPC codes can only be fixed to $[2s, s]$, which greatly reduces the choice of LRPC codes, cannot meet the diversity of needs.

The new cryptography scheme based on the improvement of Cubic Simple Matrix and LRPC codes can be optimized by combining the improvement of Cubic Simple Matrix with LRPC codes to construct core mapping. Because the core mapping of the improvement of Cubic Simple Matrix consists of a flat matrices E_1, whose dimensions are $s \times u$, where $u \geq s$, then the generator matrix G_{LRPC} which operates with $s \times (u + s)$ dimension matrix $[E_1 \mid E_2]$ is also $s \times (u + s)$ dimension, that is, the choice of LRPC codes is $[u + s, s]$, and $u + s \geq 2s$.

Compared with the combination of Cubic Simple Matrix scheme and LRPC code, the anti-quantum cryptography scheme based on the improvement of Cubic Simple Matrix and LRPC code enlarges the selection range of LRPC code, provides flexible selectivity, and meets more application scenarios and requirements.

5.2 Ciphertext Expansion Rate

In the Cubic Simple Matrix scheme, the amount of ciphertext is usually twice that of plaintext, because the dimensions of two matrices in public key mapping are $s \times s$, $s^2 = n$, where n represents the size of plaintext, so the size of ciphertext is $2n$. In the Improvement of Cubic Simple Matrix, Sect. 3.1 shows that the amount of ciphertext is larger than or equal to twice the amount of plaintext, so as to meet the diversity of needs and break the limitation of the ratio between plaintext to ciphertext.

In reference [17], it is pointed out that in the construction of LRPC codes, the size of the generator matrix G_{LRPC} constructed in the form of DC-LRPC is about $s^2 \log_2 q$, while the size of the matrix G_{LRPC} as the private key is not large, and the error matrix E is expressed in the form of a cyclic matrix, which occupies little space. Therefore, the key size of anti-quantum cryptography based on the improvement of Cubic Simple Matrix and LRPC codes the will not increase significantly. Because the transposition of the generator matrix of LRPC code is multiplied by the matrix in the core mapping of

Cubic Simple Matrix improved version, the matrix of $s \times s$ dimension is finally obtained, which is the same as plaintext. Compared with Cubic Simple Matrix scheme, the amount of ciphertext is reduced by 50%, and the compactness of ciphertext is improved. The comparison of the length of ciphertext between Cubic Simple Matrix scheme and anti-quantum cryptography scheme based on the improvement of Cubic Simple Matrix and LRPC code is shown in Table 1 below.

Table 1. Comparison of ciphertext lengths

Scheme	Parameter/($\mathbb{F} = GF(q)$, s, n, m, d, r)	PK/KB	SK/KB	X/bit	Y/bit
Cubic simple matrix	($GF(2^8)$), 6, 36, 72, 0, 0)	1120	50	201	402
	($GF(2^8)$), 8, 64, 128, 0, 0)	5991	152	512	1024
Scheme of this paper	($GF(2^8)$), 6, 36, 72, 5, 4)	1206	63	201	201
	($GF(2^8)$), 8, 64, 128, 5, 4)	6061	160	512	512

(PK represents public key, SK represents private key, X represents plaintext, Y represents ciphertext.)

Table 1 shows that compared with Cubic Simple Matrix scheme, the anti-quantum cryptography scheme based on the improvement of Cubic Simple Matrix and LRPC code reduces the ciphertext expansion rate by 50% without significantly increasing the lengths of encryption keys.

6 Conclusion

This paper studies the code-based cryptography and multivariable cryptography deeply, and combines them to design an anti-quantum cipher scheme based on multivariable and LRPC codes. This scheme combines the LRPC code in coding with the improvement of Cubic Simple Matrix in multivariable public key cryptography. Because the improved version of Cubic Simple Matrix is used, it guarantees the security and is more flexible in selecting LRPC code. Generating a fixed row relationship of matrices. Compared with Cubic Simple Matrix scheme, the anti-quantum cryptography scheme based on the improvement of Cubic Simple Matrix and LRPC code reduces the ciphertext expansion rate by 50%, and improves the compactness of ciphertext. At the same time, it combines the advantages of code-based cryptography and improves the security, which can effectively resist rank attack and algebraic attack. The disadvantage of anti-quantum cryptography schemes based on the improvement of Cubic Simple Matrix and LRPC code is that the ratio of ciphertext to plaintext is fixed to one-to-one, which makes the scheme less flexible. Further improvements are needed to expand the ratio range between plaintext and ciphertext to meet more requirements and adapt to more scenarios.

Acknowledgment. This work was supported by the National Natural Science Foundation of China (No. 61572521).

References

1. Shor, P.: Polynomial-time algorithms for prime factorization and discrete logarithms on a quantum computer. SIAM J. Comput. **26**(5), 1484–1509 (1994)
2. Bernstein, D.J., Buchmann, J., Dahmen, E. (eds.): Post Quantum Cryptography. Springer, Heidelberg (2009). https://doi.org/10.1007/978-3-540-88702-7
3. Bogdanov, A., Eisenbarth, T., Rupp, A., Wolf, C.: Time-area optimized public-key engines: *MQ*-cryptosystems as replacement for elliptic curves? In: Oswald, E., Rohatgi, P. (eds.) CHES 2008. LNCS, vol. 5154, pp. 45–61. Springer, Heidelberg (2008). https://doi.org/10.1007/978-3-540-85053-3_4
4. Matsumoto, T., Imai, H.: Public quadratic polynomial-tuples for efficient signature-verification and message-encryption. In: Barstow, D., et al. (eds.) EUROCRYPT 1988. LNCS, vol. 330, pp. 419–453. Springer, Heidelberg (1988). https://doi.org/10.1007/3-540-45961-8_39
5. Patarin, J.: Hidden fields equations (HFE) and isomorphisms of polynomials (IP): two new families of asymmetric algorithms. In: Maurer, U. (ed.) EUROCRYPT 1996. LNCS, vol. 1070, pp. 33–48. Springer, Heidelberg (1996). https://doi.org/10.1007/3-540-68339-9_4
6. Fell, H., Diffie, W.: Analysis of a public key approach based on polynomial substitution. In: Williams, H.C. (ed.) CRYPTO 1985. LNCS, vol. 218, pp. 340–349. Springer, Heidelberg (1986). https://doi.org/10.1007/3-540-39799-X_24
7. Tao, C., Diene, A., Tang, S., Ding, J.: Simple matrix scheme for encryption. In: Gaborit, P. (ed.) PQCrypto 2013. LNCS, vol. 7932, pp. 231–242. Springer, Heidelberg (2013). https://doi.org/10.1007/978-3-642-38616-9_16
8. Ding, J., Petzoldt, A., Wang, L.-C.: The cubic simple matrix encryption scheme. In: Mosca, M. (ed.) PQCrypto 2014. LNCS, vol. 8772, pp. 76–87. Springer, Cham (2014). https://doi.org/10.1007/978-3-319-11659-4_5
9. Mceliece, R.J.: A public-key cryptosystem based on algebraic. Coding Thv **4244**, 114–116 (1978)
10. Niederreiter, H.: Knapsack-type cryptosystems and algebraic coding theory. Probl. Control Inf. Theor. **15**(2), 159–166 (1986)
11. Li, Z., Yang, Y., Li, Z.: Design of public key cryptosystem based on QC-MDPC code. Comput. Appl. Res. **32**(03), 881–884 (2015)
12. Becker, O.: Symmetric unique neighbor expanders and good LDPC codes. Discrete Appl. Math. **211**, 211–216 (2016)
13. Gaborit, P., Ruatta, O., Schrek, J., Zémor, G.: New results for rank-based cryptography. In: Pointcheval, D., Vergnaud, D. (eds.) AFRICACRYPT 2014. LNCS, vol. 8469, pp. 1–12. Springer, Cham (2014). https://doi.org/10.1007/978-3-319-06734-6_1
14. Han, Y., Lan, J., Yang, X.: Signcryption scheme based on LRPC code and multivariable. J. Crypt. **3**(01), 56–66 (2016)
15. Wang, H.Z., Shenc, X., Xuz, Q., et al.: Multivariate public-key encryption scheme based on error correcting codes. China Commun. **8**(4), 23–31 (2011)
16. Han, Y., Lan, J., Yang, X., Wang, J.: Multivariable encryption scheme combined with low rank error correction coding. J. Huazhong Univ. Sci. Technol. (Nat. Sci. Ed.), **44**(03), 71–76 (2016)
17. Faugere, J.C.: A new efficient algorithm for computing Gröbner bases without reduction to zero (F5). In: Proceedings of the 2002 International Symposium on Symbolic and Algebraic Computation, pp. 75–83. ACM (2002)
18. Faugere, J.C.: A new efficient algorithm for computing Gröbner bases (F4). J. Pure Appl. Algebra **139**, 61–88 (1999)

19. Ding, J., Buchmann, J., Mohamed, M.S.E., Mohamed, W.S.A.E., Weinmann, R.-P.: Mutant XL. In: Talk at the First International Conference on Symbolic Computation and Cryptography (SCC 2008), Beijing (2008)
20. Goubin, L., Courtois, N.T.: Cryptanalysis of the TTM cryptosystem. In: Okamoto, T. (ed.) ASIACRYPT 2000. LNCS, vol. 1976, pp. 44–57. Springer, Heidelberg (2000). https://doi.org/10.1007/3-540-44448-3_4
21. Coppersmith, D., Stern, J., Vaudenay, S.: Attacks on the birational permutation signature schemes. In: Stinson, Douglas R. (ed.) CRYPTO 1993. LNCS, vol. 773, pp. 435–443. Springer, Heidelberg (1994). https://doi.org/10.1007/3-540-48329-2_37
22. Ding, J., Schmidt, D.: Rainbow, a new multivariable polynomial signature scheme. In: Ioannidis, J., Keromytis, A., Yung, M. (eds.) Applied Cryptography and Network Security ACNS 2005. LNCS, vol. 3531, pp. 164–175. Springer, Heidelberg (2005). https://doi.org/10.1007/11496137_12

Survey on Blockchain Incentive Mechanism

Jiyue Huang[1,2], Kai Lei[1,2], Maoyu Du[1,2], Hongting Zhao[1,2], Huafang Liu[2], Jin Liu[2], and Zhuyun Qi[1,2(✉)]

[1] Shenzhen Key Lab for Information Centric Networking & Blockchain Technology (ICNLAB), School of Electronics and Computer Engineering, Peking University, Shenzhen, China
{huangjiyue,mydu}@sz.pku.edu.cn, leik@pkusz.edu.cn, dove710@foxmail.com
[2] PCL Research Center of Networks and Communications, Peng Cheng Laboratory, Shenzhen, China
{liuhf,liuj01,qizy}@pcl.ac.cn

Abstract. The current research on the blockchain includes study on network architecture and the incentive. In this paper, an introduction to the architecture of information technology was given and the goal and research status of the incentive layer of blockchain were illustrated with digital economy development as the backdrop. The existing issuance of token was elaborated from computation, storage and transmission, three core facets of network technology. The development of the token allocation and the path was analyzed to confirm its rationality. It discusses the possible directions and challenges of the future research on blockchain incentive.

Keywords: Blockchain · Incentive mechanism · Digital economy

1 Introduction

Information technology architecture has been growing from large-scale centralized [11] to client/server (C/S) architecture, and to cloud computing [20]. Along the process, although the centralized architecture remains a shared feature of these three sectors, it is still entangled with the high maintenance cost and doubt on its security and reliability. As a key technology of distributed decentralized system and aims to alleviate security risk caused by centralization, the blockchain has shown promising prospects on academic and industrial application and attracted widespread attention. Blockchain can be defined as a novel distributed infrastructure and computing paradigm [32]. It uses blockchain data structure as storage tool, generates and updates data through distributed node consensus algorithms [23], secures data transmission and access via cryptography, and programs and can manipulates data through the smart contract composed of automated scripts.

© Springer Nature Singapore Pte Ltd. 2019
X. Cheng et al. (Eds.): ICPCSEE 2019, CCIS 1058, pp. 386–395, 2019.
https://doi.org/10.1007/978-981-15-0118-0_30

As the entity integrating P2P network and asymmetric encryption, blockchain is characterized by its decentralization, traceability, anti-counterfeiting, anti-tampering, and automatic execution of contacts, but the decentralized consensus and reliable basis for providing fine-grained distribution of network by tokens at the data level actually serve as the two key features distinguishing it from other systems. The consensus organizes a large scale of consensus nodes to implements the data validation and accounting of shared ledgers. In the decentralized system, it is essentially a crowd-sourcing process between consensus nodes [7,18,30], and as the consensus node itself is self-interested, maximizing its own interest is the fundamental goal of its participation in data verification and accounting [29].

Focusing on the incentive of blockchains, as the foundation of the number of consensus nodes in the network and a key sector of the prospective development of network ecosystem, main contributions of this paper are as follows:

1. To abstract the essence of blockchain and to encourage readers to have a clearer understanding on blockchain.
2. Systematically overviewed the existing work of the blockchain incentive mechanism and analyze the basic principles and key issues on basis of mechanisms of issuing and allocating.
3. To summarize and forecast the challenges and opportunities of the research on blockchain incentive to provide enlightenment and reference for future research [22].

2   Economics of Blockchain

2.1   Digital Economy

As early as in 1976, Hayek, a rugged liberalist, analyzed the fragility of currency system based on public credit in his book of Denationalization of Money [14], argued merits of issuing currencies by different entities (banks) and forecasted a prospect that issuing money is independent from government authority. Since the beginning of 21st century, with the advancement of computer and Internet communication technology, decentralized P2P communication technology and blockchain technology have emerged. The successful emergency of blockchain currencies such as Bitcoin, filecoin, and Ripple XRP provides denationalized money forecasted by Hayek with technological feasibility.

The basic economics of blockchains concerns why distributed solutions that is technically feasible may become more cost-effective than centralized solutions [6]. And the answer is that they run along three index cost curves: (1) Moore's Law - the cost of tackling digital information, namely speed, will be halved every 18 months; (2) Kryder's Law - the cost of storing digital information, namely memory, will be halved every 12 months; (3) Nielsen's Law - the cost of transporting digital information, namely bandwidth, will be halved every 24 months (Wiles 2015). The blockchain establishes reliable trust between nodes in the network, which frees the value transferring process from the interference of the intermediary, improving the efficiency of value interaction and reducing the cost, and becoming the cornerstone of building the value Internet.

However, existing blockchain currencies (such as Bitcoin) are issued by non-official organizations and gain no government credit endorsement. Most of them are characterized by decentralization, deflation, long transaction delay, low transaction security, and large primary participation [13], which is significantly different from other existing currencies, so the blockchain technology under the digital economy will inevitably contain the differentiation and improvement in terms of technological setups [32].

2.2   Governance-Centered Approach of Organization

As discussed in the previous section, the value of blockchain is not only enboded in the innovation of information and communication technology (ICT). As a distributed ledger, data in blockchain circulates throughout the network and transactions are broadcasted over all nodes, making blockchain naturally integrates the distributed system and system value more enmboded in a new type of economic organization and governance model, which provides reliable evidence to the fine-grained economic distribution system that is non-repudiation, difficult to falsify, and based on the data value.

Blockchain is building a novel institution, making new types of contracts and organizations possible. More detailedly, this new organizational model is an implementation of transaction cost economics (TCE). North argues that institutions, understood as the set of rules in a society, are the key to determining transaction costs [26]. In this sense, institutions that facilitate low transaction costs, boost economic growth. Therefore, organizations and markets serve as the alternative economic systems of economic coordination. Williamson [31] believes that the form of organization largely depends on the necessity of controlling opportunism. The ultimate reason for opportunism lies on the intent and ability of agents to use trust which is called "pursuit of self-interest" by Williamson.

According to Williamson's opportunistic ideas, the consensus mechanism of the blockchain controls opportunism to some extent by establishing the trust between strange nodes. Opportunism is inversely related to the degree of the market. When opportunism is controlled to a certain boundary, the organizational field will shrink, and the market will grow to a true distributed autonomous organization. And, of course, the formation of the organization must depend on the technical architecture of the blockchain system.

2.3   Incentive Layer of Blockchain Systems

The common intelligent transportation systems (ITS) blockchain system architecture consists of seven layers. From the bottom up, they are [24]: 1. Physical layer: It encapsulates various physical entities involved in ITS (equipment, vehicles, assets, and other objects). The key technology is the Internet of Things [33], which has enhanced device security and data privacy when integrated with the blockchain. 2. Data layer: It encapsulates the chain structure of the underlying data block, and related asymmetric public and private key data encryption and time stamp technology, serving as the basic security guarantee; 3. Network layer: It includes the P2P networking mechanism, data transmission mechanism and data verification mechanism, which means the blockchain can network automatically [27]. 4. Consensus layer: As the core of the blockchain, it encapsulates various consensus mechanism algorithms of network nodes, determining the security and reliability of the whole system. 5. Incentive layer: It integrates economic factors including the issuance mechanism and distribution mechanism of economic incentives into the blockchain technology system and promotes the development of the system in a virtuous circle; 6. Contract layer: It encapsulates various scripts, algorithms, and smart contracts and works as the foundation of the blockchain programmable feature [3,8]; 7. Application layer: It is used to deploy various scenarios and cases of the blockchain .

The incentive layer is located between the consensus layer and the contract layer, especially closely related with the consensus system [2]. Two major goals of incentive layer are: 1. To attract as many nodes to become data nodes in the network. Data flow in the network works as a foundation. Only when there are more data nodes included in the network, frequent network transactions, and real and rich transaction data, can the value of blockchain acts as a "distributed ledge" be reflected; 2. To encourage more nodes to enter the "mining circle" as a miner and participate in the consensus to win the right to serialize transactions. As the best among current consensus algorithm in terms of tackling double-spending issue, the PoW also has to run with the consensus of more than 51% of honest nodes and more nodes means higher reliability. A example more detailedly, Eyal et al. describes miners are motivated to (1) include transactions in their microblocks [9], (2) extend the heaviest chain, and (3) extend the longest chain. Unlike in Bitcoin, the latter two points are not identical. Eyal et al. mainly trade-offs between throughput and latency.

3   A Taxonomy of Incentive Focus

It turns out that the incentive economy model of digital currency is the core of the blockchain industry ecology. Two goals of the incentive mechanism are based on the token of blockchain system, which is achieved by the following two strategies: One focus on the issue mechanism of the token. It assesses what behaviors of the network node deserve newly issued coins and directly demonstrates the major functional application scenarios and development directions of the blockchain system. The incentive mechanism of the issuance responds to the main ecological demands of the networks . The other focuses on the allocation mechanism of the token. It restricts the distribution of existing tokens in the system through evaluating the nodes contributing to the network, the contribution of each node and the random factors that guarantee fairness. This one puts more emphasis on creating a good ecosystem among the participating network nodes, thereby indirectly enhance the network value, attracting nodes to participate in the network and in the consensus, and improving network reliability and the quality of the consensus.

The issuing and allocating mechanism of economic incentives are mainly considered in public chain scenarios. In public chains, it is of great significance to motivate the nodes that follow the rules to participate in the accounting and to punish the nodes violating the rules, ensuring the whole system develops into a virtuous circle. However. in the private chain, incentives are not that necessarily, because the nodes participating in the accounting often complete the game outside the chain, and participate in accounting through coercion or voluntary.

Take the Bitcoin as an example. The nodes joining the network win the billing rights through the proof of work (PoW). The Bitcoin rewards make them proactive "miners", and under the incentive mechanism, the competition for computing power drives miners to constantly optimize their "mining machines". This definitely promotes the further optimization of blockchain systems.

Obviously, the incentive mechanism encourages the nodes to participate in the maintenance of the safe operation of the block chain system and protects the general ledger from tampering through the means of economic balance. It strongly correlates with the consensus mechanism is a strong correlation and is the long-term impetus to maintain the operation of the blockchain networks.

3.1   On Issuing Mechanism

As mentioned above, the true meaning of the token is to encourage the nodes to maintain the reliability of the system consensus and to improve the ecological value of the blockchain system. Existing researches on issuance mechanism of the blockchain token aim at the three major issues of computer network: computation, storage and transmission, and have made great progress.

Calculation. The most typical example of calculating the contribution is the bitcoin system using the PoW consensus mechanism. The network solves the hash function of the specific difficulty through each miner node, and evaluates the power contribution of the node when participating in the consensus. The computing power at the single node does not exceed the total system budget. At 51% of the force, the system can effectively alleviate the very important double flower problem in the digital currency. Although the bitcoin system has aroused heating attention over the computer science field and the economic and financial circle, as a prototype of the blockchain economy proposed by Nakamoto, scholars have optimized and adapted this in later research.

Eyal et al. [10] shows that the Bitcoin mining protocol is not an incentive for an attack: the collaborating miners receive more than their fair share. Rational miners prefer to join selfish miners [1,4], and the size of the collusion group will increase until it becomes a majority. It prohibits selfish mining by pools that command less than 1/4 of the resources. This threshold is lower than the wrongly assumed 1/2 bound, but better than the current reality where a group of any size can compromise the system. Similar improvements are generally made in the trade-off of efficiency, safety and fairness [1,15].

Storage. Compared with the computing power used to calculate the contribution to solve hash function in the PoW system, with the development of centralized cloud storage, more shared storage resources in the blockchain network can be utilized, and the contribution can be evaluated based on the provided storage service.

Compared with the computing power used to calculate the hash contribution in the PoW system, with the development of centralized cloud storage, the shared storage resources utilization in the blockchain network is higher, and the contribution can be evaluated based on the provided storage service. . A typical example of a storage-based blockchain system is filecoin[1], which is built on a blockchain and a decentralized storage network with native tokens. Customers spend tokens to store data and retrieve data, while miners earn tokens by providing storage and retrieval data to form a supply and demand trading platform for storage and retrieval.

Three major roles in the Filecoin network are the customer, storage miner and search miner. When the customer publishes the demand for storing and retrieving data, the storage miner who is required to provide collateral priced in proportion to storage space provides storage and earns Filecoin,. Then the network will use the Proof-of-Space-time (PoSTs) to confirm whether the miner has complied with the commitment to store the data of the customer, while the search miner offers the data retrieval service and earns Filecoin without providing collateral. The market is operated by a network that uses PoSTs (Proof-of-Space-time) and PoRep (Proof-of-Replication) to ensure that miners properly store the data as they promised.

Transmission. The incentive on transmission aims to encourage data nodes (such as IoT devices) in the network to share real data and encourage intermediate nodes to forward data actively and honestly [16,28]. For example, Credit-Coin [21], trading and accounting information is tamper-proof and oriented to the vehicular announcement network, encourages users to share traffic information. It implements conditional privacy that allows its Trace Manager to track the identity of the malicious user in an anonymous announcement of a related transaction. Therefore, CreditCoin can encourage users to forward announcements anonymously and reliably. Similarly, to tackle potential selfish behaviors or collusion through distributed P2P applications, He et al. [16] propose a type of secure verification and pricing strategy to balance energy and bandwidth consumption by nodes forwarding files, delivering messages or uploading data.

In fact, giving incentive to transmission aims to keep nodes honest and active. On the one hand, to make efforts on honesty is to construct good economic conditions, ensuring nodes cannot change the effective transmission of the network

---

[1] https://filecoin.io/.

for personal will, which is one of the core issues of decentralization to a certain extent. On the other hand, to reward active nodes is based on the "self-interest" principle so that the nodes can achieve the approximate balance between pay and benefit as much as rules permit, which requires the Nash equilibrium to find and prove the balance point.

3.2 On Allocating Mechanism

The blockchain is not only the innovation of network technology, but also the innovation of new economic ecology. In the blockchain, the "transparent" data transmission coordinates with the distributed general ledger. Compared to the conventional architecture where IP end-to-end slave connection to transmission, this is capable to assess the contribution of the nodes more clearly. Generally, the distribution mechanism recognized by most people needs to follow three rules: 1. To ensure the reasonable evaluation on the data provider and revenue is still valid after forwarding. 2. The reward for the maintenance of real data and active forwarding by the intermediate routing node. 3. The requesting node obtains real data and the value is reasonably evaluated.

Take the bitcoin allocation as an example. In the original allocation model proposed by Nakamoto, the first miner succeed in digging for the right to accounting can get a bitcoin, while the rest all fail and can only expect the next turn although they paid the computing power too. Because of this, many small nodes have a low probability of receiving rewards. Even with the network publishing transaction data, these nodes are not motivated enough to be a miner node, which appears to contradict the reciprocal economic structure equipped with multi-accessed network and fine-grained sharing in the future. Currently, there are more than a dozen of enhanced fine-grained distribution [19] based on bitcoin [25]. The combination of ByzCoin and PBFT [5] is stuck to the transaction fee of BitCoin so that more miners can get participated and share mining reward through dynamic formation representing the recently successful block miners' hash capability ratio consensus group to achieve the Byzantine consensus while retaining the open membership of Bitcoin. Its incentive mechanism focuses on motivating intermediate nodes, and each node involved in the transaction transfer receives a portion of the transaction fee. In addition, the combination of incentives and intelligent routing can reduce communication and storage costs.

4 Challenges and Opportunities

Conventional IP network architecture can only met the demand of end-to-end data transmission. Its inconsistency causes many problems. And the exchange of data between various types of networks suffers most. The Internet with TCP/IP as its core is confronted increasingly serious technical challenges, exposing inadequacy in terms of scalability, security, reliability, flexibility and mobility. Blockchain is not an original technological innovation, but an alloy of P2P networking, digital encryption and distributed storage. In the existing network architecture, there are identity being bound to location, control being bound to transmission and security (identity) being unbound to data. The merit of the blockchain lies on that the network is equipped with the freedom to get real data, value, identity, and transaction which is credible, distinguishable, and traceable meanwhile. This is the advanced nature of the network architecture, and it is a new economic model from the data value level.

Although the text above divide the incentive into the issuance of token and allocation on the basis of the focus of previous work, the two segments are both indispensable and run jointly. For conducting solid reform on economic structure, the incentive still requires research on these fields: 1. To build quantified link between data and its value. With the support of economics and Game Theory, data value can become market-oriented. 2. To study the incentive on basis of consensus compatibility that works as its indispensable core [17]; 3. Based on the mathematics, to quantify token allocation in a fine-grained path and to balance the fairness stimulus brought by random fators and the rule-guided consensus orientation.

Research on the incentive of blockchian is still in primary stage and confronted various challenges in noted areas including evaluation on data value, the balance between random factors and allocation rules, and how to conduct valid confirmation with real transaction data still inadequate in the system [12]. It is hoped that more favorable results can be worked out with the improvement of blockchain and network structure and increasing emergence of real data. Integrating old-fashion technologies, blockchain is illuminating the future.

5 Conclusion

This paper starts with the novel economic organization model in the era of digital economy and focuses on the objectives and existing researches on the incentive layer of the blockchain. For the incentive, it begins with the three aspects that the network pays most attention to: calculation, storage and transmission, introduces cases on the token issuance, analyzes how the issuance impacts the orientation of the blockchain application, and illustrates the optimization on the distribution and rationalization mechanism. Finally, we illustrates that the research on the blockchain incentive is still in a primary stage and look into possible developments, opportunities and challenges of future research, hoping to provide reference and ideas for the later research.

Acknowledgement. This work has been financially supported by Shenzhen Project (Grant No.: JCYJ20170306091556329 and No.: ZDSYS20180205183142).

References

1. Andrychowicz, M., Dziembowski, S., Malinowski, D., Mazurek, L.: Secure multiparty computations on bitcoin. In: Security and Privacy, pp. 76–84 (2014)
2. Anjana, P.S., Kumari, S., Peri, S., Rathor, S., Somani, A.: An efficient framework for concurrent execution of smart contracts (2018)
3. Atzei, N., Bartoletti, M., Cimoli, T.: A survey of attacks on ethereum smart contracts (SoK). In: Maffei, M., Ryan, M. (eds.) POST 2017. LNCS, vol. 10204, pp. 164–186. Springer, Heidelberg (2017). https://doi.org/10.1007/978-3-662-54455-6_8
4. Babaioff, M., Dobzinski, S., Oren, S., Zohar, A.: On bitcoin and red balloons. In: ACM Conference on Electronic Commerce, pp. 56–73 (2012)
5. Castro, M., Liskov, B.: Practical byzantine fault tolerance, pp. 173–186 (1999)
6. Davidson, S., De Filippi, P., Potts, J.: Economics of blockchain. Social Science Electronic Publishing (2016)
7. Doan, A.H., Ramakrishnan, R., Halevy, A.Y.: Crowdsourcing systems on the worldwide web. Commun. ACM 54(4), 86–96 (2011)
8. Ersoy, O., Ren, Z., Erkin, Z., Lagendijk, R.L.: Transaction propagation on permissionless blockchains: incentive and routing mechanisms. In: CRYPTO Valley Conference on Blockchain Technology (2018)
9. Eyal, I., Gencer, A.E., Renesse, R.V.: Bitcoin-NG: a scalable blockchain protocol. In: Usenix Conference on Networked Systems Design and Implementation, pp. 45–59 (2016)
10. Eyal, I., Sirer, E.G.: Majority is not enough: bitcoin mining is vulnerable. In: International Conference on Financial Cryptography and Data Security, pp. 436–454 (2014)
11. Feng, Y.U.: Research on it architecture of large-scale securities centralized transaction system. Comput. Eng. 32(9), 247–250 (2006)
12. Gervais, A., Karame, G.O., Glykantzis, V., Ritzdorf, H., Capkun, S.: On the security and performance of proof of work blockchains. In: ACM Sigsac Conference on Computer and Communications Security, pp. 3–16 (2016)
13. Halpin, R., Moore, R.: Developments in electronic money regulation - the electronic money directive: a better deal for e-money issuers? Comput. Law Secur. Rev. Int. J. Technol. Pract. 25(6), 563–568 (2009)
14. Hayek, F.: The Denationalization of Money. Institute of Economic Affairs (1976)
15. He, Y., Chen, M., Ge, B., Guizani, M.: On WiFi offloading in heterogeneous networks: various incentives and trade-off strategies. IEEE Commun. Surv. Tutorials 18(4), 2345–2385 (2016)
16. He, Y., Li, H., Cheng, X., Liu, Y., Yang, C., Sun, L.: A blockchain based truthful incentive mechanism for distributed P2P applications. IEEE Access 6(99), 27324–27335 (2018)
17. Kiayias, A., Russell, A., David, B., Oliynykov, R.: Ouroboros: a provably secure proof-of-stake blockchain protocol. In: International Cryptology Conference, pp. 357–388 (2017)
18. Kittur, A., Chi, E.H., Suh, B.: Crowdsourcing user studies with Mechanical Turk. In: CHI 2008 Proceeding of the Twenty-Sixth Sigchi Conference on Human Factors in Computing Systems, pp. 453–456 (2008)
19. Kokoriskogias, E., Jovanovic, P., Gailly, N., Khoffi, I., Gasser, L., Ford, B.: Enhancing bitcoin security and performance with strong consistency via collective signing. Appl. Math. Model. 37(8), 5723–5742 (2016)

20. Kumar, K., Lu, Y.H.: Cloud computing for mobile users: can offloading computation save energy? Computer **43**(4), 51–56 (2010)
21. Li, L., et al.: CreditCoin: a privacy-preserving blockchain-based incentive announcement network for communications of smart vehicles. IEEE Trans. Intell. Transp. Syst. **19**(7), 2204–2220 (2018)
22. Liu, J., Li, M., Wang, J., Wu, F., Liu, T., Pan, Y.: A survey of MRI-based brain tumor segmentation methods. Tsinghua Sci. Technol. **19**(6), 578–595 (2014)
23. Luu, L., Teutsch, J., Kulkarni, R., Saxena, P.: Demystifying incentives in the consensus computer. In: ACM SIGSAC Conference on Computer and Communications Security, pp. 706–719 (2015)
24. Modiri, N.: The ISO reference model entities. IEEE Network **5**(4), 24–33 (2002)
25. Nakamoto, S.: Bitcoin: a peer-to-peer electronic cash system. Consulted (2008)
26. North, D.: Transaction Costs, Institutions, and Economic Performance. ICS Press (1992)
27. Okada, H., Yamasaki, S., Bracamonte, V.: Proposed classification of blockchains based on authority and incentive dimensions. In: International Conference on Advanced Communication Technology, pp. 593–597 (2017)
28. Sompolinsky, Y., Zohar, A.: Secure high-rate transaction processing in bitcoin. In: Böhme, R., Okamoto, T. (eds.) FC 2015. LNCS, vol. 8975, pp. 507–527. Springer, Heidelberg (2015). https://doi.org/10.1007/978-3-662-47854-7_32
29. Swan, M.: Blockchain thinking: the brain as a DAC (decentralized autonomous organization). IEEE Technol. Soc. Mag. **34**(4), 41–52 (2015)
30. Wang, Y., Cai, Z., Yin, G., Gao, Y., Tong, X., Wu, G.: An incentive mechanism with privacy protection in mobile crowdsourcing systems ⋆. Comput. Netw. **102**, 157–171 (2016)
31. Williamson, O.E.: The economic institutions of capitalism firms, markets, relational contracting. Soc. Sci. Electron. Publ. **32**(4), 61–75 (1998)
32. Yuan, Y., Wang, F.Y.: Blockchain: the state of the art and future trends. Acta Autom. Sinica (2016)
33. Yuan, Y., Wang, F.Y.: Towards blockchain-based intelligent transportation systems. In: IEEE International Conference on Intelligent Transportation Systems, pp. 2663–2668 (2016)

Blockchain-Based Secure Authentication Scheme for Medical Data Sharing

Xu Cheng[1,2] , Fulong Chen[1,2] , Dong Xie[1,2] , Hui Sun[1,2] ,
Cheng Huang[1] , and Zhuyun Qi[3(✉)]

[1] School of Computer and Information, Anhui Normal University,
Wuhu 241002, China
[2] Anhui Provincial Key Lab of Network and Information Security,
Wuhu 241002, China
[3] School of Electronic and Computer Engineering, Peking University,
Shenzhen 518000, China
qizy@pcl.ac.cn

Abstract. Data security is vital for medical cyber physical system (MCPS). The decentralization feature of blockchain is helpful to solve the problem that the secure authentication process is highly dependent on the trusted third party and implement data security transmission. In this paper, the blockchain technology is used to describe the security requirements in authentication process. A network model of MCPS based on blockchain is proposed. Through analysis of medical data storage architecture, data was ensured not to be tampered and trackable. The security threat was eliminated by bilinear mapping in the authentication process of medical data providers and users. The credibility problem of the trusted third party was avoided and the two-way authentication was realized between the hospital and blockchain node. The security analysis and performance test were carried out to verify the security and related performance of the authentication protocol. The results show that the MCPS based on blockchain realizes medical treatment data sharing, and meets safety requirements in the security authentication phase.

Keywords: Cyber Physical Systems · Data security sharing · Blockchain · Decentralization feature · Security authentication

1 Introduction

Medical Cyber Physical Systems (MCPS) is a Cyber Physical Systems (CPS) for embedded patient systems and independent network systems with control devices used in modern medical fields to display patient information [1]. It passes embedded systems, distributed computing and wireless, wired communication networks to monitor and control the physiological dynamics of patients [2]. This requires MCPS to have independent and complete security authentication [3] and access control mechanism [4] to verify each device's and user's identity and services. Setting access permissions for each device and each user, building the first security barrier between the MCPS.

© Springer Nature Singapore Pte Ltd. 2019
X. Cheng et al. (Eds.): ICPCSEE 2019, CCIS 1058, pp. 396–411, 2019.
https://doi.org/10.1007/978-981-15-0118-0_31

Traditional identity authentication technologies have gone through software certification to hardware certification, from single-factor authentication to multi-factor authentication, from static authentication to dynamic authentication. Later, researchers have proposed new ones based on existing identity authentication schemes. Lee said that the traditional public key infrastructure (PKI) design is not optimal and contains security vulnerabilities, which requires a lot of work to improve PKI [5]. Fromknecht and other cryptocurrencies provided by Bitcoin and Namecoin Consistency guarantees to build a PKI that ensures identity retention, called Certcoin, without central authority [6]. So a secure distributed dictionary data structure is needed to provide effective support for key lookups. Zhang et al. targeting data source security certification problem, data source certification through front-end blockchain and back-end trusted hardware SGX [7]. Liu et al. designed a new bilinear pair defined on the elliptic curve for the wireless body area network. The certificateless signature scheme [8], but their signature scheme cannot resist against simulated attack, and then they proceed to the previous scheme based on their own work. And proposed an optimized anonymous authentication scheme [8]. Lin et al. said that the decentralization advantage of blockchain makes traditional authentication reformed. The problem of CA untrustworthiness in PKI environment is always an authentication problem in distributed environment. CA attack malicious CA issues and false certificates can cause major security problems in information systems [9]. AI-Bassam can only issue certificates for user identity for current X.509 certificate standards, but can't sign fine-grained identity attribute information certificates. Based on the smart contract, it is improved and the attribute information is authenticated. If the user identity information is authenticated, the identity corresponding attribute is also trustworthy, and the trust between the user identity and the user attribute is transmitted [10]. Alexopoulos et al. ledgers implemented using blockchain technology to protect the authentication of TM (Trust Management) systems and model their systems [11]. Perera et al. based on multiple user activity authentications, need to verify the identity of multiple principals, introducing sparse representation-based identification steps and verification for this purpose [12]. Lin et al. introduced a new TCMA scheme for transmitting closed undirected graphs, which only needs to use node signatures (for example, for identifying nodes book), using a trapdoor hash function efficiently update allows signer certificate, without re-signing node [13].

The traditional authentication theory and technology are basically based on the trusted third-party trust mechanism, in this the third-party credibility has been questioned, and also includes multiple heterogeneous networks, multiple types of nodes and different user nodes in the MCPS complex environment. Security authentication between device nodes research on multiple identity authentication technologies is relatively rare and it is difficult to realize data security sharing. Therefore, we propose a blockchain-based medical data sharing security authentication scheme.

The structure of this paper is organized as follows. The second section discuss related work, including blockchain introduction, security authentication requirements, network model and mathematical background knowledge of the authentication process. The third section introduces the storage method and storage security analysis of MCPS medical records. The fourth section describes the security authentication scheme proposed in this paper. The fifth section analyzes the security scheme proposed and compares the storage overhead and computation cost. Finally, conclusions will in the last section.

2 Knowledge Background

In MCPS, the first step is to ensure the security of medical data storage through blockchain technology, and to realize the secure sharing of medical data. In order to ensure the security and reliability of communication and transaction, it is necessary to clarify the security requirements and build a network model, correctly determine the identity of two parties, and determine each item, service initiator identity and recipient identity. This paper solves the security threat of MCPS security authentication communication process to some extent by introducing bilinear mapping and intractable problems.

2.1 Blockchain Technology

Blockchain technology [14] is derived from bitcoin. The essence of technology is a distributed account book database. Based on cryptography and P2P networks, the data of a specific structure is organized into blocks in a certain way, and then these data are organized into a data chain in a chronological order in a chain structure, and the integrity and unforgeability of the data are ensured through cryptography and consensus mechanisms. The dynamic generation of block data is realized based on a certain business logic using an automatically executable mechanism, verification, synchronization of distributed computing paradigm. Blockchain data structure shown in Fig. 1.

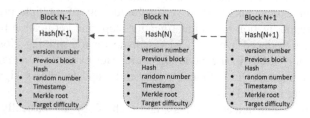

Fig. 1. Data structure of blockchain

2.2 Security Requirements and Network Model

Safety Requirements. In MCPS, information communication is required between the blockchain, hospitals and users. Therefore, the MCPS authentication scheme is subject to many threats. In order to ensure secure communication against of the MCPS, the security authentication scheme should be able to withstand multiple attacks. According to previous work [15–18], MCPS security authentication program should meet the following safety requirements.

(1) Mutual authentication: In MCPS, only the authorized users have the right to access medical data. The MCPS security authentication scheme must provide mutual authentication between the blockchain center, user and medical record provider.

(2) Anonymity: In order to ensure patient privacy, it must be ensured that no one can intercept the patient's identity information and related medical data during data transmission.

(3) Untrackability: Anonymity guarantees the privacy of users, but location privacy is also important. Therefore, the authentication scheme must provide non-traceability, that is no one including administrators and medical data providers can track patient behavior.

(4) Session key security: To ensure the confidentiality, integrity and non-repudiation of medical data transmitted in the MCPS, a shared session key must be generated between the medical consortium chain and the medical data provider. The session key is used to ensure secure communication between the two parties in the subsequent process, the authentication scheme needs to ensure the security of the session key.

(5) Perfect forward secrecy: When the attacker obtains the shared key of the transmission process, in this case, medical data transmitted in the MCPS can be accessed by decrypting the intercepted message. In order to ensure patient privacy, the MCPS security certification is required to be improved. Forward secrecy, even if the attacker obtains the key of both parties, can not disclose medical data.

(6) Logged out: When the key between the user and the medical data provider reaches the specified time, the blockchain center should be able to provide the logout operation. The logout operation here does not perform delete operation on the data like database. Because blockchain can't delete the data stored in the block, it records the operation of data on blockchain, which is a new transaction record.

System Model. The system model is shown in Fig. 2. The authentication process includes consortium blockchain, hospital, medical equipment, patients, medical staff and others. Figure 2 simplifies the complexity of MCPS, but still includes the main entities required for MCPS security certification. In network model the specific definition and description of the entity are shown in Table 1.

In MCPS, the hash of the medical data of each hospital and the location index of the summary and medical data in the cloud storage are stored in the medical consortium chain in cipher text. The storage process will be described in detail in the next section. The node classification in the medical consortium chain is shown in Table 2.

2.3 Prerequisite Knowledge

Bilinear Mapping. G_1 is an additive group with a prime order of q, and G_2 is a multiplicative group with the same order. The definition map e: $G_1 \times G_1 \to G_2$ is a bilinear pair that satisfies the following three properties:

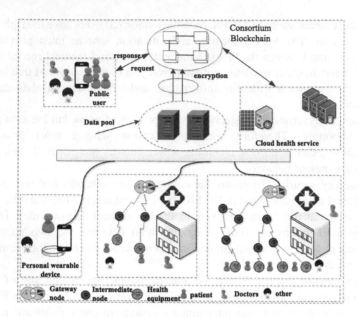

Fig. 2. System model

Table 1. Entity definitions and descriptions

Definition	Description
User Node, UN	Including patients, doctors, nurses and other user nodes. Other users are those who can only access medical data
Device Node, DN	Including personal health wearable devices, wireless sensors, electronic medical diagnostic equipment, rehabilitation equipment, hospital terminals, network equipment, etc.
Medical consortium chain, MC	Refers to the blockchain platform, responsible for registration, initialization, and key management of user nodes and device nodes
Cloud Node, CN	Provide data storage and other services for different types of users

Table 2. Node classification and descriptions

Definition	Description
Super Node, SN	The core node of the medical chain is responsible for system parameter generation, key generation, distribution, network configuration, system error correction and other tasks
Master Node, MN	Storage service and storage content retrieval service that carries the entire network block data
Light Node, LN	The bearer of the medical alliance chain network service is responsible for network optimization and data transmission

(a) Bilinear: For $\forall g_1, g_2 \in G_1$, $\forall a, b \in Zq$, Zq denotes $\{0, 1, \ldots q - 1\}$, and $e(ag_1, bg_2) = e(g_1, g_2)^{ab}$.

(b) Nondegenerate: $\exists g_1 \in G_1$ make $e(g_1, g_1) \neq 1$.

(c) Computability: Any two random $g_1, g_2 \in G_1$, $e(g_1, g_2)$ could be calculated efficiently in polynomial time.

Intractable Problems. (1) Discrete logarithm problem (DL): For $\forall g_1, g_2 \in G_1$, it is difficult to find $x \in Z_q$ such that $g_2 = xg_1$.

(2) Computational Diffie-Hellman problem (CDH): For $\forall g_1, g_2 \in G_1$, Unable to calculate the value of xyP when $g_1 = xP, g_2 = yP, x, y \in Z_q$ in polynomial time.

3 Secure Storage of Medical Records

3.1 Medical Record Storage Based on Blockchain Technology

In MCPS, patient medical records (outpatient medical records, hospital records, body temperature sheets, medical orders, laboratory tests, medical imaging data, special examination consent, surgical consent, surgery and anesthesia records, pathological data, nursing records and other medical records) are important materials. If these medical records can be shared with patients, hospitals, and researchers [19], it will promote the medical industry's tremendous progress. This section uses blockchains combined with cloud storage. The technology realizes the secure storage and data sharing of medical records. The chain structure of hash medical data stored in each hospital and the location index of abstract and medical data in the cloud storage is called the medical consortium chain, and in architecture of the medical consortium chain is shown in Fig. 3.

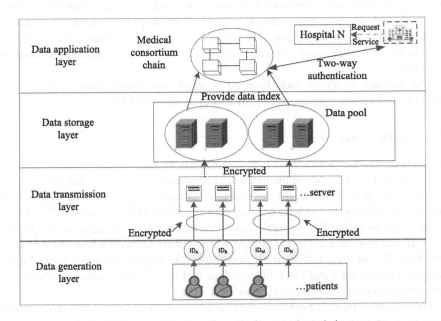

Fig. 3. Architecture of medical consortium chain

The process of publishing medical records: Due to limited storage capacity of blockchains, all hospitals store non-public medical records in cipher text, and hashes of medical records, abstracts and their location in cloud storage are stored in medical consortium chain on the master node MN.

Each hospital generates a hash of medical record M, and encrypts the medical record digest and medical record in cloud storage location L with public key pubkey of each hospital to send the medical consortium chain. The specific algorithm is as follows:

Input: Medical records of each hospital
Output: Trading results of medical data
Step 1: Generate the corresponding $E_{pubkey}\{Mac \|H(M)\| L\}$
Step 2: Save the information generated by Step 1 on the medical consortium chain
Step 3: Process transactions and broadcast to the medical consortium chain network
Step 4: Record and output the transaction results of medical data

The blockchain system has a certain technical structure. This section is based on the technical architecture of the blockchain system. The medical consortium chain is divided into data layer, network layer, consensus layer, contract layer, and the application layer; the medical consortium chain data storage depends on the data layer, summaries of all medical records and other information will be chronologically recorded in the medical alliance chain. Any medical records can be traced back to the source; network layer relies on the P2P network to communicate with each other. This is a typical decentralized distributed network, relying on maintaining a common ledger to achieve communication consistency. The consensus layer relies on multiple consensus mechanisms to allow all nodes to agree on the validity of the medical consortium chain data in a decentralized network. The contract layer encapsulates various scripts, algorithms, and smart contracts is the basis of the programmable characteristics of the medical consortium chain. The application layer encapsulates the application scenario of the secure smart medical, which can realize the management of data and authority.

The detailed design of each layer of medical consortium chain is given in the following blockchain infrastructure [20] model as shown in Fig. 4.

3.2   Security Analysis

The security analysis for medical record storage is as follows.

Tamper-proof: The data in the medical consortium chain is open and tamper-proof. It is time-stamped according to a certain time sequence. It has strong traceability, verifiability and relying on the consensus mechanism DPOS. That allows all the nodes to communicate in the P2P network. Each block saves the hash value of the previous block. If you want to modify the computing power of the node where the data needs more than 51% of the whole network, it is difficult to implement, and if hashed data is changed, original data and other hashed data will also change, which guarantees the inevitable modification of medical records.

Privacy protection: The patient's medical record hash value is stored in the medical consortium chain in an encrypted manner. Some nodes on the block chain account

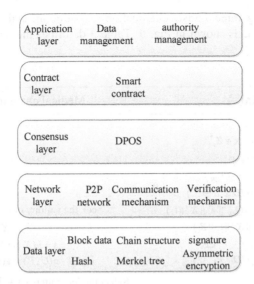

Fig. 4. Technical framework of the medical consortium chain

book contain the patient's identity information, and most of the data are stored in the cloud storage under the chain in ciphertext form. Patients have access control rights to medical data. They can authorize some third parties to access medical records or revoke their access rights. In order to ensure the authenticity of data, each hospital must disclose its identity information. Only hash of medical records is saved in the medical consortium chain. Hash value, abstracts, medical records are located in the cloud storage and processed by encryption. Without the private key of each hospital, it is impossible to obtain real data of any medical records.

Secure storage: Data security is essential. The medical records in this paper are stored in cloud storage under the chain through encryption, and the distributed storage characteristics of cloud storage can also ensure the secure storage of data. And blockchain can realize medical record sharing. In data sharing process, the party that needs medical data obtains the medical record through security authentication, and through hash calculation, hash value on the blockchain can be compared to verify the correctness of the data, whether the data is tampered with.

4   Security Authentication Scheme for MCPS

In order to realize the sharing of medical records in the MCPS, when the medical record related information is stored in the medical consortium chain, when **A** hospital needs a medical record of the **B** hospital, it will send its own data demand request to the medical consortium chain, for the R. In medical consortium chain, the medical data of each hospital is encrypted and stored in the master node MN of the medical consortium chain. The super node SN of the medical consortium chain is responsible for the related parameter generation and key management. In the security authentication scheme of this paper, including three Parts: Initialization, registration, and authentication. The authentication scheme is shown in Fig. 5. Table 3 describes the notations used in this work.

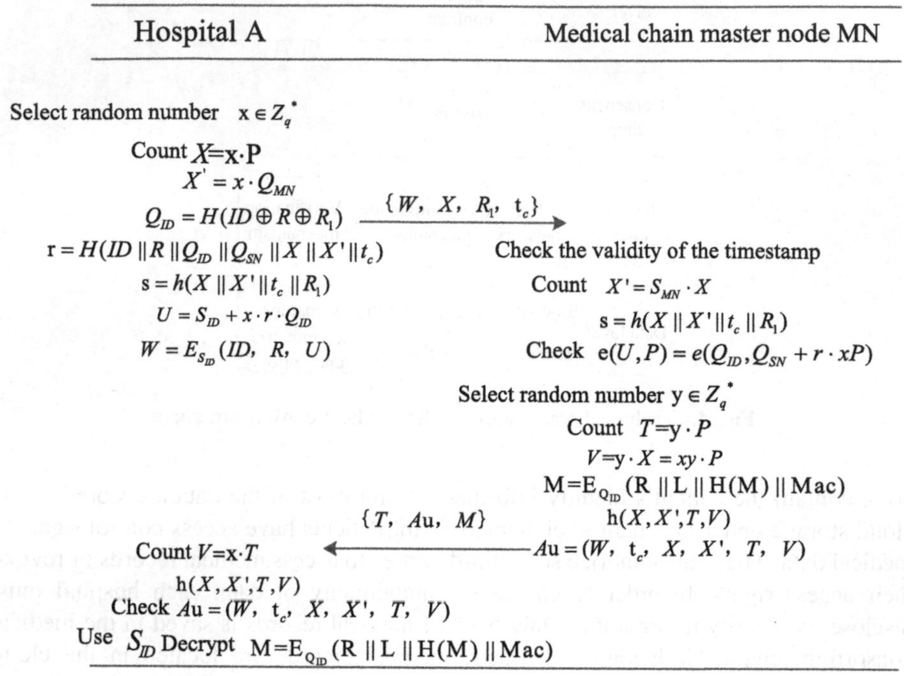

Hospital A	Medical chain master node MN

Select random number $x \in Z_q^*$

Count $X = x \cdot P$

$X' = x \cdot Q_{MN}$

$Q_{ID} = H(ID \oplus R \oplus R_1)$ $\xrightarrow{\{W,\ X,\ R_1,\ t_c\}}$

$r = H(ID \| R \| Q_{ID} \| Q_{SN} \| X \| X' \| t_c)$ Check the validity of the timestamp

$s = h(X \| X' \| t_c \| R_1)$ Count $X' = S_{MN} \cdot X$

$U = S_{ID} + x \cdot r \cdot Q_{ID}$ $s = h(X \| X' \| t_c \| R_1)$

$W = E_{S_{ID}}(ID,\ R,\ U)$ Check $e(U, P) = e(Q_{ID}, Q_{SN} + r \cdot xP)$

Select random number $y \in Z_q^*$

Count $T = y \cdot P$

$V = y \cdot X = xy \cdot P$

$M = E_{Q_{ID}}(R \| L \| H(M) \| Mac)$

   $\{T,\ Au,\ M\}$ $h(X, X', T, V)$

Count $V = x \cdot T$ $\xleftarrow{\hspace{2cm}}$ $Au = (W,\ t_c,\ X,\ X',\ T,\ V)$

$h(X, X', T, V)$

Check $Au = (W,\ t_c,\ X,\ X',\ T,\ V)$

Use S_{ID} Decrypt $M = E_{Q_{ID}}(R \| L \| H(M) \| Mac)$

Fig. 5. Authentication process

Table 3. Symbols and descriptions

Notations	Description
q	Large prime number
G_1	Addition group of order q
G_2	Multiplicative group of order q
e	Bilinear pair
P	Generator of group G_1
h	S Secure hash function h : $\{0,1\}^* \times G_2 \rightarrow Z_q^*$
H	Secure hash function H : $\{0,1\}^* \times G_1 \rightarrow G_1$
E(.)	Some encryption algorithm
t_c	Current timestamp
H(M)	**B** hospital hash of medical data
L	Location index of **B** hospital data in cloud storage
Mac	**B** hospital medical data summary

Initialization. (a) G_1 is an additive group with prime q-order, G_2 is a multiplicative group with prime q-order. Given the security parameter l, medical consortium chain super-node SN generates its public-private key pair (Q_{SN}, S_{SN}), where $Q_{SN} = S_{SN} \cdot P$, and issues system parameters $\{l, G_1, G_2, q, P, e, H, h, Q_{SN}\}$.

(b) The medical consortium chain MN generates its public-private key pair (Q_{MN}, S_{MN}), where $Q_{MN} = S_{MN} \cdot P$.

Registration. (a) **A** hospital generates its unique identification code ID and random number R_1 and records the required **B** hospital data request as R, and sends $\{ID, R, R_1\}$ to the super node SN of medical consortium chain.

(b) The medical consortium chain super node SN receives $\{ID, R, R_1\}$, calculates $Q_{ID} = H(ID \oplus R \oplus R_1)$, $S_{ID} = Q_{ID} \cdot S_{SN}$, and sends $\{S_{ID}, Q_{ID}\}$ to hospital **A** through a secure channel.

(c) When **A** hospital receives $\{S_{ID}, Q_{ID}\}$ and stores it securely.

Authentication. (a) **A** hospital chooses the random number $x \in Z_q^*$, and calculates $X = x \cdot P$, $X' = x \cdot Q_{MN}$, $Q_{ID} = H(ID \oplus R \oplus R_1)$, $r = H(ID \| R \| Q_{ID} \| Q_{SN} \| X \| X' \| t_c)$, $s = h(X \| X' \| t_c \| R_1)$, $U = S_{ID} + x \cdot r \cdot Q_{ID}$, $W = E_{S_{ID}}(ID, R, U)$, **A** hospital sends $\{W, X, R_1, t_c\}$ to the master node of the medical consortium chain MN.

(b) When the medical consortium chain master node MN receives $\{W, X, R_1, t_c\}$, it first checks the validity of the timestamp, and rejects if it's invalid; if it's valid, the MN calculates: $X' = S_{MN} \cdot X$, $s = h(X \| X' \| t_c \| R_1)$, and decrypts with the public key of hospital **A** to obtain ID, R, U. Verification e $(U, P) = e (Q_{ID}, Q_{SN} + r \cdot xP)$ is established. If not, reject the request. Otherwise, the MN selects the random number $y \in Z_q^*$, and calculates $T = y \cdot P$, $V = y \cdot X = xy \cdot P$, $M = E_{Q_{ID}}(R \| L \| H(M) \| Mac)$, and calculates the session key $= h(X, X', T, V)$, the authentication key Au $= (W, t_c, X, X', T, V)$, and the MN sends $\{T, Au, M\}$ to the hospital **A**.

(c) When **A** hospital receives $\{T, Au, M\}$, calculate $V = x \cdot T$ and $h(X, X', T, V)$, and verify whether Au $= (W, t_c, X, X', T, V)$ is established. If it is not established, reject the request; if it is established, use the private key of **A** hospital S_{ID} to decrypt M to get R, L, H (M), Mac. At this time, **A** hospital and The primary node of the medical consortium chain, MN, obtains mutual authentication, and can obtain information such as location index and data summary of medical data of hospital **B**.

5   Safety and Performance Analysis

5.1   Correctness Analysis

Deductive analysis of the verification formula used in this scheme.

After the medical master node MN receives $\{W, X, R_1, t_c\}$ sent by the hospital **A**, it needs to verify the authorization information and correctness of the verification formula is derived as follows.

$$e(U,P) = e(Q_{ID}, Q_{SN} + r \cdot xP)$$
$$e(U,P) = e(S_{ID} + x \cdot rQ_{ID}, P)$$
$$= e(Q_{ID} \cdot S_{SN} + x \cdot rQ_{ID}, P)$$
$$= e(Q_{ID} \cdot S_{SN}, P)e(x \cdot rQ_{ID}, P)$$
$$= e(Q_{ID}, P \cdot S_{SN} + x \cdot rP)$$
$$= e(Q_{ID}, Q_{SN} + r \cdot xP)$$

Through the above derivation results the correctness of this program is proved.

5.2   Safety Certification and Analysis

Safety Certificate. Based on the capabilities and network model of the attacker in the MCPS, the security model in this section is based on a game between the attacker A and the challenger C.

Theorem 1: The security authentication scheme of MCPS proposed in this chapter satisfies the unforgeability of the message under the adaptive selection message attack in the random prediction model.

Proof: It is assumed that an attacker has the ability to forge legitimate request information $\{W, X, R_1, t_c\}$ and then construct another challenger C. By running A as a subroutine, C can solve the Diffie-Hellman problem (CDH) with a non-negligible probability. C performs the following steps:

Initialization: Challenger C sets $Q_{SN} = S_{SN} \cdot P$ and generates the system public parameter $\{l, G_1, G_2, q, P, e, H, h, Q_{SN}\}$.

h Query: C maintains a list of $\{W, X, R_1, t_c\}$, initialized to null. After receiving A's request for message R, C first checks if there is a tuple $\{W, X, R_1, t_c\}$ in the list. If it already exists, C sends $\{W, X, R_1, t_c\}$ to A; otherwise, C randomly selects $t_c \in Z_q^*$, adds $\{W, X, R_1, t_c\}$ to the list and sends $\{W, X, R_1, t_c\}$ to A.

Signature Query: When C receives A request for a message, C selects the random number $x \in Z_q^*$ and calculates, $X = x \cdot P$, $X' = x \cdot Q_{MN}$, $Q_{ID} = H(ID \oplus R \oplus R_1)$, $r = H(ID \| R \| Q_{ID} \| Q_{SN} \| X \| X' \| t_c)$, $s = h(X \| X' \| t_c \| R_1)$, $U = S_{ID} + x \cdot r \cdot Q_{ID}$, $W = E_{S_{ID}}(ID, R, U)$. Because the message satisfies the equation: $e(U, P) = e(Q_{ID}, Q_{SN} + r \cdot xP)$.

Therefore, the request message $\{W, X, R_1, t_c\}$ obtained by A through the above request is valid.

Output: Finally, A outputs the message $\{W, X, R_1, t_c\}$, and C verifies whether the message is valid by the following equation.

$$e(U, P) = e(Q_{ID}, Q_{SN} + r \cdot xP)$$

If the equation does not hold, C terminates the game process. According to the fork lemma [21], if A selects a different h query request, then A can output another valid message signature R' and satisfy $e(U', P) = e(Q_{ID}, Q_{SN} + r' \cdot xP)$.

Let $Q_{ID} = a \cdot P$, $Q_{SN} = b \cdot P$.

According to the above two formulas:

$$e(r' \cdot U - r \cdot U', P) = \frac{e(r' \cdot U, P)}{e(r \cdot U', P)}$$

$$= \frac{e(U, P)^{r'}}{e(U', P)^{r}} = \frac{e(Q_{ID}, Q_{SN} + r \cdot xP)^{r'}}{e(Q_{ID}, Q_{SN} + r' \cdot xP)^{r}}$$

$$= \frac{e(Q_{ID}, r' \cdot (Q_{SN} + r \cdot xP))}{e(Q_{ID}, r \cdot (Q_{SN} + r' \cdot xP))}$$

$$= \frac{e(Q_{ID}, r' \cdot Q_{SN})}{e(Q_{ID}, r \cdot Q_{SN})}$$

$$= \frac{e(a \cdot P, r' \cdot b \cdot P)}{e(a \cdot P, r \cdot b \cdot P)}$$

$$= \frac{e(r' \cdot a \cdot b \cdot P, P)}{e(r \cdot a \cdot b \cdot P, P)} = e((r' - r) \cdot a \cdot b \cdot P, P)$$

And $a \cdot b \cdot P = (r' - r)^{-1}(r' \cdot U - r \cdot U')$.

C outputs $(r' - r)^{-1}(r' \cdot U - r \cdot U')$ as a solution to the given CDH problem instance. However, the CDH problem is actually difficult to understand. Therefore, under the adaptive selection message attack in the random prediction model, the scheme of this paper is resistant to message forgery.

Security Analysis. In this section, we can prove that the proposed safety certification scheme meets the safety requirements of Sect. 2.2.

(1) Mutual authentication: In MCPS, only valid A hospitals have access to medical data. The registrant A hospital generates a legal $\{W, X, R_1, t_c\}$ and responds to the message $\{T, Au, M\}$ to ensure the legality of the identity of both parties, thereby completing mutual authentication.

(2) Anonymity: In order to ensure patient privacy, medical data of each hospital is encrypted, the attacker can't obtain patient privacy. Then A hospital needs to calculate $X = x \cdot P$, $Q_{ID} = H(ID \oplus R \oplus R_1)$, $r = H(ID \parallel R \parallel Q_{ID} \parallel Q_{SN} \parallel X \parallel X' \parallel t_c)$, $s = h(X \parallel X' \parallel t_c \parallel R_1)$, $U = S_{ID} + x \cdot r \cdot Q_{ID}$, when implementing two-way authentication. When the attacker obtains this information, it needs to solve the CDH intractable problem, so the security authentication scheme can ensure anonymity.

(3) Untraceability: In this authentication scheme, random numbers x and y are generated in each authentication process of A hospital and MN. Therefore, it can ensure that A hospital avoids illegal molecular tracing when it authenticates with MN of the main node of the medical chain.

(4) Session key security: During the authentication process, the session key $h(X, X', T, V)$ and the authentication key $Au = (W, t_c, X, X', T, V)$ is generated, in which $X = x \cdot P$, $T = y \cdot P$, $V = y \cdot X = xy \cdot P$ involve DL problems, so the session key security can be guaranteed.

(5) Perfect forward secrecy: When the attacker obtains the session key $h(X, X', T, V)$ of transmission process, in this case attacker needs to obtain $V = y \cdot X = xy \cdot P$ through $X = x \cdot P$ and $T = y \cdot P$, which involves the CDH problem, so the security authentication scheme proposed in this paper can provide Perfect forward secrecy.

(6) Logged out: When the key between the user and medical data provider reaches the specified time, super node in blockchain can provide logout function, and solve the key expiration problem by sending a logout operation to the super node.

5.3   Storage Overhead

Storage overhead is a problem that must be considered in any authentication scheme. It involves the key and related parameter storage. In this paper, length of G_1 is 1024 bits, G_2 is 512 bits, P is 512 bits, and the length of q is 160 bits. Timestamp t_c is 32 bits, R is 64 bits, random number R_1 is 32 bits and A hospital ID is 32 bits. First, hospital A needs to store $\{S_{ID}, R, R_1\} = 1024 + 64 + 32 = 1120$ bits. Medical consortium chain MN only is the MN private key $S_{MN} \in Z_q^* = 160$ bits needs to be stored. The storage overhead of the WBAN anonymous authentication scheme proposed in this paper and Liu [8] is shown in Table 4, where n denotes the client's number in the WBAN.

Table 4. Comparison of storage overhead

	Liu et al.'s first plan	Liu et al. improved scheme	This article
Client (A hospital)	2112 bits	2112 bits	1120 bits
AP (MN node)	(160+512n) bits	(160+1536n) bits	160 bits

5.4   Calculation Overhead

Liu et al.'s [8] scheme is based on the scheme of bilinear pairing operation. This section analyzes the computational cost of this paper and compares it with the Liu et al. Because the client mobile device has limited storage and computing power, it is only considered. The client calculates the cost. The symbols used in this section are defined as follows.

(1) TG_e: Execution time of the bilinear pairing operation.
(2) TG_{mul}: The execution time of the bilinear pairing scalar multiplication operation.
(3) TG_H: The bilinear pairing hash function maps to the execution time of a point operation.
(4) TG_{add}: Execution time of the bilinear pairing point addition operation.
(5) TG_{exp}: Execution time of the bilinear paired modular exponentiation operation.
(6) T_h: The execution time of a generic hash function operation.

Table 5. Run time based on pairing operation

TG_e	TG_{mul}	TG_H	TG_{add}	TG_{exp}	T_h
5.32 s	2.45 s	0.89 s	<0.01 s	1.25 s	<0.01 s

Table 6. Computation cost comparison

Liu et al.'s first plan	Liu et al. improved scheme	This article
$4TG_{mul} + 1TG_{add} +$ $1TG_{exp} + 2T_h \approx 11.05S$	$4TG_{mul} + 2TG_{add} +$ $1TG_{exp} + 2T_h \approx 11.05S$	$4TG_{mul} + 1TG_H +$ $1TG_{add} + 4T_h \approx 10.69S$

Xiong [22] and others used 4 KB RAM, 128 KB ROM and 7.3828-MHz ATmega128L microcontroller to implement bilinear pairing time widely used in wireless sensor networks on MICAz, in which RAM and ROM usage of each operation Both are obtained using the TinyOS library. The pairing-based operation time is shown in Table 5 above. This paper uses their experimental results to evaluate the computational cost of the client. According to the experimental results of Xiong et al. the cryptographic operations and corresponding pairing operations are listed. Execution time, and compared with the Liu et al. scheme, the results of Table 6 show that the security authencation scheme of this paper can reduce the computational cost and is more effective than other schemes.

6 Conclusion

At present, blockchain technology is widely used. The existing blockchain structure can't be directly applied to the medical field. In this paper, we combines cloud storage and cryptography to analyze the feasibility of blockchain security authentication in medical information physical fusion system. Realize medical data sharing. Based on the characteristics of the blockchain we designed a medical data security storage model. Which is based on the network entity model, a security authentication scheme was designed to avoid the problem of over-reliance on third-party trusted centers to reach safety requirements. Through the analysis of the safety performance and calculation cost of the proposed scheme, the safety and efficiency of the scheme is further demonstrated.

Acknowledgement. This work has been financially supported by the National Natural Science Foundation of China through the research projects (Grant No. 61572036) and Shenzhen Key Fundamental Research Projects (Grant No. JCYJ20170306091556329).

References

1. Lee, I., Sokolsky, O.: Medical cyber physical systems. In: IEEE International Conference and Workshops on Engineering of Computer Based Systems, pp. 743–748. IEEE (2010)
2. Haro, A., Flickner, M., Essa, I.: Detecting and tracking eyes by using their physiological properties, dynamics, and appearance In: IEEE Conference on Computer Vision & Pattern Recognition, pp. 163–168. IEEE (2010)
3. Saltzer, J.H., Schroeder, M.D.: The protection of information in computer systems. IEEE CSIT Newslett. **63**(9), 1278–1308 (2005)
4. Ouaddah, A., Mousannif, H., Ouahman, A.A.: Access control models in IoT: the road ahead. In: 2015 IEEE/ACS 12th International Conference of Computer Systems and Applications (AICCSA), pp. 1–2. IEEE (2015)
5. Lee, E.A.: Cyber physical systems: design challenges. In: 11th IEEE Symposium on Object Oriented Real-Time Distributed Computing (ISORC), pp. 363–369. IEEE (2008)
6. Fromknecht, C., Velicanu, D., Yakoubov, S.A.: Decentralized public key infrastructure with identity retention. IACR Cryptology ePrint Archive, pp. 1–16 (2014)
7. Zhang, F., Cecchetti, E., Croman, K., et al.: Town crier: an authenticated data feed for smart contracts. In: The ACM Conference on Computer and Communications Security, pp. 1–13. ACM (2016)
8. Liu, J., Zhang, Z., Chen, X., et al.: Certificateless remote anonymous authentication schemes for wireless body area networks. IEEE Trans. Parallel Distrib. Syst. **25**(2), 332–342 (2013)
9. Lin, J.Q., Jing, J.W., Zhang, Q.L., et al.: Recent advances in PKI technologies. J. Cryptol. Res. **2**(6), 487–496 (2015)
10. Al-Bassam, M.: SCPKI: a smart contract-based PKI and identity system. In: ACM Workshop on Blockchain, Cryptocurrencies and Contracts, pp. 35–40. ACM (2017)
11. Alexopoulos, N., Daubert, J., Mühlhäuser, M., et al.: Beyond the hype: on using blockchains in trust management for authentication. In: 2017 IEEE Conference on Trustcom/BigDataSE/ICESS, pp. 546–553. IEEE (2017)
12. Pramuditha, P., Patel, V.M.: Face-based multiple user active authentication on mobile devices. IEEE Trans. Inf. Forensics Secur. (TIFS) **14**(5), 1240–1250 (2019)
13. Lin, C., He, D., Huang, X., et al.: A new transitively closed undirected graph authentication scheme for blockchain-based identity management systems. IEEE Access **6**, 28203–28212 (2018)
14. Liang, X., Shetty, S., Tosh, D., et al.: ProvChain: a blockchain-based data provenance architecture in cloud environment with enhanced privacy and availability. In: IEEE/ACM International Symposium on Cluster, pp. 468–477 (2017)
15. Movassaghi, S., Abolhasan, M., Lipman, J., et al.: Wireless body area networks: a survey. IEEE Commun. Surv. Tutorials **16**(3), 1658–1686 (2014)
16. Ramli, S.N., Ahmad, R., Abdollah, M.F., et al.: A biometric-based security for data authentication in Wireless Body Area Network (WBAN). In: International Conference on Advanced Communication Technology, pp. 998–1001 (2013)
17. Rivest, R.L.: A method for obtaining digital signatures and public-key cryptosystems. Commun. ACM **26**(2), 96–99 (1978)
18. Truong, T.T., Tran, M.T., Duong, A.D.: Improvement of the more efficient and secure ID-based remote mutual authentication with key agreement scheme for mobile devices on ECC. In: International Conference on Advanced Information Networking & Applications Workshops, pp. 698–703 (2012)
19. Yang, J.J., Li, J.Q., Niu, Y.: A hybrid solution for privacy preserving medical data sharing in the cloud environment. Future Gener. Comput. Syst. **43–44**(45), 74–86 (2015)

20. Jabbar, K., Bjørn, P.: Infrastructural grind: introducing blockchain technology in the shipping domain. In: ACM Conference on Supporting Groupwork, pp. 297–308. ACM (2018)
21. Pointcheval, D., Stern, J.: Security arguments for digital signatures and blind signatures. J. Cryptol. **13**(3), 361–396 (2000)
22. Xiong, X., Wong, D., Deng, T.: TinyPairing: a fast and lightweight pairing-based cryptographic library for wireless sensor networks. In: Proceedings of Wireless Communication Network Conference (WCNC 2010), pp. 1–6 (2010)

Encryption Algorithm Based NTRU in Underwater Acoustic Networks

Chunyan Peng[✉] and Xiujuan Du

Qinghai Normal University, Xining 810008, China
745691l@qq.com

Abstract. Underwater acoustic networks (UANs) adopt acoustic communication. The opening and sharing features of underwater acoustic channel make communication in UANs vulnerable to eavesdropping and interfering. The applications of UANs such as underwater military, underwater warning and energy development are very demanding for the security level. Quantum computing poses a threat to security of the traditional public key cryptosystem such as large integer factorization and discrete logarithm. To solve these problems, a public key encryption algorithm is proposed based on number theory research unit (NTRU) for underwater acoustic networks. The traditional NTRU encryption scheme was improved and a new public key cryptosystem was provided. The algorithm combined the encryption algorithm with the identity of the node. Experimental results show that the key generation speed is very fast, and the speed of encryption and decryption is faster than RSA, ECC, ElGamal and other public key cryptosystems. It verifies that the encryption algorithm can resist quantum computing attacks.

Keywords: Underwater acoustic networks · Network security · NTRU · Public key encryption

1 Introduction

Underwater Acoustic Networks (UANs) are a novel type of underwater network systems, which emphasize on effective safeguarding of the national marine rights and interests [1, 2]. UANs have been applied to many fields such as monitoring underwater environment, exploring underwater resource, collecting oceanic data, and preventing disasters etc. Even though UANs have a slice of shared properties with terrestrial sensor networks, such as limited power energy, UANs are significantly different from terrestrial sensor networks in a multitude of aspects, such as narrow bandwidth, long propagation delay, node passive mobility, and high error probability. RF signal in terrestrial wireless sensor networks at a node's maximum transmission power is not able to spread more than 1 m in underwater environment [3], laser and radio waves cannot satisfy long distance communication in water either. UANs adopt acoustic communication, which can meet with long distance transmission [4, 5]. RF signal propagates at a speed of $3*10^8$ m/s, but acoustic signal propagates at speed of 1500 m/s in water, which is much lower than the speed of RF. All these factors such as path loss, noise, multi-path, Doppler spread gives rise to higher bit-error in acoustic channels [6].

© Springer Nature Singapore Pte Ltd. 2019
X. Cheng et al. (Eds.): ICPCSEE 2019, CCIS 1058, pp. 412–424, 2019.
https://doi.org/10.1007/978-981-15-0118-0_32

UANs also suffer from rigid resource constraints, such as limited battery life and computational power. Acoustic communication is characterized by limited bandwidth, long propagation delay and low data rate, and the open acoustic channel makes UANs more vulnerable to jamming attacks or DoS attacks. Therefore, these characteristics of UANs make the existing work in terrestrial sensor networks unsuitable for UANs and bring about many security challenges. UANs also require security mechanisms and algorithms to maintain the confidentiality and integrity of important messages. The messages include nodes' information, routing items, and data. The nodes' information can be encrypted in the transmitting process, which is a common method in the field of information security.

In 2016, the author proposed a lightweight cryptography scheme [7] based on symmetric encryption for UANs, which can be used in some applications where security level is not high, but in the research fields such as underwater military, underwater warning and energy development. Where it is relevant to military and economic interests and more sensitive information, a higher security level public key encryption algorithm is needed in UANs.

At present, most public key cryptosystems such as RSA and Elliptic Curve have been widely used in many security research fields. However, with the rapid development of modern computer computing power and network technologies such as cloud platforms, the key length of the encryption algorithm has been keeping increasing to provide better Security. The increase of the key length means that the existing public key encryption scheme needs to be expanded into a larger finite field, which requires more exponential operations and more costs in the Encryption and decryption processes. At present, the new public key encryption system is developing in two aspects. One is to find difficult problems with more complexity, such as the elliptic curve public key system relying on discrete logarithm problem. The other is to find some more simple operations in order to reduce the complexity of encryption and decryption processes, such as cryptography based on lattice theory, which basically rely on the addition and multiplication of linear operations. With the quantum algorithm of factorization proposed by Shor in 1994 [8], it is possible to quickly factorize arbitrarily large certificates, which greatly shorts the time of the RSA in the encryption and decryption process. The original security of RSA is degraded in the quantum computing environment. In 1999, Shor et al. subsequently proposed a quantum algorithm for solving large integer decomposition problems and discrete logarithm problems [9]. Therefore, it is an urgent problem to find a post-quantum cryptosystem in UANs that can resist quantum attacks.

The remainder of this paper is organized as follows. In Sect. 2, the related work along with motivation is presented. In Sect. 3, the detailed explanation of lattice is discussed. Section 4 presents our proposed encryption scheme. Section 5 analyzes the security of the encryption algorithm based NTRU in underwater acoustic networks. Section 6 concludes the paper.

2   Related Works

In order to make a better analysis of the backpack password system and the RSA public key cryptosystem, lattice theory is introduced into the cryptography. Ajtai M. and Dwork C. constructed the first password system in 1997 [10], and then the NTRU cryptosystem was provided in 1998 [11]. Because there were many difficulties to proof that NTRU is a secure algorithm, NTRU hasn't got the recognition and application. NTRU has been drowned by the mainstream public cryptosystem based on integer factorization and discrete logarithm for a long time. In 2009, a fully homomorphic cryptographic scheme based on lattice theory [12] emerged, and lattice cryptography has been widely developed. In 2015, the report of the National Institute of Standards and Technology proposed four anti-quantum cryptography systems, named lattice-based cryptography system, code-based cryptography system, multivariable cryptography and signature based on hash algorithm [13]. As a kind of public key cryptosystem that can resist quantum attacks, the lattice cryptography is considered as the most powerful competitor to the standard of post-quantum cryptography algorithm. Lattice cryptography has developed rapidly in the recent years, which mainly focuses on lattice cryptography's zero-knowledge proof, lattice encryption, lattice signature scheme and lattice key exchange. Based on the post-quantum public key cryptosystem, the identity-based non-certificate digital signature scheme is constructed by using the error-correcting code theory. If the lattice cryptosystem can be used in the underwater environments, the underwater acoustic communication would resist the attack of quantum computing and realize the secure communication of the underwater acoustic networks, which can provide security guarantee for the underwater acoustic networks.

Security algorithm is the key to design safe underwater acoustic networks and plays an important role in realizing a reliable and secure communication environment. Quantum computing poses a threat to the security of the traditional public key cryptosystem based on large integer factorization and discrete logarithm. The quantum public key cryptosystem can resist the attack of quantum computing, which can be used for underwater military, underwater early warning, energy development and other fields with high security requirements. Ajtai M. proposed two difficult problems based on lattice theory [14] in 1996. One is the shortest vector problem and the other is the closest vector problem. The public key cryptosystem has been decrypted by the polynomial quantum algorithm so far, therefore the public key cryptosystem based on lattice hard problems can resist quantum attack. The NTRU encryption algorithm in lattice theory only uses simple modular multiplication and inversion of modular computation. It is much faster than RSA, ECC, ElGamal and other public key cryptography systems in the processes of key generation, the encryption and the decryption. Moreover, our proposed encryption algorithm can resist the attack of quantum computing and it is suitable for the nodes of underwater acoustic network with limited resources. The next chapter will further propose a new NTRU public key encryption scheme for underwater acoustic networks.

3 Lattice

3.1 The Definition of Lattice

Lattice theory is usually defined to different forms according to different analyzation fields. In the field of computer information security and engineering, the lattice is commonly defined as follows [15]:

Definition 1. $v_1, \ldots, v_n \in \mathbb{R}^m$ is set to be a set of linearly independent vectors. The generated lattice L refers to the set of vectors $v_1, \ldots, v_n$ which composed of linear combinations of vectors $v_1, \ldots, v_n$, and the coefficients are all in $\mathbb{Z}$, that is

$$L = \{a_1 v_1 + a_2 v_2 + \cdots + a_i v_i | a_1, a_2, \cdots, a_i \in \mathbb{Z}\} = \left\{ \sum_{i=1}^{d} a_i v_i | a_i \in \mathbb{Z} \right\},$$ such a set L is

called a lattice. Any set of linearly independent vectors that can generate a lattice are called the basis of the lattice, and the number of vectors in the set is called the dimension of the lattice, denoted as $\dim(L) = n$.

Suppose $v_1, \ldots, v_n$ is a basis of a lattice L, and $w_1, \ldots, w_n \in L$ is another set of vectors in L. According to the definition of lattice, we can indicate each w_j of them by a linear combination of basis vectors as the following formula (1).

$$
\begin{aligned}
w_1 &= a_{11} v_1 + a_{12} v_2 + \cdots + a_{1n} v_n \\
w_2 &= a_{21} v_1 + a_{22} v_2 + \cdots + a_{2n} v_n \\
&\vdots \\
w_n &= a_{n1} v_1 + a_{n2} v_2 + \cdots + a_{nn} v_n
\end{aligned}
\tag{1}
$$

As we can see all of coefficients a_{ij} are integers from the definition of the lattice.

Suppose we want to express v_i with w_j in reverse, then we need to compute the invert with the following matrix,

$$
A = \begin{bmatrix}
a_{11} & a_{12} & \cdots & a_{1n} \\
a_{21} & a_{22} & \cdots & a_{2n} \\
\vdots & & & \\
a_{n1} & a_{n2} & \cdots & a_{nn}
\end{bmatrix}
\tag{2}
$$

If you use a integer coefficients w_j linear combination to express v_i, then the numbers in A^{-1} would be all integers. Therefore,

$$1 = \det(I) = \det(AA^{-1}) = \det(A)\det(A^{-1}) \tag{3}$$

Here, $\det(A)$ and $\det(A^{-1})$ are both integers, so definitely we can get $\det(A) = \pm 1$. In reverse, if $\det(A) = \pm 1$ and all are integers in A, then according to the adjoint matrix theory, the matrix A^{-1} must be composed of integers. A lattice in which the coordinates of all vectors are integers is called an integer lattice. Equivalently, while $m \geq 1$, the integer lattice is an addition subgroup of $\mathbb{Z}^m$.

3.2 The NP Problems of Lattice

Lattice problem has a long history, which can be traced back to more than 200 years ago, Gauss studied the shortest vector algorithm of two-dimensional lattice. The key problem is to find a lattice Vector to meet minimize feature needs, the most classic of the issue is the Shortest Vector Problems (SVP) and the closest problems Vector Problem (CVP) [16–18].

Definition 2. Finding the shortest non-zero vector $v \in L$ in the lattice aims to minimize the Euclidean norm $||v||$, which is called the shortest vector problem.

Definition 3. Given a vector $w \in \mathbb{R}^m$ that is not in the lattice L, and looking for a vector $v \in L$ that is closest to w, therefore there is a vector $v \in L$ that can minimize the Euclidean norm $||w - v||$, which called the nearest vector problem.

There may be more than one shortest non-zero vector or the nearest non-zero vector in the lattice. The problems of SVP and CVP are very esoteric and hard problems. As the dimension of the lattice increases, the calculation becomes more difficult. CVP is a NP-hard problem. SVP is also under the specific "random reduction assumption". It is also a NP-hard problem. In some practical application fields, CVP can generally be considered to SVP problem with a higher dimension. For example, an SVP problem that is used to solve a $n + 1$ dimension of a backpack cryptosystem can be quickly converted to a n dimensional CVP.

You can use the method of lattice reduction to find a set of high-quality bases of the lattice. The basic hard problems in the above two lattices can be transformed into some problems with similar hardness degree by a certain lattice reduction algorithm. Common approximate problem-solving method is a small integer solution problem SIS, which is defined as the following Definition 4.

Definition 4. Given a integer q, a matrix $A \in \mathbb{Z}_q^{n \times m}$, a real number β, to find a non-zero vector $e \in \mathbb{Z}^m$ that satisfies $Ae = 0 \bmod q$, and $||e|| \leq \beta$. Here, q, m, β can be represented polynomial that generated with n, A is randomly and evenly distributed.

It can be seen from the literature [19] that since A is randomly selected uniformly, the SIS problem of small integer problems is an NP-hard problem when selecting appropriate parameters. The theory has been proofed by Micciancio and Regev.

3.3 NTRU Cryptography Analysis Based on the Lattice

The main idea of the NTRU cryptosystem is like other cryptosystems. The main idea is to construct a lattice to store the public key or ciphertext according to some form, and then solve the lattice problem to get the private key or plaintext. The following discusses how to extract a private key (f, g) from a public key h. For public parameters (N, p, q) and public keys $h(x) = h_0 + h_1 x + \cdots + h_{N-1} x^{N-1}$, we can use the following row vector as the base generator L.

$$M = \begin{bmatrix} 1 & 0 & \cdots & 0 & h_0 & h_1 & \cdots & h_{N-1} \\ 0 & 1 & \cdots & 0 & h_{N-1} & h_0 & \cdots & h_{N-2} \\ \vdots & \vdots & & \vdots & \vdots & \vdots & & \vdots \\ 0 & 0 & \cdots & 1 & h_1 & h_2 & \cdots & h_0 \\ 0 & 0 & \cdots & 0 & q & 0 & \cdots & 0 \\ 0 & 0 & \cdots & 0 & 0 & q & \cdots & 0 \\ \vdots & \vdots & & \vdots & \vdots & \vdots & & \vdots \\ 0 & 0 & \cdots & 0 & 0 & 0 & \cdots & q \end{bmatrix} \tag{4}$$

The dimension of this lattice L is $2N$, that it has $2N$ coordinates in each row vector. If the polynomial is represented as a vector, the lattice vector can be considered as a vector of two connected polynomials. Now the question is whether the result can be a vector in the lattice L if the private key f and g is connected together.

Because

$$h = f * g \pmod{q} \tag{5}$$

exists the polynomial u, and we have

$$f * h = g + qu \tag{6}$$

and

$$(f, -u) \cdot M = (f, g) \tag{7}$$

Here, the vector (f, g) can be denoted by the base M, so (f, g) is the vector indeed of the lattice L. Now we get to analyze the length of the vector (f, g), assuming in the NTRU system the public parameters $d \approx N/3$, $q \approx 2N$, then the length of the vector (f, g) can be estimated as the following formula (8).

$$\| (f, g) \| \approx \sqrt{4d} \approx 1.155\sqrt{N} \tag{8}$$

The Gaussian expectation of the shortest vector in the lattice is:

$$\sigma(L) = \sqrt{\frac{2N}{2\pi e}} (\det L)^{1/2N} = \sqrt{\frac{2N}{2\pi e}} (q^N)^{1/2N} \approx 0.484N \tag{9}$$

When N is quite a large number, the length of the vector (f, g) will be much smaller than the expected value, and the shortest vector of the lattice L has a high probability of being a vector (f, g) or its cyclic form. The shortest vector of the lattice L is obtained and the private key is also found.

4   The NTRU Public Key Encryption Scheme in UANs

Each node in the underwater has its own identity information $id_A, id_B \in \{0, 1\}^k$ and each node knows the identity information of the next hop node. The NRTU public key encryption scheme can encrypt the data and prevent the adversary from illegally stealing the message.

4.1   Parameter Generation

$KeyGen(N, p, q)$: Usually N is a large prime number, here N represents the length of the message block. p is a small prime number, and q is much larger than p, and $gcd\ (p, q) = 1$. Randomly select the polynomial $f, g \in R$ of small integer coefficients, and then compute the polynomial inverse f of modulo q and p.

$$
\begin{aligned}
F_p &\equiv f^{-1} \bmod p \\
F_q &\equiv f^{-1} \bmod q \\
h &= F_q * g (\bmod q)
\end{aligned}
\tag{10}
$$

To generate the private key (f, F_p) for the current communication and use the polynomial h as the public key.

In addition, here an available hash collision function $H(x) : \{0, 1\}* \rightarrow \{0, 1\}^t$ is introduced, and $H(x)$ can output the fixed length t bits.

4.2   Encryption Process

The length of the plaintext m is $N - t$ bits, and the following operations are performed:

(1)   To calculate $s \leftarrow H(id_A \parallel m \parallel id_B)$, "$\parallel$" indicates a bit string concatenation operation.
(2)   To select a one-time random number $u \in Z^R$, the length of u is N bits, and then calculate $m' = (m \parallel s) \oplus u$, here "$\oplus$" indicates an exclusive OR operation.
(3) $Enc(m', r, p, q)$: The plaintext m' is a polynomial with $\bmod p$ coefficients, and a polynomial r with small integer coefficients is randomly selected. The ciphertext c is calculated by the following formula (11).

$$
c \equiv (pr * h + m') \bmod \ p.
\tag{11}
$$

In the encryption process, the identity information of the plaintext and the node is randomly perturbed by $H(x)$ and the exclusive OR operation "$\oplus$" to make it have a random distribution characteristic to prevent the leakage of plaintext information.

4.3   Decryption Process

(1)   The destination node receives c and u, then decrypts the decryption algorithm $Dec(c, f, F_p)$, and calculates m'.

$$a \equiv f * c (\bmod q)$$
$$m' \equiv a * F_p (\bmod p) \tag{12}$$

Here, the destination node should appropriately select the coefficients a within the appropriate range, which cannot lead to decryption failure.

(2) The destination node executes the computation $m' \oplus u$, and then obtains a bit string of length N, taking the previous $N - t$ bits as the decrypted plaintext m'', and the following t bits as the decrypted s'.

(3) The destination node gets s by computing $H(idA \parallel m'' \parallel idB) \to s$. If $s = s'$, which indicates that the decryption is successful, it can receive the correct plaintext $m'' = m$. Otherwise, the message cannot be received because it may be lost or falsified.

5 Security Analysis

5.1 apprCVP NP

The NTRU algorithm uses polynomial algebra and modules of two different Numbers. Its security mainly depends on polynomial, lattice base reduction algorithm of mixed different modules operation, hardness in finding the maximum lattice and the nearest vector CVP. If the security assumption of a cryptographic system based on lattice theory can eventually be reduced to solve the difficult problem of apprCVP, then the cryptographic system is secure.

Theorem 1: BaBai nearest vector algorithm. Suppose $L \subset \mathbb{R}^m$ is a lattice, $v_1, v_2, \cdots, v_n$ is the basis of this lattice L, $w \in \mathbb{R}^m$ is any vector, if the orthogonality of the vectors in the basis is very good, then the following algorithm can be used to solve CVP. The algorithm Babai is showed as Fig. 1.

Input: $v_1, v_2, \cdots, v_n$ and vector w
Output: the closest lattice vector V
1    $w = t_1 v_1 + t_2 v_2 + \cdots + t_n v_n, t_1, t_2, \cdots, t_n \in \mathbb{R}$
2    $a_i = round(t_i), i = 1, 2, 3, \cdots, n$
3    $v = a_1 v_1 + a_2 v_2 + \cdots + a_n v_n$
4   **return** v

Fig. 1. The algorithm Babai

Theorem 2: LLL-apprCVP algorithm. There is a constant C, so that for any lattice L of n dimensions, when given a set of basis $v_1, v_2, \cdots, v_n$, the algorithm as following Fig. 2 can solve the apprCVP problem in C^n.

Input: $v_1, v_2, \cdots, v_n$ and vector w

Output: the closest lattice vector $\mathcal{V}$

---

1 $k = 2$

2 $v_1^* = v_1$

3   Loop while $k \le n$

4   ┃ For $j = 1$ to $k - 1$

5   ┃ $v_k = v_k - \left[\mu_{k,j}\right] v_j$

6   ┃ End for

7   ┃ If $\left\| v_k^* \right\|^2 \ge \left(\dfrac{3}{4} - \mu_{k,k-1}^2 \right) \left\| v_{k-1}^* \right\|^2$ then $k = k+1$

8   ┃   Else swap (v_{k-1}, v_k)

9   ┃ $k = \max(k-1, 2)$

10 End loop

11 LLL $\left(\{v_1, v_2, \cdots, v_n\}\right)$

12 $w = t_1 v_1 + t_2 v_2 + \cdots + t_n v_n, t_1, t_2, \cdots, t_n \in \mathbb{R}$

13 For $i = 1$ to n

14 $a_i = round(t_i)$

15 End for

16 $v = a_1 v_1 + a_2 v_2 + \cdots + a_n v_n$

17 return $\mathcal{V}$

Fig. 2. The algorithm of apprCVP

When combining the Fig. 1 and the Fig. 2, an algorithm can be formed to solve the problem of apprCVP, which can prove that the NTRU algorithm is secure.

5.2 Security Analysis of the Algorithm

If apprCVP is a difficult problem, the algorithm is secure.

Proof: After the destination node receives the ciphertext sent by the source node and decrypts it, $m' = (m \parallel H(idA \parallel m \parallel idB)) \oplus u$ can be obtained and executed $m' \oplus u$ to get $m \parallel H(id_A \parallel m \parallel id_B)$. If m is correct, the authentication information is correct when the identity information of both the node id_A and id_B are correct and belong to the legitimate nodes. Assume that the attacker had intercepted cipher c, and c will be altered into ciphertext c', at this time to get m' here by the calculation $c' \equiv (pr * h + m') \bmod p$, taking the previous $N - t$ bits as decrypted plaintext m'', and taking the rear t bits as decrypted s'. Then the destination node computes $H(idA \parallel m'' \parallel idB) \rightarrow s$ again, because of $s \neq s'$, the verification is failed. That is showed clearly the $N - t$ bits are regarded as decrypted plaintext m'' and the rear t bits are regarded as decrypted c', which is not correct.

5.3   Eavesdropping Attacks

In addition, when we suppose the adversary eavesdrop all communications from the source node to the next-hop node, and get the ciphertext c. The adversary attempts to obtain the plaintext m by calling decryption algorithm $Dec(c, f, F_p)$. The adversary chooses a coefficient of a within the appropriate scope. According to $a \equiv f * c(\mathrm{mod}\ q)$, $m \equiv a * F_p(\mathrm{mod}\ p)$ and tries to calculate m. The destination node computes $m \oplus u$ again and gets a bit string of length N to take out m'' and s'. Then the destination node computes $H(idA \parallel m'' \parallel idB) \rightarrow s$ once more to judge $s = s'$, thus to obtain the plaintext m. In this process, firstly, the adversary does not have the private key of the next hop node, so the CVP problem and m cannot be solved normally. Secondly, even if the adversary eavesdrops on random Numbers u in the process of listening, the adversary cannot get any information about m in the whole process of data transmission. Therefore, this scheme can resist the attack of information eavesdropping.

5.4   Replay Attacks

Assume that adversary has eavesdropped into a complete communication process, and he tried to pretend to be the source node sends ciphertext c and u again. Obviously the next-hop node can decrypt the ciphertext correctly, but because in the scheme u is stochastic and u is already existed in the next node, at this time the decryption is illegal and immediately the communication is terminated.

5.5   Complexity Analysis of the Algorithm

In this scenario, there are two extra execution $H(x)$ and two extra exclusive or operation in the computation, the time complexity is equivalent to NTRU. The times of encryption and decryption to a message block with length N is $O(N^2)$. However, the Fast Fourier Transform (FFT) is commonly introduced to the NTRU cryptosystem in the reference [20] written by Cooley J W. and Tukey J W. in 1965. When $n = 2^m$, we can make use of Discrete Fourier Transform (DFT) algorithm to reduce amount of calculation $O(N \log_2 N)$. We can use the length N as arithmetic of transformation, and the circular convolution of the coefficients of sum calculated f and g by FFT, the operations of multiplication and addition are reduced from the original computational complexity $O(N^2)$ to $O(N \log_2 N)$.

5.6   Performance Analysis of the Algorithm

The most important method to attack the NTRU public key cryptosystem is lattice reduction. That is the most famous is the standard LLL lattice reduction algorithm. Suppose the actual shortest vector length is t bits in the lattice L and the expected shortest vector length is d bits, then to set $c = t/d$. When c is becoming smaller, it is easier to find the nearest vector for the lattice reduction algorithm, and the NTRU is less secure. However, when c is close to 1, the cost of finding the nearest vector such as the needed time and the complexity will be greatly increased in the lattice reduction algorithm. It is proofed when we select $100 < N < 600$, $64q \leq 256$, $c = 0.18 \sim 0.25$, we can obtain enough high security theoretical and practical in general [11]. The security of NTRU is usually divided into three levels [21]. The security performance of these three levels is as follows and different parameter selection and security levels in NTRU public key cryptosystem are showed in the Table 1.

Table 1. Different security levels of NTRU public key cryptosystem (The hardware environment under test is used 200 MHz Pentium Processor)

Level	N	p	q	d_f	d_g	d_r	c	Time (s)
General	107	3	64	15	12	5	0.258	780230 (9days)
High	167	3	128	61	20	18	0.23	$1.198 * 10^{10}$ (380years)
Highest	503	3	256	216	72	55	0.18	$1.969 * 10^{35}$ ($62 * 10^{27}$ years)

(1) $N = 107$ is the general safety level, which can be used in common occasions with low requirements for safety performance,
(2) $N = 167$ is the high safety level, which is suitable for occasions with high requirements on safety performance,
(3) $N = 503$ is the highest safety level, which is suitable for special occasions with high requirements on safety performance.

5.7 Performance Analysis of the Algorithm

NTRU cryptosystem involves only simple addition and multiplication of polynomials, and there are no complicated operations such as complex modulus in the public key cryptography algorithm. Therefore, the public key cryptosystem NTRU is higher efficient in speed and the key is shorter compared with RSA, ECC, ELGamal. The cost of memory and processor requirements for underwater sensor nodes is lower and easier to the hardware of the limited resources of underwater acoustic networks. According to the underwater environment and data confidentiality, the above three security levels can be freely selected. If we use the recommended parameter set, that is, when $N = 263$, $p = 3$, $q = 128$, $d = 83$, the security level is higher than the 1024-bit RSA algorithm, and the signature time and authentication time of the NTRU algorithm are lower than the RSA-1024 algorithm. The encryption and decryption speed and key generation time of NTRU and RSA cryptosystems in the 90 MHz Pentium system respectively is showed in Table 2.

Table 2. The Comparison of NTRU and RSA

Level		Encryption speed (blocks/s)		Decryption speed (blocks/s)		Key generation time (s)	
RSA	NTRU	RSA	NTRU	RSA	NTRU	RSA	NTRU
512 bit	Low	370	1818	42	505	0.45	0.108
768 bit	Medium	189	649	15	164	1.5	0.1555
1024 bit	High	116	103	7	19	3.8	0.8571

It can be seen from the above analysis that NTRU public key cryptosystem is faster in encryption and decryption speed than RSA public key cryptosystem under the same security level. The key generation speed of NTRU is also faster than RSA, which is more suitable for underwater acoustic networks with limited transmission bandwidth, computation and storage resources.

6 Inclusion

This chapter puts forward an improved new type based on NTRU public key cryptographic algorithms. The algorithm depends on the theory of CVP hard problem Compared with the RSA public key cryptosystem, our provided algorithm implements the NTRU encryption scheme based on the identities of the modes. The algorithm in encryption and decryption processes is faster to produce the key and the key is shorter than RSA. It is proved that the new algorithm has the advantages of quantum security and it is more suitable for the high level of security requirements of applications in the field of underwater acoustic networks.

Acknowledgements. This work is supported by Key lab of IoT of Qinghai (No. 2017-ZJ-Y21), the National Social Science Foundation of China (No. 18XMZ050, No. 15XMZ057), the National Science Foundation of China (No. 61751111), Qinghai Office of Science and Technology (No. 2019-ZJ-7086, No. 2018-SF-143, No. 2015-ZJ-718).

References

1. Hu, Z., Wang, C., Zhu, Y., Kong, D.: Signal detection for the underwater acoustic voice communication. In: Proceedings of the International Symposium on Test and Measurement 2003, Washington, pp. 1–5. IEEE (2003)
2. Sozer, E.M., Stojanovic, M., Proakis, J.G.: Underwater acoustic networks. IEEE J. Oceanic Eng. **25**(1), 72–83 (2000)
3. Kilfoyle, D.B., Baggeroer, A.B.: The state of the art in underwater acoustic telemetry. IEEE J. Oceanic Eng. **25**(1), 4–27 (2000)
4. Caiti, A., Munafo, A.: Adaptive cooperative algorithms for AUV networks. In: IEEE International Conference on Communications Workshops (ICC 2010), pp. 1–5 (2010)
5. Guo, Z., Luo, H., Hong, F., Yang, M., Ni Lionel, M.: Current progress and research issues in underwater sensor networks. J. Comput. Res. Dev. **47**(3), 377–389 (2010)
6. Wei, Z., Yang, G., Cong, Y., Dong, J.: Analysis of security and threat of underwater wireless sensor network topology. In: Proceedings of the ICCEE 2010, Chengdu, pp. 506–510 (2010)
7. Peng, C., Du, X., Li, K., Li, M.: An ultra lightweight encrypted scheme in underwater acoustic networks. J. Sens. **3**, 1–10 (2016)
8. Shor, P.W.: Scheme for reducing decoherence in quantum computer memory. Phys. Rev. A **52**(4), R2493 (1995)
9. Shor, P.W.: Polynomial-time algorithms for prime factorization and discrete logarithms on a quantum computer. SIAM Rev. **41**(2), 303–332 (1999)
10. Ajtai, M., Dwork, C.: A public-key cryptosystem with worst-case/average-case equi-valence. In: The 29th Annual ACM Symposium on Theory of Computing, pp. 284–293. ACM, New York (1997)

11. Hoffstein, J., Pipher, J., Silverman, Joseph H.: NTRU: a ring-based public key cryptosystem. In: Buhler, Joe P. (ed.) ANTS 1998. LNCS, vol. 1423, pp. 267–288. Springer, Heidelberg (1998). https://doi.org/10.1007/BFb0054868

12. Gentry, C.: Fully homomorphic encryption using ideal lattices. In: STOC 2009, vol. 9, no. 4, pp. 169–178 (2009)

13. Peikert, C.: A Decade of Lattice Cryptography. Now Publishers Inc. (2016)

14. Ajtai, M.: Generating hard instances of lattice problems. In: The 28th Annual ACM Symposium on Theories of Computing, pp. 99–108. ACM, New York (1996)

15. Regev, O.: Lattice-based cryptography. In: Dwork, C. (ed.) CRYPTO 2006. LNCS, vol. 4117, pp. 131–141. Springer, Heidelberg (2006). https://doi.org/10.1007/11818175_8

16. Boas, P.V.E.: Another NP-complete partition problem and the complexity of computing short vectors in lattices. Technical report, Department of Mathematics, University of Amsterdam 81–04 (1981)

17. Khot, S.: Hardness of approximating the shortest vector problem in lattices. J. ACM (JACM) 52(5), 789–808 (2005)

18. Dinur, I.: Approximating SVP$_\infty$ to within almost-polynomial factors is NP-hard. In: Bongiovanni, G., Petreschi, R., Gambosi, G. (eds.) CIAC 2000. LNCS, vol. 1767, pp. 263–276. Springer, Heidelberg (2000). https://doi.org/10.1007/3-540-46521-9_22

19. Micciancio, D., Regev, O.: Worst-case to average-case reductions based on Gaussian measures. SIAM J. Comput. 37(1), 267–302 (2007)

20. Cooley, J.W., Tukey, J.W.: An algorithm for the machine calculation of complex Fourier series. Math. Comput. 19(90), 297–301 (1965)

21. Vredendaal, C.V.: Reduced memory meet-in-the-middle attack, against the NTRU private key. LMS J. Comput. Math. 19(A), 43–57 (2016)

Knowledge-Enhanced Bilingual Textual Representations for Cross-Lingual Semantic Textual Similarity

Hsuehkuan Lu[1]([⊠]), Yixin Cao[2], Hou Lei[1], and Juanzi Li[1]

[1] Department of CST, Tsinghua University, Beijing 100084, China
{lxk16,houlei,lijuanzi}@mails.tsinghua.edu.cn
[2] School of Computing, National University of Singapore, Singapore, Singapore
caoyixin2011@gmail.com

Abstract. Joint learning of words and entities is advantageous to various NLP tasks, while most of the works focus on single language setting. Cross-lingual representations learning receives high attention recently, but is still restricted by the availability of parallel data. In this paper, a method is proposed to jointly embed texts and entities on comparable data. In addition to evaluate on public semantic textual similarity datasets, a task (cross-lingual text extraction) was proposed to assess the similarities between texts and contribute to this dataset. It shows that the proposed method outperforms cross-lingual representations methods using parallel data on cross-lingual tasks, and achieves competitive results on mono-lingual tasks.

Keywords: Text and knowledge representations ·
Cross-lingual representations · Cross-lingual semantic textual similarity

1 Introduction

Measures of semantic relatedness or similarity are widely adopted in a variety of applications: for example, assessment of machine translation systems [16, 25], question answering [31, 37, 44], text summarization [30], document clustering [20] and information retrieval [19]. Take search engine as an example [8], a user searching for *apple* on the Web might be interested in the sense associated with computer and not in the sense of fruit. Moreover, several researches (e.g., [2, 10]) extend measures of semantic relatedness or similarity to cross-lingual settings, which explore the potential applications of cross-lingual NLP tasks. For example, extracting parallel texts (i.e., sentences, documents) for cross-lingual NLP tasks [7, 34], cross-lingual information extraction [13, 39], and plagiarism detection [33].

The challenges of measuring cross-lingual semantic similarity can be summarized to following three key points: (1) dependence on human-annotated similarity data and multi-lingual parallel data, (2) measurement of imbalanced texts, and (3) measurement of cross-lingual related texts. First, the

© Springer Nature Singapore Pte Ltd. 2019
X. Cheng et al. (Eds.): ICPCSEE 2019, CCIS 1058, pp. 425–440, 2019.
https://doi.org/10.1007/978-981-15-0118-0_33

approaches [17,32,45] develop STS systems requiring large amount of similarity labels, which restricts the applicability of minor languages with insufficient annotations. Next, in order to extend mono-lingual STS model to cross-lingual setting, it requires parallel corpus for model to learn cross-lingual information, where existing parallel corpus is either expensive to obtain or limited in narrow domains [15]. Second, existing STS methods learn from text pairs in close length, but it is insufficient to cope with actual application scenario. Because it is important to measure the similarity of texts containing imbalanced information. For example, on Web the search terms are usually far shorter than articles, and it is difficult to measure the similarity of search terms and related articles. To the best of our knowledge, there is no public evaluation dataset for imbalanced texts. At last, existing STS methods are incapable to assess the semantic similarity of cross-lingual texts. In mono-lingual setting, the dissimilarity of text pairs can be graded by supervised models. STS tasks first proposed in SemEval workshop [3] in 2012, which grades similarity from 0 (completely dissimilar) to 5 (completely equivalent). By fitting models on annotated similarity labels, it is easy to assess the similarity of text pairs. However, such data is rare in multi-lingual setting, especially in minor languages, and it troubles the measurement of cross-lingual texts.

To address the issues mentioned above, we propose a framework that integrates the information of text and Knowledge Base (KB) entities so as to model cross-lingual semantic relatedness of texts. Different to STS tasks, our method is capable to assess the relatedness between imbalanced texts. In order to validate our idea, we design a dataset of cross-lingual text extraction for evaluation, which simulates the actual scenario that the texts collected from various domains are imbalanced (in which the length of parallel text pairs might vary greatly). As Gabrilovich and Søggard [14,38] proposed, KB entities have been conventionally used to model semantics of texts. Thus, we hypothesize that combining KB entities with texts can enhance the semantic representations of texts [9,41,42]. Our novel bilingual embeddings learning model based on *comparable bilingual documents*, which aligned text pairs contain similar topics but not strictly equivalent. The representative works [11,15,18,29,36] learn cross-lingual embeddings from parallel sentences, while our method requires only weak cross-lingual supervision.

In this paper, we present a novel model that encodes bilingual texts and KB entities jointly to enhance semantic textual representations with comparable bilingual documents. Besides, in order to relieve the reliance of parallel corpus, we propose a distantly-supervised method to construct comparable bilingual corpus from Wikipedia. As to the evaluation, we contribute a new English-Chinese dataset for cross-lingual text extraction, which established by 15 people both fluent in English and Chinese. We validate our method on our proposed dataset, and public STS tasks (SemEval-2014, SemEval-2017), in which our method outperforms other cross-lingual embeddings methods in all cross-lingual datasets and several mono-lingual datasets.

Our contributions can be summed up to these following four points:

- We propose a distantly supervised method to extract the comparable bilingual texts with labeled entities from Wikipedia in two different language pairs, English-Chinese and English-Spanish.
- We manually annotate a new dataset for cross-lingual text extraction on the extracted English-Chinese data.
- We propose a novel method that encodes bilingual texts and KB entities jointly to capture cross-lingual semantic relatedness. The main strengths of our model are relatively simple and efficient to other representative cross-lingual embeddings methods [11,15,18,36] trained on sentence-level alignment data and require only weak cross-lingual supervision (i.e., comparable bilingual documents).
- We validate our model on 2 tasks separately, cross-lingual information extraction and STS tasks (mono-lingual and cross-lingual settings). Our method outperforms other baselines on all cross-lingual datasets and several mono-lingual datasets. The results support that our method is capable to measure cross-lingual semantic relatedness, and semantic similarity of texts.

2   Related Work

Joint learning representations of words and entities receives more and more attention [9,42], while most of the works target on mono-lingual settings. As summarized by Ruder [35], supervised cross-lingual embeddings methods are classified into three categories depending on different types of cross-lingual supervision: (1) methods using **parallel corpus with aligned words** [4,24,26] as bilingual regularization term. (2) methods using **parallel sentences** [11,15,18, 29,36] as semantic composition of words. (3) methods using **bilingual lexicon** [12,27] as bootstrap to learn the mapping of cross-lingual embeddings.

All of these works require parallel corpus, which is usually expensive to get. Several works attempt to tackle this issue by using unsupervised learning or comparable corpus. Artetxe [5,6] proposes the methods using small bilingual lexicon as bootstrap and applying the strategy recursively to obtain more bilingual translation pairs. In the work of Søggard [38], the work obtain dense multi-lingual word embeddings by factorizing co-occurrence matrix counted on Wikipedia comparable corpus. Another method to obtain cross-lingual embeddings by constructing "pseudo bilingual documents" [39,40].

Chandar and Hermann [11,18] train on bilingual parallel sentences, and use only cross-lingual objective to learn bilingual embeddings in the unified vector space. Their methods design to induce only cross-lingual embeddings, but ignore the impact of mono-lingual information. Afterwards, Gouws and Mogadala [15, 29] improve the previous methods by integrating mono-lingual objective into the models. Most recently, Schwenk [36] borrows the idea from neural machine translation model, and focuses on generating the intermediate representations instead of building machine translation system.

Besides, Yamada and Cao [9, 42] propose the method to jointly embed words and entities using Skip-Gram based model, and applied the embeddings to named entity disambiguation. The work [43] propose the method to jointly embed texts and entities into a unified vector space and model text with linear transformation of word vectors. The main difference to the previous work is that it models texts instead of words. Inspired by their works, our work is based on this hypothesis, and extends to the research of cross-lingual textual applications.

3　Dataset

At first, we begin with introducing our works on dataset, including automatic extraction method and evaluation dataset. Our method is based on texts containing KB entities, and we hypothesize that KB entities among texts are able to enhance semantic textual representations. Thus we use Wikipedia, which contains abundant resource of multi-lingual texts and manually-annotated entities, as our corpus. This section includes two subsections: (1) **construction of comparable corpus** and (2) **evaluation dataset for cross-lingual information extraction**.

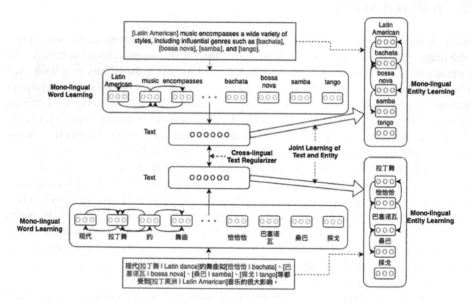

Fig. 1. The overall framework of our method. The inputs of our model are pairs of comparable texts T containing entities E, where entities are marked red in brackets. There are three main modules for our method, including (1) mono-lingual word, and entity learning, (2) cross-lingual text learning, and (3) joint learning of text and entity. (Color figure online)

3.1   Construction of Comparable Corpus

We choose Wikipedia, the April 2017 multi-lingual dump[1] as our corpus. Among all the languages, we use Chinese (Zh), Spanish (Es), and English (En) as our target languages, and we build En-Zh and En-Es bilingual corpus. The cost of acquiring parallel corpus is quite expensive, and especially insufficient in minor languages. To address this issue, we propose a **distantly supervised** method to build comparable bilingual texts on parallel concepts in Wikipedia, and the processing of data preparation includes multi-lingual parallel entity lexicon construction, and comparable texts extraction.

Before introducing our method, we first clarify the definition of comparable corpus. For example, given two parallel sentences in English: *"A boy is playing guitar."* and *"There is a kid playing guitar."* It's obvious to identify that these two sentences are semantically identical, and this data type is adopted frequently in cross-lingual related tasks. By contrast, give another two comparable sentences in English: *"A boy is playing guitar."* and *"He is singing on the road."* These two sentences are both related to music activity, but not semantically identical. This type of data alignment is so-called "comparable".

First, we build a multi-lingual parallel entity lexicon, which contains equivalent entities in different languages. Among the multi-lingual Wikipedia dump, we then pick out parallel document pairs regarding equivalent entities, so that the document pairs indicate to the same topic. Based on the parallel documents, we apply a straightforward strategy to extract highly related texts.

For instance in Fig. 1, a pair of comparable sentences describing *Latin Music*, where KB entities such as location *Latin American*, music types *samba*, *rumba*, and *tango* are parallel. In this example, we observe that sentences containing identical KB entities are likely to be semantically similar. Based on this assumption, we are able to crawl comparable texts from the multi-lingual descriptions of same entity on Wikipedia. Due to the fact that the parallel texts in Wikipedia are usually long, and may contain redundant textual information. For example, the description of *Latin Music*[2] on Wikipedia includes contents of history, definition, international organizations, and etc. Thus the textual information might cover too many domains, and it is difficult to capture the central idea of texts. So we apply distantly-supervised method to extract comparable texts from parallel mentions. We assume that the texts with same entities across different languages would be semantically similar. Based on this hypothesis and with the assist of multi-lingual entity lexicon, we extract texts containing same entities as comparable texts across different languages.

3.2   Cross-Lingual Text Extraction Dataset

Our another contribution is to manually annotate a dataset for evaluating STS application. Based on the results extracted by our method, we assume that the

---

[1] The Wikipedia dump was downloaded from the website: https://dumps.wikimedia. org.

[2] https://en.wikipedia.org/wiki/Latin_music.

bilingual aligned text pairs are semantically similar, and then we randomly sample 1,000 text pairs from Chinese-English comparable corpus, which includes 397,164 text pairs. Afterwards, we hire 15 people fluent in both English and Chinese to distinguish if the given text pair on the 1,000 sampled dataset is semantically identical. In every candidate, we add 20 texts containing same entities as negative samples. Due to the expensive cost of manual labor, we label only the first 450 pairs of the dataset, but leave all dataset for evaluation. We use a simple three labels in our annotation, which are 0 (semantically dissimilar), 1 (semantically related), 2 (semantically identical). Among all 450 samples, the total number of label 2 is 363 (80.7%), label 1 is 81 (18%), and label 0 is 6 (1.3%). The results show the feasibility of our distantly-supervised method, in which most of the texts are related (98.7%) with only small portion of unrelated texts (1.3%).

Based on the constructed dataset, we validate our method by applying STS to text extraction problem. Given two texts T^{l1} and T^{l2} in language 1 and 2 respectively, we then calculate the textual similarity for this given pair $sim(T^{l1}, T^{l2})$ and rank all similarities. In evaluation, the metric we use in this work is *Precision@N*. If the gold answer appeared in first N texts, then we consider it as a successful retrieval.

4 Methods

We propose a method that jointly embed bilingual texts and entities into a unified vector space. The overall framework is as shown in Fig. 1, including three modules: (1) mono-lingual word, entity learning, (2) cross-lingual text regularizer, and (3) joint learning of text and entity. In this work, we simply use averaging word vectors as text representations. Our data and embeddings are publicly available for further research.[3]

We denote training corpus as $\mathcal{D}$, containing multi-lingual texts T set. Given a text $T = (w_1, w_2, ..., w_N)$ as a sequence of words w containing entities $E = (e_1, e_2, ..., e_M)$ as a sequence of entities e. Following are the definition of representations of text $v_T \in \mathbb{R}^d$:

$$v_T = \frac{\sum_{w \in T} v_w}{N} \tag{1}$$

In order to jointly embed texts and entities to a unified vector space, we first assign same dimensions for both word w and entity e, and further extend the representations to cross-lingual fashion. Given two comparable texts $T_{l1} = (w_1^{l1}, w_2^{l1}, ..., w_{N_1}^{l1})$ containing entities $E_{l1} = (e_1^{l1}, e_2^{l1}, ..., e_{M_1}^{l1})$, and $T_{l2} = (w_1^{l2}, w_2^{l2}, ..., w_{N_2}^{l2})$ containing entities $E_{l2} = (e_1^{l2}, e_2^{l2}, ..., e_{M_2}^{l2})$. Where the subscripts $l1, l2$ stand for language 1, and 2 respectively.

---

[3] https://github.com/hsuehkuan-lu/KEBTR

4.1   Mono-Lingual Representation Learning

In the first module, mono-lingual word, entity learning, we apply Skip-Gram model with negative sampling (SGNS) [28] to train mono-lingual embeddings. As to the second module, we choose squared distance as cross-lingual regularizer. As stated in Gouws [15], in a aligned sentence pair, the words in $l1$ sentence can be potentially aligned to the words in $l2$ sentence. At last, we design a probability function $P(e|T)$ to force text T and entities e appeared in T to come closer. The idea is similar to Skip-Gram model, while in this function, we replace center and target word with text and entity respectively. Following are the mono-lingual SGNS models for text T and entity E, where x is input word(w)/entity(e), and then we apply SGNS models to different languages accordingly:

$$\mathcal{L}_m = -\frac{1}{N} \sum_{t=1}^{N} \sum_{-c \leq j \leq c, j \neq 0} \log P(x_{t+j}|x_t) \tag{2}$$

Since the partition function in the denominator of the softmax is expensive to compute, negative sampling approximates the softmax to make it computationally more efficient. Negative sampling defined as follows:

$$P(x_{t+j}|x_t) = \log \sigma(v'_{x_{t+j}}{}^{\top} v_{x_t}) +$$
$$\sum_{i=1}^{k} \mathbb{E}_{x_i \sim P_n(x)} [\log \sigma(-v'_{x_i}{}^{\top} v_{x_t})] \tag{3}$$

where σ is the sigmoid function $\sigma(x) = \frac{1}{1+e^{-x}}$, v_{x_i} and v'_{x_i} are the word and context word embeddings of word/entity x_i respectively, k is the number of negative samples, and P_n is the noise distribution.

4.2   Cross-Lingual Representation Learning

While training mono-lingual word and entity representations, we optimize cross-lingual text objective simultaneously. Inspired by Gouws [15], which states that more frequently word-pairs appear in parallel sentences might be more similar. Thus we design an attention method to assign weights to each words while training. Following are the cross-lingual objective functions:

$$\mathcal{L}_c = ||v_{T_{l1}} - v_{T_{l2}}||^2 \tag{4}$$

We assume that by adding this attention weights can better align similar words, and we adopt cosine similarity as our attention weights. With given $T_1 = (x_1, x_2, ..., x_N)$ and $T_2 = (y_1, y_2, ..., y_M)$:

$$a(w, T) = \max_{j \in T} \cos(w, j)$$
$$A_{12} = (a(x_1, T_2), a(x_2, T_2), ..., a(x_N, T_2)) \tag{5}$$
$$A_{21} = (a(y_1, T_1), a(y_2, T_1), ..., a(y_M, T_1))$$

Based on the Eq. 5, we then multiply the words representations with calculated attention weights to get a new attentive cross-lingual loss.

$$\mathcal{L}_c^{att} = ||v_{A_{12}T_{l1}} - v_{A_{21}T_{l2}}||^2 \qquad (6)$$

4.3   Joint Representation Learning of Text and Entity

Our goal is to embed text and entity representations in a same vector space. We assume that the distributions of words and entities in a same text should be close, and based on this hypothesis, we attempt to maximize the probability $P(e|T)$ to force the distributions of words and entities close. Thus, we design a negative sampling method to distinguish e from noises randomly drawn from all entity sequences and make noise distribution P_n of e as uniform distribution, then select k negative entity sequences from all training samples. Following is the definition of $P(e|T)$:

$$P(e|T) = \log \sigma(v_E{}^\top v_T) + \sum_{i=1}^{k} e_i \sim P_n(e)[\log \sigma(-v_{e_i}{}^\top v_T)] \qquad (7)$$

Based on the Eq. 7, we design the objective function for correlation of texts and entities:

$$\mathcal{L}_{et} = - \sum_{e \in E, (T,E) \in \mathcal{D}} \log P(e|T) \qquad (8)$$

$\mathcal{D}$ consists of (T, E) pairs, where (T, E) is a pair of word and entity sequence among a same text T. Finally, we optimize the objective functions proposed (Eqs. 2, 4, 6 and 8) above, and learn bilingual text and entity representations jointly.

$$\mathcal{L} = \mathcal{L}_m^{l1} + \mathcal{L}_m^{l2} + \mathcal{L}_{et}^{l1} + \mathcal{L}_{et}^{l2} + \alpha \mathcal{L}_c(\alpha \mathcal{L}_c^{att}) \qquad (9)$$

where the superscript of objective $\mathcal{L}$ stands for the language $l1, l2$, which is designed especially for mono-lingual embeddings learning. In cross-lingual regularizer term, we add an α term to control the impact of cross-lingual information, and in this work we simply let α equals 1. The overall objective function as shown in Eq. 9 including cross-lingual text embeddings learning, monolingual embeddings learning, and correlation of text and entity.

5   Experiments

For cross-lingual evaluation, we design experiments on 2 datasets in 2 different language pairs (Chinese-English, Spanish-English). The cross-lingual datasets are manually-annotated Chinese-English Wikipedia text extraction corpus, and SemEval-2017 Task 1 (Track 4) [2]. Besides, we also experiment on 2 mono-lingual STS datasets to prove that our model is capable to capture mono-lingual semantic information. The mono-lingual datasets are SemEval-2014 English and Spanish STS tasks.

Table 1. Statistics of extracted Wikipedia corpus.

		Text amount	Vocabulary size	Entity size
English-Spanish	English	362,397	283,298	208,618
	Spanish	362,397	292,055	148,959
English-Chinese	English	397,164	306,968	236,928
	Chinese	397,164	395,010	124,942

5.1 Setup

We begin this section with the hyper-parameters of our method. The number of dimensions d for both word and entity is 300, training epoch is 10, mini-batch size is 100, the size of entity negative samples is 8, the size of skip-gram negative samples is 8, and the window size for skip-gram model is 8. For each text, we limit the maximum numbers of words and entities to 300 and 30, and discard the tokens over the limited length. All our embeddings are initialized randomly from uniform distribution.

In the stage of preprocessing, we first build the vocabulary table for each language pairs, and the minimum counts of words and entities are set to be 30 and 5. Then we lower all the tokens and filter common words for English and Spanish corpus; as to Chinese corpus, we use Jieba[4] to do tokenization. The statistics of English-Chinese (En-Zh) and English-Spanish (En-Es) training datasets $\mathcal{D}$ are shown in Table 1. In this work, we assign different embeddings for word and entity separately, thus we create two embedding matrix for word and entity. After extracting annotated entity tokens, we remove all entity tags and convert entities to regular words.

The training per model took around 14–15 h for using a single GTX 1080 Ti GPU. In this work, Adam algorithm [22] is adopted as stochastic gradient descent optimizer, and the learning rate is fixed at 0.001.

5.2 Baselines

We compare our method with following five cross-lingual representations models as baselines. The five baselines include (1) BiCVM [18], (2) BilBOWA [15], (3) BWE Skip-Gram [39], (4) VecMap [5], and (5) LASER [36].

The methods can be classified into two categories based on the source of cross-lingual supervision: supervised method, and weakly-supervised (unsupervised) method. For supervised methods, we treat comparable texts as same as parallel data, and force models to learn cross-lingual embeddings. To make fair comparisons, we use the public toolkits for implementation of these methods, and train on the Wikipedia corpus we build.

---

[4] https://github.com/fxsjy/jieba.

Supervised Methods:

- **BiCVM and BilBOWA.** We train the model with provided toolkit in default setting, and only modify dimension d to 300 so as to make comparison.
- **LASER.** In this baseline, there is no public training module for it, so we use the trained model to obtain the sentence embeddings in cross-lingual STS tasks. The embedding size d of trained model is 1024, and Chinese model is not provided.

Weakly-Supervised (Unsupervised) Methods:

- **BWE Skip-Gram.** In this method, we first randomly shuffle our comparable texts and merge to a pseudo bilingual corpus same as the work proposed. Then we use word2vec toolkit provided by Google to train the cross-lingual word embeddings.
- **VecMap.** We first use word2vec toolkit to train word embeddings, and then learn cross-lingual word embeddings with its public unsupervised mapping toolkit.

As to the text representations in all the methods we compared, we adopt averaging word vectors as our text representations (except LASER), and discard all the out-of-vocabulary words. In this work, we do not consider the impact of word weighting, instead we simply consider each word appearing in a text contributes equally to the text. Similar to other STS works, we use cosine similarity as our similarity metric, and assume that more semantically similar texts receive higher scores.

5.3   Cross-Lingual Text Extraction

The first evaluation task we use is an application of cross-lingual STS. As stated in Sect. 3, we rank the computed similarities to see if the gold answer in *Top-N*, and adopt *Precision@N* as evaluation metric.

Table 2. Text Extraction on Wikipedia corpus. The metric is using *Precision@N* to predict golden answer in *Top-N* from 1,000 texts. Both directions En-to-Zh and Zh-to-En are experimented, and all scores reported in the table are in %.

Method	En-to-Zh			Zh-to-En		
	P@1	P@5	P@10	P@1	P@5	P@10
BiCVM	8.82	15.98	20.39	21.86	42.62	49.18
BilBOWA	0.83	2.20	2.75	0.00	0.82	1.91
BWE Skip-Gram	11.57	37.47	52.89	20.49	55.74	69.67
VecMap	17.63	38.57	49.59	7.65	21.86	31.42
Our method	44.90	64.46	74.66	43.72	66.39	77.05
Our method (att)	**48.76**	**69.97**	**78.79**	**45.08**	**69.40**	**81.15**

As shown in Table 2, our method significantly outperforms all other baselines by 30% and 25% for En-Zh and Zh-En in $P@1$. The closest result reported in the table is BWE Skip-Gram, which is proposed to apply in comparable documents. The results of the other methods BiCVM, and BilBOWA indicate that these two models are incapable to learn cross-lingual information from coarse-grained cross-lingual supervision.

5.4   Semantic Textual Similarity

STS task measures the quality of embeddings by comparing the evaluated score with human judged score [21,23,43]. In this task, we focus on evaluating cross-lingual STS tasks, and also evaluating mono-lingual STS for English and Spanish tracks individually. We apply our learned text representations to these tasks, and compare with other cross-lingual representations methods on the public STS datasets. The STS datasets we use include SemEval-2014 (STS-14) [1] English and Spanish mono-lingual tracks, and SemEval-2017 (STS-17) English-Spanish cross-lingual track [10]. All the textual similarity scores are simply computed by unsupervised cosine similarity, and the generated scores are evaluated with human annotated scores using Pearson correlation r and Spearman correlation p. In all the STS datasets we select to evaluate, the ratings rated by human range from 0 (semantically unrelated) to 5 (semantically identical).

Cross-Lingual Semantic Textual Similarity. Cross-lingual STS task can be deemed as an extension of mono-lingual STS task. The task is able to describe the cross-lingual semantic information captured by model. In Table 3, our method achieves the state-of-the-art results on STS-17 dataset. Our method outperforms the second best method (LASER) with 4% on track 4a and 0.15% on track 4b, and largely outperforms the third best method with 12% and 3% respectively.

Table 3. Pearson-r and Spearman-p correlations of our models with the state-of-the-art cross-lingual embeddings models on SemEval-2017 English-Spanish STS task. All the scores are in % and best scores, in terms of r, are marked in bold.

Method	Track 4a	Track 4b
BiCVM	10.19 (10.33)	−2.65 (−1.55)
BilBOWA	−12.45 (−16.90)	−1.42 (−5.78)
BWE Skip-Gram	32.24 (33.35)	7.99 (5.91)
VecMap	32.20 (32.82)	12.55 (9.40)
LASER	40.87 (39.95)	15.27 (19.90)
Our method	43.80 (45.25)	14.31 (13.85)
Our method (att)	**44.99 (46.02)**	**15.41 (13.92)**

Mono-Lingual Semantic Textual Similarity. In STS 2014 English track, there are 6 sub-datasets among it, including deft-forum, deft-news, headlines, images, OnWN, and tweet-news. The numbers of sentence pairs in each dataset range from 300 to 750 at most, and there are 3,750 in total. For the fair comparison, we reported the scores from work by Yamada [43] including methods Skip-grams, CBOW, Skip-thought [23], Siamese CBOW [21], and NTEE(paragraph) [43]. In addition to the 5 baselines, we also include 2 cross-lingual representations methods, VecMap and BWE Skip-Gram as baselines. As to the Spanish track in STS 2014, there is no existing distributed representations work evaluating on it. This dataset includes 2 sub-datasets, which are news, and Wikipedia.

Table 4. Pearson-r and Spearman-p correlations of our models with the state-of-the-art models on SemEval-2014 English STS task. All the scores are in % and best scores, in terms of r, are marked in bold.

Method	News	Forum	OnWN	Tweet	Images	Headlines
Skip-gram	59.06 (56.78)	31.93 (38.10)	58.48 (66.76)	63.36 (65.44)	51.31 (52.88)	57.90 (55.44)
CBOW	57.37 (55.77)	33.39 (35.07)	60.68 (68.87)	68.97 (66.15)	50.56 (52.13)	40.31 (39.10)
Skip-thought	46.17 (47.62)	37.36 (37.37)	46.82 (51.61)	51.38 (52.97)	42.57 (42.33)	40.31 (39.10)
Siamese CBOW	59.13 (57.54)	40.82 (41.88)	60.73 (65.54)	73.15 (71.28)	64.97 (64.84)	63.64 (62.60)
NTEE	72.00 (69.00)	47.00 (47.00)	**75.00 (78.00)**	**73.00 (78.00)**	**77.00 (74.00)**	65.00 (61.00)
BWE Skip-Gram	58.42 (56.08)	27.31 (30.54)	68.56 (72.10)	54.66 (50.15)	72.88 (69.25)	48.87 (45.45)
VecMap	70.95 (68.68)	31.24 (32.00)	74.75 (78.59)	72.03 (67.94)	72.69 (70.28)	61.23 (58.40)
Our method	**72.87 (67.07)**	46.15 (46.15)	73.33 (73.73)	64.04 (58.77)	72.93 (69.81)	64.06 (59.17)
Our method (att)	72.57 (67.40)	**48.05 (47.81)**	72.62 (74.20)	67.22 (61.88)	74.57 (71.18)	**66.63 (61.81)**

In Table 4 evaluating on STS-14 English track, our method achieves 3 best results among all 6 datasets over 1–2%, while among the rest of the datasets we still hold slight differences. In comparison with other methods, the most close method is NTEE, which also takes account of knowledge entities in the model. To analyze why our method scores less than NTEE, a possible reason might be the quantity of training corpus. Because our method emphasizes on cross-lingual tasks, we design to extract comparable texts from the Wikipedia corpus and which largely shrinks the scale of corpus. As reported by Yamada [43], their work use 4.4 million articles for training, while in our work only 0.36 million articles are used for training. Although the significant difference in corpus scales (11 times less), our method is still capable to capture mono-lingual semantic relatedness.

As to the evaluation on Spanish track, the results are reported in Table 5, which indicate that our method is able to perform well on different domains. Though LASER achieves the best result on news sub-dataset, it performs poorly on another sub-dataset (Wikipedia) with about 16% in difference. We assume that our training corpus extracted from Wikipedia crosses multiple domains, so that our method can take advantage of abundant semantic information across domains.

Table 5. Pearson-r and Spearman-p correlations of our models with the state-of-the-art models on SemEval-2014 Spanish STS task. All the scores are in % and best scores, in terms of r, are marked in bold.

Method	Wikipedia	News
BiCVM	54.15 (50.80)	71.97 (72.91)
BilBOWA	59.06 (57.75)	68.96 (75.71)
BWE Skip-Gram	71.99 (64.26)	77.63 (77.65)
VecMap	72.95 (67.85)	77.83 (79.24)
LASER	66.65 (63.81)	**82.71 (84.09)**
Our method	77.96 (68.81)	80.45 (77.27)
Our method (att)	**78.11 (68.83)**	81.50 (78.77)

6   Conclusion

In this work, we propose a novel method to jointly embed text and entity so as to obtain better semantic representations. We use distantly-supervised method to extract comparable data, and compare with other cross-lingual representations methods. We manually annotate a cross-lingual text extraction dataset, and contribute a novel task evaluating text similarity. We validate our idea on STS tasks, including both mono-lingual and cross-lingual tasks. The experimental results support our idea of enhancing semantic representations by integrating KB entities to texts. Our method is also capable to apply in different languages and domains.

In further work, we can apply other sentence representations methods to encode text, since it might be able to discover more complicated semantic information. Our method can learn semantically expressive representations, and performs well on STS tasks with simple setting. Additionally, our method should be able to transfer to other semantic-related NLP tasks smoothly.

Acknowledgement. The work is supported by NSFC key project (U1736204, 61533018, 61661146007), Ministry of Education and China Mobile Joint Fund (MCM20170301), and THUNUS NExT Co-Lab.

References

1. Agirre, E., et al.: Semeval-2014 task 10: multilingual semantic textual similarity. In: Proceedings of the 8th International Workshop on Semantic Evaluation (SemEval 2014), pp. 81–91. Association for Computational Linguistics (2014)
2. Agirre, E., et al.: Semeval-2016 task 1: semantic textual similarity, monolingual and cross-lingual evaluation. In: Proceedings of the 10th International Workshop on Semantic Evaluation (SemEval-2016), pp. 497–511 (2016)

3. Agirre, E., Cer, D., Diab, M., Gonzalez-Agirre, A.: Semeval-2012 task 6: a pilot on semantic textual similarity. In: *SEM 2012: The First Joint Conference on Lexical and Computational Semantics - Volume 1: Proceedings of the Main Conference and the Shared Task, and Volume 2: Proceedings of the Sixth International Workshop on Semantic Evaluation (SemEval 2012), pp. 385–393. Association for Computational Linguistics, Montréal, 7–8 June 2012

4. Ammar, W., Mulcaire, G., Tsvetkov, Y., Lample, G., Dyer, C., Smith, N.A.: Massively multilingual word embeddings. CoRR abs/1602.01925 (2016)

5. Artetxe, M., Labaka, G., Agirre, E.: Learning bilingual word embeddings with (almost) no bilingual data. In: Proceedings of the 55th Annual Meeting of the Association for Computational Linguistics (Volume 1: Long Papers), pp. 451–462 (2017)

6. Artetxe, M., Labaka, G., Agirre, E.: A robust self-learning method for fully unsupervised cross-lingual mappings of word embeddings. In: Proceedings of the 56th Annual Meeting of the Association for Computational Linguistics (Volume 1: Long Papers), pp. 789–798 (2018)

7. Aziz, W., Specia, L.: Fully automatic compilation of Portuguese-English and Portuguese-Spanish parallel corpora. In: STIL (2011)

8. Bollegala, D., Matsuo, Y., Ishizuka, M.: Measuring semantic similarity between words using web search engines. In: WWW 2007: Proceedings of the 16th International Conference on World Wide Web, pp. 757–766 (2007)

9. Cao, Y., Huang, L., Ji, H., Chen, X., Li, J.: Bridge text and knowledge by learning multi-prototype entity mention embedding. In: Proceedings of the 55th Annual Meeting of the Association for Computational Linguistics (Volume 1: Long Papers), pp. 1623–1633 (2017)

10. Cer, D., Diab, M., Agirre, E., Lopez-Gazpio, I., Specia, L.: Semeval-2017 task 1: semantic textual similarity multilingual and crosslingual focused evaluation. In: Proceedings of the 11th International Workshop on Semantic Evaluation (SemEval-2017), pp. 1–14. Association for Computational Linguistics (2017)

11. Chandar, A.P.S., et al.: An autoencoder approach to learning bilingual word representations. CoRR abs/1402.1454 (2014)

12. Faruqui, M., Dyer, C.: Improving vector space word representations using multilingual correlation. In: Proceedings of the 14th Conference of the European Chapter of the Association for Computational Linguistics, pp. 462–471. Association for Computational Linguistics (2014)

13. Franco-Salvador, M., Rosso, P., Navigli, R.: A knowledge-based representation for cross-language document retrieval and categorization. In: Proceedings of the 14th Conference of the European Chapter of the Association for Computational Linguistics, April 2014

14. Gabrilovich, E., Markovitch, S.: Computing semantic relatedness using Wikipedia-based explicit semantic analysis. In: Proceedings of the 20th International Joint Conference on Artificial Intelligence, pp. 6–12 (2007)

15. Gouws, S., Bengio, Y., Corrado, G.: BilBOWA: fast bilingual distributed representations without word alignments. In: Bach, F., Blei, D. (eds.) Proceedings of the 32nd International Conference on Machine Learning, vol. 37, pp. 748–756, 07–09 July 2015

16. He, H., Gimpel, K., Lin, J.: Multi-perspective sentence similarity modeling with convolutional neural networks. In: Proceedings of the 2015 Conference on Empirical Methods in Natural Language Processing, pp. 1576–1586. Association for Computational Linguistics (2015)

17. He, H., Lin, J.: Pairwise word interaction modeling with deep neural networks for semantic similarity measurement. In: Proceedings of the 2016 Conference of the North American Chapter of the Association for Computational Linguistics: Human Language Technologies, pp. 937–948. Association for Computational Linguistics (2016)
18. Hermann, K.M., Blunsom, P.: Multilingual models for compositional distributional semantics. In: Proceedings of ACL, June 2014
19. Hliaoutakis, A., Varelas, G., Voutsakis, E., Petrakis, E., Milios, E.: Information retrieval by semantic similarity. Int. J. Semant. Web Inf. Syst. **2**, 55–73 (2006)
20. Huang, A.: Similarity measures for text document clustering. In: Proceedings of the 6th New Zealand Computer Science Research Student Conference, pp. 49–56 (2008)
21. Kenter, T., de Rijke, M.: Short text similarity with word embeddings. In: Proceedings of the 24th ACM International on Conference on Information and Knowledge Management, pp. 1411–1420 (2015)
22. Kingma, D.P., Ba, J.: Adam: a method for stochastic optimization. CoRR abs/1412.6980 (2015)
23. Kiros, R., et al.: Skip-thought vectors. In: Cortes, C., Lawrence, N.D., Lee, D.D., Sugiyama, M., Garnett, R. (eds.) Advances in Neural Information Processing Systems, vol. 28, pp. 3294–3302. Curran Associates, Inc. (2015)
24. Klementiev, A., Titov, I., Bhattarai, B.: Inducing crosslingual distributed representations of words. In: COLING (2012)
25. Lavie, A., Denkowski, M.J.: The meteor metric for automatic evaluation of machine translation. Mach. Transl. **23**(2–3), 105–115 (2009)
26. Luong, M.T., Pham, H., Manning, C.D.: Bilingual word representations with monolingual quality in mind. In: NAACL Workshop on Vector Space Modeling for NLP, Denver, United States (2015)
27. Mikolov, T., Le, Q.V., Sutskever, I.: Exploiting similarities among languages for machine translation. CoRR (2013)
28. Mikolov, T., Sutskever, I., Chen, K., Corrado, G.S., Dean, J.: Distributed representations of words and phrases and their compositionality. In: Burges, C.J.C., Bottou, L., Welling, M., Ghahramani, Z., Weinberger, K.Q. (eds.) Advances in Neural Information Processing Systems, vol. 26, pp. 3111–3119 (2013)
29. Mogadala, A., Rettinger, A.: Bilingual word embeddings from parallel and non-parallel corpora for cross-language text classification. In: Proceedings of the 2016 Conference of the North American Chapter of the Association for Computational Linguistics: Human Language Technologies, pp. 692–702. Association for Computational Linguistics (2016)
30. Mohammad, S.M., Hirst, G.: Distributional measures as proxies for semantic relatedness (2012)
31. Mohler, M., Bunescu, R., Mihalcea, R.: Learning to grade short answer questions using semantic similarity measures and dependency graph alignments. In: Proceedings of the 49th Annual Meeting of the Association for Computational Linguistics: Human Language Technologies, HLT 2011, vol. 1, pp. 752–762 (2011)
32. Pang, L., Lan, Y., Guo, J., Xu, J., Wan, S., Cheng, X.: Text matching as image recognition. In: Proceedings of the Thirtieth AAAI Conference on Artificial Intelligence, Phoenix, Arizona, USA, 12–17 February 2016, pp. 2793–2799 (2016)
33. Potthast, M., Barrón-Cedeño, A., Stein, B., Rosso, P.: Cross-language plagiarism detection. Knowl.-Based Syst. **45**, 45–62 (2011)
34. Resnik, P., Smith, N.A.: The web as a parallel corpus. Comput. Linguist. **29**, 349–380 (2003)

35. Ruder, S.: A survey of cross-lingual embedding models. CoRR abs/1706.04902 (2017). http://arxiv.org/abs/1706.04902
36. Schwenk, H., Douze, M.: Learning joint multilingual sentence representations with neural machine translation. In: ACL workshop on Representation Learning for NLP (2017)
37. Severyn, A., Moschitti, A.: Learning to rank short text pairs with convolutional deep neural networks. In: Proceedings of the 38th International ACM SIGIR Conference on Research and Development in Information Retrieval, SIGIR 2015, pp. 373–382 (2015)
38. Søgaard, A., Agić, Ž., Martínez Alonso, H., Plank, B., Bohnet, B., Johannsen, A.: Inverted indexing for cross-lingual NLP. In: Proceedings of the 53rd Annual Meeting of the Association for Computational Linguistics and the 7th International Joint Conference on Natural Language Processing (Volume 1: Long Papers), pp. 1713–1722 (2015)
39. Vulić, I., Moens, M.F.: Bilingual word embeddings from non-parallel document-aligned data applied to bilingual lexicon induction. In: Proceedings of the 53rd Annual Meeting of the Association for Computational Linguistics and the 7th International Joint Conference on Natural Language Processing (Volume 2: Short Papers), pp. 719–725. Association for Computational Linguistics (2015)
40. Vulic, I., Moens, M.: Bilingual distributed word representations from document-aligned comparable data. J. Artif. Intell. Res. **55**, 953–994 (2016)
41. Wang, Z., Zhang, J., Feng, J., Chen, Z.: Knowledge graph and text jointly embedding. In: Proceedings of the 2014 Conference on Empirical Methods in Natural Language Processing (EMNLP), pp. 1591–1601 (2014)
42. Yamada, I., Shindo, H., Takeda, H., Takefuji, Y.: Joint learning of the embedding of words and entities for named entity disambiguation. In: Proceedings of The 20th SIGNLL Conference on Computational Natural Language Learning, pp. 250–259 (2016)
43. Yamada, I., Shindo, H., Takeda, H., Takefuji, Y.: Learning distributed representations of texts and entities from knowledge base. Trans. Assoc. Comput. Linguis. **5**, 397–411 (2017)
44. Yang, L., Ai, Q., Guo, J., Croft, W.B.: aNMM: ranking short answer texts with attention-based neural matching model. In: Proceedings of the 25th ACM International on Conference on Information and Knowledge Management, CIKM 2016, pp. 287–296 (2016)
45. Yin, W., Schütze, H., Xiang, B., Zhou, B.: ABCNN: attention-based convolutional neural network for modeling sentence pairs. Trans. Assoc. Comput. Linguis. **4**, 259–272 (2016)

Predicting the Hot Topics with User Sentiments

Qi Guo, Jinhao Shi, Yong Liu$^{(\boxtimes)}$, and Xiaokun Li$^{(\boxtimes)}$

Heilongjiang University, Harbin, China
{liuyong123456,lixiaokun}@hlju.edu.cn

Abstract. Social applications such as Weibo have provided a quick platform for information propagation, which have led to an explosive propagation for hot topic. User sentiments about propagation information play an important role in propagation speed, which receive more and more attention from data mining field. In this paper, we propose an sentiment-based hot topics prediction model called PHT-US. PHT-US firstly classifies a large amount of text data in Weibo into different topics, then converts user sentiments and time factors into embedding vectors that are input into recurrent neural networks (both LSTM and GRU), and predicts whether the target topic could be a hot spot. Experiments on Sina Weibo show that PHT-US can effectively predict the hot topics in the future.

Social applications such as Weibo provide a platform for quick information propagation, which leads to an explosive propagation for hot topics. User sentiments about propagation information play an important role in propagation speed, and thus receive more attention from data mining field. In this paper, a sentiment-based hot topics prediction model called PHT-US is proposed. Firstly a large amount of text data in Weibo was classified into different topics, and then user sentiments and time factors were converted into embedding vectors that are input into recurrent neural networks (both LSTM and GRU), and future hotspots were predicted. Experiments on Sina Weibo show that PHT-US can effectively predict hot topics in the future.

Keywords: Social networks · User sentiment · Hot topics · Recurrent neural networks

1 Introduction

In recent years, social networking applications have rapidly emerged and become popular in society, which providing an extremely fast way for the propagation of information online.

In social network, when a user posts a message and passes it to the user's neighbor, it will have an impact on its neighbors to make him do the same or similar actions. Information generated by users in social networks, is continuously transmitted through such a reaction chain. We call such propagation process an information cascade [1]. When an event occurs and attracts the attention of some groups, people use social platforms to express their related sentiments and speeches, which forming a topic. The subjective speeches and behaviors of analytic, critical, etc., which are carried out around the target topic, constitute the public opinions [2]. The hot trend of the topic can

© Springer Nature Singapore Pte Ltd. 2019
X. Cheng et al. (Eds.): ICPCSEE 2019, CCIS 1058, pp. 441–453, 2019.
https://doi.org/10.1007/978-981-15-0118-0_34

accurately reflect the attention of the social groups and mainstream views, including the sentimental polarity of support, neutrality or opposition. They are of great significance to the research work of public opinion monitoring, opinion mining and personalized recommendation. In particular, when the occurrence of an event has a great impact on the user's sentiments, the user is often more probabilistically motivated to publish content related thereto, prompting the continuous propagation of the event, that is, the user's sentiments held by the information are closely related to the formation of hot topics.

Nowadays, the research related to information propagation prediction is mainly divided into three modes: standard machine learning, point stochastic process and deep learning. The information propagation is predicted by manual extraction of propagation features, learning point process rate function and learning structure hidden features [3]. However, the current research does not consider: (1) different types of networks correspond to different stochastic rate functions, such as academic information is normally distributed generally, and information propagation in social network is as the feature of chi-square or power-law distribution; (2) The social network information propagation is affected by various factors such as text content [4], structural features, and the outside world, therefore it does not have periodicity and similarity; (3) strong sentimental fluctuations caused by the user's sentimental influence on the information itself or the information source user, which often prompt users to retweet the information, so that the information is further spread at the user node, and eventually evolve into a hot topic.

Users often use Weibo or other social applications to post information related to an event and express their own sentimental views. This paper proposes a hot topic prediction model PHT-US based on users sentiments. The main research work is as follows:

1. The PHT-US model can more effectively use the historical action log to automatically learn the characteristics of the message propagation process in a data-driven manner with strong flexibility, which could avoiding the heavy dependence of the manually defined features and the directionality of the random rate function.
2. GRU (Gated Recurrent Unit) and LSTM (Long Short-Term Memory) [5] with gate mechanism can alleviate the gradient disappearance in RNN process and improve the accuracy of model prediction.
3. The PHT-US model uses the form of user sentiment vector and time vector input to effectively integrate time and sentiment into the model, which improves the prediction effect of the model effectively.

2   Related Work

At present, research work on information propagation prediction can be divided into popularity prediction and explosive prediction. Among them, the popularity prediction can be used as a classification problem, and the number of times of retweeting the topic is comprehensively predicted according to various data such as information propagation mode, time, and text. When the number of retweeting at a certain time reaches the threshold, it can be considered a hot topic.

Traditional methods of information propagation prediction rely on the manual definition of the characteristics of the information propagation process, followed by the establishment of classification or regression models, and the results are derived from a standard machine learning framework. Hong et al. [6] defined information propagation prediction as a popular classification problem with regarding "whether information is retweeted" as a measure of popularity, classifying the popularity level according to the amount of information retweeting. He manually selected information texts and topologies, structure, time [7], etc. as features to predict the popularity of information through the spread. Shen et al. proposed an RPP model based on the self-enhanced Poisson process [8], getting rid of the serious dependence of standard machine learning prediction on the manually defined features, and considering the phenomenon that the information is more and more spread during the propagation process [9], And the decline effect of information influence over time [10, 11], modeling the learning rate function for the propagation of a single information arrival process, and thus predicting the popularity of the trend. Subsequently, Bao et al. proposed a probabilistic model based on the self-excited Hawkes process (SEHP) [12]. Unlike the RPP model, it aggregates the retweeting before a certain moment in the information retweeting process into a single triggering effect. Thus, the RPP model has been optimized and improved. Li et al. started with the topological structure of information propagation and proposed end-to-end deep learning method DeepCas [13]. By learning the representation method of independent cascade graph under the social network structure, the future scale of cascade is predicted.

At the same time, consider the close relationship between user sentiments and the results of information propagation, Bollen et al. [14] calculated the time series by considering the negative and enthusiasm of sentiments to predict industrial development. Wu et al. [15] made a prediction on the number of tweets in Twitter, and found that the negative sentiments in the information are directly related to the retweeting rate; Kamkarhaghighi et al. [16] used the stock market-related text sentiment in the social network to predict the stock market, which took a good performance boost for fluctuating trends.

In this paper, the users' sentiments and time factors are vectorized as features. While detecting the changes in users' sentiments, the trend of the target topic's propagation cascade could also been observed. And finally, we could predict the popularity of different categories of topics effectively.

3 PHT-US Model Description

3.1 Problem Definition

Hot topic predictions can be defined as a classification problem. Given the action log D (User, Item, Time), the retweeted threshold E and the observed period T, at the time T_0, a Weibo message D_i with the largest number of retweeting in the topic is selected, and according to the existing action log D_i, this message can be predicted the propagation cascade in the future. If the retweeting threshold E is the smaller one comparing with the prediction retweeting times, the corresponding topic the message could be regarded as a hot spot.

3.2 Model Definition

The sentimental topic-based hot topic prediction model PHT-US is shown in Fig. 1. The model is divided into two parts: the classification module and the prediction module. The classification module classifies the Weibo datasets according to the existing setting categories. The prediction module performs the sentiment and time vectorization of the already classified Weibo text data. After training them in the recurrent neural network, the predicted results could be generated finally.

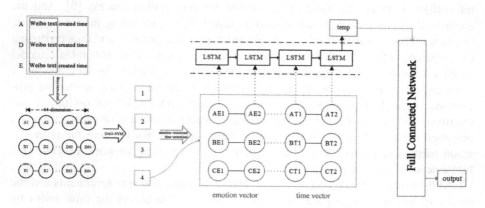

Fig. 1. Model structure of hot topic prediction problem based on sentiment

Classification Module

The classification module mainly includes three parts: Chinese word segmentation, vectorized representation and text classification.

Chinese Word Segmentation

In social networks, the information published by users is often colloquial, and often with network terms such as popular words and emoji, which interferes with topic classification. The Chinese character sequence in the text data is divided into independent phrases, which can remove unnecessary interference words and obtain high frequency and representative phrases. This experiment uses the API interface provided by the NLPIR2016 Chinese word segmentation system of Beijing Institute of Technology to set the text information segmentation and labeling, and then performs secondary processing to delete redundant symbols, stop words, and remove text noise.

Vectorized Representation

The word2vec tool is used to process the phrase after the word segmentation into a K-dimensional space vector, so that the vector operation between the word and the word corresponds to the semantics. At the same time, short texts such as Weibo related to an event, the similar words trained are more related words at the same level. Selecting a vector with a lower dimension to process the text will result in a lower discrimination between words. Choosing a dimension that is too high can easily cause the matrix to be sparse. In summary, in order to better represent the text, this experiment sets $K = 64$, which is to process the text content into a 64-dimension vector.

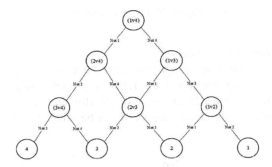

Fig. 2. DAG-SVM tree

Text Categorization
Integrate all the words of a Weibo and adjust the vector to a uniform length. According to the vector similarity between them, the topic classification is performed by the DAG-SVM algorithm [17]. As shown in Fig. 2, the algorithm is an improved algorithm for SVM binary classification. It can be classified by only k-1 decision functions. Compared with similar algorithms, it has higher fault tolerance and faster running speed.

Prediction Module
The prediction module mainly includes three parts: sentiment vector acquisition, time vectorization and cyclic neural network detection.

Sentimental Vector Embedding
The texts, pictures, expressions, etc. posted by users in social networks can reflect user sentiments. This experiment focuses on mining user sentiments from text. By segmenting the text (NLPIR2016), denoising, matching with the multidimensional sentiment dictionary of DUTIR Lab, and calculating according to the obtained sentiment categories and levels, the final sentiment vector is obtained. After sentimental algorithm mining, each Weibo will be represented by 7 dimensions of sentiment vector $e = (e_1, e_2, e_3, e_4, e_5, e_6, e_7)$. Corresponding to music, good, angry, sad, fearful, evil, and shocked. Each dimension represents the corresponding sentimental intensity.

Time Vectorization
In the process of Weibo information propagation, the time series feature is the key information for LSTM to predict the number of retweeting. As shown in Fig. 3,

considering the time value and the unit map, the vector value of each unit value at a certain time t is taken to obtain a time vector corresponding to the time.

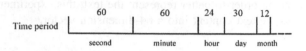

Fig. 3. Time vectorization

Recurrent Neural Networks Prediction
In the process of Weibo message propagation, the information cascade structure could be build, base on the user nodes and the retweeting relationship between the users. After randomly walks of the information cascade structure according to the Markov chain (Fig. 4), a series of isometric sequences can be obtained.

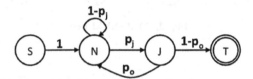

Fig. 4. State transition diagram based random walk

Assuming that the node v currently traveling in the N state, the transition probability p that it continues to activate a certain neighboring node u of V can be expressed by the *Eq.* (1).

$$p(u \in N(v)|v) = \frac{sc_t(u) + \alpha}{\sum_{s \in N(v)}(sc_t(s) + \alpha)} \tag{1}$$

Where α is a smoothing constant, $N(v)$ represents the neighbor set of the v-node, and $sc_t(u)$ is a scoring function, which is defined according to the degree of the u-node, the degree, and the weight of the edge (v, u). The transition probability setting during the random walking process, could effectively express the graph structure while the trivialness of traveling the graph structure could be optimized.

When predicting information propagation, we input the time series of information retweeting as a model, so RNN is more suitable for processing. As shown in Fig. 5, LSTM acts as a recurrent neural network with a gate mechanism. Each unit is equipped with an input gate (i_t), an forgetting gate (f_t), and an output gate (o_t). The information is completed through the three control gates. The selective storage and delivery solves the problem of gradient disappearance of RNN.

Xt is the input of the t_{th} time of the RNN, corresponding to the t_{th} element in the input sequence X, which contains the time vector and the user sentiment vector. I is the index operation, v_t is the input vector obtained after x_t is indexed, U is the input weight

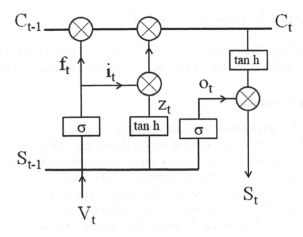

Fig. 5. LSTM unit structure

matrix, W_r is the cyclic weight matrix, and s_t is the output of the hidden layer in the t_{th} time step of the RNN. Calculated as follows:

$$V_t = I(x_t) \tag{2}$$

$$s_t = f(Uv_t + W^r s_{t-1}) \tag{3}$$

To formalize the forward propagation calculation process of LSTM, we use U_i, W_i, and b_i to represent the input weight matrix, the cyclic weight matrix, and the bias matrix of the input gate, respectively. The forward propagation calculation process of the LSTM layer is as follows:

$$i_t = \sigma(U_i v_t + W_i s_{t-1} + b_i) \tag{4}$$

$$f_t = \sigma(U_f v_t + W_f s_{t-1} + b_f) \tag{5}$$

$$o_t = \sigma(U_o v_t + W_o s_{t-1} + b_o) \tag{6}$$

$$z_t = g(U_z v_t + W_z s_{t-1} + b_z) \tag{7}$$

$$c_t = i_t \circ z_t + f_t \circ c_{t-1} \tag{8}$$

$$s_t = o_t \circ h(c_t) \tag{9}$$

The training results of the recurrent neural network then enter the fully connected layer. The fully connected neural network maps it to the tag space of the sample for classification. The vector obtained through learning is converted into a scalar through the fully connected layer, and then compared with the retweeting threshold E to finally determine whether the Weibo can become popular.

4   Experimental Results and Analysis

This section uses Sina Weibo data to train and evaluate the PHT-US model, and then compares it with existing related models to give experimental results and analysis.

4.1   Data Preprocessing

The datasets of this experiment were all obtained by crawling on Sina Weibo. Randomly select 100 users as seed nodes and set the distance length to 3 to crawl. A total of 1.7 million users and 40 million directed edges are obtained, and each user has 200 edges. For each user, collect the most recent 1000 Weibo (including Weibo and retweeting) and get a total of 1 billion Weibo data. The data set screened 1 billion Weibo according to Weibo retweeting and user behavior, and selected 300,000 Weibo data that are meaningful for research. Each data set contains the original Weibo data and all its retweeting, and each message is retweeted 80 times on average. Then, four types of events are determined, the initial message set of each event is filtered, and the text message collection is traversed according to the Weibo ID to obtain interaction information. After preprocessing, the statistics of data are shown in Table 1:

Table 1. Training datasets

Dataset	User number	Edge number	Log number
Technology	242681	2858047	399892
Education	393031	5030796	628851
National defense	336799	4468093	519617
Entertainment	453862	8175471	857914

In order to train the model, this chapter divides the data set into three parts: training set, verification set and test set, with a ratio of 7:1:2. We plotted the frequency

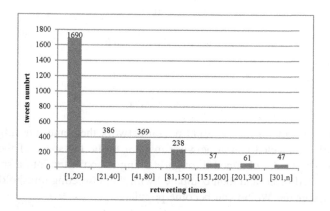

Fig. 6. Weibo test datasets frequency distribution histogram

distribution histogram for the test set, as shown in Fig. 6. We sorted Weibo by the number of retweeting and selected the top 10% as a hot topic.

In the experiment, the data processing algorithm is written in Python and executed in PyCharm. The computer processor used is an Intel® CoreTM i7 CPU with 64 GB of RAM, 2 TB of hard drive and Windows 10 operating system.

4.2 Evaluation Criteria

This experiment defines hot topic prediction in social networks as a binary classification problem, citing the precision and the recall in the confusion matrix to evaluate the PHT-US model experimental results as F_1-score. After determining the four parts of the confusion matrix, we divide the evaluation of the model into actual real values and predicted values. In the process of information propagation, the action log depicts the behavior of the user at a certain moment, so we call it the actual real value; and the predicted value is the user behavior prediction result obtained by training the RNN model with the real datasets. For the relationship between predicted and true values, we give a more intuitive definition: event occurrence is defined as 1; event does not occur as defined as 0. Furthermore, we get four permutations and combine them into a two-dimensional matrix as shown in Table 2:

Table 2. Confusion matrix

		Truth	
		1	0
Estimate	1	TP	FP
	0	FN	TN

The confusion matrix can be used to describe the relationship between the four cases more accurately, and then obtain the common machine learning and model evaluation indicators such as the precision (P) and the recall (R). Therefore, we used the precision (P), the recall (R) and F_1-score to evaluate the results of the PHT-US model experiment.

$$P = \frac{TP}{TP + FP} \tag{10}$$

Precision (P): is a measure of accuracy, which is the proportion of messages that are predicted to be popular in the experiment and is actually a popular Weibo, which itself is for the prediction result of the algorithm.

$$R = \frac{TP}{TP + FN} \tag{11}$$

Recall (R): is a measure of coverage, indicating how much of the actual popular Weibo is predicted to be a popular Weibo in the experiment. It is itself for the real sample, it shows how many positive examples in the sample are accurately predicted.

$$F_1 = \frac{2PR}{P+R} \tag{12}$$

F_1-Score: As a comprehensive indicator, it is to comprehensively evaluate the accuracy of a model prediction in order to balance the impact of the precision and the recall.

4.3   Comparison Models

In this section, we use the Sina Weibo dataset to verify the predictive effect of the model on hot topics by setting up different contrast experiments. Before that, we first studied the effect of the retweeting threshold on the accuracy of model prediction. The time vector and the sentiment vector are input as features, and different retweeting thresholds are set to train the model. The experimental results are evaluated by F_1-score, and the curve of F_1-score with the retweeting threshold is obtained.

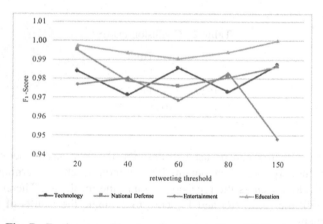

Fig. 7. F_1 changes with retweeting threshold on Weibo datasets

As can be seen from Fig. 7, as the set retweeting threshold increases, F_1-Socre fluctuates up and down, that is, the experimental result is affected by the retweeting threshold. At the same time, the types of data sets are different, and the experimental results generated are different. After the retweeting threshold increased to 80, the F_1-score dropped significantly. This is because in the test set, the probability distribution of the number of retweeting times of the message is concentrated in a small retweeting interval, and when the retweeting threshold is too large, part of its popular information with a small number of pre-transmissions are ignored. Combined with Fig. 2, we set the retweeting threshold $E = 40$ for subsequent experiments.

Table 3. Model evaluation on Weibo dataset – with time vector and sentiment vector

	Technology dataset			National defense dataset			Education dataset			Entertainment dataset		
	P	R	F_1	P	R	F_1	P	R	F_1	P	R	F_1
PHT-US-LSTM	0.990	1	0.994	0.957	1	0.978	0.976	1	0.987	0.987	1	0.993
PHT-US-GRU	0.958	0.985	0.971	0.996	0.962	0.978	0.965	0.996	0.988	0.987	1	0.993

We selected the RNN-GRU and RNN-LSTM neural networks, input time and sentiment vectors, and train the four types of Weibo datasets to obtain the comparison results of the two models, as shown in Table 3.

We use the sentiment vector as the user vector, sort the action log by the periodicity of time, divide the action log into the training set, the intermediate set and the test set according to the ratio of 7:1:2, and use the action log to integrate the time vector to perform randomization. Walk around and get the sequence. As shown in Table 3, the Precision index of the PHT-US-GRU model is 0.97 on average, and the F_1-Score is above 0.95. The overall PHT-US-LSTM index is less than or equal to the PHT-US-GRU model. Therefore, the overall prediction result of the PHT-US-GRU model is better than the PHT-US-LSTM model.

Next, we select the DeepCas [1] model to set up a comparison experiment to detect the predictive performance of the PHT-US-GRU model on hot topics. The results are shown in Table 4.

Table 4. Model comparison evaluation on Weibo Technology dataset-DeepCas and PHT-US

	Precision	Recall	F_1
DeepCas-GRU	0.732	0.971	0.833
PHT-US-GRU	0.958	0.985	0.971

Table 5. Model comparison evaluation on Weibo datasets with sentiment vector

	Technology datasets			National defense datasets		
	Precision	Recall	F_1	Precision	Recall	F_1
PHT-US-LSTM	0.980	0.980	0.970	1	0.952	0.975
PHT-US-GRU	1	0.928	0.962	0.983	0.989	0.985

Finally, only the sentiment vector is used as a feature for comparison experiments. The experimental results are shown in Table 5. It can be seen that the influence of the sentiment vector on the prediction is different from the influence of the sentiment vector and the time vector on the prediction, but the difference is not large. The PHT-US-GRU model and the user sentiment vector optimal effect F_1-score reached 0.978, and the performance displayed on the two data sets was basically the same. Experiments show that the effect of using the time vector is slightly better than using the graph structure to obtain the propagation sequence.

5   Conclusion

Researches on information propagation in social networks is one of the hotspots in the field of data mining field. This work proposes the assumption of hot topic prediction and implements the corresponding prediction model PHT-US. Through extracting the user's sentiments from the huge Weibo text data, embedding the time vector, and then training the RNN to predict the target topic's propagation cascade in the future, in order to judge the hot trend of the target topic in the future. Compared with other models, the PHT-US model has good data adaptability. Secondly, according to the characteristics of neural network, the method of time vectorization and user sentiment vectorization has been improved, which depend on the adopted way to embed time and user sentiment. The time periodicity and user sentiment are reasonably integrated into the model, which expanding the diversity of the model input. The experimental results on the Sina Weibo datasets show that the PHT-US model can effectively predict whether the target topic could become popular in the future.

Acknowledgment. This work was supported by the National Natural Science Foundation of China (No. 61602159), the Natural Science Foundation of Heilongjiang Province (No. F201430), the Innovation Talents Project of Science and Technology Bureau of Harbin (No. 2017RAQXJ094), and the fundamental research funds of universities in Heilongjiang Province, special fund of Heilongjiang University (No. HDJCCX-201608).

References

1. Bikhchandani, S., Hirshleifer, D., Welch, I.: A theory of fads, fashion, custom, and cultural change as informational cascades. J. Polit. Econ. **100**(5), 992–1026 (1992)
2. Xian-Yill, C., Ling-ling, Z., Qian, Z., et al.: The framework of network public opinion monitor-ing and analyzing system based on semantic content identification. J. Convergence Inf. Technol. **5**(10), 1–5 (2010)
3. Lipton, Z.C., Berkowitz, J., Elkan, C.: A critical review of recurrent neural networks for sequence learning. arXiv preprint arXiv:1506.00019 (2015)
4. Li, C., Guo, X., Mei, Q.: Joint modeling of text and networks for cascade prediction. In: Twelfth International AAAI Conference on Web and Social Media (2018)
5. Hochreiter, S., Schmidhuber, J.: Long short-term memory. Neural Comput. **9**(8), 1735–1780 (1997)
6. Hong, L., Dan, O., Davison, B.D.: Predicting popular messages in Twitter. In: Proceedings of the 20th International Conference Companion on World Wide Web, pp. 57–58. ACM (2011)
7. Suh, B., Hong, L., Pirolli, P., et al.: Want to be retweeted? Large scale analytics on factors impacting retweet in twitter network. In: 2010 IEEE Second International Conference on Social Computing, pp. 177–184. IEEE (2010)
8. Shen, H., Wang, D., Song, C., et al.: Modeling and predicting popularity dynamics via reinforced poisson processes. In: Twenty-Eighth AAAI Conference on Artificial Intelligence (2014)
9. Crane, R., Sornette, D.: Robust dynamic classes revealed by measuring the response function of a social system. Proc. Natl. Acad. Sci. **105**(41), 15649–15653 (2008)

10. Ulrich, R., Miller, J.: Information processing models generating lognormally distributed reaction times. J. Math. Psychol. **37**(4), 513–525 (1993)
11. Wang, D., Song, C., Barabási, A.L.: Quantifying long-term scientific impact. Science **342** (6154), 127–132 (2013)
12. Bao, P., Shen, H.W., Jin, X., et al.: Modeling and predicting popularity dynamics of microblogs using self-excited hawkes processes. In: Proceedings of the 24th International Conference on World Wide Web, pp. 9–10. ACM (2015)
13. Li, C., Ma, J., Guo, X., et al.: DeepCas: an end-to-end predictor of information cascades. In: Proceedings of the 26th International Conference on World Wide Web, International World Wide Web Conferences Steering Committee, pp. 577–586 (2017)
14. Bollen, J., Mao, H., Pepe, A.: Modeling public mood and emotion: Twitter sentiment and socio-economic phenomena. In: Fifth International AAAI Conference on Web-Logs and Social Media (2011)
15. Wu, B., Shen, H.: Analyzing and predicting news popularity on Twitter. Int. J. Inf. Manag. **35**(6), 702–711 (2015)
16. Kamkarhaghighi, M., Chepurna, I., Aghababaei, S., et al.: Discovering credible Twitter users in stock market domain. In: 2016 IEEE/WIC/ACM International Conference on Web Intelligence (WI), pp. 66–72. IEEE (2016)
17. Chen, P., Liu, S.: An improved DAG-SVM for multi-class classification. In: 2009 Fifth International Conference on Natural Computation, vol. 1, pp. 460–462. IEEE (2009)

Research on Cross-Language Retrieval Using Bilingual Word Vectors in Different Languages

Yulong Li and Dong Zhou[✉]

School of Computer Science and Engineering,
Hunan University of Science and Technology, Xiangtan 411201, Hunan, China
liyulong198@gmail.com, dongzhou1979@hotmail.com

Abstract. Bilingual word vectors have been exploited a lot in cross-language information retrieval research. However, most of the research is currently focused on similar language pairs. There are very few studies exploring the impact of using bilingual word vectors for cross-language information retrieval in long-distance language pairs. In this paper, it systematically analyzes the retrieval performance of various European languages (English, German, Italian, French, Finnish, Dutch) as well as Asian languages (Chinese, Japanese) in the adhoc task of CLEF 2002–2003 campaign. Genetic proximity was used to visually represent the relationships between languages and compare their cross-lingual retrieval performance in various settings. The results show that the differences in language vocabulary would dramatically affect the retrieval performance. At the same time, the term by term translation retrieval method performs slightly better than the simple vector addition retrieval methods. It proves that the translation-based retrieval model can still maintain its advantage under the new semantic scheme.

Keywords: Cross-language information retrieval · Bilingual word embedding · Genetic proximity · Language pairs

1 Introduction

With the popularity of the Internet and the acceleration of globalization, people are increasingly interested in information written in different languages. Monolingual retrieval is certainly not enough to meet the daily needs of users nowadays [1, 2]. Cross-language Information Retrieval (CLIR) has received more and more attention. This technique was first proposed by Salton and Gerard [3], referring to the process of constructing queries in one language and retrieving documents in other languages. Although the study of CLIR has been in the academic world for more than 50 years, the retrieval effect is still far from satisfactory. One of the main reasons is that the traditional information retrieval methods are based on sparse document representation [4]. Due to differences between different languages, there is a huge vocabulary gap that leads to poor performance in CLIR. Another reason is that the digital resources of languages are not equal. Most of CLIR studies focus on resource-equal language pairs or closely related language pairs, there are few studies on comparing CLIR using different language pairs.

© Springer Nature Singapore Pte Ltd. 2019
X. Cheng et al. (Eds.): ICPCSEE 2019, CCIS 1058, pp. 454–465, 2019.
https://doi.org/10.1007/978-981-15-0118-0_35

At present, the research on CLIR methods mainly focus on the following two aspects [1]. One is translation-based retrieval [5, 6]. It can be divided into three categories. (1) Translating queries into target language [7]. (2) Translating documents into source language [8]. (3) Translating both queries and documents into an intermediate language [9]. Another is based on the semantic matching [10, 11]. This type of methods maps queries and documents into a semantic space for retrieval. Translation-based methods require bilingual texts and/or other bilingual signals as the basis. Although it performs well under the resource-rich languages, it is not satisfactory in resource-lean languages. In contrast, the semantic-based methods do not need the process of translation, thus avoiding the errors introduced by translation. It retains the semantics of user requirements and retrieving documents to a greater extend, making the retrieval results more accurate. In recent years, semantic matching based retrieval models have received a lot of attention, in particular with the development of cross-language word vectors [12].

The study of cross-Lingual word vectors [13] can be divided into two categories: supervised training [14] and unsupervised training [15]. Supervised training can be further divided into three categories according to the different granularities of training corpus. (1) Word-based level [14]. (2) Sentence-based level [16]. (3) Document-based level [17]. Compared with supervised training, unsupervised training can obtain cross-language word vectors without any parallel or comparable corpora. However, due to the large amount of computation during training, it cannot be used under large-scale word vectors at present On the contrast, supervised training methods are more effective in similar language pairs, the performance is much worse in the long-distance language pairs [18].

Correlation between language pairs can be analyzed in several ways. Nerbonne and Hinrichs [19] classified language relevance by pronunciation, syntax, and semantics. Litschko et al. [4] distinguished the relationship between language pairs by language family, but they had not given a clear definition, and their experiments only focused on three language pairs. The study of CLIR has been moved to more language pairs, it is necessary to have a more specific representation of the relationships between languages [20].

Although there are literatures studying the performance of using bilingual language vectors in different languages in CLIR [21], they only compared a small number of language pairs [22], most of which only use the bilingual word vectors in Machine Translation Task. Few researchers compared similar language pairs with long-distance language pairs [23]. Our research explores bilingual word vectors in cross-language retrieval for different language pairs [18] (closely-related language pairs and long-distance language pairs). At the same time, the genetic proximity in languages is used to quantitatively measure the distance between languages [24].

2 Related Work

2.1 Bilingual Word Vectors

Mikolov et al. [25] proposed the most popular word2vec model. It was then found that although the word vectors in different languages are independent, the distribution patterns are very similar. It is then found that the word vector representation of the source language can be mapped to the word vector representation of the target language by a linear transformation. This linear transformation can be obtained by using a known bilingual dictionary [26]. The bilingual word vectors obtained 10%–20% improvements in the machine translation task compared to the previous methods [27].

Faruqui and Dyer [28] pointed out that linear discriminant analysis can solve the problem of learning mapping between two word vectors in different languages. The resulting bilingual word vectors improve the accuracy of machine translation by 10%.

Vulic and Moens [29] proposed a bilingual word vectors training method by using comparable Wikipedia[1] corpora. It solves the problem of training bilingual word vectors with insufficient resources. The bilingual word vectors they obtained achieved good performance in the English-Dutch retrieval by using the CLEF 2002 and 2003 test collections[2]. In particular, results in both monolingual and cross-language retrieval are higher than those obtained by the bilingual topic models [30] and the traditional language models.

Smith et al. [14] trained bilingual word vectors by using bilingual dictionary as others before [26], but they proved that the linear mapping between two monolingual word vectors should be orthogonal. Based on that truth, they could use Singular Value Decomposition (SVD) to get the mapping matrix. The experimental results show that their methods performed better than Canonical Correlation Analysis (CCA) [28], which had gotten higher accuracy as well as the method proposed by Mikolov et al. [26].

Conneau et al. [15] proposed a method of training bilingual word vectors without any bilingual dictionary and parallel corpora. They first used the generative adversarial networks [31] as the training model to get the ideal linear mapping, then a lot of word pairs could be induced by looking for the nearest words between two languages. Finally they finished their training steps by using the method from Smith et al.'s paper [14]. The experimental results showed that the cross-language word vectors they trained in the word translation task have exceeded the accuracy obtained by using supervised training method, which has a milestone significance.

Litschko et al. [4] proposed a cross-language retrieval framework that does not require any bilingual data. The vectors of queries and documents are constructed by mapping them into the readily trained bilingual word vectors. The relevance between queries and documents are acquired according to the vector similarities. They conducted cross-language retrieval in the CLEF test collections by using English-Dutch, English-Italian and English-Finnish language pairs. They found that cross-language search that does not require any bilingual training data can achieve good performance.

2.2 Language Similarity

Nerbonne and Hinrichs [19] systematically analyzed several common concepts of similarity between languages, including pronunciation, syntax, and semantics. In terms of pronunciation, Laver and John [32] proposed the basic concept of "phonetic

---

[1] http://www.wikipedia.org.

[2] http://www.clef-initiative.eu/.

similarity" to measure the pronunciation differences between different languages. Miller and Nicely [33] established a more complex analysis to construct a segmental aggregation confusion matrix. Albright and Hayes [34] proposed a speech learning model based on "minimal generalization". In terms of syntax, Comrie [35] proposed syntactic typology as an important field of linguistics, aiming to identify interrelated grammatical features in all languages. Homola and Kubon [36] used the relationships in the dependency tree to analyze the syntax, and proposed to use the edit distance to simulate the degree of syntactic differences. In terms of semantics, Firth [37] proposed that the meaning of a word should be related to its adjacent word. Zesch et al. [38] proposed that the judgment of semantic similarity could be judged by the similarity between pairs of words. Dridan and Bond [39] used a recursive semantic framework to obtain a more appropriate method for measuring the similarity of sentences.

Vincent proposes a sophisticated model for comparison between languages[3]. The criteria for comparison are based on the basic vocabularies between languages and the ability to generate an automatic language classification of language and sublanguage systems. Finally, they use genetic proximity to express the similarities between languages.

3   Experiments

3.1   Experimental Data

Our experiments used the CLEF 2002 and 2003 test collections. The document set includes approximately one million news reports. The statistics of documents, non-duplicate words and languages used in the experiment are listed in Table 1.

For bilingual dictionaries we use the same method as in Smith et al.'s paper [14] to extract 100,000 pairs of words from Google Translate[4] as a dictionaries to train the supervised bilingual word vectors.

Table 1. Statistics of test collections

Language	Document	Unique words
English	166,753	235,710
Dutch	190,000	29,600,000
French	87,191	479,682
German	225,371	1,670,316
Italian	108,000	17,100,000
Finnish	55,000	9,250,000

---

[3] https://www.elinguistics.net/.

[4] https://translate.google.com/.

3.2   Document and Query Processing

We process documents and queries by a normal procedure including word segmentation, stop words and stemming, etc. We use *Mecab*[5] for Japanese word segmentation, *jieba*[6] for Chinese word segmentation. Based on past experiences, we use the *title* and *description* fields of CLEF topics, documents are selected according to the guideline of CLEF for a specific language.

3.3   Language Pairs and Monolingual Vector

We have selected a number of language pairs, including common European language pairs such as English to German, French, Italian, Dutch, Finnish respectively, and long-distance language pairs such as English to Japanese and English to Chinese. At the same time, we use the fasttext[7] word vectors pre-trained by Facebook[8] research as the monolingual vectors in all of our experiments.

3.4   CLIR Model

We used the same methods in Litschko et al.'s paper [4]. We named the supervised model as CLIR-sup and unsupervised model as CLIR-unsup. The experimental settings are consistent with the Smith et al.'s paper [14] and the Conneau et al.'s paper [15].

3.5   Experimental Runs

Our experiments are divided into two groups. The first group of experiment uses different queries to retrieval English test collections, and the queries' languages are made of German (DE), Italian (IT), French (FR), Dutch (NL), Finnish (FI), Chinese (ZH), Japanese (JP). Another uses English queries to retrieval test collections in different languages, such as German, Italian, French, Dutch, Finnish.

We use the same framework proposed by Litschko et al. [4] as our cross-language search framework. It mainly includes three cross-language retrieval models: BWE-Agg-Add, BWE-Agg-IDF and TbT-QT. Some introductions about these methods are shown as follow:

(1) BWE-Agg-Add, Use bilingual word vectors to replace words that appear in queries and documents, then the vector representation of the queries and documents is the sum of the word vectors.
(2) BWE-Agg-IDF, In the process of adding the word vectors, the inverse document frequency of the word is added.
(3) TbT-QT, Use bilingual word vector translation to query and then search.

---

[5] https://github.com/taku910/mecab.

[6] https://github.com/fxsjy/jieba.

[7] https://fasttext.cc.

[8] https://github.com/rlitschk/UnsupCLIR.

In our article, we will use method1, method2 and method3 instead. We compared six combination of methods, including two models that generate bilingual word vectors (CLIR-sup and CLIR-unsup) and three cross-language search models (method1–method3).

4 Results

First, in order to visually show the differences between different language pairs, we selected the different queries corresponding to a same query id in the CLEF 2003 ad-hoc task, as shown in Table 2. We find that the above-mentioned European language pairs are partially similar in vocabularies, especially in English and Dutch. The Finnish language is slightly different from other European languages, while the vocabularies of Chinese, Japanese and European languages are very different.

Table 2. Different representations of a same CLEF query.

Language	Query
EN	Letter Bomb for Kiesbauer Find information on the explosion of a letter bomb in the studio of the TV channel PRO7 presenter Arabella Kiesbauer.
NL	Bombrief voor Kiesbauer Zoek informatie over de explosie van een bombrief in de studio van presentatrice Arabella Kiesbauer van TV-zender PRO7.
FR	Une lettre piégée pour Arabella Kiesbauer Trouvez des informations sur l'explosion d'une lettre piégée dans le studio de la présentatrice Arabella Kiesbauer de la chaîne de télévision PRO 7.
IT	Lettera Bomba per Kiesbauer Recupera le informazioni relative all'esplosione di una lettera bomba nello studio della presentatrice della rete televisiva PRO7.
DE	Briefbombe für Kiesbauer Finde Informationen über die Explosion einer Briefbombe im Studio der Moderatorin Arabella Kiesbauer beim Fernsehsender PRO7.
FI	Kirjepommi Kiesbauerille Etsi tietoja juontaja Arabella Kiesbauerille osoitetun kirjepommin räjähdyksestä TV-kanava PRO7:n studiossa.
ZH	給Kiesbauer的郵件炸彈找出在PRO7電視台主持人Arabella Kiesbauer身上發生的郵件炸彈爆炸案資訊。
JP	キ〖スバウア〖案ての緘絵曲冏テレビ渡PRO7の皇柴莢アナベラ゛キ〖スバウア〖のスタジオでの緘絵曲冏曲券に簇する擾鼠を玫したい。

In order to more scientifically show the gaps between language pairs, we list the genetic proximity between the language pairs in this experiment according to the comparative criteria of the comparative linguistics website (see Fig. 1).

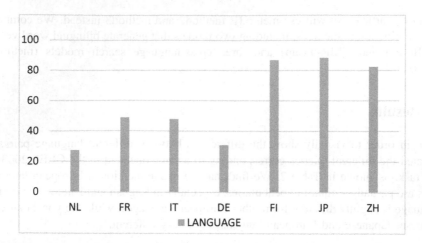

Fig. 1. The x-axis represents different languages, and the y-axis represents the degree of genetic proximity (larges numbers mean bigger differences). The genetic proximity is measured between English and languages presented in x-axis.

The values of Genetic proximity are different, and the correlations between languages are also different. Vincent divides the relationships between languages into five degrees according to the genetic proximity values, as shown in Table 3.

Table 3. Degrees of language similarities.

Genetic proximity	Degree
1–30	Highly related languages
30–50	Related languages
50–70	Remotely related languages
70–78	Very remotely related languages
78–100	No recognizable relationship

Below we show the performance of a number of cross-language search results by using supervised and unsupervised cross-language word vectors in different languages, measured by the mean average accuracy (MAP). The results are shown in Table 4. We can easily find that, when using queries in different languages to retrieve English test collection, no matter which method do we choose, the better performance always belongs to similar languages pairs. The highest scores are obtained in NL-EN language pair by using CLIR-sup embeddings together with method1 and method2. For method3, the highest scores are acquired in FR-EN language pair, which is also a similar language pair.

Similar Language Pairs vs Long-Distance Language Pairs. We use genetic proximity value of 70 as the boundary, defining that values less than 70 are considered to be similar language pairs, and greater than 70 is considered to be long-distance language pairs. As is shown in Fig. 1, it is clearly that these three language pairs {ZH-EN, JP-EN, FI-EN} should belong to long-distance language pairs, and the other language pairs used in our experiment are members of similar language pairs. We find that when use either supervised or supervised bilingual word vectors in the long-distance languages for cross-language retrieval, the results are much worse than those obtained for similar language pairs.

Table 4. X-en results of CLIR.

Emb	Method	Language pair						
		NL-EN	FR-EN	IT-EN	DE-EN	FI-EN	ZH-EN	JP-EN
CLIR-sup	method1	0.159	0.150	0.129	0.154	0.115	0.002	2.836E-04
	method2	0.175	0.158	0.163	0.159	0.116	3.000E-04	1.625E-04
	method3	0.179	0.250	0.209	0.157	0.136	0.022	9.010E-05
CLIR-unsup	method1	0.159	0.167	0.136	0.123	0.105	2.521E-04	1.146E-04
	method2	0.180	0.210	0.201	0.138	0.119	2.010E-04	1.312E-04
	method3	0.175	0.232	0.201	0.210	0.135	2.238E-02	6.102E-05

method3 vs method1&method2. It is shown that method3 performs better than method1 and method2, no matter by using supervised bilingual word vectors and unsupervised bilingual word vectors. This proves that the translation retrieval model based on bilingual word vectors is still effect in the new semantic space.

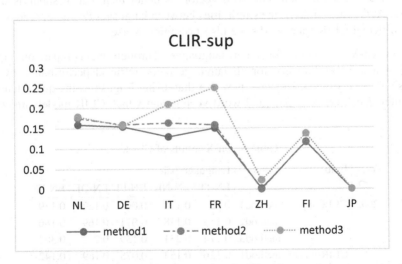

Fig. 2. Performance of using supervised bilingual embeddings in cross-language retrieval under different language queries

In order to better demonstrate the relationships between language pairs and cross-language retrieval, we visualize the data in Figs. 2 and 3. The x-axis represents different languages. We have followed their language relevance from English to the left. The right is arranged in order, and the y-axis represents the retrieved MAP value.

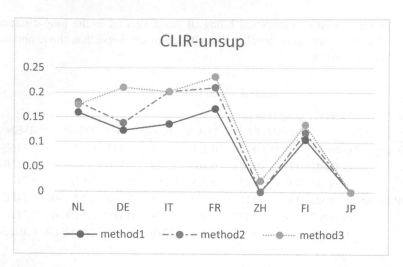

Fig. 3. Performance of using unsupervised embedding in cross-language retrieval under different language queries

CLIR-sup vs CLIR-unsup. We find that under different language pairs, the performance of unsupervised bilingual word vectors is better than that of supervised word vectors. At the same time, as the similarities between languages decrease, search results under different CLIR frameworks are also becoming worse.

X-en vs en-X. X stands for other languages. Through the comparison between Tables 4 and 5, we find that for all language pairs, retrieval performance by using English as the document language is worse than using English as the query language, no matter which types of bilingual word vectors and which CLIR models utilized.

Table 5. en-X results of CLIR.

Year	Emb	Method	Language pair				
			EN-IT	EN-NL	EN-FI	EN-DE	EN-FR
2002	CLIR-sup	method1	0.12	0.138	0.06	0.132	0.139
		method2	0.152	0.178	0.071	0.169	0.186
		method3	0.274	0.231	0.139	0.222	0.303
	CLIR-unsup	method1	0.126	0.153	0.078	0.169	0.142
		method2	0.161	0.203	0.102	0.208	0.203
		method3	0.256	0.257	0.144	0.227	0.32
2003	CLIR-sup	method1	0.202	0.184	0.21	0.18	0.159
		method2	0.233	0.226	0.197	0.221	0.215
		method3	0.32	0.256	0.185	0.224	0.327
	CLIR-unsup	method1	0.212	0.197	0.239	0.214	0.199
		method2	0.253	0.25	0.223	0.238	0.251
		method3	0.32	0.299	0.242	0.25	0.337

5   Conclusions

In this paper we have systematically compared the CLIR retrieval methods by using bilingual word vectors in different language pairs. We used genetic proximity as a measure for the correlation between languages. Finally, we find that when use either supervised or unsupervised bilingual word vectors under similar language pairs for cross-lingual retrieval, the results are always better than those obtained for long-distance language pairs. At the same time, we have also shown the effect using term by term CLIR method under different language pairs. No matter which bilingual word vectors we choose, this method always performs better than the other two methods. We would like to try to use some complex neural networks combining with a new bilingual word vector method in the future, so the documents' and queries' semantics could be detected completely. In addition, How to improve the performance of bilingual word vectors in long-distance language pairs is also a problem worthy of further study.

Acknowledgement. The work described in this paper was supported by National Natural Science Foundation of China under Project No. 61876062, Scientific Research Fund of Hunan Provincial Education Department of China under Project No. 16K030, Hunan Provincial Natural Science Foundation of China under Project No. 2017JJ2101, Hunan Provincial Innovation Foundation for Postgraduate under Project No. CX2018B671.

References

1. Sharma, V.K., Mittal, N.: Cross lingual information retrieval (CLIR): review of tools, challenges and translation approaches. In: Satapathy, S.C., Mandal, J.K., Udgata, S.K., Bhateja, V. (eds.) Information Systems Design and Intelligent Applications. AISC, vol. 433, pp. 699–708. Springer, New Delhi (2016). https://doi.org/10.1007/978-81-322-2755-7_72
2. Hajič, J., Homola, P., Kuboň, V.: A simple multilingual machine translation system. In: Proceedings of the MT Summit IX, pp. 157–164 (2016)
3. Salton, G.: Experiments in multi-lingual information retrieval. Cornell University (1972)
4. Litschko, R., Glavaš, G., Ponzetto, S.P., Vulić, I.: Unsupervised cross-lingual information retrieval using monolingual data only. arXiv preprint arXiv:1805.00879 (2018)
5. Zhou, D., Truran, M., Brailsford, T., Wade, V., Ashman, H.: Translation techniques in cross-language information retrieval. ACM Comput. Surv. **45**, 1 (2012)
6. Zhou, D., Lawless, S., Wu, X., Zhao, W., Liu, J.: A study of user profile representation for personalized cross-language information retrieval. Aslib J. Inf. Manag. **68**, 448–477 (2016)
7. Gao, J., Nie, J.-Y., Xun, E., Zhang, J., Zhou, M., Huang, C.: Improving query translation for cross-language information retrieval using statistical models. In: Proceedings of the 24th Annual International ACM SIGIR Conference on Research and Development in Information Retrieval, pp. 96–104. ACM (2001)
8. Oard, D.W.: A comparative study of query and document translation for cross-language information retrieval. In: Farwell, D., Gerber, L., Hovy, E. (eds.) AMTA 1998. LNCS (LNAI), vol. 1529, pp. 472–483. Springer, Heidelberg (1998). https://doi.org/10.1007/3-540-49478-2_42
9. Gollins, T., Sanderson, M.: Improving cross language retrieval with triangulated translation. In: Proceedings of the 24th Annual International ACM SIGIR Conference on Research and Development in Information Retrieval, pp. 90–95. ACM (2001)

10. Zhou, D., Wu, X., Zhao, W., Lawless, S., Liu, J., Engineering, D.: Query expansion with enriched user profiles for personalized search utilizing folksonomy data. IEEE Trans. Knowl. **29**, 1536–1548 (2017)
11. Zhou, D., Zhao, W., Wu, X., Lawless, S., Liu, J.: An iterative method for personalized results adaptation in cross-language search. Inf. Sci. **430**, 200–215 (2018)
12. Vulić, I., De Smet, W., Moens, M.-F.: Cross-language information retrieval with latent topic models trained on a comparable corpus. In: Salem, M.V.M., Shaalan, K., Oroumchian, F., Shakery, A., Khelalfa, H. (eds.) AIRS 2011. LNCS, vol. 7097, pp. 37–48. Springer, Heidelberg (2011). https://doi.org/10.1007/978-3-642-25631-8_4
13. Ruder, S., Vulić, I., Søgaard, A.: A survey of cross-lingual word embedding models. arXiv preprint arXiv:1706.04902 (2017)
14. Smith, S.L., Turban, D.H., Hamblin, S., Hammerla, N.Y.: Offline bilingual word vectors, orthogonal transformations and the inverted softmax. arXiv preprint arXiv:1702.03859 (2017)
15. Conneau, A., Lample, G., Ranzato, M.A., Denoyer, L., Jégou, H.: Word translation without parallel data. arXiv preprint arXiv:1710.04087 (2017)
16. Zou, W.Y., Socher, R., Cer, D., Manning, C.D.: Bilingual word embeddings for phrase-based machine translation. In: Proceedings of the 2013 Conference on Empirical Methods in Natural Language Processing, pp. 1393–1398 (2013)
17. Vulić, I., Moens, M.-F.: Bilingual distributed word representations from document-aligned comparable data. J. Artif. Intell. Res. **55**, 953–994 (2016)
18. Adams, O., Makarucha, A., Neubig, G., Bird, S., Cohn, T.: Cross-lingual word embeddings for low-resource language modeling. In: Proceedings of the 15th Conference of the European Chapter of the Association for Computational Linguistics: Volume 1, Long Papers, pp. 937–947 (2017)
19. Nerbonne, J., Hinrichs, E.: Linguistic distances. In: Proceedings of the Workshop on Linguistic Distances, pp. 1–6 (2006)
20. Lazaridou, A., Dinu, G., Baroni, M.: Hubness and pollution: delving into cross-space mapping for zero-shot learning. In: Proceedings of the 53rd Annual Meeting of the Association for Computational Linguistics and the 7th International Joint Conference on Natural Language Processing (Volume 1: Long Papers), pp. 270–280 (2015)
21. Levy, O., Søgaard, A., Goldberg, Y.: A strong baseline for learning cross-lingual word embeddings from sentence alignments. arXiv preprint arXiv:1608.05426 (2016)
22. Pan, X., Zhang, B., May, J., Nothman, J., Knight, K., Ji, H.: Cross-lingual name tagging and linking for 282 languages. In: Proceedings of the 55th Annual Meeting of the Association for Computational Linguistics (Volume 1: Long Papers), pp. 1946–1958 (2017)
23. Upadhyay, S., Faruqui, M., Dyer, C., Roth, D.: Cross-lingual models of word embeddings: an empirical comparison. arXiv preprint arXiv:1604.00425 (2016)
24. Song, Y., Upadhyay, S., Peng, H., Roth, D.: Cross-lingual dataless classification for many languages. In: IJCAI, pp. 2901–2907 (2016)
25. Mikolov, T., Sutskever, I., Chen, K., Corrado, G.S., Dean, J.: Distributed representations of words and phrases and their compositionality. In: Advances in Neural Information Processing Systems, pp. 3111–3119 (2013)
26. Mikolov, T., Le, Q.V., Sutskever, I.: Exploiting similarities among languages for machine translation. arXiv preprint arXiv:1908.00879 (2013)
27. Haghighi, A., Liang, P., Berg-Kirkpatrick, T., Klein, D.: Learning bilingual lexicons from monolingual corpora. In: Proceedings of ACL-2008: Hlt, pp. 771–779 (2008)
28. Faruqui, M., Dyer, C.: Improving vector space word representations using multilingual correlation. In: Proceedings of the 14th Conference of the European Chapter of the Association for Computational Linguistics, pp. 462–471 (2008)

29. Vulić, I., Moens, M.-F.: Monolingual and cross-lingual information retrieval models based on (bilingual) word embeddings. In: Proceedings of the 38th International ACM SIGIR Conference on Research and Development in Information Retrieval, pp. 363–372. ACM (2015)
30. Vulić, I., De Smet, W., Moens, M.-F.: Cross-language information retrieval models based on latent topic models trained with document-aligned comparable corpora. Inf. Retrieval **16**, 331–368 (2013)
31. Goodfellow, I., et al.: Generative adversarial nets. In: Advances in Neural Information Processing Systems, pp. 2672–2680 (2014)
32. Laver, J., John, L.: Principles of Phonetics. Cambridge University Press, Cambridge (1994)
33. Miller, G.A., Nicely, P.E.: An analysis of perceptual confusions among some English consonants. J. Acoust. Soc. Am. **27**, 338–352 (1955)
34. Albright, A., Hayes, B.: Rules vs. analogy in English past tenses: a computational/ experimental study. Cognition **90**, 119–161 (2003)
35. Comrie, B.: Language Universals and Linguistic Typology: Syntax and Morphology. University of Chicago Press, Chicago (1989)
36. Homola, P., Kubon, V.: A translation model for languages of acceding countries. In: Broadening Horizons of Machine Translation and its Applications. Proceedings of the Ninth EAMT workshop, Foundation for International Studies, University of Malta, Valletta, pp. 90–97 (2004)
37. Firth, J.R.: Selected Papers of JR Firth, pp. 1952–1959. Indiana University Press (1968)
38. Zesch, T., Müller, C., Gurevych, I.: Extracting lexical semantic knowledge from Wikipedia and wiktionary. In: LREC, pp. 1646–1652 (1968)
39. Dridan, R., Bond, F.: Sentence comparison using robust minimal recursion semantics and an ontology. In: Proceedings of the Workshop on Linguistic Distances, pp. 35–42. Association for Computational Linguistics (2006)

Word Segmentation for Chinese Judicial Documents

Linxia Yao[1], Jidong Ge[1(✉)], Chuanyi Li[1], Yuan Yao[1], Zhenhao Li[1], Jin Zeng[1], Bin Luo[1], and Victor Chang[2,3]

[1] State Key Laboratory for Novel Software Technology, Nanjing University, Nanjing, China
gjdnju@163.com
[2] International Business School Suzhou, Xi'an Jiaotong-Liverpool University, Suzhou, China
[3] Research Institute of Big Data Analytics, Xi'an Jiaotong-Liverpool University, Suzhou, China

Abstract. Word segmentation is an integral step in many knowledge discovery applications. However, existing word segmentation methods have problems when applying to Chinese judicial documents: (1) existing methods rely on large-scale labeled data which is typically unavailable in judicial documents, and (2) judicial document has its own language features and writing formats. In this paper, a word segmentation method is proposed for Chinese judicial documents. The proposed method consists of two steps: (1) automatically generating some labeled data as legal dictionaries, and (2) applying a hybrid multi-layer neural networks to do word segmentation incorporating legal dictionaries. Experiments are conducted on a dataset of Chinese judicial documents showing that the proposed model can achieve better results than the existing methods.

Keywords: Chinese word segmentation · Knowledge discovery · Judicial documents

1 Introduction

In 2013, China Judgement Online System have been open to the public. Until now, it has already included more than 49 million electronic judicial documents and has become the largest sharing website about judicial documents in the world, making knowledge discovery under legal big data become possible. Word segmentation of Chinese judicial documents is an essential pre-processing step for the upper application of natural language processing in Chinese judicial documents. Nowadays, there is no specific word segmentation methods on Chinese judicial documents. Instead, researchers use mainstream methods and tools for Chinese word segmentation, which have good results on common documents but perform less desirably on Chinese judicial documents. The latter has its own language features and writing formats, which involves large scale legal terms that general word segmentation method cannot recognize.

© Springer Nature Singapore Pte Ltd. 2019
X. Cheng et al. (Eds.): ICPCSEE 2019, CCIS 1058, pp. 466–478, 2019.
https://doi.org/10.1007/978-981-15-0118-0_36

In this paper, Our main contributions lie in:

(1) Propose Model I that can generate a corresponding legal dictionary. Which includes the LawRank based on the improved PageRank and can solve the current lack of labeled data and can extract legal related keywords as legal dictionary.
(2) Propose a template creation rule under the study of linguistic structure of judicial documents. Correspondingly, dict feature template was created based on the legal dictionary generated from model I. POS-tag feature templates was created based on the statistics on part of speech of documents.
(3) Propose Model II that can integrate the templates created for each word into hybrid multi-layer neural networks based on its language's own semantic rules to do word segmentation. Which can handle legally unregistered words.
(4) Extensive experiments on large scale judicial documents are conducted to compare the proposed method with state-of-the-art methods, and the results show that the proposed method significantly outperforms the competitors in terms of F-measure.

2 Related Work

Initially, Chinese word segmentation was regarded as a linguistics task. Recently, most existing methods based on the statistical methods relied on large-scale corpus have become the de facto standard. In statistical methods, we usually need to manually label each word or each character in a given text or sentence [1]. Based on these labels, various models have been used, such as maximum entropy [2], SVM [3], HMM [4], and Condition Random Fields (CRF) [5].

Recently, methods on machine learning [6, 7], especially deep learning [16–18], have been applied to Chinese word segmentation [21, 22] and achieved remarkable results. At the same time, there are many word segmentation methods that combine the above two or more methods [19, 20] and have achieved excellent results on different data sets [8, 9]. The statistical method is based on the frequency and other characteristics of statistical words to do knowledge discoveries and special treatments are needed for different scenes and applications, which is used as a basis for word segmentation [10, 11]. The word segmentation method of machine learning has been hotter with the increase of text datasets and the rise of machine learning algorithms in recent years [14, 15].

3 Methods

For the problems on word segmentation for Chinese judicial documents mentioned above, corresponding methods were proposed to solve these problems. The workflow of the methods can be shown in the Fig. 1.

For the problem 1: existing methods rely on large-scale labeled data which is typically unavailable in judicial documents. Model I was proposed to generate a corresponding legal dictionary. Which includes the LawRank based on the improved PageRank algorithm. By calculating the LawRank value of each candidate word, after PNS and GMS, the initial word segmentation results can be obtained. Then through the information entropy and mutual information, the legal related keywords can be obtained as legal dictionary.

For the problem 2: judicial document has its own language features and writing formats involving. For each word in the documents, two different templates were proposed. The Pos-tag feature template was based on the statistics on part of speech of documents. Dict feature template was created based on the legal dictionary generated from model I. Then combining the two templates, a ten-dimensional feature vector for each word is generated. This vector incorporates the linguistic information of the legal language and will play a large role in our Model II.

For the problem 3: existing methods or tools for Chinese word segmentation, which have good results on common documents but perform less desirably on Chinese judicial documents. In resent research, the model Bi-LSTM-CRF has a good effect on word segmentation. Hence, we also used Bi-LSTM-CRF as our network layers. The difference is that we use two Bi-LSTMs, which are called MainLSTM and SuperLSTM. Also, we integrate the legal lexicon into neural networks. The input of SuperLSTM is the ten-dimensional feature vector constructed which includes the semantic information of the judicial document. The input of MainLSTM is the word embedding based on the toolkit Word2Vec [13].

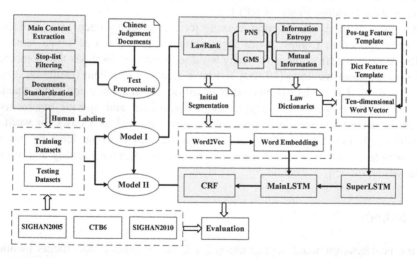

Fig. 1. Overview of the workflow of this paper, gray sections are the technology designed by each module, the dotted areas are the details involved

3.1   Model I

LawRank Layer. In this paper, different from original PageRank algorithm [23], improvements have been made to apply to text segmentation on legal field, which we called LawRank. The specific algorithm is as follows.

$$R(u) = c_1 \sum_{v \in B_u} \frac{R(v)}{N_v} + c_2 \sum_{v \in F_u} \frac{R(v)}{M_v} \tag{1}$$

$$R = c_1 \left(R^T A \right)^T + c_2 A R \tag{2}$$

Where u represents a potential word node, F_u represents a collection of potential words pointed to by u, and B_u represents a collection of potential words pointing to u. c_1, c_2 are feedback coefficients that are used to balance LawRank length. If c_1 is larger, the long-term LawRank value is relatively high. For a relational matrix A (m is the number of potential words w), if there is an edge from u to $v(u \neq v)$, $A_{u,v} = 1$ otherwise $A_{u,v} = 0$.

Positive and Negative Segmentation (PNS). Positive and negative segmentation sets up a sliding window of size t. We cut the candidate word's left side and right side, and slide the sliding window to the right of the candidate word in the right direction. In contrast, backward matching is the sliding window sliding from the right side to the left side, the example see Fig. 2.

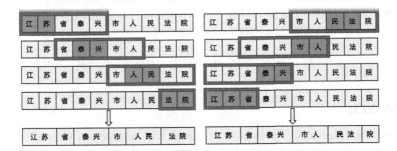

Fig. 2. Example of "江苏省泰兴市人民法院" (People's Court of Taixing City, Jiangsu Province). The left is positive segmentation results, and the right is negative segmentation results. The orange box in the figure represents the sliding window, and the green section represents the candidate with the largest LawRank value in the sliding window. (Color figure online)

Global Maximum Segmentation (GMS). The next part is to compare the results obtained from positive and negative cuts, where the same cuts are directly retained as the result of candidate words. Figure 3 shows an example.

Fig. 3. Example of "江苏省泰兴市人民法院" (People's Court of Taixing City, Jiangsu Province) global maximum segmentation, the orange box in the figure represents the matching parts of positive segmentation and negative segmentation. (Color figure online)

After using the LawRank algorithm to segment words, the results of the preliminary segmentation are not accurate, and there are still some problems in the candidate words. The first is the problem of the individual words being separated out. For a single word c in the candidate word list and, H represents information entropy, note that the former candidate word is p and the latter candidate word is n. the following conditions can be satisfied as follows:

$$H_L(pc) + H_R(pc) > H_L(p) + H_R(p) \tag{3}$$

$$H_L(cn) + H_R(cn) > H_L(n) + H_R(n) \tag{4}$$

$$H_L(pc_1) + H_R(pc_1) + H_L(c_2n) + H_R(c_2n) \\ > H_L(p) + H_R(p) + H_L(n) + H_R(n) + H_L(c_1c_2) + H_R(c_1c_2) \tag{5}$$

Then, the information entropy of the two new words pc_1 and c_2n after splitting and merging is greater than the old word before splitting. Hence, the word c_1c_2 is selected to be split and merged. Based on the methods above, keywords extraction should be mentioned on the agenda. The specific sorting criteria are:

$$score(word) = [I(word) + H_L(word) + H_R(word)] * len(word) \tag{6}$$

3.2   Model II

In this section, Inspired by the approaches above and the study of judicial documents, we consider incorporating the legal dictionaries and the unique language feature of Chinese judicial documents into Bi-LSTM-CRF based models.

Word Vector Construction. Under the research of Zhang et al. [8], we customize related templates for judicial documents. For a given text, we have prepared a dictionary of corresponding legal documents. First, based on the model, we make out some words and we can get corresponding dictionaries. For the above rules, we customize the Dict feature templates. For the type of 1-gram of the i-th word, which templates are w_i. For the type of 2-gram, the templates of the i-th word are $w_{i-1}w_i$, w_iw_{i+1}. For the type of 3-gram, the templates are $w_{i-2}w_{i-1}w_i$, $w_iw_{i+1}w_{i+2}$.

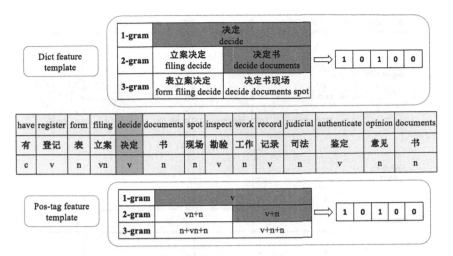

Fig. 4. Example of word feature construction, the word with orange shadow is the word w_i, the modules with green shadow express whether w_i meet the requirements of Dict feature template or POS-tag feature template. (Color figure online)

Also, legal vocabularies are relatively few in terms of emotional vocabulary, and nouns account for most of them. Based on the basis of this research, we customize the following POS-tag template. That is if the POS-tag feature is N, N+N, A+N, N+V+N, V+V+V, then the Binary value is 1. If not, then 0. Also, the details of defining word feature vector can be seen in following rules:

Rule 1: Binary value will be 1 if the word legal dictionary contains, else will be 0 for the first 5 binary values.

Rule 2: Binary value will be 1 if adjacent combinations of the word meets the requirements of POS-tag feature template, else will be 0 for the last 5 binary values.

Based on the rules above, we can generate a ten-dimensional vector. Figure 4 is an example, showing the construction of a feature word vector.

LSTM Layer. Inspired by Zhang et al. [8] and Ha et al. [30], we take hypernetwork into consideration which is an approach Ha et al. proposed to use one network to generate the weights for another network. Different from the network proposed by Zhang et al. [25] The network flowchart can be seen in Fig. 5. The input of SuperLSTM is the word feature vector we proposed based on judicial documents. Based on the feature vectors we proposed above, the boundary candidates could be received. Under the above study of baseline model, traditional LSTM have constraints on weight-sharing. The input of SuperLSTM layer is the feature vector f_i indicating some information of the words vocabulary and dictionaries from the text. The output of this layer is the weight of the input for the layer2.

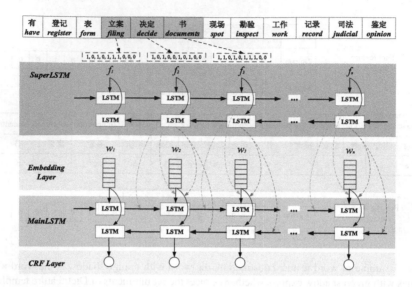

Fig. 5. The structure of Model II, the w_i represent word embeddings, f_i represent word feature vectors. Red and green arrows are for parameter generation. The bottom module is the CRF layer to represent the last layer of data processing and training. (Color figure online)

In contrast to traditional LSTM, the SuperLSTM can dynamically generate weights which are more competitive compared to some state-of-the-art models. In SuperLSTM, we allow W_h and W_x to float over time by using a smaller hypernetwork to generate these parameters of MainLSTM at each step. More concretely, the parameters W_h, W_x, b of MainLSTM will have different time steps, so h_t can now be computed as:

$$h_t = \phi(W_h(z_h)h_{t-1} + W_x(z_x) + b(z_b)) \tag{7}$$

$$W_h(z_h) = \langle W_{hz}, z_h \rangle \tag{8}$$

$$W_x(z_x) = \langle W_{xz}, z_x \rangle \tag{9}$$

$$b(z_b) = W_{bz}z_b + b_0 \tag{10}$$

Where $W_{hz} \in \mathrm{R}^{N_h \times N_h \times N_z}$, $W_{xz} \in \mathrm{R}^{N_h \times N_x \times N_z}$, $W_{bz} \in \mathrm{R}^{N_h \times N_z}$, $b_0 \in \mathrm{R}^{N_h}$ and z_h, z_x, $z_z \in \mathrm{R}^{N_h}$. Vector z_h, z_x, z_z can generate the weights of the four gates which are linear projection of hidden states of the SuperLSTM.

4 Experiments

This section, for assessing the knowledge we have discovered. We give the experimental results for five commonly used same domain datasets, four cross domain datasets and three type of cases on judicial documents for studying. Then, we will discuss the performance of our models on CWS.

4.1 Datasets

To measure the effectiveness of our model, we choose the datasets that usually used for evaluating CWS tasks. For the evaluation of cross domain, the datasets we choose are SIGHAN2010 [32], which contains data from literature, computer, medical, and financial areas, see Table 1.

Table 1. Datasets we choose from cross domain

Datasets		Literature	Computer	Medical	Financial
Training sets	Sentence	15.3K	24.3K	17.1K	16.4K
	Word	0.99M	1.11M	1.32M	1.32M
	Character	1.09M	2.01M	1.41M	1.76M
Testing sets	Sentence	1.40K	2.32K	1.53K	1.59K
	Word	1.01M	0.07M	1.06M	1.23M
	Character	0.98M	0.12M	1.13M	1.45M
OOV rates		6.8%	8.9%	9.0%	5.4%

However, the problem we want to solve is segmentation on Chinese judicial documents. Hence, we prepare a serious of judicial documents from the type of civil, criminal and administrative. Table 2 presents the size we choose for testing, training and corresponding OOV rates.

Table 2. Datasets we choose from judicial documents

Datasets		Civil	Criminal	Administrative
Training sets	Sentence	809.1K	765.3K	691.0K
	Word	6.34M	5.76M	4.97M
	Character	10.7M	6.44M	6.45M
Testing sets	Sentence	16.7K	12.8K	11.3K
	Word	1.91M	1.03M	0.91M
	Character	3.04M	1.99M	1.78M
OOV rates		5.4%	5.6%	4.9%

4.2 Experiments

At present, there have been many efforts to use novel BLSTM [31, 29] and adaptive language model [28] for word segmentation and have achieved good success. The training data used legal documents. We selected 70% as the training set, 10% as the development set, and 20% as the test set.

Embedding Layer Processing. For embedding layer, in Model II, it contains two different sections and with different input data. For embedding layer, based on the result from Model I, initial segmentation will be obtained and then we use the word2vec [13] toolkit to do train and fine tuned in the training process. Following previous work [27], we also used bi-gram character embedding, which were initialized by averaging the embedding of two contiguous characters. For embedding layer II, the input is not the judicial documents, but the word feature vector t_i we constructed before the details shown as follows.

We counting the corresponding number of occurrences in legal documents. From the results, we can know the nouns in judicial documents are the most common and numerous, the next is verbs. At the same time, in order to confirm our thoughts, we do experiments also on gold corpora and compare it with the results of judicial documents. We can know that in general documents, there is little difference between the number of nouns, verbs and adjectives. Especially the adjectives in judicial documents contain few.

Based on this knowledge discovery, we also counted out the probabilities of the nouns and their surrounding words appear together. We set the sliding window *win* to 2 and 3, according to the size of *win*. Figure 6 shows an example of the process of part-of-speech combination statistics in the sentence of judicial documents.

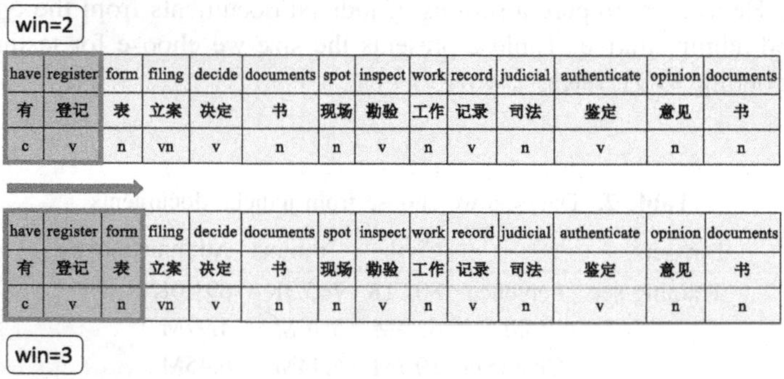

win=2

have	register	form	filing	decide	documents	spot	inspect	work	record	judicial	authenticate	opinion	documents
有	登记	表	立案	决定	书	现场	勘验	工作	记录	司法	鉴定	意见	书
c	v	n	vn	v	n	n	v	n	v	n	v	n	n

have	register	form	filing	decide	documents	spot	inspect	work	record	judicial	authenticate	opinion	documents
有	登记	表	立案	决定	书	现场	勘验	工作	记录	司法	鉴定	意见	书
c	v	n	vn	v	n	n	v	n	v	n	v	n	n

win=3

Fig. 6. An example about the work of sliding window, the section with orange shadow is the size of sliding window, the arrow represents the direction of the sliding window flow (Color figure online)

The results showed that the nouns generally appeared together with adjectives or verbs, which results are shown in the Fig. 7.

Figure 7 shows that when the size of sliding window is 2, the combination of "N+N" accounts for the largest proportion, next are the combination of "V+N" and "A+N". For the remaining 18 cases add up to only 6.45%, certainly considering it as negligible. For example, the percentage of combination "A+A+A" and "A+V+A" is zero. Based on this result, we create POS-tag template based on the judicial documents, the details can be seen in the introduction section of Model II.

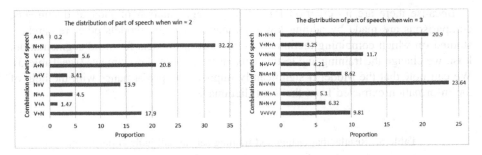

Fig. 7. The left and the right figure is the distribution of part of speech when the size of sliding window is 2 and 3

Network Configuration. The parameters are very important for neural networks, the details of parameters can be seen in Table 3.

Table 3. Parameters of our Model

Initial learning rate	$\alpha = 0.02$
Embedding size	$d = 150$
Hidden unit number1	$H_1 = 128$
Hidden unit number2	$H_2 = 128$
Dropout rate	$p = 0.2$
Batch size	$B = 128$
Gradient clipping	$c = 5$

Evaluations on Legal Area. Also, there have been many studies dedicated to integrate dictionaries which contain human knowledge discoveries into existing models. Table 4 shows the details. For Comprehensive evaluations on judicial documents, the first comparison we taken is making the training sets and testing sets are all from standard datasets. The datasets we choose are SIGHAN2010 [32], which contains data from literature, computer, medical, and financial areas. To compare with the cross-domain in public datasets, our model we use the corresponding dictionaries. The result shows that the word segmentation effect is not better than other studies. The reason that our model adds the feature which only judicial documents have and do not apply to general corpus.

Table 4. Results on cross-domain evaluation

Model	Literature	Computer	Medicine	Finance
Jiang et al. [24]	95.5	91.2	93.3	93.2
Liu et al. [12]	–	93.4	91.7	–
Zhao et al. [32]	94.6	95.1	93.9	95.9
Zhang et al. (Model I) [8]	94.4	94.4	93.9	95.7
Zhang et al. (Model II) [8]	94.7	94.7	94.1	96.0
Model II+domian dic	91.4	90.9	91.6	89.9

Because there are no criterion for the word segmentation of judicial documents, comparisons are made by adding law dictionaries into the models other researchers studied on which combining dictionaries into existing models for word segmentation. Last, we change the training and testing sets all from judicial documents (see Table 5). We can note that the Model II significantly improve the performance with the help of the information extracted from the legal dictionary.

Table 5. Results that testing and training sets are all judicial documents

Training set	Judicial documents
Testing set	
Jiang et al.+lawdic [24]	94.8
Liu et al.+lawdic [12]	92.0
Zhao et al.+lawdic [32]	93.9
Zhang et al. (Model I)+lawdic [8]	94.8
Zhang et al. (Model II)+lawdic [8]	94.9
Model II+lawdic	**95.1**

We also investigate the effect of the dictionary size. We randomly select 70% to 95% of the words from the law dictionary. Results show that the performance of the proposed Model Improves gradually with an increase in the dictionary size. Therefore, we could infer that we will get better results if we can obtain a dictionary containing more words.

This proves that incorporating dictionaries into neural network models can significantly boost the performance of the CWS. In particular, Model II performs better on judicial documents, possibly because Model II presented a more flexible way to extract features for each word w_i based on the legal dictionaries and contexts. Moreover, no corresponding development set is provided for these datasets. For a fair evaluation, we do not select some data from the testing sets as a development sets, but instead of using the models trained on the training set for a fixed epoch for testing. In particular, our dictionaries contain most legal words of testing sets. In order to prove that our models could profit from legal dictionaries, we filter partial legal dictionaries words appearing in the external dictionary as specific dictionaries and only use them during testing.

5   Conclusions

In this paper, two different models are proposed to generate legal dictionaries and do word segmentation based on the knowledge discoveries of judicial documents. As for the result, combining with linguistic features of judicial documents, the POS-tag and Dict feature templates are proposed respectively. Form the templates, a 10-dimensional binary word feature vector is constructed for each fine-grained word in the judicial documents. In order to better use the word feature vector, SuperLSTM and MainLSTM are built. This multi-layer neural networks take Bi-LSTM structure which can not only consider past information but also future information. In this layer, we incorporate word vectors created based on legal lexicons and legal statistical features as input for this layer. Finally, through the concat layer, the structure of the two layers is connected, and as the input of the last CRF layer, which has also achieved great success in sequence labeling and text segmentation. The experiment found that the field of legal documents can have a good segmentation effect.

In the future, the models will be further improved and the training speed will be increased. For the construction of the word vector template, the template we made at present is still relatively simple. It only extracts some features of legal documents. In truth, there are other knowledge can be discovered in legal area based on the large scale judicial documents. We will further study the other features of the judicial documents language and improve the word vector template architecture.

Acknowledgments. This work was supported by the National Key R&D Program of China (2016YFC0800803). Jidong Ge is the corresponding author.

References

1. Xue, N.: Chinese word segmentation as character tagging. IJCLCLP **8**(1) (2003)
2. Berger, A., Pietra, V., Pietra, S., Pietra, V.J.D.: A maximum entropy approach to natural language processing. Comput. Linguist. **22**(1), 39–71 (1996)
3. Boser, B., Guyon, I., Vapnik, V.: A training algorithm for optimal margin classifiers. In: Proceedings of the 1992 5th Computational Learning Theory/ACM, COLT 1992, pp. 144–152 (1992)
4. Eddy, S.: Hidden Markov models. Curr. Opin. Struct. Biol. **6**(3), 361–365 (1996)
5. Lafferty, J., McCallum, A., Pereira, F.: Conditional random fields: probabilistic models for segmenting and labeling sequence data. In: ICML, pp. 282–289 (2001)
6. Mikolov, T., Chen, K., Corrado, G., et al.: Efficient estimation of word representations in vector space. In: First International Conference on Learning Representations, pp. 1–12 (2013). https://arxiv.org/abs/1301.3781
7. Chen, X., Qiu, X., Zhu, C.: Long short-term memory neural networks for Chinese word segmentation. In: Proceedings of the 2015 Conference on Empirical Methods in Natural Language Processing, EMNLP 2015, pp. 1197–1206 (2015)
8. Zhang, Q., Liu, X., Fu, J.: Neural networks incorporating dictionaries for Chinese word segmentation. In: AAAI (2018)
9. Zhang, J., Huang, D., Han, X.: Rules-based Chinese word segmentation on MicroBlog for CIPS-SIGHAN on CLP2012. In: Proceedings of the 2012 2th CIPS-SIGHAN Joint Conference on Chinese Language Processing, JCCLP 2012, pp. 74–78 (2012)
10. Lin, Q., Chang, C.: 基於特製隱藏式馬可夫模型之中文斷詞研究 (Chinese Word Segmentation using Specialized HMM). ROCLING (2006). (in Chinese)
11. Zeng, D., Wei, M., Chau, M., Wang, F.: Domain-specific Chinese word segmentation using suffix tree and mutual information. Inf. Syst. Frontiers **13**(1), 115–125 (2011)
12. Chen, M., Xu, Z., Weinberger, K., et al.: Marginalized denoising autoencoders for domain adaptation. In: 29th International Conference on Machine Learning, pp. 1–8 (2012). https://icml.cc/2012/papers/416.pdf
13. Mikolov, T., Chen, K., Corrado, G., et al.: Efficient estimation of word representations in vector space. arXiv preprint arXiv:1301.3781 (2013)

14. Lample, G., Ballesteros, M., Subramanian, S., et al.: Neural architectures for named entity recognition. arXiv preprint arXiv:1603.01360 (2016)
15. Low, J., Ng, H., Guo, W.: A maximum entropy approach to Chinese word segmentation. In: Proceedings of the 2005 4th SIGHAN Workshop on Chinese Language Processing, CLP 2005 (2005)
16. Xue, N., Converse, S.: Combining classifiers for Chinese word segmentation. In: Proceedings of the 2002 1st SIGHAN Workshop on Chinese Language Processing-Volume 18, ACL 2002, pp. 1–7. Association for Computational Linguistics (2002)
17. Chen, X., Shi, Z., Qiu, X., et al.: Adversarial multi-criteria learning for Chinese word segmentation. arXiv preprint arXiv:1704.07556 (2017)
18. Nuo, M., Liu, H., Long, C., et al.: Tibetan unknown word identification from news corpora for supporting Lexicon-based Tibetan word segmentation. In: Proceedings of the 53rd Annual Meeting of the Association for Computational Linguistics and the 7th International Joint Conference on Natural Language Processing, ACL 2015, vol. 2, pp. 451–457 (2015)
19. Liang, Z., Xu, B., Zhao, J., et al.: Chinese new words detection using mutual information. In: Proceedings of 2012 International Conference on Trustworthy Computing and Services, ICTCS 2012, pp. 341–348 (2012)
20. Nakagawa, T.: Chinese and Japanese word segmentation using word-level and character-level information. In: Proceedings of 2004 the 20th International Conference on Computational Linguistics, ACL 2004, p. 466. Association for Computational Linguistics (2004)
21. Shiraz, M., Gani, A., Shamim, A., et al.: Energy efficient computational offloading framework for mobile cloud computing. J. Grid Comput. 13(1), 1–18 (2015)
22. Zhao, H., Kit, C.: Integrating unsupervised and supervised word segmentation: the role of goodness measures. Inf. Sci. 181(1), 163–183 (2011)
23. Yu, H., Yang, Z., Tan, L., et al.: Methods and datasets on semantic segmentation: a review. Neurocomputing 304, 82–103 (2018)
24. Jiang, W., Sun, M., Lü, Y., et al.: Discriminative learning with natural annotations: word segmentation as a case study. In: Proceedings of the 2013 51th Annual Meeting of the Association for Computational Linguistics, ACL 2013, pp. 761–769 (2013)
25. Zhang, M., Zhang, Y., Fu, G.: Transition-based neural word segmentation. In: Proceedings of the 2016 54th Annual Meeting of the Association for Computational Linguistics, ACL 2016, pp. 421–431 (2016)
26. Zhang, H., Liu, Q.: ICTCLAS. Institute of Computing Technology, Chinese Academy of Sciences (2002). http://www.ict.ac.cn/freeware/003_ictclas.Asp
27. Pei, W., Ge, T., Chang, B.: Max-margin tensor neural network for Chinese word segmentation. In: Proceedings of the 2014 52th Annual Meeting of the Association for Computational Linguistics, ACL 2014, pp. 293–303 (2014)
28. Teahan, W., Wen, Y., McNab, R., et al.: A compression-based algorithm for Chinese word segmentation. Comput. Linguist. 26(3), 375–393 (2000)
29. Cai, D., Zhao, H.: Neural word segmentation learning for Chinese. arXiv preprint arXiv: 1606.04300 (2016)
30. Ha, D., Dai, A., Le, Q.: Hypernetworks, arXiv preprint arXiv:1609.09106 (2016)
31. Huang, S., Sun, X., Wang, H.: Addressing domain adaptation for Chinese word segmentation with global recurrent structure. In: Proceedings of the 2017 8th International Joint Conference on Natural Language Processing, EIJCNLP 2017, pp. 184–193 (2017)
32. Zhao, H., Liu, Q.: The CIPS-SIGHAN CLP2010 Chinese Word Segmentation Backoff. In: CIPS-SIGHAN Joint Conference on Chinese Language Processing (2010)

Visual Sentiment Analysis with Local Object Regions Attention

Guoyong Cai[1(✉)], Xinhao He[1], and Jiao Pan[2]

[1] Guangxi Key Lab of Trusted Software,
Guilin University of Electronic Technology, Guilin, China
ccgycai@guet.edu.cn
[2] Guilin Kaige Information Technology Co. Ltd., Guilin, China

Abstract. Human action recognition has gained popularity because of its worldwide applications such as video surveillance, video retrieval and human–computer interaction. This paper provides a comprehensive overview of notable advances made by deep neural networks in this field. Firstly, the basic conception of action recognition and its common applications were introduced. Secondly, action recognition was categorized as action classification and action detection according to its respective research goals. And various deep learning frameworks for recognition tasks were discussed in detail and the most challenging datasets and taxonomies were briefly reviewed. Finally, the limitations of the state-of-the-art and promising directions of the research were briefly outlined.

Keywords: Action recognition · Deep neural network · Action classification · Action detection

1 Introduction

Nowadays, more and more social media users tend to use images to express emotions or opinions. Compared with text, images are easier to express personal emotions intuitively. Therefore, the visual sentiment analysis of images has gained considerable attention and research [1, 2]. Visual sentiment analysis is a task to study the emotional response of humans to visual stimuli such as images and videos [3]. The key challenge of visual sentiment analysis is the giant gap between sentiment space and visual feature space.

Feature engineering methods have been studied extensively to construct image sentiment features in previous literatures, such as colors, textures and shapes [4–6]. Recently, deep neural network learning has achieved great success in computer vision because of its ability to learn abstract and robust features [7–9]. In particular, Convolution Neural Networks (CNN) can automatically learn robust features from large-scale image data and demonstrate excellent performance. CNN is widely used in image-related tasks such as image classification and object detection, therefore, methods based on CNN have also been proposed for predicting image sentiment [10].

However, visual sentiment analysis is more challenging than conventional recognition tasks due to a higher level of subjectivity in the emotion recognition process, and

© Springer Nature Singapore Pte Ltd. 2019
X. Cheng et al. (Eds.): ICPCSEE 2019, CCIS 1058, pp. 479–489, 2019.
https://doi.org/10.1007/978-981-15-0118-0_37

almost all these approaches have been trying to reveal sentiment from the global perspective of the whole image. Little attention has been paid to research the sentimental response from local objects' regions. Therefore, it may lead to low accuracy performance on sentiment prediction.

In order to make prediction more accurate, we proposed a novel visual sentiment analysis method that integrating local objects' region features with global features to enhance sentiment classification performance. The main contributions is that the proposed method can selectively focus on sentimentally important objects' regions through object detection model and transfer learning strategy to learn more discriminative representation for visual sentiment classification.

2 Related Work

Feature engineering methods for visual sentiment analysis are categorized mainly into two types: feature selection and feature extraction. Lv et al. introduced color features for expressing emotion based on SIFT features of three RGB color channels and combined together to form a 384-dimensional C-SIFT features for predicting image sentiment [11]. Roth et al. analyzed image sentiment by extracting texture features of an image, using support vector machine (SVM) to classify sentiment polarity of an image [12]. Borth and Yuan et al. respectively, proposed using visual entities and attributes extraction to obtain semantic features to overcome sentiment gap between low-level visual features and high-level emotional semantics. Borth et al. constructed a large visual sentiment ontology library composed of 1,200 adjective noun pairs (ANP) [13]. Using this ontology library, the authors proposed SentiBank and MVSO emotion detectors to extract middle-level representation of the input images, which is treated as image features to train sentiment classifiers. Yuan et al. applied a similar strategy to Borth et al., but 102 pre-defined scene attributes are used instead of ANP as the middle-level representation [2].

More recently, researchers began to use deep models to automatically learn sentiment representations from large-scale image data and obtained better results. For example, Chen et al. studied the classification of visual sentiment concepts, trained models on the large dataset given in [13] to obtain an upgraded version of SentiBank, called DeepSentibank [14]. You et al. defined a CNN architecture for visual sentiment analysis to solve the problem of training on large-scale and noisy datasets, the authors used a progressive training strategy to fine-tune the network architecture, called PCNN [15]. Campos et al. used transfer learning strategy to fine-tune the image classification network with the Flickr dataset and applied it for image sentiment analysis [16, 17].

Although the above methods have achieved some positive performance, they basically considered to extract features only from whole images, and did not pay enough attention to local objects' regions expressed prominent emotion. Sun et al. used deep model to automatically discover local regions that contain objects and use them for visual sentiment analysis [18]. Li et al. proposed a context-aware classification model that considered both the local context and local-global context [19]. Different from existing research, this paper mine sentiment information from whole images and the local objects' regions simultaneously. The paper mainly focuses on the following points: First, we obtaining a localized region with accurate positioning and carrying emotional objects; Second, the feature fusion method is used to consider both the whole image and the local regions at the COIS architecture.

3 COIS Framework Description

The overall framework of the proposed COIS method is illustrated in Fig. 1. The goal is to learn discriminative sentiment representations from full images and local regions containing salient objects which may also express emotions. The framework consists of four parts: global feature extraction from whole images, local object regions detection, local features extraction from local objects' region, global features and local features integrating for visual sentiment classification. The global feature representation from the whole image is extracted by VGGNet-16 [20], as shown in (a) of Fig. 1; Faster R-CNN [21] is a popular and excellent object detection model, in order to extract the local objects' region features, COIS framework tunes a pre-trained Faster R-CNN model to detect objects' regions, as shown in (c) of Fig. 1 by using an affective image dataset as shown in (b) of Fig. 1; The global features and the local objects' region features are combined and used to train a sentiment classifier, which is shown in (d) of Fig. 1.

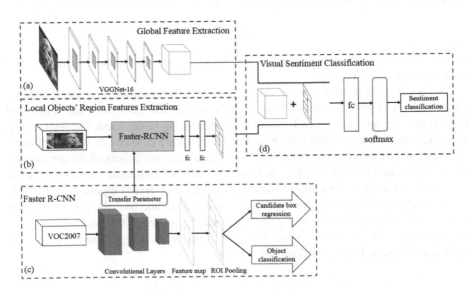

Fig. 1. Overview of the proposed COIS method.

3.1   Local Objects' Regions Generation

Local objects' region usually contain fine-grained information of objects in the image. We use Faster R-CNN to generate local regions for providing localized information. The input image is first fed into Faster R-CNN to generate a multi-channel feature map, and a series of candidate boxes is obtained. We get local objects' regions by comparing the overlap ratio of each candidate box with the ground truth label of the object detection image. A transfer learning strategy is applied to overcome the difference between the object detection dataset and the affective image dataset. That is, Faster R-CNN model is pre-trained on the object detection dataset PASCAL VOC 2007, transfer the parameters of the learned object region detection model to be further tuned by the affective image dataset. The following is a detailed description of how to use the Faster R-CNN to generate objects' region candidate boxes.

The input of the candidate box generation network is an image, and the output is a set of candidate boxes. Suppose the input image size is $M \times N$. The image is fed into convolutional layer to obtain a response convolutional feature map $F \in R^{w \times h \times n}$, where w and h are the width and height of F, and n is the number of channels of the convolutional feature map. Let the size of the convolution feature map be $(M/16) \times (N/16)$, which means the width, the height of the input image and the output feature map are both scaled to 1/16. To generate the candidate boxes, Faster R-CNN utilizes a two-layer CNN on the convolution feature map, the first layer contains c filters $g \in R^{a \times c}$ of size $a \times a$, the filter g slides on the input convolution feature map F to generate a lower-dimensional feature $F' \in R^{w' \times h' \times l}$, which is calculated as Eq. (1):

$$F' = \delta(g * F + b) \tag{1}$$

where $*$ is a convolution operation, $b \in R$ is the bias, and R is real number set, $\delta(\cdot)$ is a nonlinear activation function.

For each position on F', we considering k possible candidate box sizes to better detect objects of different sizes. Supposing W and H are the spatial size(width and height) of F', respectively, then WHk candidate boxes can be obtained. The feature map F' is delivered to two parallel fully connected layers. One is to decide whether there is an object in the candidate box; The other is to predict the coordinates of the center point of the candidate box and the size, as shown by the two rightmost branches in (c) of Fig. 1. Therefore, for k candidate boxes, the classification layer outputs $2k$ probability scores to evaluate whether there is an object in the candidate box, the regression layer outputs the candidate box center point coordinates and the width and height of the candidate box. The weighted expression of the classification layer and the regression layer loss function is given in Eq. (2):

$$L(\{p_i\}, \{t_i\}) = \frac{1}{N_{cls}} \sum_i^k L_{cls}(p_i, p_i^*) + \lambda \frac{1}{N_{reg}} \sum_i^k p_i^* L_{reg}(t_i, t_i^*) \tag{2}$$

where p_i denotes the prediction result of the candidate box i, $p_i^* = 1$ if the candidate box i is a positive sample, that is, the object exists in the candidate box, and $p_i^* = 0$ otherwise, that is, the candidate box is the background. N_{cls} represents the number of all candidate boxes generated by a Minibatch. t_i represents the size of the candidate box, t_i^* is the ground truth label corresponding to t_i, and λ is the hyper-parameter.

Since determining whether an object is in the candidate box belongs to the two-class classification problem, L_{cls} adopts Log Loss, which is commonly used for the two-class problem, and the calculation formula is shown in Eq. (3):

$$\ell(\theta) = p_i^* \log p_i + \left(1 - p_i^*\right) \log(1 - p_i) \tag{3}$$

L_{reg} uses the common loss function Smooth L_1 Loss, which measures the degree of deviation between the predicted value and the ground truth label, Smooth L_1 Loss is given in Eq. (5):

$$L_{reg}\left(t_i, t_i^*\right) = \sum smooth_{L1}\left(t_i - t_i^*\right) \tag{4}$$

$$smooth_{L1}(x) = \begin{cases} 0.5x^2 & if\ |x| < 1 \\ |x| - 0.5 & otherwise \end{cases} \tag{5}$$

3.2 Local Objects' Region Features Extraction

Let $L = \{r_1, \cdots, r_n\}$ be the generated candidate boxes set containing objects. The candidate boxes set L is projected onto the convolution feature map $F \subset R^{w \times h \times n}$, then we extract local objects' region features, thereby avoiding the missing image information caused by cropping or scaling the candidate box, and reducing the time spent on a large number of convolution operations [22]. Any one of the candidate box sets $r_i = \{(x_i, y_i)\}_{i=1}^n$ is used as a sample generated in the affective image, as shown by the orange rectangular box in Fig. 2(a), where x_i is usually expressed as a four-dimensional vector, which represents the coordinates of the center point of the candidate box and the width and height, respectively, $y_i \in \{0, 1\}$ represents the sentiment label corresponding to the object in the generated box. For each sample, we divided a generated box into m different granularities to obtain the semantic information of multiple levels, as shown in Fig. 2(b) where three sizes of granularities are applied. Then, the maximum pooling operation is performed on each of the divided sub-blocks to obtain a series of distinguishing feature maps $\{f_1, f_2, \cdots, f_d\}$, and d represents the number of divided sub-blocks, the calculation formula is shown in Eq. (6):

$$f_i = G_{max}\left(b_j\right) \tag{6}$$

where b_j denotes a sub-block after division, f_i denotes a feature map corresponding to the sub-block b_j, and $G_{max}(\cdot)$ denotes a maximum pooling operation. Finally, the feature maps of all the sub-blocks are added to obtain a local feature vector L_{f_i} of fixed dimension, which is specifically expressed as Eq. (7):

$$L_{f_i} = \sum_{i=1}^{d} f_i \tag{7}$$

Let L_{f_i} be a feature vector of a local region detected in an image, thus all detected local regions can be represented as a feature vector set $\{L_{f_1}, \cdots, L_{f_n}\}$, where n represents the number of detected local regions.

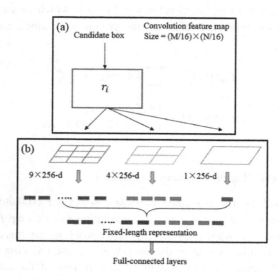

Fig. 2. Overview of local objects' region features extraction.

3.3   Global Feature Extraction

The global feature is an important factor related to the sentiment representation of an image, and typically includes the overall appearance information of the image and the contextual information surrounding the object in the image. We uses the VGGNet-16 shown in Fig. 3 to extract global features. VGGNet-16 consists of 5 convolutional blocks and 3 fully connected layers, this deep neural network jointly developed by Oxford University and DeepMind has a deeper network structure and unified network configuration than ordinary convolutional neural networks. It allows for more nonlinear transformations while reducing parameters, resulting in better feature extraction capabilities.

Specifically, we extract the global feature of an image from the last fully connected layer $fc7$ of VGGNet-16, and get a 4096-dimensional feature vector, denoted as G_f, as shown in the rightmost side of Fig. 3.

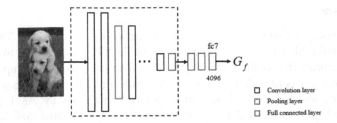

Fig. 3. Overview of VGGNet-16.

3.4 Image Sentiment Classification

Global feature representations and local objects' region features are G_f and $\{L_{f_1}, L_{f_2}, \cdots, L_{f_n}\}$, respectively. We select the top 5 objects detected to represent important local region information. Therefore each image can ultimately be represented as a set of feature vectors $U = \{G_f, L_{f_1}, \cdots, L_{f_n}\}$. We fuse these two type of features by the following Eq. (8):

$$\varphi(U) = G_f \oplus L_{f_1} \oplus \cdots \oplus L_{f_n} \tag{8}$$

where $\oplus$ represents concat two type of features, which are expressed by tensors.

For supervised image sentiment classification, the role of sentiment label in the training process cannot be ignored. Therefore, we assign a same sentiment polarity label to the local regions of the corresponding image. After the integrated joint feature vector $\varphi(U)$ is obtained, it is delivered to the fully connected layer and classified into the output category by softmax. To measure the model loss, we uses cross entropy to define the loss function. The softmax layer maps the joint feature vector $\varphi(U)$ to the output category by assigning a corresponding probability score q_i. If the number of sentiment categories is s, then q_i is given in (9):

$$q_i = \frac{\exp(\varphi(U)_i)}{\sum_i \exp(\varphi(U)_i)}, i = 1, \ldots, s \tag{9}$$

$$\ell = -\sum_i h_i \log(q_i) \tag{10}$$

where ℓ is the cross entropy loss of the network and h_i is the true emotional label of the image.

4 Experiments

In this section, we evaluate our method against other methods of sentiment classification by global feature of image to demonstrate the performance of the COIS for visual sentiment analysis.

4.1 Datasets

We evaluate our method on two public datasets including TwitterI [13], TwitterII [15]. TwitterI is assembled from social media Twitter in 881 images with two sentiment polarity categories (i.e. positive and negative), which are based on group intelligence strategy. Twitter II is provided by You *et al.* which contains 1269 images from social media Twitter and labeled with sentiment polarity categories by AMT participants. Both TwitterI and TwitterII datasets are split randomly into 80% training and 20% testing sets.

4.2 Experiment Settings

Our framework is implemented using Linux-Ubuntu14.04, Python 2.7, Tensorflow 1.3.0, and the development tool is PyCharm. All of our experiments are performed on an NVIDIA Tesla P100-PCIE GPU with 64 GB on-board memory. The framework used to extract feature of images is based on COIS. To conduct fair comparison, we keep most of the settings same as previous study [13, 14, 20, 23]. The image size is 224×224. The image is padded by 2 pixels on each side, filled with 0 value. In order to supervised training to obtain a local objects' regions, we labeled affective images artificially with five types of objects such as people, vehicles, etc. The annotation tool is called ImageLab. At this point, the dataset contains both the sentiment label and the object detection label (including the center point coordinates and width and height of the object's candidate box).

We train the model using MomentumOptimizer with a momentum of 0.9. The learning rate of the convolutional layers are initialized as 0.001. Our method uses the Dropout strategy [23] and the L2 paradigm to reduce over-fitting, the Dropout value is set to 0.5. Cross entropy is chosen as the loss function. The total number of iterations are 100 epochs. In addition, for local objects' regions features, we adjusts the pooling layer of detection branch. The pool kernel adopts 3×3, 2×2, 1×1 to adapt to the datasets of this paper.

4.3 Comparison Method

For traditional methods, we compares the SentiBank [13] model. For the basic CNN methods, we compared DeepSentiBank [14], and two classical deep learning methods pre-trained on ImageNet and fine-tuned on the affective datasets: AlexNet [23], VGGNet [20]. The specific instructions are as follows:

SentiBank: A visual concept detector library that can be used to detect the presence of 1,200 adjective noun phrase pair (ANPs) in an image, which is treated as an image middle-level representation for visual sentiment analysis [13].
DeepSentiBank: A visual sentiment concept classifier trained on large datasets using deep convolutional neural networks, an improved version of SentiBank [14].
ImageNet-AlexNet: A deep learning architecture AlexNet pre-trained on ImageNet and fine-tuned it on the affective image dataset for visual sentiment analysis [23].
ImageNet-VGGNet16: The same idea as the ImageNet-AlexNet, the difference is that the network is replaced by VGGNet-16 [20].

Local regions-Net: A method only considering local objects' region feature with Faster R-CNN and fully connected layers to learn sentiment representation.

4.4   Experiment Result

Table 1 shows that the classification accuracy(%) of the proposed method and the comparison method on the TwitterI, TwitterII datasets. As shown in Table 1, our proposed method COIS achieves 75.81% accuracy on TwitterI and 78.90% accuracy on TwitterII compared with 66.63% and 65.93% accuracy on TwitterI and TwitterII for SentiBank method. The COIS method outperforms Deepsentibank by 5% and 8% on TwitterI and TwitterII datasets. It is suggests that our proposed COIS method can learn more discriminative representation for visual sentiment analysis.

In addition, compared with ImageNet-AlexNet and ImageNet-VGGNet16, our proposed method improved by about 10% accuracy on TwitterI and TwitterII under similar parameter size. The result suggests that the effectiveness of combining local objects' regions representation.

Table 1. Classification accuracy of different methods on TwitterI and TwitterII.

Method	Accuracy@ Twitter I	Accuracy@ Twitter II
SentiBank	66.63%	65.93%
DeepSentibank	71.25%	70.23%
ImageNet-AlexNet	65.80%	67.88%
ImageNet-VGGNet	67.49%	68.79%
Local regions-Net	61.54%	58.63%
COIS	**75.81%**	**78.90%**

5   Conclusions

Visual sentiment analysis is gaining more and more attention. Considering that the emotions generated by images are not only from the whole image, but also from the local objects' regions in the images. This paper presents a novel COIS method for visual sentiment analysis. In the COIS method, the Faster R-CNN model is used to detect the objects' regions in an image, and then the fully connected network is used to learn the sentiment representation of the local objects' region of the image and integrate it with the global feature of the image to obtain a more discriminative sentiment representation. The performance of the proposed method is evaluated on two real datasets, and the experimental results show that the proposed COIS method is better than the method of learning emotional representation only from the whole image.

However, in this study, only the local region sentiment of the objects contained in the images are used to enhance the visual sentiment analysis, but other regions that do not contain objects are ignored. In the future work, we will consider the method of weakly supervised learning to discover more emotional regions in an image and consider better strategy of feature fusion to further improve the performance of COIS.

Acknowledgment. This work is supported by The National Science Foundation of China (#61763007), The Nature Science Foundation of Guangxi in China (#2017JJD160017), The Science and Technology Project of Guilin City (20170113-6).

References

1. Jin, X., Gallagher, A., Cao, L., et al.: The wisdom of social multimedia: using flickr for prediction and forecast. In: International Conference on Multimedia, pp. 1235–1244, DBLP, Firenze (2010)
2. Yuan, J., Mcdonough, S., You, Q., et al.: Sentribute: image sentiment analysis from a mid-level perspective. In: International Workshop on Issues of Sentiment Discovery and Opinion Mining, pp. 1–8 (2013)
3. Yang, J., She, D., Lai, Y.K., et al.: Weakly supervised coupled networks for visual sentiment analysis. In: Proceedings of the IEEE Conference on Computer Vision and Pattern Recognition, pp. 7584–7592 (2018)
4. Wang, X., Jia, J., Hu, P., et al.: Understanding the emotional impact of images. In: ACM International Conference on Multimedia. ACM (2012)
5. Cheng, Y.C., Chen, S.Y.: Image classification using color, texture and regions. Image Vis. Comput. **21**, 759–776 (2003)
6. Iqbal, Q., Aggarwal, J.K.: Retrieval by classification of images containing large manmade objects using perceptual grouping. Pattern Recogn. **35**, 1463–1479 (2002)
7. Karpathy, A., Toderici, G., Shetty, S., et al.: Large-scale video classification with convolutional neural networks. In: IEEE Conference on Computer Vision and Pattern Recognition. IEEE Computer Society, pp. 1725–1732 (2014)
8. Chen, M., Zhang, L., Allebach, J.P.: Learning deep features for image emotion classification. In: IEEE International Conference on Image Processing. IEEE, pp. 4491–4495 (2015)
9. Szegedy, C., Liu, N.W., Jia, N.Y., et al.: Going deeper with convolutions. In: IEEE Conference on Computer Vision and Pattern Recognition (CVPR). IEEE Computer Society (2015)
10. You, Q., Yang, J., Yang, J., et al.: Building a large scale dataset for image emotion recognition: the fine print and the benchmark. In: Thirtieth AAAI Conference on Artificial Intelligence. AAAI Press, pp. 308–314 (2016)
11. Lv, P.: Research on image emotion categorization. Yanshan University (2014)
12. Yanulevskaya, V., Gemert, J.C.V., Roth, K., et al.: Emotional valence categorization using holistic image features. In: IEEE International Conference on Image Processing. IEEE (2008)
13. Borth, D., Ji, R., Chen, T., et al.: Large-scale visual sentiment ontology and detectors using adjective noun pairs. In: ACM International Conference on Multimedia, pp. 223–232. ACM (2013)
14. Chen, T., Borth, D., Darrell, T., et al.: DeepSentiBank: visual sentiment concept classification with deep convolutional neural networks. Comput. Sci. 675–678 (2014)
15. You, Q., Luo, J., Jin, H., et al.: Robust image sentiment analysis using progressively trained and domain transferred deep networks, pp. 381–388 (2015)
16. Campos, V., Salvador, A., Giro-I-Nieto, X., et al.: Diving deep into sentiment: understanding fine-tuned CNNs for visual sentiment prediction. In: International Workshop on Affect & Sentiment in Multimedia, pp. 57–62. ACM (2015)
17. Campos, V., Jou, B., Giró-I-Nieto, X.: From pixels to sentiment: fine-tuning CNNs for visual sentiment prediction. Image Vis. Comput. **65**, 15–22 (2017)

18. Sun, M., Yang, J., Wang, K., et al.: Discovering affective regions in deep convolutional neural networks for visual sentiment prediction. In: IEEE International Conference on Multimedia and Expo, pp. 1–6. IEEE (2016)
19. Li, B., Xiong, W., Hu, W., et al.: Context-aware affective images classification based on bilayer sparse representation. In: ACM International Conference on Multimedia, pp. 721–724. ACM (2012)
20. Simonyan, K., Zisserman, A.: Very deep convolutional networks for large-scale image recognition. Comput. Sci. 1409–1556 (2014)
21. Ren, S., He, K., Girshick, R., et al.: Faster R-CNN: towards real-time object detection with region proposal networks. In: International Conference on Neural Information Processing Systems, pp. 91–99. MIT Press (2015)
22. He, K., Zhang, X., Ren, S., et al.: Spatial pyramid pooling in deep convolutional networks for visual recognition. IEEE Trans. Pattern Anal. Mach. Intell. **37**, 1904–1916 (2014)
23. Krizhevsky, A., Sutskever, I., Hinton, G.E.: ImageNet classification with deep convolutional neural networks. In: International Conference on Neural Information Processing Systems, pp. 4–12. Curran Associates Inc. (2012)

TPOS Tagging Method Based on BiLSTM_CRF Model

Lili Wang[1], Ziyan Chen[1], and Hongwu Yang[1,2,3](✉)

[1] College of Physics and Electronic Engineering, Northwest Normal University,
Lanzhou 730070, China
yanghw@nwnu.edu.cn

[2] Engineering Research Center of Gansu Province for Intelligent Information
Technology and Application, Lanzhou 730070, China

[3] National and Provincial Joint Engineering Laboratory of Learning Analysis
Technology in Online Education, Lanzhou 730070, China

Abstract. Part of speech (POS) tagging determines the attributes of each word, and it is the fundamental work in machine translation, speech recognition, information retrieval and other fields. For Tibetan part-of-speech (TPOS) tagging, a tagging method is proposed based on bidirectional long short-term memory with conditional random field model (BiLSTM_CRF). Firstly, the designed TOPS tagging set and manual tagging corpus were used to get word vectors by embedding Tibetan words and corresponding TPOS tags in continuous bag-of-words (CBOW) model. Secondly, the word vectors were input into the BiLSTM_CRF model. To obtain the predictive score matrix, this model using the past input features and future input feature information respectively learned by forward long short-term memory (LSTM) and backward LSTM performs non-linear operations on the softmax layer. The prediction score matrix was input into the CRF model to judge the threshold value and calculate the sequence score error. Lastly, a Tibetan part of speech tagging model was got based on the BiLSTM_CRF model. The experimental results indicate that the accuracy of TPOS tagging model based on the BiLSTM_CRF model can reach 92.7%.

Keywords: Deep learning · Word embedding model · Tibetan part-of-speech

1 Introduction

POS tagging is to determine the attributes of each word, such as noun, verb, adjective and other part of speech. As a basic research in the field of information processing and a very important basic work in natural language processing, it is commonly used in machine translation, speech recognition, information retrieval and other fields. At present, the technical route of TPOS tagging mainly adopts the rule-based method and the combination of rule-based with statistical method.

The rule-based approach tries to find a suitable set of annotation rules so that the corpus to be annotated can get the correct annotation. First, according to the tagged corpus, we summarize the context rules, and establish the separate tagging rules based on TPOS ambiguity. Then, we search for a dictionary when tagging, and if a word possesses more than one part of speech, we can disambiguate it by the rules.

© Springer Nature Singapore Pte Ltd. 2019
X. Cheng et al. (Eds.): ICPCSEE 2019, CCIS 1058, pp. 490–503, 2019.
https://doi.org/10.1007/978-981-15-0118-0_38

The method of combining statistics with rules is the main method of part of speech tagging in Tibetan. In 2003, paper [1] first discussed the issue of TPOS tagging in Tibetan. In 2006, paper [2] proposed a set of basic markers for the part of speech in Tibetan. In 2010, the paper [3] put forward a research method of automatic POS Tagging Based on Tibetan corpus. This research mainly aims at finding the difference between Tibetan and English-Chinese, so as to have a focus on analyzing the morphological characteristics of Tibetan, and ultimately get the TPOS tagging set. The paper trains the model of binary grammar through manual tagging corpus, and realizes TPOS Tagging based on statistical method through algorithm, but this paper does not give exact experimental results. In 2012, hidden Markova model was adopted in the paper [4] to achieve Tibetan part-of-speech tagging, and 40 thousand words of corpus were used as training corpus to carry out part-of-speech tagging for 20 articles, with an accuracy rate of 84%. In 2013, the paper [5] proposed a maximum entropy POS tagging method based on syllable features. This method takes maximum entropy model as the basic framework, defines and chooses feature templates according to the word-formation characteristics and statistical analysis results of Tibetan, and studies the maximum entropy POS tagging model based on linguistic features. Experiments show that the syllable feature improves the effect of POS tagging in Tibetan and reduces the error rate by 6.4%. In 2014, the paper [6] proposed a discriminant part of speech tagging method based on perceptron training model. In the same year, the paper [7] adopted the method of combining maximum entropy with conditional random field. In 2015, a part of speech prediction study based on TPOS tagging was proposed in the paper [8]. They selected part of the corpus of Tibetan primary and secondary school textbooks and constructed a corpus with Tibetan character markers, word boundary markers and POS markers, which realized the integration of word segmentation and part-of-speech marking. In the same year, the paper [9] proposed an open source tip-las system for Tibetan word segmentation and TPOS tagging. However, the process of building rules often requires a large amount of linguistic knowledge of Tibetan, and the conflict between rules needs to be dealt with. Moreover, the rule building process is time-consuming and laborious, with poor portability.

With the rise of deep learning, neural networks have achieved good results in natural language processing, but there is no application of neural networks in TPOS tagging. For timing data, the bidirectional long-short term memory (BiLSTM) model [10] can combine two groups of long-short term memory (LSTM) layers, which allows the current part of speech to contain both historical and future information, and is more conducive to marking the current part of speech. In order to optimize the output results, it is desirable to learn the optimal label sequence from the whole sequence and get the optimal results by following a layer of conditional random field model (CRF). As a corollary, this paper proposes a TPOS tagging method based on bidirectional long-short term memory network with conditional random field model (BiLSTM_CRF), which combines BiLSTM and CRF.

The rest of this paper is organized as follows: Section 2 introduces the specific application of BiLSTM_CRF model in Tibetan part-of-speech tagging, including the design of part-of-speech tagging set, word embedding and the introduction of BiLSTM_CRF model. Section 3 introduces relevant evaluation experiments. Section 4 summarizes the work of this paper.

2   The Framework of BiLSTM_CRF Model

Figure 1 shows the flow chart of TPOS tagging. The specific implementation steps are as follows:

- We label the Tibetan corpus.
- We read the Tibetan POS tagged corpus data firstly and get two lists of word and tag, in which, word saves the participle and tag saves the corresponding tag. Then we construct a dictionary to count the words' frequency and arrange the words according to their occurrence frequency. Finally we build two dictionary lists, word_to_id and tag_to_id, which store the ID of words and tags respectively. The tag of unknown words is UNKNOWN = "*".
- We read the training set, validation set and test set data respectively, and convert the word list and tag list of the data set into their corresponding word_to_id and tag_to_id list.
- The dictionary table and TPOS tags are constructed by CBOW model in word2vec toolkit to obtain the word vector of word and TPOS tagging.
- We build the BiLSTM_CRF model through the tensorflow framework.
- We initialize the model, then input training samples, construct an iterator, and each time return the word and label pairs that read the size of the input batch as the input of the BiLSTM_CRF model.
- After learning the features of the forward LSTM model and the backward LSTM model in BiLSTM model, the score matrix is computed through the softmax layer and transmitted it to the CRF model to calculate the sequence score errors, so as to determine whether the errors meet the preset threshold. If satisfied, it saves network parameters. If not, the network's weight will be updated automatically after the new batch data is input, and the above work will be repeated until the model converges.
- When the training is finished, we save the optimized network model, then input the test Tibetan word segmentation corpus, load the model parameters, and finally get the TPOS tagging results.

2.1   Design of TPOS Tagging Set

The choice of POS tagging set is the primary task of POS tagging research. At present, the TPOS tagging set is not uniform in Tibetan information processing field. We define three main principles for the modern TPOS marker set for information processing.

- Grammatical function is the main basis for the division of TPOS. The meaning of Tibetan words does not serve as the main basis for classifying parts of speech, but sometimes it also serves as a reference.
- The classification in the part of speech tag set should cover all the vocabulary in modern Tibetan. In order to meet the needs of computer processing of real text, the symbols in marker set should cover not only words in linguistic sense, but also smaller units than words, such as function words, morpheme words, non-morpheme words, and larger language units than words, such as idioms, idioms, abbreviations, abbreviations, punctuation marks and non-Tibetan symbols.

- We allow both classes when we divide parts of speech. The labeling method of concurrent words is to connect the parts of speech it has concurrently with "/", such as ང་ཚོ། denoting noun-verb concurrent case auxiliary words, ཞིང་ཆེན། denoting noun-verb-adjective concurrent category words, etc.

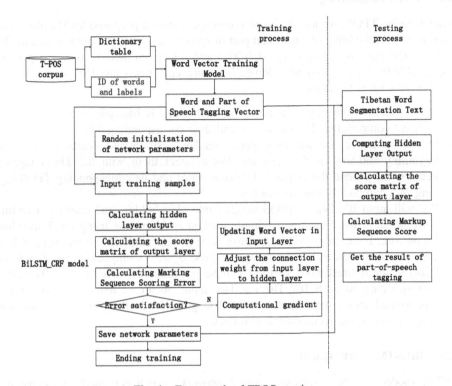

Fig. 1. Framework of TPOS tagging.

For studying the TPOS, we combine the research objectives to choose three commonly used Tibetan books as the benchmarks, which are the Tibetan-Chinese Dictionary, the newly compiled Tibetan Dictionary and the Tibetan Verb Dictionary. We identify and merge the vocabulary through the design procedure, and finally construct a part of speech database containing more than 70,000 words. Meanwhile, the Tibetan edition of the People's Daily of China, the China Tibetan Broadcasting Network and some modern Tibetan articles are selected as the basic materials. After summarizing and analyzing, a total of 120,000 words are labeled manually. We designed 21 parent classes and 35 child classes, as shown in Table 1. We labeled the data manually according to the following guidelines:

- Grammatical function as the main basis for classification.
- Every word should have a label.
- The number of labels of a word is equal to the number of segments of the word.

- If a stem can be classified as an adverb or an adjective, we consider it an adjective.
- If a stem always appears as an adverb, we mark it as an adverb.
- All loanwords and complex Sanskrit characters are labeled as unknown words.

2.2 Word Embedding

After labeling TPOS, we use the word vector space model proposed by Mikolov et al. to realize the embedding of words and part of speech tags [11]. The open source toolkit includes Continuous bag-of-words (CBOW) and skip-gram models. In this paper, we use CBOW model for word embedding. The word vector contains 156783 words. The training process is as follows:

- Input features: The input feature set of each entry is basically composed of two different components, Tibetan words and their corresponding TPOS tags.
- In each sentence, we add two special markers <start> and <end> to mark the beginning and end of the sentence. We connect them with the TPOS tagging vectors, and extend the embedded Tibetan word and its corresponding TPOS tagging in a completely unsupervised way.
- We will input ID data of TPOS tagging into placeholders and convert them into corresponding word vectors. A simple method is defined by using the Tensorflow framework. First, an embedding matrix is generated randomly in the form of [vocab_size, size], that is, the dictionary size [vocab_size] multiplied by the dimension [size] of the defined word vector. Then Search method is used to find the vector corresponding to each ID. The one-Hot input vector of length [vocab_size] is transformed into a word vector of fixed length size. In the process of backward propagation, word vectors are also trained.

2.3 BiLSTM_CRF Model

LSTM model is a time recursive neural network, which can model long-distance dependent information [12]. The model structure is shown in Fig. 2. The information is input from the input gate and connected by a cell unit which is circularly connected. The cell unit is used to control the information flowing to the input gate and to control the forgetting gate of the ell state before the forgetting gate. The calculation expressions of each unit at each time are shown in expressions (1), (2), (3), (4), (5).

$$i_t = \sigma(w_i \cdot [h_{t-1}, x_t] + b_i]) \tag{1}$$

$$f_t = \sigma(w_f \cdot [h_{t-1}, x_t] + b_f) \tag{2}$$

$$o_t = \sigma(w_o \cdot [h_{t-1}, x_t] + b_o) \tag{3}$$

$$c_t = f_t \times c_{t-1} + i_t \times (\tanh(w_c \cdot [h_{t-1}, x_t] + b_c)) \tag{4}$$

$$h_t = o_t \times \tanh(c_t) \tag{5}$$

i_t, f_t, o_t, c_t respectively represents the output of t-Time input gate, forgetting gate, output gate and cell state. h_t and x_t stand for hidden layer vectors and input vectors at t time. σ represents the sigmoid activation function and it can output values between 0 and 1 to describe how much each part can pass. 0 stands for "no quantity is allowed to pass" and 1 stands for "any quantity is allowed to pass", w and b stand for weight matrix and bias vector separately.

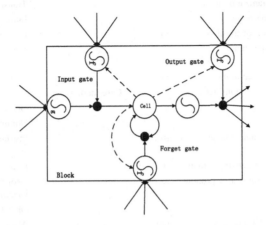

Fig. 2. Framework of LSTM model

Although LSTM network has a good performance in POS tagging, this model advances from left to right and ignores the influence of the latter part of speech on the former part of speech. Therefore, in order to get better context information of current words, we use BiLSTM network to build model. This model combines forward LSTM and backward LSTM models. In addition to using past input features and sentence-level markup information, it can also use future input features. The features learned from BiLSTM model are input into the prediction TPOS tagging information on the softmax layer. The input parameter labels of softmax function are shaped as [batch_-size, num_classes], and each element type is float64. The label vectors corresponding to each sample correspond to multiple labels, that's to say, the label of each sample is not limited to one, and the probability distribution corresponding to each category is found. Then choose the label output with the highest part of speech score. At the same time, we use back propagation algorithm to train our network in our model, and update parameter information α, as shown in expression (6). Where x is the input vector, y is the marker corresponding to the input vector, and ρ is the maximum probability calculated by the softmax function. We estimate the value through Stochastic Gradient Descent (SGD), that is, all the parameter information in the network.

$$\alpha \leftarrow \alpha + \gamma \frac{\partial \ln \rho(y|x, \alpha)}{\partial \alpha} \qquad (6)$$

Table 1. The table Tibetan part of speech tag

Parent class	Child class	Pos markers	Parent class	Child class	Pos markers
Noun		n	Classifier		q
	Common noun	nc	Adverb		d
	Name	nr	Conjunction		c
	Place name	ns	Verb		v
	Organization name	no		Transitive verb	vt
	Time noun	nt		Intransitive verb	vi
	Noun of location	nl		Existential verbs	ve
	Proper noun	nz		Verb of judgment	vj
	Rhetoric terms	ne		Other verbs	vo
	Sanskrit transliteration	nf	Case-marker		p
Adjective		a		Genre	pg
Pronoun		r		Nominative	pa
	Personal pronoun	rh		Instrumental case	pi
	Interrogative pronoun	rw		Hour format	pt
	Demonstrative pronoun	rd		Geographical location	pl
Auxiliary word		u		Object lattice	pd
	Modal auxiliary	um		Ablative case	pc
	Modal auxiliary	uv		Comparison lattice	pn
	Auxiliary auxiliary words	ug		Self character	ps
	Omission auxiliary words	ue	Volume labeling		t
	Auxiliary particle	uw	Nominalization markers		h
	Metaphor auxiliary words	uo	Plurality marking		f
	Result auxiliary word	ub	Statement label designator		y
	Supplementary auxiliary words	ua	Interjection		e
	Causal auxiliary	us	Idiom		l
An onomatopoeia		o	Syllabic word		s
Abbreviation		j	Unknown symbol		*
Punctuation		w	Numeral		m

2.4   CRF Model

The CRF model has a long history for sequential tagging tasks [13]. It mainly considers the whole tag of a given observation sequence to learn the conditional probability distribution jointly and decode the optimal tag sequence of the input sentence. It uses adjacent information to predict the current tag. The CRF model predicts the distribution of each tag, and then uses beam decoding to find the optimal tag sequence. The CRF layer can effectively utilize the tag information at sentence level, which is represented by the line connecting the continuous output layer.

For sequential tagging tasks, we need to consider the correlation between community tags. It is beneficial to decode the optimal tag chain of a given input sentence commonly. Hence, we use conditional random fields to jointly model tag sequences instead of decoding each tag independently. The input parameters of CRF layer are the output of BiLSTM layer, a matrix P of $n * m$, where n is the number of Tibetan words and m is the type of label. Definition p_{ij} is the transition score from the i tag to the j tag. For a predicted tag sequence $y = y_1, y_2 \ldots y_n$. Its probability is shown in Eq. (7).

$$score(x, y) = \sum_{i=1}^{n} P_{i,y_i} + \sum_{i=1}^{n+1} A_{y_{i-1},y_1} \tag{7}$$

The whole sequence probability consists of two parts. One is the P matrix output from BiLSTM layer, the other part is the transfer matrix A of CRF layer. y_0 and y_n in Eq. (7) are the end and beginning markers of predictive sentences.

The likelihood function of maximizing markers for a training sample x, y^x in CRF level training is shown in Eq. (8)

$$\log(p(y^x|x)) = score(x, y^x) - \log(\sum_{y'} e^{score(x,y')}) \tag{8}$$

Where y' represents the real markup value. In the prediction process, Viterbi algorithm of dynamic programming is used to solve the optimal path, as shown in Eq. (9).

$$y^* = \arg\max_{y'} score(x, y') \tag{9}$$

3   Experiments

3.1   Introduction of Corpus

The corpus we use combined the Tibetan edition of the People's Daily of China, the Tibetan Broadcasting Network of China and some modern Tibetan articles. The corpus contains novels, news, humanities, ecological environment, legal documents, history, blogs, texts and other topics, so as to cover more fields of words and solve the problem of unknown words as far as possible. After word segmentation, we summarized, sorted and analyzed a total of 26,228 sentences and 353,932 words, and took them as the research objects to carry out manual tagging. We used the part of speech tagging set we designed to complete the part of speech tagging with the maximum matching method. Then 15 Tibetan native speakers independently completed TPOS proofreading, summarized the proofreading results, and discussed the controversial labeling results to determine the final results.

We show the percentage distribution of POS tags in the tagged corpus in Fig. 3, and we can see that some tags are less distributed than columns, especially in the child class TPOS tagging category, in which interrogative pronouns, demonstrative pronouns, other verbs, instrumental case, subordinate case and normalization mark are seldom distributed. The label distribution of nouns, case markers, verbs, transitive verbs, genus, punctuation marks, auxiliary words, conjunctions and adjectives decreased in turn. We can clearly see that the number of parent class TPOS tags is significantly more than that of child class TPOS tags. Also, this skewed label distribution affects the performance of automatic sequential label tasks.

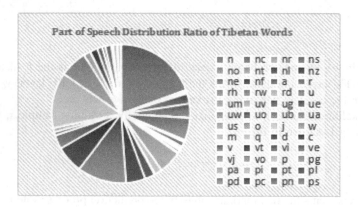

Fig. 3. Distribution ratio of part of speech tagging in Tibetan words

After tagging the corpus, we randomly divided the corpus into training set, verification set and test set, each accounting for about 80%, 10% and 10% words, as shown in Table 2. We used training set and verification set to conduct preliminary experiments and then used test set to evaluate the final experimental results. What need to point out is, many words in test set data partitioning do not exist in training set. The out-of-vocabulary (oov) words are counted regardless of the POS tags.

Table 2. Data set distribution table

Train	Sentence	20793
	words	281960
dev	Sentence	2607
	words	35200
	oov	985
test	Sentence	2828
	words	36772
	oov	1139

3.2 Design of Experimental Parameters

The window size of the word vector training part set as 5 that is, 5 word parts before and after the current part of speech are considered. If there is not enough 10 POS, 0 vector is added as a supplement. If more than 10 POS are added, 1 vector is added as a supplement. The pre-training word vector and the random initial word vector have a great influence on the accuracy of the lexical TPOS tagging.

All neural networks used in this paper shared a common SGD forward and backward training process. We chose the BiLSTM_CRF model to illustrate the training algorithm. All training data was first batch processed, and each batch contained a list of sentences determined by the batch size parameter. Using a batch size of 100 means that the total length of sentences included cannot be greater than 100. For each batch, the BiLSTM_CRF model has an LSTM layer forward and backward with a size of 100. We found that adjusting this dimension had no significant effect on the model. We run the BiLSTM model to get all the scores for all tag locations, then run the CRF layer forward and backward to calculate the network output and state transition matrix to propagate errors from the output to the input, including the inverse of the forward and reverse states of the LSTM pass to pass. Finally, we updated the network parameters including the state transition matrix and the BiLSTM model parameters. We used a learning rate of 0.1 to train the model, and the hidden layer size was set to 100 because the hidden layer size did not have a large impact on model performance. We used a stochastic gradient descent algorithm with a gradient clipping of 5.0 and an initial gradient descent rate of 0.5. Dropout network is used to prevent over-fitting, and the retention rate is 0.8.

The experimental framework used in this paper is Tensorflow. The accuracy of the most widely used evaluation index was used to evaluate the result of part-of-speech tagging. Its definition is shown in the following formula (10), which represents the ratio of correctly labeled words in all test sets, in which the correctly labeled words are the marks that match the marks manually labeled. Where, n denotes the number of Tibetan words correctly labeled and m denotes the total number of Tibetan words.

$$p = \frac{n}{m} \times 100\% \tag{10}$$

3.3   Experimental Design

Experiment 1. The effect of different models on the accuracy of lexical TPOS tagging.

In this paper, in order to compare the effects of LSTM, BiLSTM, LSTM_CRF and BiLSTM_CRF models on TPOS tagging results when handling the random initialization word vector and pre-training word vector, we used the same experimental parameters and the same experimental data for TPOS tagging. The experimental results are shown in Table 3. The difference in results is entirely due to the different network structures and whether or not the pre-trained word vectors are used.

Table 3. Comparison of different model results

Model	Initialization word vector	Pre-training word vector
LSTM	85.7%	86.4%
BiLSTM	86.3%	88.5%
LSTM_CRF	89.7%	90.9%
BiLSTM_CRF	91.3%	92.7%

From Table 3, we can clearly see that after adding the pre-training word vector to the BiLSTM model, the accuracy rate is increased by 2.2%. In the BiLSTM_CRF model, the accuracy rate is increased by 1.4% after adding the pre-training word vector. The experimental results show that using pre-trained word vectors can improve the accuracy of Tibetan word tagging.

Table 3 also shows that the P of LSTM_CRF model is 4.0%, which is higher than that of LSTM model when the word vector is randomly initialized. BiLSTM_CRF model was 5.0% more accurate than BiLSTM model. When pre-trained word vectors were added, the accuracy of LSTM_CRF model was 4.4% higher than that of LSTM model, and the accuracy of BiLSTM_CRF model was 4.2% higher than that of BiLSTM model. The experimental results show that adding CRF model in the decoding part can improve the accuracy of TPOS tagging.

When the processor is Core i5-7500 and the running memory is 16G, the BiLSTM_CRF model is compared with the traditional CRF model and HMM model which are the most widely used in TPOS tagging. It can be seen from Table 4 that the HMM model runs at a high speed, 16.5 min faster than the BiLSTM model, but the P is reduced by 9.9%. Although the accuracy of CRF model is 0.8% higher than that of HMM model, it is still lower than that of BiLSTM_CRF model, which can explain the adaptability of BiLSTM_CRF model in TPOS tagging. In addition, although the BiLSTM_CRF model is slow in training, it is faster than the HMM model in the part-of-speech tagging test of the same sentence, and higher accuracy can be achieved without any features, because it can capture non-lexical relations and dialectical trends and model them well.

Table 4. Comparison of experimental results between CRF model HMM model and BiLSTM_CRF model

Model	P	Operation hours
CRF	83.6%	82 min
HMM	82.8%	76 min
BiLSTM_CRF	92.7%	94.5 min

Experiment 2. Analysis of experimental results of BiLSTM_CRF model.

From experiment 1, it can be seen that the BiLSTM_CRF model performs well in the tagging of Tibetan words. Therefore, we further analyze the error rate generated by the BiLSTM_CRF model by using the confusion matrix. We found that the most common mistake was to confuse nouns with case markers by 16.8%. The second kind of mistake is the confusion of case mark and noun, which accounts for 32.7% of the systematic errors. The reason for this error may be that some nouns can be both nouns and case markers, or it may be that hidden words outside the vocabulary prefer the noun tag because it is more common. A third mistake is confusion between nouns and verbs, and vice versa. The fourth common mistake is to confuse an adjective with a noun. In Table 5, we listed the partial confusion error rate of BiLSTM_CRF model.

Table 5. Analysis of error types

Error type	Error rate	example
n→p	16.8%	ང་ཚོ།
p→n	15.9%	ཁོ་ཚོ།
n→v	12.5%	ལས་ཀ།
a→n	10.1%	འདི་ཟད།
v→n	9.7%	རི་ཚོ་འཁྲིད་བ།
pa→f	5.6%	ཁོ་ཚོ།
n→a	4.3%	དཀར་པོ་ནི།

Experiment 3. Comparison of the experimental results of BiLSTM_CRF model with other scholars.

To further illustrate the applicability of BiLSTM_CRF model in Tibetan part-of-speech tagging, we compare the effect of BiLSTM_CRF model with other parts-of-speech tagging methods proposed by other scholars. Although the methods, corpus and part-of-speech Dictionary of the existing Tibetan part-of-speech tagging system are different, there is no unified and standardized corpus, and it is difficult to make a strict comparison. The comparison in Table 6 shows that the BiLSTM_CRF model adopted in this paper is adaptable to Tibetan word segmentation. At present, there is no literature on part-of-speech tagging in Tibetan based on neural network, so, compared with the traditional method, the effect of this method is similar or better than the traditional method. Finally, this paper realized Tibetan word segmentation without using other language rules to assist.

Table 6. Comparison of the experimental results of BiLSTM_CRF model with other scholars

Model	p	Model	p
Paper [4]	84%	Paper [8]	91.6%
Paper [5]	90.94%	Paper [9]	93.90%
Paper [6]	98.26%	BiLSTM_CRF	92.7%
Paper [7]	87.76%		

4 Conclusion

In TPOS tagging, we propose a method based on BiLSTM_CRF model. First and foremost, we design a TPOS tagging set with 21 parent-class and 35 child-class classes. On top of this, we use CBOW model to insert word vectors into the labeled Tibetan words and their corresponding parts of speech to obtain word vectors, and then it is input into the BiLSTM_CRF model. At this time, the model first runs BiLSTM model to get all the scores of all tag positions, then runs the CRF layer forward and backward to propagate errors from output to input by calculating network output and state transition matrix, and updates network parameters to get the optimal model. In fact, by comparing LSTM, BiLSTM, LSTM_CRF and BiLSTM_CRF models, the accuracy of BiLSTM_CRF model is the highest. The common errors include the confusion of nouns and case particles, verbs and nouns, which can be improved by adding experimental data.

References

1. Di, J.: On syntactic chunks and formal markers of Tibetan. In: National Joint Conference on Computational Linguistics (2003)
2. Rang-jia, C., Tai-jia, J.: On the code classification of parts of speech in Tibetan corpora. In: Academic Conference on the 25th Anniversary of the Chinese Information Society of China (2006)
3. Jun-fen, S., Kong-yu, Q., Tai, B.: Research on automatic part-of-speech tagging in Tibetan corpus based on HMM. J. Northwest Natl. (Nat. Sci.) 30(1), 42–45 (2009)
4. Duo-Xi, Z.J., Cai-Rang, A.J.: Research and implementation of Tibetan part of speech tagging based on HMM. Comput. CD Softw. Its Appl. 12, 100–101 (2012)
5. Hong-Zhi, Y., Ya-Chao, L., Kun, W.: Fusion of syllable features for Tibetan part of speech based on maximum entropy model. J. Chin. Inf. Process. 27(5), 160–166 (2013)
6. Que-Cai-Rang, H., Qun, L., Hai-Xing, Z.: Discriminative Tibetan part-of-speech tagging with perceptron model. J. Chin. Inf. Process. 28(2), 56–60 (2014)
7. Cai-Jun, K.: Research on Tibetan word segmentation and part of speech tagging (2014)
8. Cong-Jun, L., Hui-Dan, L., Jian, W.: Research on tagging of Tibetan syllables. J. Chin. Inf. Process. 29(5), 211–216 (2015)
9. Ya-Chao, L., Jing, J., Yang-Ji, J., et al.: TIP-LAS: an open source toolkit for Tibetan word segmentation and pos tagging. J. Chin. Inf. Process. 29(6), 203–207 (2015)
10. Ren, Y., Teng, C., Li, F., et al.: Relation classification via sequence features and bi-directional LSTMs. Wuhan Univ. J. Nat. Sci. 22(6), 489–497 (2017)

11. Mikolov, T., Chen, K., Corrado, G., et al.: Efficient estimation of word representations in vector space. Comput. Sci., 1–12 (2013)
12. Soutner, D., Müller, L.: Application of LSTM neural networks in language modelling. In: Habernal, I., Matoušek, V. (eds.) TSD 2013. LNCS (LNAI), vol. 8082, pp. 105–112. Springer, Heidelberg (2013). https://doi.org/10.1007/978-3-642-40585-3_14
13. Krishnapriya, V., Sreesha, P., Harithalakshmi, T.R., et al.: Design of a POS tagger using conditional random fields for Malayalam. In: First International Conference on Computational Systems and Communications (2015)

Judicial Case Screening Based on LDA

Jin Xu[1], Tieke He[1], Hao Lian[1], Jiabing Wan[1], and Qin Kong[2](✉)

[1] State Key Laboratory for Novel Software Technology, Nanjing University,
Nanjing 210093, China
[2] JinLing College, Nanjing University, Nanjing 210093, China
qinkong@vip.126.com

Abstract. Under the background of judicial responsibility system, making similar judgments according to similar cases is vital for front-line judges to solve complicated problems such as non-standard use of law and inconsistency of judicial ruling standards. In this paper, a method is proposed for judicial cases based on the LDA topic model. The case, penalty and legal provisions were set. Gibbs Sampling algorithm was employed to estimate the probability distribution of topics on the implicit topic set in a text and calculate the similarity between texts by cosine similarity. The quality of screening was used as a final evaluation indicator. The verification of massive experiments shows that the case screening method based on LDA and cosine similarity has a satisfactory effect.

Keywords: Topic model · LDA · Cosine similarity

1 Introduction

With the globalization of information, the transmission of information has become more and more frequent and the data resources in the Internet show a large number of characteristics. In order to refine high-quality information from these complex data, content-based data mining [1] and information classification technology [3] has become common methods. Among them, text classification technology [27] is an important component of text information mining. Text categorization is based on the similarity of content under the condition of a multi-label category text collection. Since the 1990s, with the maturity of machine learning technology [16], the text classification method based on machine learning [16] has broken the traditional classification model based on knowledge theory and strengthened the ability of information mining and optimization, which greatly improved the efficiency of word processing and increase the depth of text mining. This method is used in many fields such as medical care and transportation. Judicature is one of them.

For a long time, in the process of judicial activities [14] in China, we mainly filled the legal loopholes by holding trial meetings and organizing training of judges, and then solved the judicial problems faced in the process of applying the law to achieve the goal of unifying judicial standards and safeguarding

© Springer Nature Singapore Pte Ltd. 2019
X. Cheng et al. (Eds.): ICPCSEE 2019, CCIS 1058, pp. 504–517, 2019.
https://doi.org/10.1007/978-981-15-0118-0_39

judicial justice. However, due to the influence of regional humanities and other factors, in the face of the same facts, the judges of different courts and even the different judges of the same court are likely to make extremely different judgments. The above measures have not achieved the desired results. In view of the different rulings in the same case, the similar judgment undoubtedly provides a new solution to the quality of the case of the governing judge, and also makes an important contribution to the construction of judicial information. Judgments [18] mean that the same or similar judgments can be obtained in the same or similar cases. The mechanism is to find a case similar or even identical to the case being handled by the judge in a passive or active manner to achieve inspiration and expand the judgment of the judge. Inspired by this, this paper aims to find the application of text categorization methods in the judicial field, that is, to introduce a new model to assist judicial staff to efficiently classify massive judicial documents and select suitable cases as recommendations to help judges to make correct judgment and improve the quality and efficiency of the trial.

Because the judicial structure [6] has a complex structure including the facts of the case, the origin of the evidence, and the legal basis, the classification of texts based on massive judicial documents is an arduous task. In the field of natural language processing [13], text classification methods such as TF-IDF [29] (Term Frequency–Inverse Document Frequency), VSM [8] (Vector Space Model), and LSI [28] (Latent Semantic Indexing) are common. However, these methods have some inherent defects. For example, in TF-IDF, the dimension of word weight vector is too high to distinguish between synonyms and polysemy; LSI has significant effects on dimensionality reduction, but it is likely to be weak due to the neglect of some important and rare classification features. The above deficiencies will have a serious impact on the classification performance of judicial texts. At the same time, the traditional classification methods mentioned do not take into account the complex structure of the judicial case documents and the semantics of the case descriptions in the instruments, which makes them unable to meet the high requirements of accurate classification of judicial texts, so they are difficult to apply directly to the real judicial field. In view of the above problems, this paper proposes a text classification method based on LDA [10] topic model [11,24]. LDA makes up for the shortcomings of traditional methods. It uses the polynomial distribution of words to obtain the distribution of topic polynomial, which realizes the feature dimension reduction of text as a topic vector, and effectively solves the problem of data sparseness and synonym. More importantly, the parameter space of the LDA model is fixed and not affected by the size of the data set. Therefore, LDA is very suitable for document classification of massive judicial texts.

In general, this paper has made the following contributions:

– In the process of judicial research, we use the LDA topic model in the field of machine learning to find similar judicial cases. We observe the similarities between judicial texts at the topic level through synonym, polysemy, overlapping and other semantic relations to cluster texts based on topics and screen out cases of high similarity for legal staff to analyze and refer to;

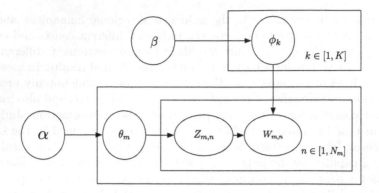

Fig. 1. LDA generation model

- Gibbs Sampling [19] is used to estimate the parameters in the text matrix, such as the subject polynomial distribution θ and the word polynomial distribution ϕ in the judicial text. The algorithm is very simple, but the calculation results are accurate, which improves the credibility of the screening results;
- We conduct a large number of simulation experiments on the judicial dataset and obtain a relatively high value of screening quality, which indicates that the method can significantly improve the text clustering effect.

   The rest of this paper is organized as follows. Section 2 introduces the LDA model framework. We mainly describe how to deploy LDA to the screening of similar cases in Sect. 3. Section 4 shows the experiment. We discuss the related work in Sect. 5. Section 6 concludes this paper.

2   LDA Model Framework

This section introduces the LDA topic model and its modeling process, followed by a detailed description of the model's training and inference processes [21].

2.1   Model Description

As we all know, the judicial case documents are complex in form and contain a lot of information. According to different information values, a judicial document can be divided into multiple topics, and each topic is a collection of data composed of several keywords with high frequency. We can understand that each judicial document is generally composed of multiple topics, one of which is composed of a number of words closely related to the subject. Therefore, constructing a topic model based on judicial knowledge can be of great help in analyzing the content of judicial cases and text classification. This paper builds an LDA topic model based on the theory of judicial knowledge. The model is shown in Fig. 1.

   In the model, we define the key data structures and corresponding symbols. For the meaning of related symbols, see Table 1.

By observing the model, we can find that the textual modeling of LDA can be divided into two physical processes: the formation of the topic corresponding to the n-th word in the m-th document and the generation of the n-th word in the m-th document. The LDA model treats each case as a collection of K-topics composed of a mixture of N_m words. Each topic generates a certain number of words with a certain probability, and each topic is generated with a certain probability. The text-to-topic follows the polynomial distribution, and the topic-to-word obeys the polynomial distribution. In other words, each text represents a probability distribution formed by some topics, and each topic represents a probability distribution composed of many words. We use LDA to model judicial cases to find the relationship between different topics and words, and to get the relevant topic distribution of the text. The process of generating each text is as follows:

1. Select the number of words N_m by obeying the Poisson distribution [9], $N_m \sim Possion(\zeta)$;
2. Select θ by obeying the Dirichlet distribution [17], $\theta \sim Dirichlet(\alpha)$;
3. For each word associated with N_m: Choose a topic by obeying the $Multinomial(\theta)$ and generate a word with the conditional polynomial probability of the topic;
4. Repeat the above operation until each word in the text is traversed.

Table 1. Symbol representation

Symbol	Description
M	The number of case texts
K	The number of topics
N_m	The number of words in the m-th text
α	Dirichlet prior parameters
β	Dirichlet prior parameters of multiple
$Z_{m,n}$	The topic of the nth word in the m-th text
$W_{m,n}$	The n-th word in the m-th text
θ_m	Topic distributions of the m-th text
ϕ_k	Word distributions for the k-th topic

2.2 Training and Inference

There are two goals in constructing an LDA model: one is to estimate the optimal parameter values of the subject polynomial distribution θ and the word polynomial distribution ϕ in the model, the other is to calculate the distribution of topics in the specified judicial text, namely model training and inference.

The process of training is to obtain samples of the topics and words in the case text set. We can estimate all the parameters in the LDA model based on the collected samples. The training process is not complicated:

1. Randomly initialize α and β;
2. Traverse each text in the training set, resample and calculate the topic distribution of each document and the topic distribution of each word according to the Gibbs Sampling algorithm. Repeat this step until Gibbs Sampling converges;
3. Construct the co-occurrence frequency matrix of topics and co-occurrence frequency matrix of words in training set respectively to end the entire training process of the LDA model.

The inference process is almost identical to the training process. We can think of the word polynomial distribution ϕ as a constant so that we only need to calculate the subject distribution θ of the specified case text during the sampling process. The forecasting process is as follows:

1. Random initialization: randomly assign a topic number to each word in the current text;
2. Rescan the current text and use Gibbs Sampling to resample each word its topic;
3. Repeat the above steps unless the result converges and calculate the topic distribution θ of the text.

The Gibbs Sampling algorithm mentioned above aims at the probability distribution of the text on the topic θ and the probability distribution of the topic on the words ϕ. The main idea of this method is: we find the joint distribution $P(w, z)$ with the subject z according to the known word vectors w converted from texts (the probability distribution of the subject z is not known at present), and calculate the conditional probability distribution $p(z_i = \frac{k}{w}, z_i)$ of the word w_i corresponding to the subject z_i, where w_i refers to the i-th word of the word vector set w, z_i is the topic associated with the word w_i in the corpus, and z_i represents the probability distribution of the topic after the word w_i is removed. Based on the above results, we get the sampling topic of the word w_i by Gibbs sampling. By analogy, we can find the corresponding topics of all words. Using the correspondence between the words and the topics obtained from all the samples, we can obtain the required θ and ϕ. The conditional probability formula for Gibbs sampling is:

$$p(z_i = \frac{k}{w}, z_{-i}) = \frac{(n_{(k,-i)}^{(t)} + \beta_t)(n_{(m,-i)}^{(t)} + \alpha_k)}{\sum_{x=1}^{V}(n_{k,-i}^{(x)} + \beta_x)\sum_{s=1}^{K}(n_{m,-i}^{(s)} + \alpha_s)} \tag{1}$$

In (1), $n_{m,-i}^{(k)}$ is the number of occurrences of the topic k in the text m in the case where the word i is removed, $n_{k,-i}^{(t)}$ implies the number of words t corresponding to the topic k after the word i is removed.

The specific application of Gibbs sampling algorithm is as follows. When we get the probability distribution of all the topics Z to which the current word belongs, we sample a new topic for the word according to this probability distribution, and then use the same method to constantly update the topic of the next word. until the subject polynomial distribution θ and the word polynomial distribution ϕ converge. At this point the algorithm stops, the parameters to be estimated are output.

3 Case Screening Based on LDA Model

The main work of this section is to introduce how to use LDA to screen out the similar case texts needed from the massive judicial texts.

In order to improve the screening quality, this paper uses the combination of LDA and cosine similarity. We use LDA to get all the other texts of the class of the specified judicial case, and then use the vector-dependent cosine distance formula to calculate the cosine similarity between these texts and the case, finally filter the top-5 texts with the highest similarity as the push. The screening process needs to go through three phases: data preprocessing [5], representation of case text information and screening of similar cases.

3.1 Data Preprocessing

The original text of the target case has a large number of party names (such as "Wang", "Liu"), time, location, etc., but these data do not provide much help for us to analyze the case information. Therefore, we need to perform data pre-processing on it to ensure the quality of experimental data and improve the accuracy of experimental results. Pre-processing includes word segmentation [25], stop word removal [15], and stem extraction. Specifically, we analyze the text's syntax and semantics to divide the sequence of words into separate words and use the stop word list to determine whether the existence of stop words in the text (if it is, delete immediately), finally extract keywords containing topic information to form new text.

3.2 Textual Representation of the Case

We count the number of occurrences of each keyword in the case and use the bag-of-words [26] to convert it into a fixed-dimensional word frequency vector to construct a text vector model. In layman's terms, we construct a word-text matrix with terms and frequencies as constituent elements. We then use the transformed text as an input to the LDA model to infer and obtain two result vectors (thematic polynomial distribution and word polynomial distribution). This result will serve as the basis for the screening of the case.

The key to using the LDA model for inference is as follows:

$$P(\frac{word}{document}) = \sum_{topics} P(\frac{word}{topics}) \times P(\frac{topics}{document}) \tag{2}$$

(2) shows that the probability of occurrence of a word w in a case document d can be calculated by $P(\frac{word}{topics})$ and $P(\frac{topics}{document})$ (the subject as a transition), where $P(\frac{word}{topics})$ and $P(\frac{topics}{document})$ can be obtained by using the current word polynomial distribution and the topic polynomial distribution, respectively.

We mentioned that the inference process is similar to the training process in Sect. 2. In order to easily understand the inference details, we will detail the training process. Now make the following settings: There are n cases in training set D, involving k different topics and m different words related to these topics. All topics constitute a set of topics $T<t_1, t_2, \cdots, t_k>$, where ti denotes the i-th topic and all words constitute a set of words $W<w_1, w_2 \cdots, w_m>$, where w_i denotes the i-th word in W.

Each topic t in the topic set T corresponds to a polynomial distribution $\phi_t<p_{w1}, p_{w2}, \cdots, p_{wm}>$ of all words in the word set W, where pwi represents the probability of the i-th word in W. Then the probability distribution of each word on all topics can be obtained, as shown in (3).

$$\begin{pmatrix} P_{w_1 t_1} & P_{w_1 t_2} & \cdots & P_{w_1 t_k} \\ P_{w_2 t_1} & P_{w_2 t_2} & \cdots & P_{w_2 t_k} \\ \vdots & \cdots & \ddots & \vdots \\ P_{w_m t_1} & P_{w_m t_2} & \cdots & P_{w_m t_k} \end{pmatrix} \tag{3}$$

In (3), $P_{w_m t_k}$ represents the probability distribution of the m-th word in W corresponding to the k-th topic and can be calculated by dividing the number of word m in the topic k by the number of all words in the topic k.

In the same way, Each case document d in D corresponds to a polynomial distribution $\theta_d<p_{t1}, p_{t2}, \cdots, p_{tk}>$ of all topics in the topic set T, where p_{ti} represents the probability of the i-th topic in T. So that the probability distribution of each topic on all documents can be obtained, as shown in Eq. (4).

$$\begin{pmatrix} P_{t_1 d_1} & P_{t_1 d_2} & \cdots & P_{t_1 d_n} \\ P_{t_2 d_1} & P_{t_2 d_2} & \cdots & P_{t_2 d_n} \\ \vdots & \cdots & \ddots & \vdots \\ P_{t_k d_1} & P_{t_k d_2} & \cdots & P_{t_k d_n} \end{pmatrix} \tag{4}$$

In (4), $P_{t_k d_n}$ shows the probability distribution of the k-th topic in T corresponding to the n-th document and can be calculated by dividing the number of topic k in the document n by the number of all topics in the document n.

3.3   Screening of Similar Cases

In Sect. 3.2, we translate the semantic relationship between the specified target case and the word into the probability distribution of the topic. Since the distribution of text on the subject is a mapping of text vectors, we can measure the similarity between the target document and the documents in the training set by calculating the polynomial distributions of the corresponding topics. This paper uses cosine similarity as the distance metric. Cosine similarity uses the cosine of the angles of two vectors in vector space as the measure of similarity between texts. The distance between text vectors is defined as follows:

$$S(d_i, d_j) = cos^2_{d_i, d_j}(\theta)$$

$$= (\frac{\sum_{x=0}^{k} P_{ix} * P_{jx}}{\sqrt{\sum_{x=0}^{k} (P_{ix})^2} * \sqrt{\sum_{x=0}^{k} (P_{jx})^2}})^2 \qquad (5)$$

$$= \frac{(\sum_{x=0}^{k} P_{ix} * P_{jx})^2}{\sum_{x=0}^{k} (P_{ix})^2 * \sum_{x=0}^{j} (P_{jx})^2}$$

In (5), θ refers to the angle between text vectors and k is the number of topics.

$S(d_i, d_j)$ is between 0 and 1. The larger its value implies that the shorter the cosine distance of the two texts and the more similar they are. When $S(d_i, d_j) = 0$, it means that the two texts are unrelated. When $S(d_i, d_j) = 1$, the two texts are identical. We use the calculation result of the vector distance from the target text to other instruments to classify the text and filter out the five case documents with the largest $S(d_i, d_j)$ value as the push.

4   Experimental Analysis and Results

In this section, we describe experimental settings such as data sets, evaluation indicators, and compare the LDA proposed with the traditional text classification method TF-IDF, then show the experimental results.

4.1   Experimental Design

In this paper, the experiment uses a collection of legal cases provided by CAIL, with a total of more than 120,000 cases, each of which contains labels for cases, penalty, and legal provisions. We divide the case set into a training set and a test set according to a certain proportion and preprocess the entire data set by the method described in Sect. 3. The processing includes stop words removal, data cleaning such as the criminal name (such as "Li") and converting the case texts in the case set into word vectors. Next, we apply LDA and TF-IDF to construct a text vector on the training set and use the cosine similarity as a measure to find the n case texts in the training set that are closest to the target text in the test set, then we filter them out. We label n cases filtered by LDA vector as $nlda$ and $ntfidf$ filtered by TF-IDF vector and finally judge between the two methods according to the evaluation index. In theory, LDA-based screening is better.

4.2   Evaluation Indicators

We define R_{case}, $R_{penalty}$, R_{laws}, Q to indicate the hit rate of the case, the hit rate of the penalty, the hit rate of the legal provisions and the quality of the screening, and define i as the target case in the test set. Among the n texts selected from the training set, the case of $ncase$ texts is the same as the case of the target text and the penalty of npenalty texts is the same as the penalty of the target text. The number of intersections between the legal provisions of the selected case i and the target case is $nlaw_i$.

$$R_{case} = \frac{ncase}{n}$$

$$R_{penalty} = \frac{npenalty}{n}$$

$$R_{laws} = \frac{\sum_{i=1}^{n} nlaw_i}{n}$$

$$Q = \sigma(P_{case} + R_{penalty} + R_{laws})$$

In the above formulas, σ symbolizes the *sigmoid function* that causes the result to fall between 0 and 1.

4.3   Experimental Results

We compare the screening effects of the two methods by the different ratios, that is, the training set VS test set is $2:1, 4:1, 4:1, 5:1, 10:1$. The results are shown in Table 2.

Table 2. Screening results based on LDA and TF-IDF

Training set VS Test set	Method	R_{case}	$R_{penalty}$	R_{laws}	Q
2:1	TF-IDF	0.3135	**0.2679**	0.2152	**0.6892**
	LDA	0.1760	**0.2831**	0.1296	**0.6430**
4:1	TF-IDF	0.2886	**0.2645**	0.2088	**0.6817**
	LDA	0.1628	**0.2688**	0.1237	**0.6354**
5:1	TF-IDF	0.2729	**0.2667**	0.2037	**0.6774**
	LDA	0.1575	**0.2679**	0.1218	**0.6335**
10:1	TF-IDF	0.3358	**0.2307**	0.2476	**0.6930**
	LDA	0.2074	**0.2332**	0.1578	**0.6453**

From the experimental data in Table 2, we can find that LDA is significantly stronger than TF-IDF from the perspective of penalty hit rate, while LDA is slightly weaker from the standpoint of text-based case and the hit rate of legal provision. Since the LDA model is based on term semantics to construct a text vector representation, the segmentation effect of the case text in the preprocessing process will directly result in the final screening. Here we give a simple example - "". There may be two outcomes after the word segmentation: " " and " ". As we all know, the machine does not process the word segmentation after fully understanding the semantics of the entire text as we humans do, so the machine segmentation processing will produce bias to a certain extent. Unlike LDA, TF-IDF builds feature vectors of text based on word frequency without considering text content. Therefore, we allow and accept that the hit rate of penalty and hit rate of legal provisions based on the LDA model will be lower. At the same time, we can find that the screening quality of the two is very close and the screening quality of LDA is almost 70, indicating that the screening effect is good.

Besides, We compare the LDA and TF-IDF in point of time consuming and memory space for the prediction phase of the model. Table 3 shows the average time they spent on the screening process at different ratios.

Table 3. Time-consuming results

Traing set VS Test set	Average time consuming	
2:1	LDA	6.3278
	TF-IDF	8.2744
4:1	LDA	5.7842
	TF-IDF	8.6458
5:1	LDA	6.5800
	TF-IDF	9.8021
10:1	LDA	8.0403
	TF-IDF	13.2525

Looking at the data in Table 3, we can know that the time consumption of LDA is significantly less than that of TF-IDF. And as the proportion continues to expand, their time-consuming gaps are also growing, which means that LDA-based inferential speed are excellent with less mental loss.

By the way, we analyze the size of the occupied memory. For TF-IDF, it is necessary to count the frequency of all words in the text to construct a feature vector matrix and the memory holds all the terms in the text. But for LDA, what stored in memory is the word-subject matrix (the matrix size is the number of words multiplied by the number of topics) and the subject-text matrix (the matrix size is the number of topics multiplied by the number of texts), where the term is keywords that are closely related to the topic, not all words. This fact makes the number of features of the LDA matrix much less than that of the TF-IDF, which greatly reduces the memory loss.

As can be seen from the above, LDA only reduces the screening accuracy of TF-IDF by a little bit on the basis of time saving and memory saving. Thus we can conclude that LDA is much more practical than TF-IDF and can be applied to the screening of judicial cases.

5 Related Work

We present some of the existing related work that recommends similar cases [22] and describe the topic model, the word bag model and the TF-IDF model.

5.1 Recommendations for Similar Cases

When encountering difficult and complicated cases, it may be difficult to obtain a satisfactory conclusion based on the legal provisions. At this time, turning to the study of similar cases is a good choice. It seems that it is extremely important to recommend appropriate cases for judges to help broaden their judgments. Collaborative Filtering Recommendations [7] (CF) is currently the most popular recommendation method [4] and has been widely used in the research community. CF mainly includes online collaborative and offline filtering. The so-called online collaboration is to find items of interest to users based on online resources, while offline filtering is to filter information that is worthless or not recommended. In general, there are three types of collaborative filtering recommendations, namely user-based collaborative filtering [20], item-based collaborative filtering [12], and model-based collaborative filtering [2]. The basic principle of user-based collaborative filtering recommendation is to compare the similarity between users. The basic principle of user-based collaborative filtering recommendation is to compare the similarity between users, find out the user groups similar to the current user interest, and use the KNN [23] algorithm to obtain the past preference information of K "neighbors" as the current user for recommendation. The idea of item-based collaborative filtering recommendation is similar. It compares the similarities between items and recommends similar items to users according to the preferences of the target users. For small-scale data, the similarity between users and the similarity between items will not change too much in a short time and can be calculated offline. However, with the expansion of the data scale, due to the large amount of calculation, it is impossible to recommend users in real time. Model-based collaborative filtering effectively solves this problem. Model-based collaborative filtering is the main idea of recommending items to target users based on existing user information and several items. Commonly used methods are: regression algorithm, graph model, implicit semantic model LFM (latent factor model). The implicit semantic model is to establish the relationship between the target user and the hidden class and the relationship between the item and the hidden class through matrix decomposition to obtain the user's preference relationship with the item. It is commonly used in the field of natural language processing to semantically analyze user behavior and recommend items to users.

5.2 Topic Model

Judging whether two documents are similar is not only a simple repetition of several words, but also a hidden semantic relationship in the text. Topic models are often used for data analysis of natural language processing (such as text mining, semantic association mining), bioinformatics research as well as modeling in an unsupervised learning manner. Different from the method of calculating the similarity between documents in the previous information retrieval, the topic model is good at automatically finding the semantic topic between words from a large-scale text set, or it can be said to obtain the probability distribution of the text on the topic.

The topic in the document is a general term for a category. It is a probability distribution of all the words in the text as a set of closely related characters. For example, regarding the topic of "fishin", we can easily think of high-frequency words such as "fishing gear" and "pond". The usual way to find topics in text is based on statistical generation: when a text has multiple topics, you can assume that a topic is selected with a certain probability and the relevant words are selected with a certain probability, finally these words form the whole document. Following this line of thought, the topic model can automatically analyze all text and semantic information, while using a mathematical framework to represent the text as a polynomial distribution of topics.

5.3   BoW Model

The bag of words is a method of text representation commonly used in the field of natural language processing and treats each document as a collection of several words. Based on this, the word set can be used as a frequency vector of a word to transform complex text content into digital information that is easy to model. The bag of words method does not take into account the word order of the text, grammar and other elements, that is, the bag of words ignores the order between the words and the words and considers each word to be independent, which greatly reduces the complexity of problems. Specifically, the bag of words model divides a document into words, and each document can be replaced with a long vector. Each dimension in the vectors represents a word and the weight of the corresponding dimension also reflects the importance of the word in the text.

5.4   TF-IDF Model

The TF-IDF (term frequency–inverse document frequency) model was proposed in 1983 and is often used for information retrieval and data mining. TF refers to the frequency of occurrence of a word in a document. The implication is that the more times a word that appears in a document, the more important it is to the document. IDF is the number of documents in which a word appears in the dataset. It can be understood that if a word appears in more documents, the word has a weaker distinction to the document and its weight in the document is smaller. The TF-IDF model mainly constructs a V-D matrix (V stands for the lexicon, which contains all possible words in the text, D represents the text set) and replaces texts of different lengths with a fixed-length matrix. TF-IDF is relatively simple but needs to consider all the elements contained in the full text. It sees each word as an independent individual, resulting in a high time complexity and spatial complexity of TF-IDF, which requires more time to train large-scale data. Moreover, TF-IDF does not care about the order between words, and does not recognize semantic relationships in documents.

6   Conclusion

In this paper, we introduce a text categorization method based on the LDA topic model to assist judicial staff to efficiently classify massive judicial documents and screen out appropriate cases as a push. In more detail, LDA uses the polynomial distribution of words to get the distribution of topic polynomial, which displays the text as a topic vector to achieve feature dimensionality reduction. And it observes the similarity between judicial texts at the topic level through semantic relations such as synonyms, polysemy and overlap.

Acknowledgement. The work is supported in part by the National Key Research and Development Program of China (2016YFC0800805) and the National Natural Science Foundation of China (61772014).

References

1. Aggarwal, C.C., Zhai, C.: Mining Text Data. Springer, New York (2012). https://doi.org/10.1007/978-1-4614-3223-4
2. Breese, J.S., Heckerman, D., Kadie, C.: Empirical analysis of predictive algorithms for collaborative filtering. In: Proceedings of the Fourteenth Conference on Uncertainty in Artificial Intelligence, pp. 43–52. Morgan Kaufmann Publishers Inc. (1998)
3. Brutlag, J.D., Meek, C.: Challenges of the email domain for text classification. In: ICML 2000, pp. 103–110 (2000)
4. Desrosiers, C., Karypis, G.: A comprehensive survey of neighborhood-based recommendation methods. In: Ricci, F., Rokach, L., Shapira, B., Kantor, P.B. (eds.) Recommender Systems Handbook, pp. 107–144. Springer, Boston, MA (2011). https://doi.org/10.1007/978-0-387-85820-3_4
5. Gutierrez-Osuna, R., Nagle, H.T.: A method for evaluating data-preprocessing techniques for odour classification with an array of gas sensors. IEEE Trans. Syst. Man Cybern. Part B (Cybern.) **29**(5), 626–632 (1999)
6. Helfer, L.R.: The politics of judicial structure: creating the United States court of veterans appeals. Conn. L. Rev. **25**, 155 (1992)
7. Herlocker, J.L., Konstan, J.A., Riedl, J.: Explaining collaborative filtering recommendations. In: Proceedings of the 2000 ACM Conference on Computer Supported Cooperative Work, pp. 241–250. ACM (2000)
8. Kansheng, S., Jie, H., Liu, H.T., Zhang, N.T., Song, W.T.: Efficient text classification method based on improved term reduction and term weighting. J. China Univ. Posts Telecommun. **18**, 131–135 (2011)
9. Le Cam, L., et al.: An approximation theorem for the poisson binomial distribution. Pac. J. Math. **10**(4), 1181–1197 (1960)
10. Li, W., Sun, L., Zhang, D.K.: Text classification based on labeled-LDA model. Chin. J. Comput.-Chin. Ed.- **31**(4), 620 (2008)

11. Lin, J., Wilbur, W.J.: Pubmed related articles: a probabilistic topic-based model for content similarity. BMC Bioinform. **8**(1), 423 (2007)
12. Linden, G., Smith, B., York, J.: Amazon.com recommendations: item-to-item collaborative filtering. IEEE Internet Comput. **1**, 76–80 (2003)
13. Manning, C.D., Manning, C.D., Schütze, H.: Foundations of Statistical Natural Language Processing. MIT Press (1999)
14. McKay, R.B.: The judiciary and nonjudicial activities. Law Contemp. Probl. **35**(1), 9–36 (1970)
15. Méndez, J.R., Iglesias, E.L., Fdez-Riverola, F., Díaz, F., Corchado, J.M.: Tokenising, stemming and stopword removal on anti-spam filtering domain. In: Marín, R., Onaindía, E., Bugarín, A., Santos, J. (eds.) CAEPIA 2005. LNCS (LNAI), vol. 4177, pp. 449–458. Springer, Heidelberg (2006). https://doi.org/10.1007/11881216_47
16. Mitchell, T., Buchanan, B., de Jong, G., Dietterich, T., Rosenbloon, P.: Machine learning annual review of computer science. J. Comput. Sci **4**, 417–433 (1990)
17. Newman, D., Smyth, P., Welling, M., Asuncion, A.U.: Distributed inference for latent Dirichlet allocation. In: Advances in Neural Information Processing Systems, pp. 1081–1088 (2008)
18. Pizarro, D.: Nothing more than feelings? The role of emotions in moral judgment. J. Theory Soc. Behav. **30**(4), 355–375 (2000)
19. Porteous, I., Newman, D., Ihler, A., Asuncion, A., Smyth, P., Welling, M.: Fast collapsed Gibbs sampling for latent Dirichlet allocation. In: Proceedings of the 14th ACM SIGKDD International Conference on Knowledge Discovery and Data Mining, pp. 569–577. ACM (2008)
20. Sarwar, B.M., Karypis, G., Konstan, J.A., Riedl, J., et al.: Item-based collaborative filtering recommendation algorithms. Www **1**, 285–295 (2001)
21. Si, X., Sun, M.: Tag-LDA for scalable real-time tag recommendation. J. Inf. Comput. Sci. **6**(2), 1009–1016 (2009)
22. Small, M.L.: How many cases do I need?' On science and the logic of case selection in field-based research. Ethnography **10**(1), 5–38 (2009)
23. Soucy, P., Mineau, G.W.: A simple KNN algorithm for text categorization. In: Proceedings 2001 IEEE International Conference on Data Mining, pp. 647–648. IEEE (2001)
24. Steyvers, M., Griffiths, T.: Probabilistic topic models. Handb. Latent Semant. Anal. **427**(7), 424–440 (2007)
25. Sun, Q., Li, R., Luo, D., Wu, X.: Text segmentation with LDA-based Fisher kernel. In: Proceedings of the 46th Annual Meeting of the Association for Computational Linguistics on Human Language Technologies: Short Papers, pp. 269–272. Association for Computational Linguistics (2008)
26. Wei, X., Croft, W.B.: LDA-based document models for ad-hoc retrieval. In: Proceedings of the 29th Annual International ACM SIGIR Conference on Research and Development in Information Retrieval, pp. 178–185. ACM (2006)
27. Weng, S.S., Tsai, H.J., Liu, S.C., Hsu, C.H.: Ontology construction for information classification. Expert Syst. Appl. **31**(1), 1–12 (2006)
28. Zelikovitz, S., Hirsh, H.: Using LSI for text classification in the presence of background text. In: Proceedings of the Tenth International Conference on Information and Knowledge Management, pp. 113–118. ACM (2001)
29. Zhang, W., Yoshida, T., Tang, X.: A comparative study of TF* IDF, LSI and multi-words for text classification. Expert Syst. Appl. **38**(3), 2758–2765 (2011)

Evaluation System for Reasoning Description of Judgment Documents Based on TensorFlow CNN

Mengting He[1,2], Zhongyue Li[1,2], Yanshu Wei[1,2], Jidong Ge[1,2(✉)], Peitang Ling[1,2], Chuanyi Li[1,2], Ting Lei[1,2], and Bin Luo[1,2]

[1] State Key Laboratory for Novel Software Technology, Nanjing University, Nanjing 210093, China
gjdnju@163.com
[2] Software Institute, Nanjing University, Nanjing 210093, China

Abstract. In order to improve the quality of the judgment documents, the state and government have introduced laws and regulations. However, the current status of trials in our country is that the number of cases is very large. Using system to verify the documents can reduce the burden on the judges and ensure the accuracy of the judgment. This paper describes an evaluation system for reasoning description of judgment documents. The main evaluation steps include: segmenting the front and back of the law; extracting the key information in the document by using XML parsing technology; constructing the legal exclusive stop word library and preprocessing inputing text; entering the text input into the model to get the text matching result; using the "match keyword, compare sentencing degree" idea to judge whether the logic is consistent if it is the evaluation of "law and conclusion"; integrating the calculation results of each evaluation subject and feeding clear and concise results back to the system user. Simulation of real application scenarios was conducted to test whether the reasoning lacks key links or is insufficient or the judgment result is unreasonable. The result show that evaluation speed of each document is relatively fast and the accuracy of the evaluation of the common nine criminal cases is high.

Keywords: Judgment documents · Reasoning description · Evaluation · Text matching · Correlation calculation

1 Introduction

The judgment document is the final product of the judicial trial. It is the main carrier that the parties can hear and see, and it is also a vivid teaching material for the legal publicity and education. It reflects not only the personal qualities of judges, but also the image of the national judiciary. The essence of the judgment document is "reasoning". The reasoning part presents the judge process in written form, which not only reflects the professional quality of the judge, but also shows the interpretation of fairness and justice by the national judicial institution. A strong rationale with pertinence, logic, and

© Springer Nature Singapore Pte Ltd. 2019
X. Cheng et al. (Eds.): ICPCSEE 2019, CCIS 1058, pp. 518–533, 2019.
https://doi.org/10.1007/978-981-15-0118-0_40

sufficiency can make the people feel fair and just, and plays a vital role in serving the lawsuit and fixing the dispute.

However, in real life, due to various reasons, such as the enormous workload of the judges, the weak sense of responsibility of some judges, and the irrationality of document drafting system, the instrument evaluation system, and the judge evaluation mechanism, many "simple and rude" judgment documents come out. These judging documents are not sufficient or even unreasonable. It is inevitable that the people will have doubts about the rationality of the trial results, which affects the credibility of the judiciary and has a bad effect on the image of the state's judicial and judicial organs. Therefore, it is very necessary to standardize the rational part of the judgment document and make the judgment document achieve the organic unity of "law" and "ration", so that the judgment result is rationalized and justified [1, 2].

In order to improve the quality of the judgment documents, the state and various levels of government have also promulgated laws and regulations to promote the openness of the Internet, requiring the court to uniformly apply the law in the trial process, effectively respond to the hot spots of disputes, and make the arguments clear and clear. As a result, the judges often need to discuss and reconfirm several times to ensure that the clues are logically clear and the results are reasonable. However, the current situation of trials in our country is that there are many cases, and one court often has to hear a number of cases every day, on average about three to six in the hands of each judge. These reasons make the work of the judge writing a very heavy burden. Therefore, the writing of the judgment part of the judgment document has become a grand project. If a system can be used to verify the reasoning description part of judgment document, it can reduce the workload of the judge and ensure the accuracy of the judgment.

The facts of the case, the quotation of the law, and the conclusion of the judgment are a logical chain connected by reference to the law. Among them, the quotation of the law is a part of the past and the future. The text of the quotation has the following characteristics: (1) In addition to the semantic conjunctions, mainly use professional legal terms, which has a high degree of specificity; (2) Summarize multiple related, possible events in one legal clause which has a high degree of abstraction. Under these characteristics, it is difficult to obtain a satisfactory effect by analyzing the relationship between the citation rule and the facts of the case, the citation rule and the judgment conclusion from the character level. It is also necessary to analyze the semantics between the texts from the semantic level of the expression. In natural language processing, there are multiple models to calculate the degree of text similarity. This paper proposes an Attention-based neural network method based on the CNN model of the convolutional neural network. The Attention mechanism is proposed by Volodymyr et al. from the inspiration of human beings to focus on a certain local area of the image when observing the image. Different weights are assigned to different regions of the image, so that the model can make more accurate judgments. This mechanism was originally used in the field of digital image processing and was subsequently widely used in natural language processing problems such as machine translation, image annotation, and relation extraction. We focus on researching an evaluation system for reasoning description of judgment document is based on the Attention-based neural network model, using referee documents and the common criminal law civil law legal documents as the data sources.

2 Related Work

TensorFlow CNN Developed by the Google Brain Group, TensorFlow [3] is an open source artificial intelligence [5] learning system for machine learning [6] and deep neural network research. The numerical calculation is performed in the data flow graph: the mathematical operation corresponds to the nodes in the data flow graph, and the interconnected multi-dimensional data array corresponds to the line connecting the nodes in the data flow graph, that is, the tensor. When all the tensors at the input are ready, the nodes will be assigned to various computing devices, performing asynchronous parallel operations. Each tensor is flowed from the input to the output in the form of an intuitive image, so the tool is called "TensorFlow" [4]. CNN, the abbreviation of Convolutional Neural Network [7], is derived from the study of cat's visual cortical cells. CNN was originally used to solve problems such as image recognition. With the development of computers, CNN now has applications in natural language processing (NLP) [8], medical discovery, and text processing. In general, a convolutional neural network consists of an input layer, a convolutional layer, a pooled layer, a fully connected layer, and an output layer. The idea of using CNN for text categorization comes from Yoon Kim's Convolutional Neural Networks for Sentence Classification. Since CNN's traditions and advantages are handled by images, TensorFlow is used for text categorization, often based on TensorFlow's TextCNN class. TextCNN made some minor adjustments to the operation of CNN to make it a CNN model for text data. According to TensorFlow's official API, users can configure various hyperparameters, enter placeholders, define embedded layers to map lexical indexes into low-dimensional vector representations, build convolutional layers, pool them, and merge the results into large ones. Feature vectors, calculation of scores and predictions, etc., complete text classification and other operations.

Gensim consists of two parts, gen-sim, the abbreviation of generate similar. Semantically, it is a free Python library that mines the semantic structure of a document by concentrating similar phrases. Only need a common text corpus, you can use gensim to discover its semantic structure, easy and efficient.

According to Gensim's official API, it can be seen that in addition to providing basic corpus processing functions such as text preprocessing, feature extraction, and topic clustering, Gensim also provides LSI, LDA, DTM, DIM and other theme models, word2vec, paragra [9].

Jieba is a lexical-based word segmentation library that supports the continuous Chinese character sequence into individual words and recombined into word sequences according to certain specifications. The main functions are as follows:

(1) Word segmentation: Jieba supports three word segmentation modes: precise mode: the most accurate separation of sentences, suitable for text analysis; full mode: scan all words in the sentence that can be worded, but there are ambiguities; search

engine mode: Based on the precise mode, the long words are split again and are suitable for search engines.

(2) Custom dictionary addition: Although Jieba has the ability to recognize new words, users can also add custom dictionaries to ensure higher word segmentation accuracy.

(3) Keyword extraction: Extract the list of words most relevant to this document from the text of the document. Jieba provides two keyword extraction modes: extraction based on TF-IDF algorithm and extraction based on TextRank algorithm.

(4) Part of speech tagging: Jieba can perform part-of-speech tagging on each word obtained after the word segmentation, using the tag method compatible with ictclas.

(5) Tokenize: Returns the starting position and ending position of the word in the original text.

(6) Parallel word segmentation: Jieba can improve the speed of word segmentation by separating the target text by line, then assigning each line of text to multiple Python processes for parallel segmentation, and finally merging the results.

(7) ChineseAnalyzer for Whoosh: Whoosh is the only native full-text search engine implemented in Python. Not only is it functionally complete, it is fast, and it is small in size. Jieba encapsulates Whoosh and provides users with the ability to search for Chinese.

3 Approach

3.1 Overview

Use XML parsing technology to extract the key elements of the document, to avoid the problem of introducing other irrelevant information in the full-length use of the document and reduce the influence of unnecessary feature vocabulary on the results of the evaluation. By calculating the document frequency, category frequency and category information entropy automatically build a legal-specific stop-loss vocabulary and combines it with the common stop-loss vocabulary commonly used in Chinese, which greatly saves the time required to manually construct the stop-loss lexicon, and reduces the influence of the irrelevant vocabulary to the final evaluation result. In order to prevent over-fitting, the obtained word list is vectored and expanded to a fixed length, and two word vectors are input into a predefined model to obtain a calculation result. These meet the requirements of judges checking that in real application scenarios whether the reasoning lacks key links or is insufficient or the judgment result is unreasonable. The detailed workflow is shown in Fig. 1. The evaluation method mainly includes the following steps:

(1) According to the extracted models of front and rear pieces of the law, the front and back parts of the law are segmented.

(2) Analyze the judgment documents and extract key information of the documents;

(3) Establish a stop-words dictionary to preprocess the text.

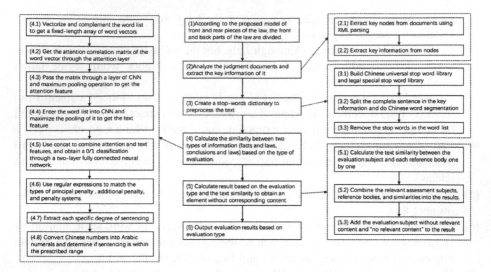

Fig. 1. Overview of our proposed method

(4) Calculate the similarity between two kinds of information (facts and laws, conclusions and laws) according to the type of assessment.

(5) Obtain elements without corresponding content according to the type of evaluation and text similarity calculation results.

(6) Output an evaluation result based on the type of evaluation.

3.2 Segmentation of the Front and Rear of the Law

Segment the legal text. The former is used as an abstract description of the possible facts to match the factual segment; the latter is used as an abstract description of the processing method to match the judgment conclusion [10].

Refer to the "Criminal Law of the People's Republic of China", "Criminal Procedure Law of the People's Republic of China", "General Principles of the Civil Law of the People's Republic of China", "General Principles of the Civil Law of the People's Republic of China" and other laws. The front and rear parts of the legal provisions appear as follows:

(1) [XXXX]的, [XXXX] ("的" is a keyword that distinguishes the front and back pieces)

(2) [XXXX]应当/可以[XXXX] ("应当", "可以" is the key to distinguish the front and rear pieces)

(3) 对/对于[XXXX], /应当/可以[XXXX]; ("对", "对于" is the key to distinguish the front and rear pieces)

(4) [XXXX]是指/是/为[XXXX]; ("是指", "是", "为" are the keywords that distinguish the front and rear parts)

In addition to simply matching keywords, the front and rear pieces themselves have a logical relationship, such as: "有XX, XX 或XX 情形之一的, XXXX",

"同时满足XX, XX 的, XXXX", these two cases correspond Two kinds of relationships: or relations and relationships, that is, "|" and "&" in computer language. In the work of segmentation, it is also necessary to distinguish between the two cases, otherwise it will affect the logical accuracy of the results of the segmentation of the front and rear parts.

3.3 Extracting Key Information

In order to obtain the content paragraphs related to the rational evaluation from the judgment documents, remove the useless data, and improve the system performance, it is necessary to extract the key information in the judgment documents. The specific steps are:

(1) Extract the start of the judgment document, the litigation record, the analysis process, the judgment result, the end of the document, and reference legal nodes. Due to the semi-structured nature of the judgment documents, a well-constructed judgment document mainly consists of the start of the document, the basic situation of the case, the original telling, the defendant's argument, the evidence, the facts ascertained, the judgment result, the reason for the judgment, the quotation of the law, the end of the document and etc. Different parts of the content have different credibility. In order to reduce the noise data, improve the accuracy of the similarity result calculation, reduce the complexity, and improve the system performance, it is necessary to extract the appropriate document information in the judgment documents. The code that uses XML parsing techniques to extract a particular node, and get the specific node content is shown in Fig. 2.

```
# The node content acquisition under the full-text node of the document
def getQWChildContent(path,childname):
    content = "
    qw = getQW(path)
    for qwchild in qw:
        if qwchild.tag == childname:
            content += qwchild.attrib['value']
    return content
```

Fig. 2. Code of getting specific nodes' content

(2) Use the XML parsing technique [10] to extract the information needed for the rational evaluation from the specific node obtained in the previous step. For example, the case number, case type and other information are extracted from the first node, and the information of the party, the charge, the case, the facts, the accusation, etc. are extracted from the litigation record node, and the referee is extracted from the end node. Information such as time, judges, etc., extracts information such as citation rules from the reference legal node.

3.4 Text Preprocessing

In order to remove the noise data and improve the training effect of the theme model, the data needs to be pre-processed [11, 12] before calculating the similarity. The specific steps are as follows:

(1) First construct the stop word library commonly used in Chinese: including punctuation, sequence number, modal particle, semantic conjunction, etc.; then construct a legal proprietary stop word library: calculate each occurrence in all documents under a specific case. The word frequency of each word, calculate their category information entropy, sort in descending order according to document frequency and category information entropy, and filter out words whose document frequency is greater than 2000 and whose category information entropy is greater than 2.0 as the stop word database. Then remove the stop words in the word segmentation results, including removing Chinese general stop words, legal exclusive stop words, removing words with word length less than 2, removing high frequency words, and removing low frequency words.

(2) Use the ";" and "." to segment for the obtained case facts, citation rules, and judgment conclusion information, and obtain the corresponding case fact set F = {f1, f2, …, fn}, reference law set L = {l1, l2, …, ln}and the set of decision conclusions J = {j1, j2, …, jn}. For each fi, li, ji, conduct Chinese word segmentation. By observing the analysis results, we found that most of the meaningful words basically belong to the three parts of "n", "v", "a". Therefore, we choose words with the word "n", "v", "a" to retain and converted them into a list of words.

(3) For the obtained list of words, use the built-in general Chinese stop word library and legal special stop word library to perform de-stop word processing.

3.5 Correlation Calculation

In order to distinguish the evaluation of facts and laws from the evaluation of conclusions and laws, it is necessary to select two types of text according to the type of evaluation and calculate the similarity. The method this paper proposed evaluates facts and laws, conclusions and rules from the perspective of text similarity and semantic similarity, and constructs FL model and LJ model. After the word list is vectored and input into the corresponding model, the predicted classification result can be obtained. The corresponding similarity calculation flow chart is shown in Fig. 3 [14]. Specific sub-steps include:

(1) The word list of the two paragraphs of text is input into an LSTM layer for vectorization to complement, and a fixed length word vector array is obtained, thereby capturing the word order information of the input text. To calculate semantic similarity, you need to get the word vector of two paragraphs of text. There are many ways to get word vectors. RNN (Recurrent Neural Network) is commonly used. The key to traditional RNN relies on the calculations of the previous moment. Therefore, the hidden state at the last moment can represent the word order information of the entire input. It can easily map any length of input into a fixed custom length output vector without losing word order information.

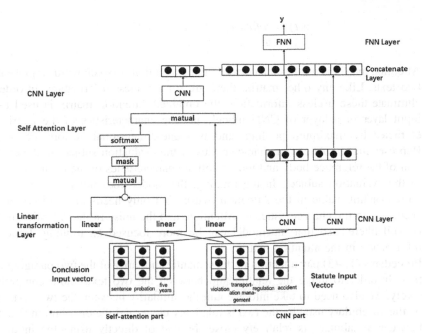

Fig. 3. Similarity calculation flow chart

However, RNN faces the problem of gradient explosion and gradient disappearance. When the input sequence is too long, the RNN cannot capture long-distance information. Therefore, a variant LSTM (Long-Short-Term Memory) of the RNN is used [13]. LSTM introduces three gates—input gates, forgetting gates, and output gates—to achieve information persistence. The forgotten gate decides which information to discard, the input gate determines what value needs to be updated, and the output gate controls what value needs to be output. Through the gate mechanism, LSTM can better capture the word order information in the input data.

(2) Although LSTM can capture the word order information of the input text, it cannot obtain the relationship between the texts. For example, when calculating the similarity between "facts and laws", we need to know which facts correspond to which of the first parts of the law, that is, which part of the law. The LSTM cannot get this part of the information, and the Attention mechanism [15] can solve this problem well. The Attention mechanism is derived from the field of digital image processing, which gives different weights to different parts of the image, enabling the model to make more accurate judgments. Therefore, construct an Attention layer, input two word vector arrays into the Attention layer, use the "dot production" method to obtain the Attention correlation matrix of the two, and perform a softmax operation on each row of the matrix to obtain a matrix; The FL model of "Fact and Law" is an example: the i-th row of the matrix, the j-th column represents the Attention weight of the i-th input word of F to the j-th input word of L, and the higher the correlation between the two words The greater the weight. The specific calculation formula is as follows:

$$\text{a}(t) = softmax(RM(1,t),\ldots,RM(|L|,t))$$
$$\alpha = [\text{a}(1),\text{a}(2),\ldots,\text{a}(|F|)]$$

(3) Attention matrix is used as the corresponding attention between words representing two texts. Like any other matrix, there are some useless information. In order to eliminate these useless information, the obtained Attention matrix is used as an input layer of a layer of CNN model, and the characteristics of the matrix are extracted by maximum pooling, and two one-dimensional vectors are output. Representing the Attention characteristics of the evaluation subject to the evaluation of the reference body and the Attention characteristics of the evaluation body to the evaluation subject. In the training, the model does not need the specific corresponding value in the Attention matrix, but only needs to be discriminated according to the characteristics of Attention, and the noise in the Attention matrix is well filtered, so that the model can be more focused on training. Important information in the attention matrix.

(4) Procedures (1)–(3) pay attention to the semantic similarity of the two paragraphs of text. In order to judge the relationship between the two texts more comprehensively, we also need to take into account the similarity between the two texts, that is, the vocabulary used by the two is relatively close, and the distribution structure between vocabulary is relatively close. Instead of directly using the input text vector for comparison, we abstract the input text into a feature vector that can represent the main information of the text, and operate on the feature vectors of the two texts. This can reduce the noise interference in the input data, and can better compare the text similarity between the two. Therefore, we use CNN to extract the text feature vector, input the word list of the two paragraphs into a CNN model and perform the maximum pooling operation to obtain two one-dimensional vectors and features.

(5) Take the FL model of "facts and rules" as an example: for facts and laws, a predecessor of a law may correspond to a variety of facts, and because of the different vocabulary of each fact, different fact types Text features may vary widely, but semantically related features are similar because they correspond to the same rule predecessor. Therefore, for text features, the aid of semantic-related features in discriminating between facts and legal relationships is more accurate. In addition, a class of facts with representative vocabulary may correspond to multiple predecessors in different legal articles, which may correspond to different semantic related features. Therefore, for semantic features, the combination of text features can be more accurately identified when discriminating between facts and laws. Based on this idea, we use the simplest concat method to combine the above four vectors, and, and input a two-layer fully connected neural network for training, get the possibility of 0/1 classification, and choose the possibility. The output is the final classification result. This can maximize the semantic related information and text feature information, so that the model can automatically find the relationship between the four vectors during training.

If the evaluation type is an evaluation between the reference law and the judgment conclusion, the operations of (6)–(8) are also required. This is because the evaluation of "laws and conclusions", in addition to the need to calculate similarities, also considers logical consistency. For example, the similarity of the text "with imprisonment for three years or less" and "sentenced to imprisonment for one year and two months" and "sentenced to five years in prison" may be very similar, but only logically consistent with the first one. Since the sentencing part of the criminal law in our country is relatively uniform, it is basically composed of one or more of the main penalty, additional penalty and penalty execution systems, supplemented by the specific degree of sentencing. Therefore, the logic is relatively simple. You can consider using the text processing library of the third party without directly relying on the idea of "matching keywords and comparing the degree of sentencing".

(6) Using regular expressions to match the types of principal, additional, and penalty systems for judgments and citations.

(7) Extracting the specific degree of sentencing for each of the cited principal, additional, and penal systems.

(8) Judgment documents use a more formal language expression. Chinese numbers are used to express the degree of sentencing. However, it is difficult to directly compare Chinese numbers with Chinese numbers. The Chinese digital expressions in the degree of sentencing need to be converted into Arabic numerals. For each of the citing laws and judgments, pairwise match keywords of the main penalty, the additional penalty and the penalty system, and for the corresponding part, judge whether the sentence in the judgment result is within the scope of the sentencing specified in the law, and return the result of the judgment.

3.6   Result Processing

In order to feed back the user's concise and understandable information, it is also necessary to synthesize the text similarity calculation result to obtain an element without corresponding content. Specific sub-steps include:

(1) Calculate the similarity between each of the items of the evaluation type and the item of the evaluation type for each item in the body of the evaluation type.

(2) If the similarity between the two is "correlated", the subject, the reference body, and the similarity of the evaluation type are combined and added to the returned result.

(3) If the similarity is "unrelated", skip the reference body and calculate the similarity and logical consistency of the evaluation subject and the next reference body.

(4) After the calculation of each evaluation subject is completed, if it is not relevant to each item in the corresponding evaluation reference, the evaluation subject and "no corresponding content" are added to the returned result.

4 Evaluation

4.1 Operation Result

Experiments were conducted on the judgment documents of nine common criminal cases, and the average time for evaluation of a single document was 9 s. The system operation result is shown in Fig. 4.

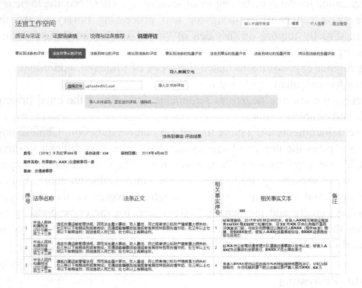

Fig. 4. Operation result

4.2 Segmentation Experiment of the Law

The Criminal Law of the People's Republic of China and the General Principles of the Civil Law of the People's Republic of China, which are often used in criminal cases and civil cases, have been divided into front and back. Based on the manual understanding, the results of the program were tested and adjusted. It can be seen that for the results of the front and back of a law, the smaller the proportion of manual adjustment or modification, the more accurate the results obtained by the program segmentation. The statistical results of manual adjustment are shown in Table 1.

It can be seen that the accuracy of using the program to segment the front and back parts is about 95%. The accuracy is high, so the results can be used for calculation of the correlation between facts and laws, laws and conclusions in the latter step.

Table 1. Statistical results of manual adjustment of the front and back parts of the law

Legal name	Total legal terms	Total number of front and back pieces obtained by segmentation	Total number of manual adjustment	Manual adjustment percentage	Segmentation accuracy
Criminal law	452	1466	64	4.37%	95.63%
General rule of the civil law	156	251	15	5.98%	94.02%

4.3 Comparison Experiment of Three Methods for Calculating Correlation

For the most frequently cited law of nine criminal cases, we randomly select the facts and conclusions that are typically related to the former and the latter respectively and calculate correlations by Gensim LDA, Gensim LSI and TensorFlow CNN respectively. Using LDA to calculate the similarity of two documents is to extract the word bag combination of two documents separately and obtain the subject of each document basing on that. The closer the distance between the subject vectors of the two articles, the higher the similarity between them. Using LSI to calculate the similarity of two documents is to generate the document matrix for two documents, and decompose and reduce the matrix by SVD (singular value decomposition) and calculate the cosine similarity between the two matrixes. The method we use CNN to calculate the similarity is to preprocess the two sentences input and extract features separately by embedding, convolution, and pooling operations, then full connect with concat method, and input the result to a fully connected neural network, let the neural network learn the relationship, and finally classify the results. There are only two classification results here: 0 or 1. 0 means there is no correlation between the two, and 1 means that the two are related. The calculated results are shown in Tables 2, 3, and 4 respectively: (The calculation results of both LDA and LSI retain four decimal places).

Table 2. Statistics of calculation results of LDA

LDA	Similarity to typical related texts		Similarity to typical irrelevant text	
	Fact	Conclusion	Fact	Conclusion
Theft	0.9817	0.8663	0.9902	0.8791
Intentional injury	0.9963	0.8891	0.9713	0.8813
Traffic accident	0.9974	0.8974	0.9966	0.8810
Rape	0.9994	0.8795	0.9958	0.8586
Robbery	0.9825	0.9004	0.9752	0.8885
Dangerous driving	0.9819	0.8942	0.9728	0.8934
Credit card fraud	0.9895	0.8752	0.9848	0.8946
Fraud	0.9922	0.8926	0.9804	0.8794
Smuggling, selling, transporting and manufacturing drugs	0.9943	0.9113	0.9618	0.8965

Table 3. Statistics of calculation results of Gensim LSI

LSI	Similarity to typical related texts		Similarity to typical irrelevant text	
	Fact	Conclusion	Fact	Conclusion
Theft	0.9998	0.4084	0.0000	0.2063
Intentional injury	0.9815	0.4115	0.0074	0.2056
Traffic accident	0.9910	0.4262	0.0000	0.2175
Rape	0.9975	0.3914	0.0009	0.1998
Robbery	0.9865	0.4635	0.0105	0.2258
Dangerous driving	0.9891	0.4250	0.0063	0.2071
Credit card fraud	0.9901	0.4469	0.0092	0.2456
Fraud	0.9886	0.4395	0.0016	0.2284
Smuggling, selling, transporting and manufacturing drugs	0.9913	0.4224	0.0004	0.2206

Table 4. Statistics of calculation results of TensorFlow CNN

CNN	Similarity to typical related texts		Similarity to typical irrelevant text	
	Fact	Conclusion	Fact	Conclusion
Theft	0	0	0	0
Intentional injury	1	0	0	0
Traffic accident	0	0	0	0
Rape	1	1	0	0.5
Robbery	1	0	0	0
Dangerous driving	0	0	0	0
Credit card fraud	1	1	0	0.5
Fraud	0	1	0	0.5
Smuggling, selling, transporting and manufacturing drugs	0	0	0	0

(1) Although the part of the instrument is often a piece of text, it often involves multiple facts, multiple articles, and multiple conclusions. Generally speaking, the facts related to a crime, the premise and the latter of the law corresponding to the crime, and the conclusions related to the crime are sentences of less than 50 words in length, that is, short texts. Therefore, in order to meet the actual situation, the data used in the experiment are also short texts of no more than 50 words in length.

(2) The data values of "similarity to typical unrelated conclusions" in the table are average values. For example, "one-year and two-month imprisonment" and "three-year imprisonment" have similarities in keywords (similarity noted as $s1k$, it can be seen that $s1k \neq 0$), but it is completely irrelevant in logic (similarity noted as $s1l$, it can be seen that $s1l = 0$), so the overall similarity $s1 = s1k + s1l$; and it is

completely irrelevant to the "life imprisonment" in terms of keywords and logic (similarity is noted as s2k and s2l respectively, it can be seen that s2k = 0, s2l = 0), the overall similarity s2 = s2k + s2l. The data for this column takes the arithmetic mean of s1 and s2, which is (s1 + s2)/2.

It can be seen from the table that the correlation between the law calculated by LDA and the typical related and irrelevant facts and conclusions exceeds 0.85. Generally speaking, if the value exceeds 0.5, it means that there is a certain relationship between the two. The degree of association. That is, for both typical irrelevant, LDA also considers the two to be related. This is because LDA is implemented by calculating the subject, and for short text, the subject may be clustered into crime, illegal, etc., so the error is large, so LDA is not suitable as a tool for calculating the similarity of short texts. At the same time, the vast majority of correlations between the laws calculated by CNN and the typical related facts and conclusions are zero, that is, for the typical correlation, CNN also considers the two to be irrelevant. This is because the CNN calculation correlation used in this paper is realized by text classification. For short text, there are few feature vectors remaining after convolution, pooling, etc. After the full connection operation, the neurons need to learn autonomously. The relationship is also very small, so the related short text does not have the same classification characteristics, so CNN is not suitable as a tool for calculating the similarity of short text.

At the same time, it can be seen from the data in the table that the result of the correlation between LSI and the facts is more accurate, and there are some problems between the calculation method and the conclusion. This is because the criminal law's laws and judgments are usually composed of the main penalty, the additional penalty, and the enforcement system. The main penalty and the additional penalty are followed by the specific degree of sentencing. For example, "the term of imprisonment for three years or less" and "the fine for five hundred yuan or less", when calculating the correlation with LSI, only the calculation of the similarity of keywords is performed, that is, the similarity of the type of sentencing is calculated. However, the specific degree of sentencing of the numbers cannot be calculated as a logical relationship. For example, "two months a year" is in line with "three years or less" and does not meet "more than ten years", resulting in its There are certain problems in the relevance of the conclusion.

5 Conclusion

This paper introduces an evaluation system for reasoning description of judicial judgment based on TensorFlow CNN. The evaluation method the system uses has the following advantages: using XML parsing technology to extract key information in the document, avoiding the problem of introducing other irrelevant information in the full-length use of the document, reducing the influence of unnecessary feature vocabulary on the result of the evaluation; by calculating the document frequency of the word, category frequency, category information entropy to automatically construct a legal-specific stop word library, and combine it with the common stop word library commonly used in Chinese, which greatly saves the time required for manually constructing the stop word library, reducing the time The impact of the related vocabulary on the final evaluation result; in order to prevent over-fitting, the obtained word list is vectored and expanded to a fixed length, and the two word vectors are input into a predefined model to obtain a calculation result.

In our work, there are still some disadvantages. We use three methods to calculate the correlation and these three methods are the most popular methods for text processing in natural language processing and artificial intelligence. However, there are still many feasible methods to calculate text relevance and we have not experimented on them. At the same time, the method used in this paper can improve the accuracy of correlation calculation by adjusting the parameter value or increasing the scale of the training set. But due to time and manpower limitations, this development work has not been repeatedly tested in these aspects.

Acknowledgements. This work was supported by the National Key R&D Program of China (2016YFC0800803).

References

1. Pollicino, O.: Legal reasoning of the court of justice in the context of the principle of equality between judicial activism and self-restraint. Ger. Law J. **5**(3), 283–317 (2004)
2. Sun, H.-P., Ma, L.-K.: Legal Reasoning in Adjudication Transparency, Justice and Credibility. Beijing University Press, pp. 5–40, (2016). (in Chinese)
3. Abadi, M., Agarwal, A., Barham, P., et al.: TensorFlow: large-scale machine learning on heterogeneous systems (2015). tensorflow.org
4. Abadi, M., Barham, P., Chen, J., et al.: TensorFlow: a system for large-scale machine learnin. In: Proceedings of the 12th USENIX Symposium on Operating Systems Design and Implementation (OSDI 2016), pp. 265–283 (2016)
5. Murray, C., Joseph, H., Feng-h, H.: Deep blue. Artif. Intell. **134**(1–2), 57–83 (2002)
6. Shalev-Shwartz, S., Ben-David, S.: Understanding Machine Learning: From Theory to Algorithms. Cambridge University Press, pp. 19–24 (2014)
7. Hijazi, S., Kumar, R., Rowen, C.: Using convolutional neural networks for image recognition, pp. 1–7. IP Group, Cadence (2015)
8. Kim, Y.: Convolutional neural networks for sentence classification. In: Proceedings of the 2014 Conference on Empirical Methods in Natural Language Processing (EMNLP), pp. 1746–1751 (2014)
9. Rehurek, R.: Gensim Documentation Release 0.8.6 (2017)
10. Bach, N.X., Nguyen, M.L., Shimazu, A.: PRE task: the task of recognition of requisite part and effectuation part in law sentences. Int. J. Comput. Process. Lang. (IJCPOL) **23**(2), 109–130 (2011)
11. Vijayarani, S., Ilamathi, J., Nithya, M.: Preprocessing techniques for text mining - an overview. J. Comput. Sci. Commun. Netw. **5**(1), 7–16 (2015)
12. Wang, Y.: Various Approaches in Text Pre-processing. Technical report, Department of Computer Science, University of Liverpool (2004)
13. Hochreiter, S., Schmidhuber, J.: Long short-term memory. Neural Comput. **9**(8), 1735–1780 (1997)

14. Wu, Y., Wu W., Li, Z., Zhou, M.: Knowledge enhanced hybrid neural network for text matching. In: Proceedings of the Thirty-Second AAAI Conference on Artificial Intelligence (AAAI-18), pp. 5586–5593 (2018)
15. Vaswani, A., Shazeer, N., Parmar, N., et al.: Attention is all you need. In: Proceedings the 30th of Annual Conference on Neural Information Processing Systems (NIPS 2017), pp. 6000–6010 (2017)

Statute Recommendation Based
on Word Embedding

Peitang Ling[1,2], Zian Wang[1,2], Yi Feng[1,2], Jidong Ge[1,2(✉)],
Mengting He[1,2], Chuanyi Li[1,2], and Bin Luo[1,2]

[1] State Key Laboratory for Novel Software Technology, Nanjing University,
Nanjing 210093, China
gjdnju@163.com
[2] Software Institute, Nanjing University, Nanjing 210093, China

Abstract. The statute recommendation problem is a sub problem of the automated decision system, which can help the legal staff to deal with the process of the case in an intelligent and automated way. In this paper, an improved common word similarity algorithm is proposed for normalization. Meanwhile, word mover's distance (WMD) algorithm was applied to the similarity measurement and statute recommendation problem, and the problem scene which was originally used for classification was extended. Finally, a variety of recommendation strategies different from traditional collaborative filtering methods were proposed. The experimental results show that it achieves the best value of F-measure reaching 0.799. And the comparative experiment shows that WMD algorithm can achieve better results than TF-IDF and LDA algorithm.

Keywords: Statute recommendation · Word embedding ·
Word mover's distance · Collaborative filtering

1 Introduction

The statutes recommendation problem refers to problem of modeling the case information and then automating the recommendation of the statutes. The purpose of the recommendation question is to save manpower and the pressure of handling the case, and to help legal staff in the process of handling the case through intelligent and automated methods.

Aiming at the statutes recommendation problem of legal judgment documents, this paper adopts the WMD algorithm based on word vector model to measure the similarity between the judgment documents, so as to use the recommendation strategy of collaborative filtering to recommend the statutes to the target judgment documents.

The main work of this paper are the following three: (1) Applying the WMD algorithm to the similarity comparison of the legal judgement documents and the statute recommendation problem, and expanding the problem scene originally used only for text classification; (2) A variety of recommendation strategies different from traditional collaborative filtering methods are proposed. (3) An improved common word similarity algorithm is proposed for the normalization of the law.

© Springer Nature Singapore Pte Ltd. 2019
X. Cheng et al. (Eds.): ICPCSEE 2019, CCIS 1058, pp. 534–548, 2019.
https://doi.org/10.1007/978-981-15-0118-0_41

This paper is organized as follows. The first section of this article introduces the issues of the article research, Sect. 2 describes related work and techniques commonly used. Section 3 details the algorithmic content used in the entire process and process of text-based similarity based on word vectors. Section 4 introduces the specific details of the experiment and analyzes the results of the experiments and comparative experiments. Section 5 summarizes and forecasts the work of this paper.

2 Related Work

There are many applications in the field of artificial intelligence applied to the legal field. In 2016, The paper published by Aletras et al. [1] introduced the modeling of 584 human rights cases through N-gram language model and Support Vector Machines (SVM) technology, achieving high accuracy. The downside is that the data set is limited to hundreds of cases with similar cases.

In 2016, Liu and others made some research on the recommendation of the law firstly [2]. They recommend the law through three-phase prediction (TPP). The limitation lies in the use of words as the analysis object, without considering the semantic information attached to the words themselves.

Bag of Words and Word Vector Technology: Bag of Words refers to a model that ignores the position of a word in a sentence and the meaning of the itself, and stores, calculates, and analyzes in a collection's data structure, just like loading words into a bag. Word vector refers to the representation, calculation and analysis of words in the form of vectors. Words express certain semantic information in the position of vector space.

One way to obtain the word vector representation of a word is to train it through a neural network model. This idea was first proposed by Xu [3]. Bengio then proposed a three-layer neural network based on the n-gram model to train the word vector [4]. Subsequently, Mikolov's skip-gram model [5] and CBOW (Continuous Bag of Words) model [6] became one of the most widely used word vector generation models (Word2Vec).

Text Similarity Measurement Technique: Text similarity algorithm is used to compare the similarity between texts. In the study of text similarity measurement, in 2015, Kusner et al. [7] proposed a new word vector model based algorithm for measuring the similarity between texts, WMD (Word Mover's Distance) algorithm. Their research shows that the classification effect of WMD algorithm is better than mainstream text analysis algorithms such as Bag of Words (BOW), TF-IDF [8] (Term Frequency–Inverse Document Frequency), BM25 Okapi [9] (a modified TF-IDF), LSI [10] (Latent Semantic Indexing), LDA [11] (Latent Dirichlet Allocation), mSDA [12] (Marginalized Stacked Denoising Autoencoder), CCG [13] (Componential Counting Grid) these algorithms.

Kusner et al. combined the word vector model with the EMD (Earth Mover's Distance) algorithm and proposed the WMD algorithm to measure the similarity between texts. EMD algorithm was first applied in computer vision problem scene [14], Due to the high computational complexity of the WMD algorithm, Kusner proposed two WMD lower bound algorithms: WCD(Word Centroid Distance) algorithms and RWMD (Relaxed Word Mover's Distance) algorithms.

The WCD algorithm measures the distance between two texts D_1 and D_2 by calculating the distance of average of the word vectors $\bar{x}$_1 of all words of D_1 and mean $\bar{x}$_2 of the word vectors of all words of D_2. The RWMD algorithm measures the distance between two texts D_1 and D_2 as shown in Eqs. 1 to 3:

$$DRWMD(D_1, D_2)_1 = \min_{T \geq 0} T_{ij} c(i,j) \quad s.t. \sum_{j=1}^{m} T_{ij} = p_i \; \forall i \in \{1, \ldots, n\} \tag{1}$$

$$RWMD(D_1, D_2)_2 = \min_{T \geq 0} T_{ij} c(i,j) \quad s.t. \sum_{i=1}^{n} T_{ij} = p'_j \; \forall j \in \{1, \ldots, m\} \tag{2}$$

$$dist(D_1, D_2) = \max\left(RWMD(D_1, D_2)_1, RWMD(D_1, D_2)_2\right) \tag{3}$$

Recommendation System and Collaborative Filtering Technology: In 2011, Wang et al. [15] applied collaborative filtering to the paper recommendation system. However, traditional collaborative filtering applications use result information to describe and compare similarities between things. Considering the separation of transaction characteristics and result information, other strategies should be adopted for collaborative filtering recommendation strategy.

3   Algorithm Design

3.1   Algorithm Framework

The algorithm takes the basic situation of the target documents set as input, and the recommended law is the output. Firstly, pre-process the basic situation text of the case, normalize the law, and then train the word vector model to obtain the word vector representation of the basic situation of the case, and then use WMD algorithm to measure the distance between the basic conditions of the case to obtain a similar set of documents, then use the verification set to determine the optimal parameters of the collaborative filtering recommendation strategy, and finally recommend laws to the instrument according to the collaborative filtering recommendation strategy. The overall process of the recommendation method is shown in Fig. 1.

(1) Obtaining the initial data set: obtaining the basic situation set $A_1 = \{a_1, a_2, \ldots, a_n\}$ of case and the referenced statutes set $F_1 = \{f_1, f_2, \ldots, f_n\}$;

(2) Pre-processing the basic situation of the case: using the word segmentation tool to segment the basic situation a_i of each case in A_1 and remove the stop words, and obtain the basic situation set of the pre-processed case $A_2 = \{a'_1, a'_2, \ldots, a'_n\}$;

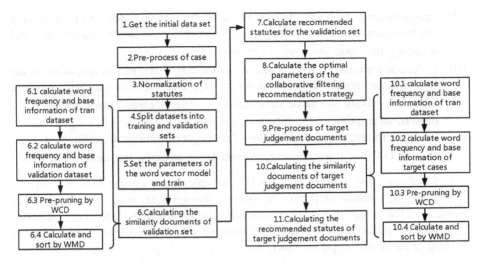

Fig. 1. Process of collaborative filtering recommendation algorithm

(3) Normalize the statutes: normalize all $statutes f_i = \{flft_1, flft_2, \ldots\}$ in F_1, get $f_i' = \{std\_flft_1, std\_flft_2, \ldots\}$, get $F_2 = \{f_1', f_2', \ldots, f_n'\}$; For a detailed description of the sub-steps, refer to Sect. 3.2.

(4) Split data set: split the case basic situation set A_2 and the law strip set F_2 into training sets $A_{2,t}$, $F_{2,t}$ and verification set $A_{2,v}$, $F_{2,v}$ according to a certain proportion;

(5) Set the word vector model parameters and train: set the window size, the minimum frequency of occurrence of the word, the length of the word vector, use the word vector model to train the case basic set $A_{2,t}$, and obtain the trained word vector model M;

(6) Calculate similar set of documents for the validation set: According to the word vector model M, the training set $A_{2,t}$, the verification set $A_{2,v}$, and the text similarity measure algorithm, obtain the basic situation of the m cases $A_{WMD,i}$ closest to the basic situation of each case a_i' in verification set; For a detailed description of the sub-steps, refer to Sect. 3.3.

(7) Calculate the set of recommended statutes for the verification set: for each case in the verification set a_i', according to $A_{WMD,i}$, $F_{2,t}$ and the collaborative filtering recommendation strategy, recommend the statute f_i'', Get the recommended statutes set $F_{3,v}$;

(8) Calculate the optimal parameters of the recommendation strategy: Calculate the accuracy, recall rate and F value(Harmonic average of accuracy and recall) of the verification set under the different parameters of the collaborative filtering recommendation strategy according to $F_{2,v}$ and $F_{3,v}$, determining the parameters of the collaborative filtering recommendation strategy with the best value of F value;

(9) Pre-processing the basic situation of the target judgment document: classifying the basic situation a_g of the target document and removing the stop words, and obtaining the basic situation of the pre-processed case a'_g;

(10) Calculate a similar set of documents of the target document: according to the word vector model M, the training set $A_{2,t}$ and the text similarity measure algorithm, obtain the basic case $A_{WMD,g}$ of the m cases closest to a'_g; For a detailed description of the sub-steps, refer to Sect. 3.3.

(11) Calculate the recommended statutes of the target document: according to $A_{WMD,g}$, $F_{2,t}$ and collaborative filtering recommendation strategy, recommend the law f''_g.

In the above steps, step (6) and step (10) are calculating similar document sets for verification set, wherein step (6) The data of the verification set can be similarly regarded as the target data, which contains four similar sub-steps. For a detailed description of the examples of the collaborative filtering recommendation strategy involved in the above steps, see Sect. 3.4.

3.2 Data Preprocessing

Most of the processing techniques are based on the words as the smallest unit of processing, so it is necessary to use word segmentation technique to segment the basic situation of the case from the judgment documents.

After the basic situation of the case was processed by word segmentation, the sequence of words obtained needs to be cleaned up to remove a series of stop words. After the stop words are removed, the basic situation of the case can be analyzed. The process is shown in Fig. 2.

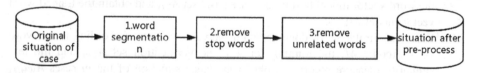

Fig. 2. Preprocessing process of case basic situation

This paper has simplified statute, only retains the articles of the law, but removes the paragraph, subparagraph, and clause. In the normalization of statutes, using the similarity algorithm based word units. Therefore, it is necessary to standardize the laws extracted from the judgment documents, and adopt automated method. The specific steps can be described as:

(1) Separating the statute name and item information of the law $flft$ into fl and ft;

(2) Remove punctuation and special characters from the statute name fl;

(3) Using the similarity algorithm in words to measure the similarity between fl and the standard statute name set $FL = \{std\_fl_1, std\_fl_2, \ldots, std\_fl_n\}$;

(4) Select the standard legal name std_fl_i with the highest similarity with the statute name fl as the result of standardization of the statute name;

(5) The paragraph, subparagraph, and clause in the information ft are removed, and only the article is retained, and std_ft is obtained as a result of standardization of the item information;

(6) The results obtained in (4) and (5) are spliced to obtain the result of normalization $std\_flft = std\_fl_i + std\_ft$;

In the step (3) the normalization of statutes above-mentioned, the similarity algorithm in word units is involved. The Levenshtein edit distance algorithm and the improved common word similarity algorithm proposed in this paper are introduced and compared below.

The Levenshtein edit distance algorithm is an algorithm proposed by Levenshtein in 1965 to compare the similarity between two strings. The edit distance refers to the minimum number of operations by converting a string into another character only by inserting, deleting, and replacing. The statute name fl and the standard statute name set $FL = \{std\_fl_1, std\_fl_2, \ldots, std\_fl_n\}$ are calculated one by one, and the distance is the smallest std_fl_i is the standard statute name with the highest similarity to the statute name fl.

Algorithm 1 Improved common word similarity algorithm
01. $sim = 0$
02. for ch in fl do
03. $point = 1$
04.   if ch in IDF_mapper then
05. $point = IDF\_mapper[ch]$
06.   end if
07.   if ch in std_fl_i then
08. sim += $point$
09.   else
10. sim -= $point$
11.   end if
12. end for
13. for ch in std_fl_i do
14. $point = 1$
15.   if ch in IDF_mapper then
16. $point = IDF\_mapper[ch]$
17.   end if
18.   if ch in f then
19. sim += $point$
20.   else
21. sim -= $point$
22.   end if
23. end for
24. sim -= abs(len(fl) − len(std_fl_i))
25. return sim

Fig. 3. The pseudocode of improved common word similarity algorithm

When manually standardizing the statute name fl, it usually pays attention to the more special words as the basis for standardization, and compares the number of characters shared by the statute name fl and a standard legal name std_fl_i, the number of non-common characters and the difference between the lengths to determine the similarity between the two. Therefore, this project draws on the TF-IDF method, the common word similarity algorithm and the process of manual standardization, proposes a new The similarity measure method (as shown in Fig. 3), with the statute name fl to be standardized, the standard legal name std_fl_i, the IDF value IDF_mapper of the character in the standard statute name set as input, the similarity sim is the output, the specific steps described as follows:

(1) Calculate the IDF value of the character: use the standard name set FL as the data set, calculate the IDF value of each character, and get IDF_mapper;
(2) Initialization similarity: the similarity sim between the statute name fl and the standardized statute name std_fl_i initialize 0;
(3) Calculate the inclusion similarity of the standard statute name: For each word ch in the standard statute name std_fl_i, if the statute name fl contains ch, then sim increases the IDF value corresponding to ch, if not, sim subtracts ch corresponding IDF value;
(4) Calculate the inclusion similarity of the statute name: For each word ch in the statute name fl, if the standard statute name std_fl_i contains ch, sim increases the IDF value corresponding to ch, if not, sim subtracts the corresponding to ch IDF value;
(5) Calculate the length similarity between the statute name and the standard statute name: sim minus the length difference between the statute name fl and the standard statute name std_fl_i.

Table 1. The comparison of legal name standardization between Levenshtein edit distance and improved common word similarity algorithm

Statute name fl	Irregular reason	Standardization result std_fl	
		L edit distance	Improved similarity
Opinions on Dealing with Several Specific Issues of Self-confidence and Standing Power	Omit the issuing authority	Reply of the Supreme People's Court on the issue of Ma Naping's divorce	Interpretation of the Supreme People's Court on Several Issues Concerning the Specific Application of Law of Self-sufficiency and contribution
Civil Law	Short name	Nurse regulations	General Law of the People's Republic of China
Criminal Law of the Republic of China	Duplicate part	National Emblem Law	Criminal Law of the People's Republic
Road Safety Law of the People's Republic of China	Some content is missing	Mine Safety Law of People's Republic of China	Road Traffic Safety Law of People's Republic of China
Marriage Law of People's Republic of China	No	Marriage Law of People's Republic of China	Marriage Law of People's Republic of China

If there is an abbreviation in the statute name *fl*, such as "the General Principles of the Civil Law of the People's Republic of China" is abbreviated as "General Principles of the Civil Law", the editing distance between the L and the standard statute name calculated by the L edit distance algorithm is 7, which is greater than the edit distance of all standard statute names less than 7 in length, so the statute name cannot be correctly normalized. But the improved common word similarity algorithm does not have such limitation, although it will also reduce the similarity due to the difference in length, due to the successful matching of the common words with high IDF values, the correct standardized results will be obtained. The standardization results of the two text similarity comparison algorithms on different statute names are shown in Table 1. As shown, it can be seen that the improved common word similarity algorithm can obtain better standardized results.

3.3 Text Similarity Measure

After pre-processing, we can use the Word2vec provided in the gensim text analysis package to train the word vector model M. After obtaining the word vector model M, the WMD algorithm can be used to measure the similarity of the basic situation of the case, and obtain the judgment documents that are similar to the basic situation of the target case, which is used as the basis for the follow-up recommendation. The WMD algorithm is used to calculate the specific case description $A_{WMD,g}$ of m the cases with the highest degree of similarity in the dataset $A_1 = \{a_1, a_2, \ldots, a_n\}$ and the basic situation of the target case, as shown in Fig. 4:

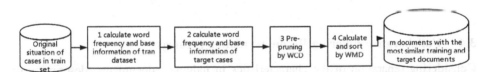

Fig. 4. The process of measuring case basic situation similarity by WMD algorithm

(1) Calculate the basic situation and word frequency of the training set case based on the word vector model: Calculate the word frequency of the basic case a_i of each case in A_1 according to the word vector model M, obtain the word frequency set $P_1 = \{p_1, p_2, \ldots, p_n\}$, and the words Replace the basic case a_i' in the form of a word vector, and obtain A_2;

(2) Calculate the basic situation and word frequency of the target case based on the word vector model: According to the word vector model M, calculate the word frequency of a_g, obtain the word frequency p_g, and replace the word with the word vector form case basic case a_g';

(3) Use WCD pre-pruning: According to the basic situation a_i' of each case in a_g', p_g and A_2 and the corresponding p_i in P_1, calculate WCD and sort, select $h = 2\ m$–$10\ m$, get $A_{WCD,g}$;

(4) Calculate WMD and sort: According to the basic situation a'_i of each case in a'_g, p_g and $A_{WCD,g}$ and the corresponding p_i in P_1, calculate WMD and sort, select top m small, get $A_{WMD,g}$;

In the above steps, the WMD is used to directly calculate the similarity after WCD pre-prune, and the RWMD pre-pruning is not continued between the two steps. Directly sort similar documents using the WMD method after sorting by WCD pre-pruning, which is faster, and WCD has better pre-pruning effect, so calculating RWMD itself is an extra step, and in the case of algorithm optimization, WMD and RWMD calculations are not much different from speed, and they are all solved based on the EMD problem, so the pre-pruning of RWMD can be directly removed in the actual situation. The average calculation time of these similarity algorithms can be seen in Table 2.

Table 2. The time of calculating m documents that are most similar to a_g by different methods while $|A_1| = 3000$

Method name	Average time (seconds)
WCD	2.34
RWMD	56.13
WMD (without pre-prune)	81.83
WMD (pre-pruning using WCD, parameter h = 2 m)	2.78
WMD (pre-pruning using WCD, parameter h = 4 m)	3.24
WMD (pre-pruning using WCD, parameter h = 10 m)	4.84

3.4   Collaborative Filtering Recommendation Strategy

Considering separation of the characteristics in judgment documents (description of information of the case, etc.) and the result information (citing the law), use WMD to measure the similarity of the basic situation characteristics of the documents, and then use a variety of strategies recommend law to the target document. After obtaining the basic situation $A_{WMD,g}$ of the top m small in a single document, a variety of recommendation strategies can be adopted to recommend the law f_g, select four descriptions are as follows:

(1) Strategy 1 (frequency top k recommendation strategy): Select top k by the number of occurrences of the law contained in the documents $A_{WMD,g}$ of *topm* in distance is recommended to the target document, and the value of k is determined by the verification set selecting best value F (the average of the harmonics of the accuracy and recall);

(2) Strategy 2 (cumulative distance reciprocal top-k recommendation strategy): Calculate the weight of the law contained in the document $A_{WMD,g}$ of the distance top

m, the weight is the reciprocal of WMD, and recommend top k to the target document. The value of k is taken from the verification set to obtain the optimal F value;

(3) Strategy 3 (threshold recommendation strategy): In $A_{WMD,g}$, the law that exceeds a certain percentage t of the number of occurrences of the law contained in the documents which are in top s by distance is recommended to the target document, and the values of s and t is derived from the verification set to obtain the optimal F value. In order to prevent the occurrence of the statute without exceeding the t, a statute that has the highest number of occurrences of the law contained in the document in top s is recommended to the target document;

(4) Strategy 4 (full recommendation strategy): In $A_{WMD,g}$, all the statutes contained in the document from the top s by distance are recommended to the target document, and the value of s is derived from the verification set.

4 Simulation Experiment

This section experiments on the statute recommendation algorithm proposed in this paper. In the experiment, two cases of civil cases were used in the case of divorce disputes and civil loan disputes, and a case of criminal cases was used as a data set for the experiment under the crime of traffic accidents. These judging documents have undergone a certain coarse-grained processing before the experiment, the basic situation of the case contains the basic situation of the case of the judgment document, and the referee analysis process segment contains the law cited in the judgment document.

4.1 Experimental Parameters

The number of judging documents used was: 12,845 for divorce disputes, 11,953 for private lending disputes, and 10,000 for traffic accidents. In the collaborative filtering recommendation process, the verification set was set to determine the optimal parameters of the collaborative filtering recommendation strategy because the parameters would affect the final experimental results. The ratio of the data set, the verification set and the test set is set to the ratio commonly used in machine learning 3:1:1. The basic cases $A_{WMD,g}$ from a single document obtained by WMD top-m, m is set to 10. Because the experiment shows that the number of similar judgements is generally less than this value, so when applying the collaborative filtering recommendation strategy, the basic situation of the subsequent case is no longer needed as the basis for recommendation. In the training parameters of the model, the set window size is 5, the minimum occurrence frequency of the word is 5, and the length of the word vector is 100.

4.2 Experiment Metrics

Precision, recall and F values are commonly used statistical evaluation indicators for evaluating the results of search, recommendation, prediction, etc.

In the experiment, the F value is used as the criterion for obtaining the optimal parameters of the verification set and the criterion for evaluating the recommendation result, because the F value has a main characteristic as the harmonic mean of the accuracy rate and the recall rate: the harmonic mean is susceptible to extreme values. In particular, the effect of extreme small values, that is, if the accuracy or recall rate has a value that is too small, the F value will become smaller, which can fully reflect the influence of the recommended method on the recommended effect.

In the scenario of recommending the law, if we pursue high accuracy, method that recommend only a few statutes with high statistical probability will get a very big advantage. This fixed recommendation strategy is not practical in practical applications. If the pursuit of excessive recall rate, the recommendation strategy of recommending a large number of statutes will get a very big advantage. In practical application, it will lead to too many invalid articles, which is not convenient to judge the case. Therefore, this paper uses the F value as the evaluation standard for the results of the recommended method.

4.3 Experiment and Results Analysis

The data set of the experimental results comes from all the judgment documents of the three cases, and the mixed case consisting of 3,000 papers randomly selected from each case. The experimental results of the four data sets under the four collaborative filtering recommendation strategies are shown in Table 3, including the parameters of the collaborative filtering recommendation strategy, the accuracy of the test set, the recall rate, and the F value when the verification set obtains the optimal F value. Among them, the strategy 1 refers to the frequency top-k recommendation strategy, and the strategy 2 refers to cumulative distance reciprocal top-k recommendation strategy, the strategy 3 refers to the threshold recommendation strategy, the strategy 4 refers to the full recommendation strategy, and Fig. 5 shows the comparison of the F values of the four data sets under the four collaborative filtering recommendation strategies.

By analyzing the data in the table, all the cases have achieved good experimental results, and the F values are all over 0.5. Among them, the traffic threshold crime policy has reached the F value of 0.799 by the threshold recommendation strategy, has achieved very good experimental results.

In the four data sets, the threshold recommendation strategy has nearly achieved the optimal accuracy, recall rate and F value compared with other collaborative filtering recommendation strategies, which shows that this recommendation strategy has advantage, which can be clearly seen in Fig. 6.

Table 3. Experimental results of four data sets under four collaborative filtering recommendation strategies

Case	Strategies	Parameter		Precision	Recall	F-value
Loan dispute	1	k = 6		0.466	0.593	0.522
	2	k = 6		0.478	0.61	0.536
	3	s = 9	t = 0.3	0.489	0.612	0.543
	4	s = 1		0.538	0.563	0.55
Divorce dispute	1	k = 2		0.721	0.6	0.655
	2	k = 2		0.722	0.601	0.656
	3	s = 10	t = 0.4	0.756	0.657	0.703
	4	s = 1		0.676	0.636	0.655
Traffic crime	1	k = 4		0.815	0.75	0.781
	2	k = 4		0.817	0.752	0.783
	3	s = 7	t = 0.5	0.825	0.775	0.799
	4	s = 1		0.821	0.758	0.788
Mixed case	1	k = 4		0.597	0.622	0.61
	2	k = 4		0.607	0.632	0.619
	3	s = 10	t = 0.4	0.672	0.625	0.648
	4	s = 1		0.63	0.58	0.604

4.4 Comparison Experiment of WCD Pre-pruning Parameter

When using the WMD algorithm to measure the degree of similarity between documents, WCD will be used for pre-pruning. In this case, the basic case of top $h = 2$ m–10 m small by WCD will be selected, and $A_{WCD,g}$ will be obtained. Taking the influence of different values on the results, the experiment selected 3,000 documents from the divorce dispute case, using the threshold recommendation strategy as the recommendation strategy for collaborative filtering. Table 5 and Fig. 6 show the experimental results when the parameters h are 2 m, 4 m and 10 m respectively and the time of m judgements that are most similar to a single document.

It can be seen from Table 4 and Fig. 6 that the WCD pre-pruning parameters h take 2 m, 4 m and 10 m respectively, the running time has a significant increase, but the result F value does not change much, which shows that the WCD method has a good pre-pruning effect, after filtering out 2 m texts, it is no longer necessary to increase the data of the basic situation of the case for WMD calculation. At the same time, because the change of the F value is not large, the comprehensive time factor and effect consideration are considered in this experiment. Other experiments use $h = 2$ m as the parameter for WCD pre- pruning by default in this paper.

Table 4. The comparison of the result and time of h with different values

h	F-value	Time (seconds)
2 m	0.703	2.78
4 m	0.713	3.24
10 m	0.714	4.84

4.5   Comparison Experiment of Different Text Similarity Measurement Methods

In order to illustrate the advantages of WMD in measuring text similarity, the experiment selected 3,000 documents from the divorce dispute case, using four different collaborative filtering recommendation strategies. Table 5 and Fig. 7 show effect of algorithms on measuring text similarity. It can be seen under these four collaborative filtering recommendation strategies, the WMD algorithm is superior to the other two text similarity recommendation algorithms, and the F value is superior to the other two algorithms in each recommendation strategy by 4% or even More. The WMD algorithm has significant advantages in measuring text similarity.

Table 5. Comparison of three text similarity comparison algorithms

Similarity algorithm	Strategies	Parameter		F-value
TF-IDF	1	k = 1		0.617
	2	k = 1		0.617
	3	s = 7	t = 0.5	0.648
	4	s = 1		0.578
LDA	1	k = 1		0.617
	2	k = 1		0.617
	3	s = 7	t = 0.5	0.623
	4	s = 1		0.527
WMD	1	k = 2		0.655
	2	k = 2		0.656
	3	s = 10	t = 0.4	0.703
	4	s = 1		0.655

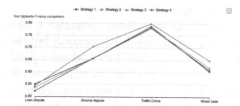

Fig. 5. Comparison of four data sets under four collaborative filtering recommendation strategies

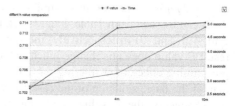

Fig. 6. The result and time comparison diagram of h with different values

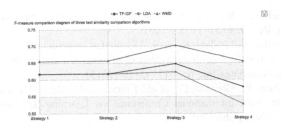

Fig. 7. F-measure comparison diagram of three text similarity comparison algorithms

5 Conclusion

This paper introduces a collaborative filtering method recommendation algorithm based on word vector model and WMD text similarity measurement algorithm. WMD method is derived from EMD method, and with the powerful expression ability of word vector, it can measure the similarity between texts and provided good raw data. A variety of collaborative filtering recommendation strategies can make up for the traditional application collaborative filtering recommendation strategy to measure the similarity between things based on the result information, and achieve a better recommendation effect.

The data set used in the experiment comes from the basic case of the judgment documents. The recommendation process uses the entire text segment as the basis to recommend statutes for the target documents. If we can split the basic situation of the case and use the split sub-text segment to establish a certain relationship with the reference law of the documents, we can recommend the law at a finer granularity, which may achieve better results.

Results of the experiment show that the threshold recommendation strategy obtains better results than the other recommendation strategies. However, the current strategy has ample space, such as combined with the above-mentioned strategy and the total citation frequency of the law, to recommend statutes, or based on other evaluation criteria to recommend statutes. The use of a more novel statute recommendation strategy in the framework of collaborative filtering is another direction to research.

Acknowledgements. This work was supported by the National Key R&D Program of China (2016YFC0800803).

References

1. Aletras, N., Tsarapatsanis, D., Preoţiuc-Pietro, D., et al.: Predicting judicial decisions of the European Court of Human Rights: a natural language processing perspective. Peerj Comput. Sci. **2**(2), e93 (2016)
2. Liu, Y.H., Chen, Y.L., Ho, W.L.: Predicting associated statutes for legal problems. Inf. Process. Manag. **51**(1), 194–211 (2015)
3. Xu, W., Rudnicky, A.I.: Can artificial neural networks learn language models? In: Sixth International Conference on Spoken Language Processing, pp. 202–205 (2000)

4. Bengio, Y., Ducharme, R., Vincent, P., Jauvin, C.: A neural probabilistic language model. J. Mach. Learn. Res. **3**, 1137–1155 (2003)
5. Mikolov, T., Sutskever, I., Chen, K., et al.: Distributed representations of words and phrases and their compositionality. In: International Conference on Neural Information Processing Systems Curran Associates Inc. pp. 3111–3119 (2013)
6. Mikolov, T., Chen, K., Corrado, G., et al.: Efficient estimation of word representations in vector space. In: First International Conference on Learning Representations, pp. 1–12 (2013). https://arxiv.org/abs/1301.3781
7. Kusner, M.J., Sun, Y., Kolkin, N.I., et al.: From word embeddings to document distances. In: International Conference on International Conference on Machine Learning JMLR.org, pp. 957–966 (2015)
8. Salton, G., Buckley, C.: Term-weighting approaches in automatic text retrieval. Inf. Process. Manag. **24**(5), 513–523 (1988)
9. Robertson, S.: Okapi at TREC-3. In: Overview of the Third Text REtrieval Conference, pp. 109–125 (1995)
10. Deerwester, S.: Indexing by latent semantic analysis. J. Am. Soc. Inf. Sci. **41**(6), 391–407 (1990)
11. Blei, D.M., Ng, A.Y., Jordan, M.I.: Latent Dirichlet allocation. J. Mach. Learn. Res. Arch. **3**, 993–1022 (2003)
12. Chen, M., Xu, Z., Weinberger, K., et al.: Marginalized denoising autoencoders for domain adaptation. In: 29th International Conference on Machine Learning, pp. 1–8 (2012)
13. Perina, A., Jojic, N., Bicego, M., et al.: Documents as multiple overlapping windows into a grid of counts. In: Advances in Neural Information Processing Systems, pp. 10–18 (2013)
14. Rubner, Y., Tomasi, C., Guibas, L.J.: A metric for distributions with applications to image databases. In: International Conference on Computer Vision, p. 59. IEEE Computer Society (1998)
15. Wang, C., Blei, D.M.: Collaborative topic modeling for recommending scientific articles. In: ACM SIGKDD International Conference on Knowledge Discovery and Data Mining. ACM, pp. 448–456 (2011)

Machine Learning

Knowledge Graph Embedding Based on Adaptive Negative Sampling

Saige Qin, Guanjun Rao[✉], Chenzhong Bin, Liang Chang,
Tianlong Gu, and Wen Xuan

Guangxi Key Laboratory of Trusted Software,
Guilin University of Electronic Tochnology, Guilin 541004, China
rao-rgj@foxmail.com

Abstract. Knowledge graph embedding aims at embedding entities and relations in a knowledge graph into a continuous, dense, low-dimensional and real-
valued vector space. Among various embedding models appeared in recent
years, translation-based models such as TransE, TransH and TransR achieve
state-of-the-art performance. However, in these models, negative triples used for
training phase are generated by replacing each positive entity in positive triples
with negative entities from the entity set with the same probability; as a result, a
large number of invalid negative triples will be generated and used in the
training process. In this paper, a method named adaptive negative sampling
(ANS) is proposed to generate valid negative triples. In this method, it first
divided all the entities into a number of groups which consist of similar entities
by some clustering algorithms such as K-Means. Then, corresponding to each
positive triple, the head entity was replaced by a negative entity from the cluster
in which the head entity was located and the tail entity was replaced in a similar
approach. As a result, it generated a set of high-quality negative triples which
benefit for improving the effectiveness of embedding models. The ANS method
was combined with the TransE model and the resulted model was named as
TransE-ANS. Experimental results show that TransE-ANS achieves significant
improvement in the link prediction task.

Keywords: Adaptive negative sampling · Knowledge graph embedding ·
Translation-based model

1 Introduction

Knowledge graph is a directed graph composing of different types of entities as nodes
and relations between entities as edges. Essentially, knowledge graph is a semantic
network that expresses the semantic relations between various types of entities. Its
basic constituent unit is a triple (h, r, t), where h is a head entity, t is a tail entity, and r
represents the relationship between head and tail. Recently, the knowledge graph has
played a vital role in data mining, artificial intelligence and other fields, and it promotes
the development of artificial intelligence applications, such as web search, and question
answering.

© Springer Nature Singapore Pte Ltd. 2019
X. Cheng et al. (Eds.): ICPCSEE 2019, CCIS 1058, pp. 551–563, 2019.
https://doi.org/10.1007/978-981-15-0118-0_42

With the advent of big data era, the scale of knowledge graph has grown rapidly, and various large-scale knowledge graphs (such as Freebase, WordNet, and NELL) have appeared. Although the scale of knowledge graph is very large, but it is still incomplete. So it is necessary to complete the current knowledge graphs. This is one of the current hot topics in knowledge graph. Recently, it is very popular to learn methods of learning vector representations of entities and relations in the knowledge graph, where embedding-based representation learning method show strong feasibility and robustness. These methods are to embed the entities and relations in knowledge graph into a continuous, dense, low-dimensional, and real-valued vector space.

Among various representational learning methods, translation-based presentation learning model achieves state-of-the-art performance. TransE is the most typical translation-based model which proposed by Bordes et al. in 2013. TransE treats relations as translations from head entity to tail entity in vector space. If a triple (h, r, t) holds, head entity vector $\mathbf{h}$, tail entity vector $\mathbf{t}$ and relation vector $\mathbf{r}$ should satisfy $\mathbf{h} + \mathbf{r} = \mathbf{t}$. TransE is extremely simple and shows excellent performance in processing large-scale data, thus set off a research boom in translation-based presentation learning.

TransH model proposes that an entity has different representations in different relations, projects entity onto a hyperplane where the relation lies, and then performs translation operations on hyperplane. Lin et al. believe that entities and relations should be in different semantic spaces, and propose TransR model. TransR projects entities from the entity space into the relation space through a projection matrix, and then establishes translation operations in the relational space. Ji et al. proposed the TranSparse model considering imbalance and heterogeneity of entities and relations. TranSparse solves this problem by using a sparse matrix instead of a dense matrix in TransR. FT model considers that translation principle of $\mathbf{h} + \mathbf{r} = \mathbf{t}$ is too strict, thus establishing a more flexible translation principle $\mathbf{h} + \mathbf{r} = \alpha\mathbf{t}$, which improves the expression ability of the translation model. DT model considers that translation principle of FT model are still too complex, and further proposes dynamic translation principle $(\mathbf{h} + \alpha_h) + (\mathbf{r} + \alpha_r) = (\mathbf{t} + \alpha_t)$, which improves the performance of the translation model.

Although the translation-based representation learning model has been improved steadily, this aspect of research still faces a common challenge. In the representation learning model, knowledge graph used for training only has valid positive knowledge (positive triples), and there is no valid negative knowledge (negative triples), which proposes a great challenge for representation learning model training. In order to get a negative triple, the existing model usually deletes the head entity (or tail entity) in the positive triple and selects one entity randomly to replace head entity (or tail entity) in the entity set. Unfortunately, this method is not ideal, because of the large number of entity sets, the number of negative triples is even larger corresponding to each positive triple. In the negative triples, most of the negative triples are very different from the positive triples, making them extremely easy to distinguish. Thus, they are an invalid negative triple (e.g., negative triples (UnitedStates, President, NewYork)). If we just randomly replace an entity with equal-probability to generate a negative triple, that usually result in the generation is easy to distinguish negative in most cases. In other words, the large number of negative triples generated are not similar to the positive triples and such negative triples do not help model training.

In this paper, we believe that generating an effective negative triple is very important for training the model, and the current method of random substitution with equal-probability is not reasonable. To that end, we propose an adaptive negative sampling method to generate a valid negative triple. We integrate adaptive negative sampling method into TransE, named TransE-ANS, and perform extensive experiments on four common datasets (i.e., WN18, WN11, FB15k and FB13), both of them obtain superior state-of-the-art performance.

The rest of this paper is organized as follows. We introduce related work in Sect. 2. We detail a limitation of equal-probability sampling and a method of adaptive negative sampling in Sect. 3. We evaluate our method through experiments in Sect. 4. Conclusions are summarized in Sect. 5.

2   Related Work

We divided the previous studies into two categories: one is a translation-based model, and the other is an embedded model.

2.1   Translation-Based Models

TransE [1] treats r as the translation operation from h to t in triple (h, r, t). TransE believes that each triple in knowledge graph should satisfy $\mathbf{h} + \mathbf{r} = \mathbf{t}$. In other words, when (h, r, t) is a positive triple, $(\mathbf{h} + \mathbf{r})$ should be close to $(\mathbf{t})$; otherwise $(\mathbf{h} + \mathbf{r})$ will go away from $(\mathbf{t})$. Therefore, score function for embedding of a triple is $f_r(h, t) = \|\mathbf{h} + \mathbf{r} - \mathbf{t}\|_2^2$. TransE is a simple and effective model. Especially, when dealing with 1-to-1 simple relations, the performance of the model is particularly outstanding, but there are some defects in dealing with complex relations of 1-to-N, N-to-1 and N-to-N. For example, there is a triple set (h_1, r, t), (h_2, r, t), ... (h_n, r, t). Embedding by TransE, we might get $h_1 = h_2 = \ldots = h_n$.

Wang et al. [2] believe that each entity should have different representations in different relations. TransH projects head entity vector and tail entity vector to a relation hyperplane by normal vector ω_r, and correspondingly obtains the projected vectors $\mathbf{h}_\perp$ and $\mathbf{t}_\perp$. Its score function is $f_r(h, t) = \|\mathbf{h}_\perp + \mathbf{r} - \mathbf{t}_\perp\|_2^2$.

TransR/CtransR [3] proposes to embed entities and relations into different semantic spaces respectively and constructs a projection matrix M_r for each relation r. Then, it uses M_r to project the entity vector into the relation space. Its score function is $f_r(h, t) = \|\mathbf{h}M_r + \mathbf{r} - \mathbf{t}M_r\|_2^2$.

TranSparse [4] mainly considers the heterogeneity and imbalance of entities and relations in knowledge graph to prevent overfitting of simple relations and underfitting of complex relations. TranSparse constructs sparse matrices $M_r^h(\theta_r^h)$ and $M_r^t(\theta_r^t)$ for head entities and tail entities adaption based on the complexity of each relation. Its score function is $f_r(h, t) = \left\|M_r^h(\theta_r^h)\mathbf{h} + \mathbf{r} - M_r^t(\theta_r^t)\mathbf{t}\right\|_2^2$.

FT [5] thought that the translation principles of the existing models $\mathbf{h} + \mathbf{r} = \mathbf{t}$ are too strict to express the complex diversity of entities and relations. Therefore, a novel and more flexible translation principle $\mathbf{h} + \mathbf{r} = \alpha\mathbf{t}$ is proposed. The FT can flexibly model complex and various entities and relations, and easily extend to a series of translation models such as TransE. FT uses the following score function $f_r(h, t) = (\mathbf{h} + \mathbf{r})^\mathrm{T}\mathbf{t} + \mathbf{h}^\mathrm{T}(\mathbf{t} + \mathbf{r})$.

Chang et al. [6] believed that FT model was still too strict in changing entities in a certain direction, and put forward a representation learning model based on the principle of dynamic translation. DT defined a new kind of translation rules $(\mathbf{h} + \alpha_h) + (\mathbf{r} + \alpha_r) = (\mathbf{t} + \alpha_t)$, relax the constraints of entities and relations. Like the FT model, the DT model is also easy to extend to a series of translation models such as TransE.

2.2 Other Models

SE [7] constructs two relation-specific matrices $\mathbf{M}_{rh}$ and $\mathbf{M}_{rt}$ respectively for head entities and tail entities. Its score function is $f_r(h, t) = \|\mathbf{M}_{rh}\mathbf{h} - \mathbf{M}_{rt}\mathbf{t}\|_1$.

LFM [8, 9] encodes each entity and sets a matrix $\mathbf{M}_r$ for each relation. It defines the following bilinear scoring function for each triple $f_r(h, t) = \mathbf{h}^\mathrm{T}\mathbf{M}_r\mathbf{t}$.

SME [10, 11] constructs four different projection matrices to capture semantic associations between entities and relations, and for each triples defines a linear score function $f_r(h, t) = (\mathbf{M}_h\mathbf{h} + \mathbf{M}_{r1}\mathbf{r} + \mathbf{b}_1)^\mathrm{T}(\mathbf{M}_t\mathbf{t} + \mathbf{M}_{r2}\mathbf{r} + \mathbf{b}_2)$, and a bilinear score function $f_r(h, t) = (\mathbf{M}_h\mathbf{h} \otimes \mathbf{M}_{r1}\mathbf{r} + \mathbf{b}_1)^\mathrm{T}(\mathbf{M}_t\mathbf{t} \otimes \mathbf{M}_{r2}\mathbf{r} + \mathbf{b}_2)$.

NTN [12] uses bilinear tensors to replace the linear transformation layer of traditional neural networks, and connects head entities and tail entities vectors in different dimensions.

3 Our Method

In this section, we first demonstrate the impact of generating a negative triple by using the same probability to replace h (or t) in a positive triple on knowledge graph embedding models. Then we propose an adaptive negative sampling principle and combine it with the typical model TransE.

3.1 The Limitations of Simple Random Sampling

Entity equal-probability sampling is used for generating negative triples in most of existing knowledge graph embedding models, that is, an entity was extracted from entity set with the same probability to replace h (or t) in positive triples. We refer to the extracted entity as a negative entity and the replaced entity as a positive entity. However, this method may suffer from a problem in generating negative triples: Negative entities are not similar to positive entities, which results in an invalid negative triple. Next, we will describe this issue in detail.

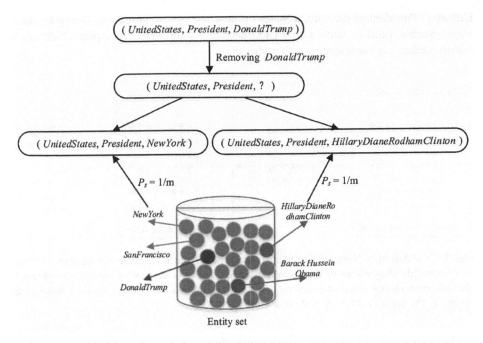

Fig. 1. The Simple Random Sampling. The circle represent entities, and the closer colors of two circles are, the more similar they are; P_s is sampling probability; m is the number of entities.

A certain similarity exists between each entity. Given a positive entity, it has multiple negative entities, and the similarity between positive entity and each negative entity is different. A negative entity extracted by simple random sampling may have a very low similarity with the corresponding positive entity, so an invalid negative triple may be generated. Invalid negative triples are little help to the effective embedding of learning knowledge graph. In the following paragraph, we use a concrete example to illustrate the importance of valid negative triple in knowledge graph embedding learning.

As shown in Fig. 1, supposing there is a positive triple (UnitedStates, President, Trump) existing in the knowledge graph. According to simple random sampling, a negative triple is generated by replacing the tail. Specifically, we get an incomplete triple (UnitedStates, President, ?) by removing Trump. Then, we extract a tail entity from entity set with simple random sampling. Assuming that a City type entity NewYork is extracted, we will get a ridiculous negative triple (UnitedStates, President, NewYork), which is an invalid negative triple, as the relation President requires a tail to be a Person type. Therefore, the generated triple is an invalid tail for the relation President by a simple type constraint judgment, which means the negative triple is invalid. In contrast, if a Person type entity Hillary is extracted, we will get a very valid negative triple (UnitedStates, President, Hillary). For anyone who does not have a detailed understanding of the United States president history, he cannot judge whether

Hillary is President of the United States because she is very similar to Trump in many ways. Such a valid negative entity can generate a valid negative triple, which has a positive effect on knowledge graph embedding.

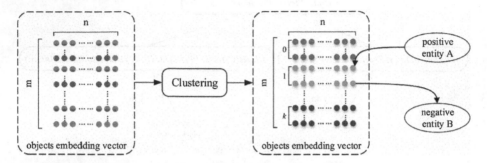

Fig. 2. The Adaptive Negative Sampling framework. m and n represent the number of entity vectors and the dimensions of the vector, respectively. Entity vectors are clustered to obtain k clusters, each cluster containing several entity vectors. When the positive entity A belongs to cluster 1, the negative entity B will come from cluster 1.

In the training of knowledge graph embedding model, we hope to obtain a negative entity that is similar to positive entity. However, this issue has seriously affected knowledge graph embedding. Therefore, we propose Adaptive Negative Sampling to solve this problem.

3.2 Adaptive Negative Sampling

A valid negative triple can help the knowledge graph embedding, so we need to get a valid negative triple. The key of obtaining a valid negative triple is extracting a negative entity similar to a positive entity. In vector space, we can determine whether they are similar by calculating the distance between two entity vectors. If the distance between two entity vectors is smaller, the more similar they are and vice versa. This motivates us the idea of clustering entities and then looking for similar entities in a cluster. Therefore, we decided to use simple and effective K-Means algorithm [13–15] to cluster entities, so that positive and negative entities come from the same cluster. We refer to the above sampling process of negative entities as Adaptive Negative Sampling (ANS). The Adaptive Negative Sampling framework is shown in Fig. 2. The followed is the K-Means clustering of entities in knowledge graph.

Given a knowledge graph $G = (E, R, S)$, where $E = \{e_1, e_2, \ldots, e_N\}$ denotes a set of entities consisting N entities in knowledge graph, $R = \{r_1, r_2, \ldots, r_M\}$ indicates a set of relations consisting M relations in knowledge graph, and $S \subseteq E \times R \times E$ represents a set of triples in knowledge graph. We aim to divide N entities into K clusters by K-Means in which each entity belongs to a cluster with the nearest mean, i.e., the sum of Euclidian distances from each entity to its cluster center is the smallest. Our target formula is as follows:

$$\arg min \sum_{i=1}^{K} \sum_{e \in C_i} \|e - c_i\|_{l1} \qquad (1)$$

where K denotes the number of clusters, e is an entity vector, c_i represents i^{th} cluster center vector, C_i is a set of entity e in i^{th} cluster. Entity set E is divided into K clusters $E_1, E_2, \ldots, E_K$ by K-Means clustering. There is a high similarity between entities in the same cluster. Given a positive entity $e \in E_k (k \in 1, 2, \ldots, K)$, a negative entity e' will be selected from the cluster E_k. Then, the obtained negative entity e' and the positive entity e will have a high similarity. Therefore, we can get a valid negative triple to help knowledge graph embedding.

Algorithm 1 Learning TransE-ANS

Require : Training sets positive triples $S = \{(h, r, t)\}$ and negative triples $S' = \{(h', r, t) | h' \in E_h\} \cup \{(h, r, t') | t' \in E_t\}$, entity set E, relation set R, margin γ, embedding dimension n, learning rate α, K-Means cluster number k, entity cluster E_i ($i = 1, 2, \ldots, k$)

Ensure : Entity and relation embeddings.

1: initialize:

2: $r \leftarrow$ uniform $(-\frac{6}{\sqrt{n}}, \frac{6}{\sqrt{n}})$ for each $r \in R$

3: $r \leftarrow r/\|r\|$ for each relation $r \in R$

4: $e \leftarrow$ uniform $(-\frac{6}{\sqrt{n}}, \frac{6}{\sqrt{n}})$ for each $e \in E$

5: $e \leftarrow e/\|e\|$ for each relation $e \in E$

6: **loop**:

7: $S_{batch} \leftarrow$ sample (S, b) // sample a minibatch of size b

8: $T_{batch} \leftarrow \emptyset$ // initialize the set of pairs of triplets

9: **for** $(h, r, t) \in S_{batch}$ **do**

10:     $(h', r, t') \leftarrow$ sample$(S'_{(h, r, t)})$ //sample a corrupted triplet

11:        $T_{batch} \leftarrow T_{batch} \cup \{(h, r, t), (h', r, t')\}$

12: **end for**

13: Update embedding w.r.t.

$$\sum_{((h,r,t),(h',r,t') \in T_{batch})} \nabla [f_r(h,t) + \gamma - f_r(h',t')]_+$$

14:     if epoch % 50 == 0 then

15:        Update E_i, // K-Means clustering

16: **end if**

17: **end loop**

3.3   TransE-ANS

In this section, we discuss the TransE-ANS model constructed by combining ANS with TransE model. TransE-ANS embeds entities and relations into the same vector space using the translation principles $\mathbf{h} + \mathbf{r} = \mathbf{t}$ of TransE. Therefore, scoring function of TransE-ANS is as follows:

$$f_r(h,t) = \|\mathbf{h}+\mathbf{r}-\mathbf{t}\|_{l1} \tag{2}$$

In TransE-ANS, we use a marginal loss function as our training target. The marginal loss function is as follows:

$$L= \sum_{(h,r,t)\in S} \sum_{(h',r,t')\in S'} [f_r(h,t)+\gamma -f_r(h',t')]_{+} \tag{3}$$

where S is a set of positive triples, $S' = \{(h',r,t)|h' \in E_h\} \cup (h,r,t')|t' \in E_t\}$, E_h is the cluster where h is located, E_t is the cluster where t is located.) is a set of negative triples, $[f_r(h,t)+\gamma -f_r(h',t')]_{+} = \max(0, [f_r(h,t)+\gamma -f_r(h',t')])$, γ is a margin. We use the stochastic gradient descent (SGD) [16] to minimize the marginal loss function.

Algorithm 1 shows the complete training process of our model. For every 50 training epochs, we cluster the entity vectors we get. In 15th line of the algorithm, assuming that we use the K-Means clustering algorithm, but obviously which can substitute it with any clustering algorithm.

3.4   Comparison of Complexity

The complexity of several knowledge graph embedding models is shown in Table 1. In general, the number of cluster centers is much smaller than entities, so the parameter complexity of TransE-ANS is same as TransE. In terms of time complexity, comparing with TransE model, the TransE-ANS model only increases the time of each K-Means clustering. Therefore, our approach has high efficiency.

Table 1. The complexity of several knowledge graphs embedded in the model.

Model	Parameters	Operations (Time)
SE	$O(N_e m + 2N_r n^2)(m = n)$	$O(2m^2 N_t)$
SME(linear)	$O(N_e m + N_r n + 4mk + 4k)(m = n)$	$O(4mkN_t)$
SME(bilinear)	$O(N_e m + N_r n + 4mks + 4k)(m = n)$	$O(4mksN_t)$
LFM	$O(N_e m + N_r n^2)(m = n)$	$O((m^2 + m)N_t)$
SLM	$O(N_e m + N_r(2k + 2nk))(m = n)$	$O((2mk + k)N_t)$
TransE	$O(N_e m + N_r n)(m = n)$	$O(N_t)$
TransH	$O(N_e m + 2N_r n)(m = n)$	$O(2mN_t)$
TransR	$O(N_e m + N_r(m + 1)n)$	$O(2mnN_t)$
CTransR	$O(N_e m + N_r(m + d)n)$	$O(2mnN_t)$
TransD	$O(2N_e m + 2N_r n)(m = n)$	$O(2nN_t)$
GTrans-SW	$O(N_e m + 3N_r n)(m = n)$	$O(nN_t)$
TransE-ANS	$O((N_e + K)m + N_r n)(m = n)$	$O(N_t + KiN_e n)$

4 Experiments and Analysis

Table 1 shows the complexity of various embedded models. In Table 1, N_e is the number of entities and N_r is the number of relations. N_t is the number of triples in the knowledge graph. m and n represent the dimension of the entity embedding space and relations embedding space, respectively. k represents the number of hidden nodes of the neural network, s denotes the number of slices of the tensor, and K denotes the number of K-Means cluster centers, i denotes the number of cluster iterations. We use link prediction and triple classification tasks to evaluate our approach. For reducing the implementation time of model, our code uses multiple threads in Linux system to achieve rapid training and testing.

4.1 Data Setting

The datasets that we use are from two widely used knowledge graphs: WordNet and Freebase. WordNet is a large knowledge graph of English vocabulary. Freebase is a large-scale knowledge graph of human knowledge that stores general facts of the world. In this experiments, we use four subsets from WordNet and Freebase, i.e., WN18 and FB15K. Table 2 shows these two datasets statistics.

Table 2. Dataset information.

Dataset	#Ent	#Rel	#Train	#Vaild	#Test
WN18	40943	18	141442	5000	5000
FB15K	14951	1345	483142	50000	59071

Table 3. Optimal parameter setting in link prediction.

Dataset	Metric	Epoch	λ	γ	n	B	K	i	D.S
WN18	Mean Rank	2000	0.001	5.5	50	100	14	20	l_1
	Hit@10	2000	0.001	3	50	100	14	20	l_1
FB15K	Mean Rank	2000	0.001	4	200	200	64	20	l_1
	Hit@10	2000	0.001	2	200	200	64	20	l_1

4.2 Link Prediction

Link prediction aims to predict a missing entity h (or t) in a triple (h, r, t) [1, 8, 12]. We call entity h (or t) missing in a test triple (h, r, t) as the correct entity, and all the entity other than the correct entity are considered as the candidate entity. Firstly, we use each candidate entity to replace h (or t) of a test triple (h, r, t) to get the candidate triple. Then, the score of the test triple and each candidate triple are calculated. Finally, the correct entity and the candidate entity are sorted in ascending order according to the score. We use the two indicators [1] as our evaluation metrics: (1) the average rank of the correct entity (Mean Rank); and (2) the proportion of the correct entity ranked in top 10 (Hits@10).

It is worth nothing that the candidate triples may exist in knowledge graph, and these candidate triples should be considered as the correct triples. Its scores are likely to be lower than the correct triples. Thus, we should filter out these candidate triples that have already appeared in train, valid and test sets. We set the filter during the test to filter out these candidate triples, and we use evaluation setting by "Filt", otherwise we use evaluation setting by "Raw".

In this task, in order to get the influence of each parameter on our model, we try various parameter settings. We select the training period epoch among {1000, 1500, 2000}, the learning rate λ among {0.01, 0.001, 0.0001}, the margin value γ among {1, 2, 2.5, 3, 3.5, 4, 4.5, 5, 5.5, 6}, the dimension n among {25, 50, 100, 200}, the nimi-batch size B among {100, 200, 500, 1000}, the number of cluster center K among {14, 32, 64}, the number of clustering iteration i among {10, 20, 50}, and the dissimilarity D.S is l1-norm. We obtained two different optimal parameter settings for Mean Rank and Hits@10 on the two datasets, as shown in Table 3. We obtained the experimental results of Mean Rank and Hits@10 by these two settings.

Table 4. Link prediction results on WN18.

Metric	Mean Rank		Hits@10	
	Raw	Filt	Raw	Filt
SE	1011	985	68.5	80.5
SME(linear/bilinear)	542/526	533/509	65.1/54.7	74.1/61.3
LFM	469	456	71.4	81.6
TransE	263	251	75.4	89.2
TransH(unif/bern)	318/401	303/388	75.4/73.0	86.7/82.3
TransR(unif/bern)	232/238	219/225	78.3/79.8	91.7/92.0
CTransR(unif/bern)	243/231	230/218	78.9/79.4	92.3/92.3
TransD(unif/bern)	242/224	229/212	79.2/79.6	92.5/92.2
TransSparse(unif/bern)	233/223	221/211	79.6/80.1	93.4/93.2
TransSparse-DT(unif/bern)	248/234	232/221	80.0/**81.4**	93.6/94.3
TransE-ANS(unif/bern)	**220/207**	**208/195**	**80.2**/80.6	**94.0/94.6**

The experimental results of WN18 and FB15K link prediction tasks are shown in Tables 4 and 5, respectively. We compare TransE-SNS with a few state-of-the-art methods in the link prediction task on WN18 and FB15K, including SE, SME, LFM, TransE, TransH, TransR/CTransR, TransD, TranSparse and TranSparse-DT. We use the results reported in their papers or [3] directly since the data set is the same.

Table 5. Link prediction results on FB15K.

Metric	Mean Rank		Hits@10	
	Raw	Filt	Raw	Filt
SE	273	162	28.8	39.8
SME(linear/bilinear)	274/284	154/158	30.7/31.3	40.8/41.3
LFM	283	164	26.0	33.1
TransE	243	125	34.9	47.1
TransH(unif/bern)	211/212	84/87	42.5/45.7	58.5/64.4
TransR(unif/bern)	226/198	78/77	43.8/48.2	65.5/68.7
CTransR(unif/bern)	233/199	82/**75**	44.0/48.4	66.3/70.2
TransD(unif/bern)	211/194	67/91	49.4/53.4	74.2/77.3
TransSparse(unif/bern)	216/190	66/82	50.3/53.7	78.4/79.9
TransSparse-DT(unif/bern)	208/**188**	58/79	**51.2/53.9**	78.4/80.2
TransE-ANS(unif/bern)	**198**/210	**56**/95	48.9/52.5	**80.1/83.0**

As shown in Table 4, TransE-ANS performance has been greatly improved compared to TransE. Although, TransE-ANS slightly behind TranSparse-DT in Hits@10 (raw, bern), TransE-ANS still achieved state-of-the-art performance in most cases. As shown in Table 5, TransE-ANS performance has still been greatly improved in FB15K compared with TransE. Regrettably, our method failed to achieve state-of-the-art performance in both Mean Rank(bern) and Hits@10(raw). We believe that there are two reasons for this result: One reason is that the data of FB15K is sparse, and there are fewer entities with Multiple identical relationships (i.e. each entity has fewer similar entities), which results in a certain number of entities with lower similarity in each cluster after clustering. Another reason is that cluster centers K is difficult to determine, and the choice of our K value has some limitations. Therefore, K-Means clustering does not classify entities well.

Tables 6 and 7 are link prediction results on FB15K by relation category. The relation categories include 1-to-1, 1-to-N, N-to-1, and N-to-N, which account for 24%, 23%, 29% and 24% of the total number of relations, respectively. Table 6 shows results of predicting head entities in different relation categories. Table 7 shows results of predicting tail entities in different relation categories.

As can be seen from Table 6, Among different categories of relations, our method always finds the best result in a certain situation. Particularly, in the N-to-N relationship, our approach achieves the most state-of-the-art performance. As shown in Table 7, we can get similar conclusions as in Table 6, and achieve state-of-the-art performance in more cases in Table 7. Overall, the performance of our approach is far superior to the baseline model TransE, and in most cases achieves the most advanced performance over all current models.

Table 6. Prediction results of head on FB15K by relation category (%).

Tasks	Predicting head (Hit@10)			
Relation category	1-to-1	1-to-N	N-to-1	N-to-N
SE	35.6	62.6	17.2	37.5
SME	35.1/30.9	53.7/69.6	19.0/19.9	40.8/38.6
TransE	43.7	65.7	18.2	47.2
TransH(unif/bern)	66.7/66.8	81.7/87.6	30.2/28.7	57.4/64.5
TransR(unif/bern)	76.9/78.8	77.9/89.2	38.1/34.1	66.9/69.2
CTransR(unif/bern)	78.6/81.5	77.8/89.0	36.4/34.7	68.0/71.2
TransD(unif/bern)	80.7/86.1	85.8/95.5	47.1/39.8	75.6/78.5
TransSparse(unif/bern)	83.2/87.1	85.2/**95.8**	51.8/44.4	80.3/81.2
TransSparse-DT(unif/bern)	83.0/**87.4**	85.7/**95.8**	**51.9**/47.7	80.5/81.6
TransE-ANS(unif/bern)	**83.4**/84.1	**88.8/95.8**	45.6/**48.4**	**83.2/85.3**

Table 7. Prediction results of tail on FB15K by relation category (%).

Tasks	Predicting tail (Hit@10)			
Relation category	1-to-1	1-to-N	N-to-1	N-to-N
SE	34.9	14.6	68.3	41.3
SME	32.7/28.2	14.9/13.1	61.6/76.0	40.8/38.6
TransE	43.7	19.7	66.7	50.0
TransH(unif/bern)	63.7/65.5	30.1/39.8	83.2/83.3	60.8/67.2
TransR(unif/bern)	76.2/79.2	38.4/37.4	76.2/90.4	69.1/72.1
CTransR(unif/bern)	77.4/80.8	37.8/38.6	78.0/90.1	70.3/73.8
TransD(unif/bern)	80.0/85.4	54.5/50.6	80.7/94.4	77.9/81.2
TransSparse(unif/bern)	82.6/87.5	60.0/57.0	**85.5**/94.5	82.5/83.7
TransSparse-DT(unif/bern)	82.8/86.7	59.9/56.3	**85.5/94.8**	82.9/84.0
TransE-ANS(unif/bern)	**87.4/88.5**	**60.8/60.5**	83.3/94.5	**83.3/85.7**

5 Conclusion

In this paper, we introduce a method named Adaptive Negative Sampling that generates valid negative triples to help knowledge graph embedding. It considers the limitations of simple random sampling. We construct TransE-ANS model by combining ANS with TransE model. Experimental results show that TransE-ANS performs better than other models in link prediction task, since it can generate a set of high-quality negative triples which are benefit for improving the effectiveness of embedding models. We will further explore the combination of different clustering algorithms and knowledge graph embedding models in the future.

Acknowledgements. This work was partially supported by the National Natural Science Foundation of China (Nos. U1501252, 61572146 and U1711263), the Natural Science Foundation of Guangxi Province (No. 2016GXNSFDA380006), the Guangxi Innovation-Driven Development Project (No. AA17202024) and the Guangxi Universities Young and Middle-aged Teacher Basic Ability Enhancement Project (No. 2018KY0203).

References

1. Bordes, A., Usunier, N., García-Durán, A., Weston, J., Yakhnenko, O.: Translating embeddings for modeling multi-relational data. In: Proceedings of NIPS, Lake Tahoe, NV, USA, pp. 2787–2795 (2013)
2. Wang, Z., Zhang, J., Feng, J., Chen, Z.: Knowledge graph embedding by translating on hyperplanes. In: Proceedings of AAAI, Quebéc city, QC, Canada, pp. 1112–1119 (2014)
3. Lin, Y., Liu, Z., Sun, M., Liu, Y., Zhu, X.: Learning entity and relation embeddings for knowledge graph completion. In: Proceedings of AAAI, Austin, TX, USA, pp. 2181–2187 (2015)
4. Ji, G., Liu, K., He, S., Zhao, J.: Knowledge graph completion with adaptive sparse transfer matrix. In: Proceedings of AAAI, Phienix, AZ, USA, pp. 985–991 (2016)
5. Feng, J., Huang, M., Wang, M., Zhou, M., Hao, Y., Zhu, X.: Knowledge graph embedding by flexible translation. In: Proceedings of KR, Cape Town, South Africa, pp. 557–560 (2016)
6. Chang, L., Zhu, M., Gu, T., Bin, C., Qian, J., Zhang, J.: Knowledge graph embedding by dynamic translation. IEEE Access 5, 20898–20907 (2017)
7. Bordes, A., Weston, J., Collobert, R., Bengio, Y.: Learning structured embeddings of knowledge bases. In: Proceedings of AAAI, San Francisco, CA, USA, pp. 301–306 (2011)
8. Jenatton, R., Roux, N.L., Bordes, A., Obozinski, G.: A latent factor model for highly multi-relational data. In: Proceedings of NIPS, Lake Tahoe, NV, USA, pp. 3167–3175 (2012)
9. Sutskever, I., Salakhutdinov, R., Tenenbaum, J.B.: Modelling relational data using Bayesian clustered tensor factorization. In: Proceedings of NIPS, Vancouver, BC, Canada, pp. 1821–1828 (2009)
10. Bordes, A., Glorot, X., Weston, J., Bengio, Y.: A semantic matching energy function for learning with multi-relational data. Mach. Learn. **94**(2), 233–259 (2014)
11. Bordes, A., Glorot, X., Weston, J., Bengio, Y.: Joint learning of words and meaning representations for open-text semantic parsing. In: Proceedings of AISTATS, La Palma, Canary Islands, pp. 127–135 (2012)
12. Socher, R., Chen, D., Manning, C.D., Ng, A.Y.: Reasoning with neural tensor networks for knowledge base completion. In: Proceedings of NIPS, Lake Tahoe, NV, USA, pp. 926–934 (2013)
13. Hartigan, J.A., Wong, M.A.: Algorithm AS 136: a K-Means clustering algorithm. J. R. Stat. Soc. **28**(1), 100–108 (1979)
14. Hamerly, G., Elkan, C.: Algorithm AS 136: a K-Means clustering algorithm. In: Proceedings of CIKM, McLean, VA, USA, pp. 600–607 (2002)
15. Celebi, M.E., Kingravi, H.A., Vela, P.A.: A comparative study of efficient initialization methods for the k-means clustering algorithm. Expert Syst. Appl. **40**(1), 200–210 (2013)
16. Duchi, J.C., Hazan, E., Singer, Y.: Adaptive subgradient methods for online learning and stochastic optimization. JMLR **12**, 2121–2159 (2011)

Multi-grained Pruning Method
of Convolutional Neural Network

Zhenshan Bao, Wanqing Zhou, and Wenbo Zhang[✉]

Faculty of Information Technology, Beijing University of Technology,
Beijing 100124, China
zhangwenbo@bjut.edu.cn

Abstract. Although the deep learning technology has shown great power in solving the complex tasks, these neural network models are large and redundant as a matter of fact, which makes these networks difficult to be placed in embedded devices with limited memory and computing resources. In order to compress the neural network to a slimmer and smaller one, the multi-grained network pruning framework is proposed in this paper. In our framework, the pruning process was divided into the filter-level pruning and the weight-level pruning. In the process of the filter-level pruning, the importance of the filter was measured by the entropy of the activation tensor of the filter. In the other process, the dynamic recoverable pruning method was adopted to prune the weights deeply. Different from these popular pruning methods, the weight-level pruning is also taken into account based on the employment of the filter-level pruning to achieve more effectively pruning. The proposed approach is validated on two representative CNN models - AlexNet and VGG16, pre-trained on ILSVRC12. Experimental results show that AlexNet and VGG16 network models are compressed $19.75\times$ and $22.53\times$ respectively by this approach, which are 2.05 and 5.89 higher than the classical approaches of dynamic Network Surgery and ThiNet.

Keywords: Pruning · Model compression · CNN · Computational · Modeling training

1 Introduction

In the past few years, deep neural networks have achieved the most advanced performance in many fields, such as computer vision, natural language processing, and speech recognition, etc. At the same time, these network structures have become more complex and redundant, and their computing power and memory requirements for computing platforms are getting higher and higher, which makes the networks difficult to be placed in devices with limited memory and computing resources [1]. For example, the famous VGG16 network [2] has 138.34 million parameters and 30.94G float operations (flops), which is hardly be placed in the mid- or low-end consumer devices.

In order to solve the above problems, many scholars have proposed a number of acceleration and compression approaches based on deep neural networks [3] to reduce the size of the network model by low-rank approximation [4–6], network pruning [7],

© Springer Nature Singapore Pte Ltd. 2019
X. Cheng et al. (Eds.): ICPCSEE 2019, CCIS 1058, pp. 564–576, 2019.
https://doi.org/10.1007/978-981-15-0118-0_43

quantization [8, 9], knowledge distillation [10, 11], and compact network design [12–14]. Among of these approaches, network pruning is more intuitive and efficient, and has the advantages of high compression ratio and low precision loss by eliminating unimportant parameters in the network. Early works focus on pruning all connections below the weight threshold and retraining the network to restore network performance [7], [15] but these approaches require constant iterations to seek optimal compression and accuracy, which results in slower convergence and longer pruning cycles. In order to speed up network pruning, another part of the scholars perform filter-level pruning on the network to make the network more "thin" and accelerate the speed of network compression [16–18]. However, the filter-level pruning approach is not completely pruning the network. Both of these approaches are pruned at a single scale and do not quickly and efficiently prune the network.

Therefore, we combine the filter-level pruning with the weight-level pruning, and use global average pooling (GAP) layer instead of fully connected layer, and propose a new multi-grained network pruning framework to achieve deeper network compression and accelerate the speed of network. The first stage is the filter-level pruning. We calculate the output map of the filter for different input feature maps at first, and calculate the global average value of each output map as the response of the filter. These different response values constitute a one-dimensional tensor. We then use the entropy of the tensor to measure the importance of the filter and eliminate unimportant filters. Additionally, in the final stage of the filter-level pruning, we use L1 regularization to make the remaining parameters in the network model closer to zero, which is beneficial to improve the efficiency of weight-level pruning. In the second stage, the weights in the sparse network model are deeply pruned by a dynamic pruning approach, which means the weights pruning is recoverable. We delete connections with values smaller than threshold A and recover connections with values bigger than the other threshold B to avoid incorrect pruning. The multi-grained network pruning framework can flexibly and effectively prune the network. The experimental results show that our approach increases the compression ratios of AlexNet and VGG16 networks from $17.70\times$ and $16.64\times$ to $19.75\times$ and $22.53\times$, respectively.

2 Related Work

A lot of researches indicated that most of the parameters in the network model are not important [19, 20]. Network pruning is a kind of methods to find unimportant parameters and remove them. According to the different pruning granularity, [21] divided the existing approaches into two categories, namely the weight-level pruning and the filter-level pruning. Figure 1 vividly shows the difference between the two approaches.

The pruning target of the weight-level pruning approach is the weights of the filter. Early works [22, 23] on pruning used the approximate second-order derivatives of the loss function to determine the saliency of the parameters, and then pruned those parameters with low saliency. However, the second-order derivative calculation of the network is too complex to be applied to deep networks. [19] proposed a deep compression framework, which applied pruning, quantization and Huffman coding to the compression of deep neural networks, and compressed the AlexNet network model by

35× without drops in accuracy. On this basis, [15] proposed a dynamic network pruning approach, which dynamically divided network connections into two categories, important connections and unimportant connections. This approach requires only a small training epochs to achieve the compression ratio of Han's approach by repeating pruning and restoring connections. [20] and [24] focused on pruning the weight matrix in a grouping manner, which effectively improved the inference speed of the network.

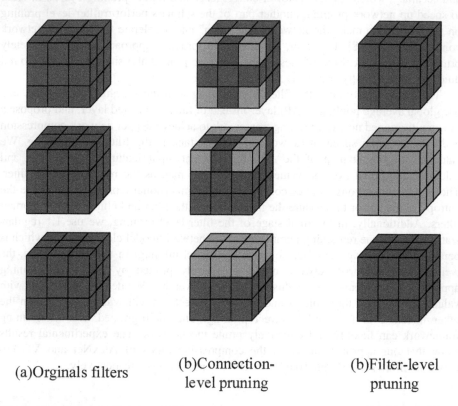

(a)Orginals filters   (b)Connection-level pruning   (b)Filter-level pruning

Fig. 1. The schematic diagram of different pruning approaches' effect for a convolutional layer which has 3 convolutional filters of kernel size 3 × 3, where a large cube represents a filter and a small cube represents a connection

The filter-level pruning makes the convolutional layer of the network thin by directly pruning the filter, it is a coarse-grained pruning approach compared with the weight-level pruning approach. These approaches reduce the network parameters of the current layer and also reduce the input channels of the next layer. For this reason, the network parameters are quickly reduced by these approaches, and the pruned network is very friendly to the program implementation. [25] proposed an approach based on magnitude-level pruning, they used the sum of the weights to measure the importance of the filter. This approach achieved 1.5× compression and 2.9× speed up on the VGG16. [26] used the APoZ to judge the importance of the filter, where APoZ meant the average percentage of zero in the feature map of the filter. However, this approach ignored the filters whose activation values are close to 0, so it retained some filters with less information. In order to deeply prune unimportant filters, [27] proposed an entropy-based pruning approach. This approach used the entropy value of the tensor, where the output feature map to measure the importance of the filter. [28] proposed a filter-level pruning approach called ThiNet, which used the statistics of the next layer to guide the pruning of the current layer. ThiNet achieved 16.63× and 2× compression with less accuracy degradation on VGG16 and ResNet, respectively. Recently, Zou [29] defined the critical points among the discriminability values of feature maps in each convolutional layer, and used these critical points to prune feature maps.

3 Design of Multi-grained Pruning Convolutional Neural Network

In this section we provide a comprehensive introduction to our approach of Multi-grained Pruning of Convolutional Neural Network. We combine filter-level pruning with weight-level pruning for deeper and faster network pruning. In addition, we apply some strategies during the pruning process, which will be given in the Experiments section. Figure 2 shows the implementation framework and details of our approach. It should be noted that we only prune the convolution layer. For the fully connected layer, we replace it with GAP layer [30], which can greatly reduce the parameter size of the fully connected layer and reduce the risk of over-fitting caused by the fully connected layer.

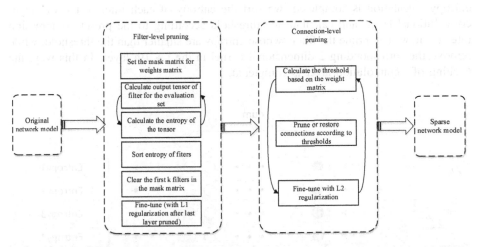

Fig. 2. The implementation framework and details of our approach:In the first stage, the importance of the filter is measured by the entropy of the activation tensor of the filter for filter-level pruning. Then, L1 regularization is introduced into the retraining process with parameters approaching zero. In the second stage, the dynamic recoverable pruning method is used to deeply prune the weight, which we call the weight-level pruning. And in the second stage, L2 regularization is introduced into the fine-tuning process to prevent over-fitting

3.1   Filter-Level Pruning

How to measure the importance of filters is the primary task that we have to figure out. As we know, filter is the basic unit to extract information from input pictures. The filter generates different response values for different input images. For different input images, the larger the difference in the response value, the stronger the distinguishing ability of filter on the input image, and the smaller the difference in response value, the weaker the ability of filter to distinguish the input image. Therefore, it seems to be a reasonable inference that if a filter is not sensitive to the input image, its influence on the network is lower, so it can be pruned.

As shown in Fig. 3, we randomly select 100 pictures as evaluation set. We calculate the mean of the feature map of the filter as its response value. Then, we get a 100-dimensional response tensor by calculating the images in the evaluation set. We should prune the filters whose values in the tensor are nearly identical and keep the filters whose values have large difference. In the information theory, entropy is used to measure the disorder of signals [27]. We use the entropy of tensor to measure the degree of difference of its elements. At first, we divide the value range of the tensor elements into m bins, then we count the number of elements in each block and calculate the probability p_j of each bin. Finally, the entropy of the tensor can be calculated as (1).

$$H_{i,k} = -\sum_{j=1}^{m} p_j \log p_j, i \in N, k \in C_i \tag{1}$$

$H_{i,k}$ is the entropy of the k-th filter of the i-th layer, N and C_i are the number of convolution layers and the number of channels in the i-th layer, respectively. After the entropy calculation is completed, we sort the entropy of each filter in the i-th layer convolutional layer, and calculate the threshold according to the preset compression rate. Then, we can prune the filters whose entropy are smaller than the threshold, while remove the corresponding 2-dimensional kernel in the i + 1th layer. In this way, the pruning of a convolution layer is completed.

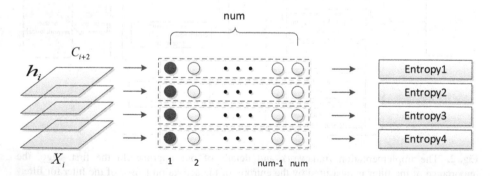

Fig. 3. The calculation process of entropy of response tensor of filters for different feature maps in convolutional layer

Then, how do we achieve it? We set a binary mask matrix T that is exactly the same as the network model. Each element in the matrix corresponds to a weight in the network model. The initial value of the matrix's elements are all set to 1. When a filter is pruned, the matrix elements corresponding to the filter are all set to 0. Therefore, for a filter $W_{i,k}$, when its input feature map is X_i, its convolution operation will changes as shown in (2).

$$f(X_i \otimes W_{i,k}) \rightarrow f(X_i \otimes W_{i,k} \odot T_{i,k}) \tag{2}$$

Where, $f(\bullet)$ is the activation function, $T_{i,k}$ is the mask matrix corresponding to $W_{i,k}$, $\otimes$ is the convolution operation and $\odot$ represent the hadamard operator.

Figure 4 shows the details of the filter-level pruning. We can see that the filter-level pruning is also beneficial to the reduction of calculation amount. Suppose w_i and h_i are the width and height of the input feature map $X_i \in \mathbb{R}^{C_i \times h_i \times w_i}$, respectively. $X_{i+1} \in \mathbb{R}^{C_{i+1} \times h_{i+1} \times w_{i+1}}$ are generated after the i-th layer convolution operation, these are also the input feature maps of the i + 1-th layer at the same time. A convolutional layer contains C_{i+1} filters, and each filter contains C_i 2-dimensional kernels $k \in \mathbb{R}^{k \times k}$. Therefore, the number of operations of the i + 1th layer convolution layer is $C_{i+1}C_ik^2h_{i+1}w_{i+1}$. When we prune a filter of the i + 1th layer, we can reduce $C_ik^2h_{i+1}w_{i+1}$ operations. At the same time, since the feature maps of the i + 2th layer are reduced, the number of the operations of the i + 1th layer additionally is reduced $C_{i+2}k^2h_{i+2}w_{i+2}$ operations.

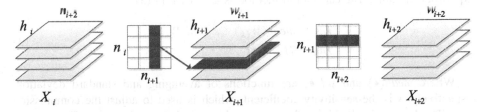

Fig. 4. The effect of filter-level pruning on the network structure

After the pruning is completed, the accuracy is dropped due to the destruction of the network structure. Therefore, we need to retrain the network to restore the accuracy. The number of iterations of the training is determined by the network type and compression ratio. For example, when the compression rate is 30% during the LeNet network pruning process. It only takes 2500 iterations to restore network performance.

3.2 Weight-Level Pruning

The redundant filters in the network are pruned in Section A, however there are still a large number of invalid connections in the remaining filters, which are the weighting parameters in the filter. The weight-level pruning is required to pruning those unimportant connections in the network. We mainly refer to the dynamic pruning approach [15], which judges whether the connection is important by judging the value of a connection, prunes the connection below the threshold Th_A, and recovers the connection above the threshold Th_B. This recoverable mechanism prevents the problem that the network cannot be recovered due to the incorrect pruning. The schematic diagram of the pruning process is shown in Fig. 5.

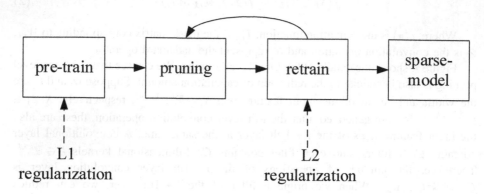

Fig. 5. Connection level pruning diagram

The threshold Th_A is determined by the mean and standard deviation of $W_{i,k}$, $\forall 0 \leq i \leq N, \forall 0 \leq k \leq C_i$, and the threshold Th_B is based on the addition of an experimental value. The calculation method is as shown in (3).

$$Th_A = mean(W_{i,k}) + s * std(W_{i,k})$$
$$Th_B = Th_A + \Delta t \tag{3}$$

Where $mean(\bullet)$ and $std(\bullet)$ are functions of averaging and standard deviation, respectively. s is the sensitivity coefficient, which is used to adjust the compression ratio. The larger s is, the higher the compression ratio is, and vice versa. The pruning and recovery of the connection is implemented by setting and clearing the corresponding elements of the mask matrix T. And then we let $M_{i,k}(p)$ be a parameter in the filter $W_{i,k}$, $\forall 0 \leq i \leq N, \forall 0 \leq k \leq C_i$, and the update strategy of mask matrix is shown in (4).

$$M_{i,k}(p) = \begin{cases} 0 & |W_{i,k}(p)| < th_A \\ M_{i,k}(p) & th_A \leq |W_{i,k}(p)| < th_B \\ 1 & |W_{i,k}(p)| \geq th_B \end{cases} \tag{4}$$

In order to improve the pruning efficiency, we use L1 regularization in the filter-level pruning, so that as many parameters as possible approach 0. In the second phase we use L2 regularization to effectively avoid over-fitting while ensuring that the network has sufficient expressive power. The update strategy for the parameters is shown in (5).

$$W_{i,k}(p) = W_{i,k}(p) - \beta \frac{\partial}{\partial W_{i,k}(p) M_{i,k}(p)} L(\partial W_{i,k}(p) \odot M_{i,k}(p)) \tag{5}$$

4 Experiments and Results

We verified the effectiveness of the proposed approach in the AlexNet and VGG16 network models. Both experiments were based on Caffe framework [31] and used the classification of the ILSVRC12 dataset in the ImageNet dataset [32] as the validation set and training set. The ImageNet dataset has 14 million images covering more than 20,000 categories, with more than one million images with category and location annotations of objects in the image. This dataset is widely used in research papers in the field of computer vision, and has become a standard data set for the performance testing of algorithms in the field of deep learning images. The experimental environment is a desktop computer with an NVIDIA GTX1080TI graphics card.

We compared our approach with the pruning approaches such as Network Pruning, Dynamic, and Entropy. The experimental results show that our approach achieves a compression ratio of 19.75× and 22.53× on the AlexNet and VGG16 networks respectively, and the precision loss of the pruned networks are within a reasonable range. Table 1 shows the results of comparative experiments between the approach and other typical approaches.

Table 1. Comparison of our methods with other network pruning methods

Method	Model	Top-1 error	Top-5 error	Compression
Network pruning	AlexNet	42.77%	19.67%	9×
	VGG16	31.34%	10.88%	13×
Dynamic	AlexNet	43.09%	19.99%	17.7×
	—	—	—	—
Entropy	—	—	—	—
	VGG16	33.60%	12.84%	16.64×
Our method	AlexNet	43.50%	19.93%	19.75×
	VGG16	33.60%	12.60%	22.53×

We will explain the performance of this approach on the AlexNet and VGG16 networks in two subsections. In order to save space, only the changes of the parameters and calculations of the VGG16 network are analyzed in detail.

4.1   AlexNet Based on ImageNet Pre-training

AlexNet is a classic of deep learning, which won the title of image classification task in the 2012 ImageNet competition. The AlexNet is an 8-layer network with 5 convolutional layers and 3 fully connected layers, including 630 million connections, 60 million parameters and 650,000 neurons. The model achieved a 42.8% top-1 error rate and a 19.7% top-5 error rate on the ImageNet dataset.

In the filter-level pruning phase, we pruned the convolutional layer of by preset compression rate. For the fully connected layer, its parameters account for more than 90% of all parameter quantities of network, yet its contribution to the network is far less than that. In addition, it increases the risk of network over-fitting. As described in [30], the global average pooling layer aggregates the spatial information of the entire feature map, so the GAP can be regarded as a regularization operation. In many experiments, global average pooling is superior to the fully connected layer under the same network architecture. Therefore, we replaced the fully connected layer with a GAP layer.

During the retraining phase, all image sizes were adjusted to 256×256, then the image size was randomly cropped to 224×224 as the input of network. Each convolution layer was fine-tuned with a learning rate of 10^{-3} to 10^{-5}, and after the last layer was pruned, L1 regularization was introduced to retrain the network. L1 regularization tends to produce a small number of features, while other features are zero. Therefore, we used L1 regularization in the retraining phase to cause more weight parameters to approach 0. By comparing the top-1 error rate and top-5 error rate of the network under different compression ratios, we found that when the compression ratio is $14.3\times$, the network can achieve better performance with less precision loss. Then, we performed weight-level pruning on the network, keeping the training batch size, basic learning rate, and learning strategy consistent with the first phase. And we used L2 regularization and 200K iterations of the convolutional layer to obtain a sparse network model. Table 2 lists the parameter changes and compression ratios of AlexNet network in different stage of network pruning.

Table 2. Parameters and accuracy of AlexNet in different stage of network pruning

AlexNet	Top-1 error	Parameters	Compression
Reference	42.8%	61M	–
Filter level pruned	43%	4.27M	14.3×
All pruned	44.5%	3.09M	19.75×

4.2   VGG16 Based on ImageNet Pre-training

We applied the same pruning approach to the VGG16 network based on the ImageNet ILSVRC12 dataset pre-training. The VGG16 network uses a small convolution kernel (all kernels size are 3×3), and increases the depth of the network to ensure that the receptive field is large enough. And the VGG16 network has good generalization capabilities for other data sets. The network model has 13 convolutional layers and three fully connected layers, achieving a 31.5% top-1 error rate and a 11.32% top-5 error rate on the ImageNet dataset.

Table 3. Parameters and accuracy of VGG16 in different stage of network pruning

VGG16	Top-1 error	Parameters	Compression
Reference	31.5%	138.34M	–
Filter level pruned	33.00%	8.39M	16.50×
All pruned	33.60%	6.09M	22.53×

Table 4. Parameter changes of VGG16 in different network pruning stages

Layer	Original parameters	Filter-level pruning		Connection-level pruning	
		Parameters		Parameters	
		Pruned	Percentage	Pruned	Percentage
Conv1–1	1.73K	0.87K	50%	0.74K	86%
Conv1–2	36.86K	18.43K	50%	15.30K	83%
Conv2–1	73.73K	36.87K	50%	28.02K	76%
Conv2–2	147.46K	73.73K	50%	45.71K	62%
Conv3–1	294.91K	73.73K	25%	45.71K	62%
Conv3–2	589.82K	147.46K	25%	87K	59%
Conv3–3	589.82K	147.46K	25%	101.74K	69%
Conv4–1	1180K	295K	25%	215.35K	73%
Conv4–2	2360K	590K	25%	371.7K	63%
Conv4–3	2360K	590K	25%	418.9K	71%
Conv5–1	2360K	1180K	50%	802.4K	68%
Conv5–2	2360K	2360K	100%	1864.4K	79%
Conv5–3	2360K	2360K	100%	1581.2K	67%
FC6	102760K	512K	0.40%	512K	100%
FC7	16780K				
FC8	4100K				
Total	138340K	8385.52K	6.06%	6090.17K	4.40%

The pre-processing of the image during the training phase is consistent with previous experiments. The first ten layers of VGG16 have more than 90% of the calculation, so they are the mainly part we need to prune. Besides the last layer (conv4–3), we fine-tuned the network with a learning rate of 10^{-3} to 10^{-5} for 1–2 epochs after pruning of every layer. After the last layer pruned, we introduced L1 regularization and use 8 epochs to fine-tune with the learning rate of 10^{-5}. At the weight-level pruning stage, the same training parameters were used for 360K iterations of the network. Table 3 lists the parameter changes and compression ratios in different stage of network pruning. It can be seen from the table that the parameter amount of the VGG16 network was reduced from 138.34M to 6.09M, and the accuracy loss is only 2.1%.

Tables 4 and 5 show the variation of the parameters and operations of the VGG16 network at different stages of pruning in detail. We can see that the parameters are mainly concentrated in the fully connected layer, while the operations are concentrated in the convolutional layer. In the filter-level pruning stage, for the first four layers, we only pruned 50% of the filters in Conv1-1 and Conv2-1, keeping the filters of Conv1-2 and Conv2-2 unchanged. For the subsequent layer, we pruned 50% of the filters separately. We ensure that the early stages of the network do not lose too much information through this strategy. And it effectively prevents the network from accuracy loss. Finally, the pruned network parameters only account for 6.06% of the original network while reducing the number of operations to 5.8G flops from 15.5.

In the weight-level pruning phase, we only prune the convolution layer and keep the GAP layer unchanged. Since most of the unimportant filters have been pruned in the previous stage, and retraining makes the parameters in the sparse network important, only about one-third of the connections were pruned at this stage. Finally, we achieve $3.78\times$ acceleration and $22.53\times$ compression on VGG16.

Table 5. Operations changes of VGG16 in different network pruning stages

Layer	Original flops	Filter-level pruning		Connection-level pruning	
		Flops		Flops	
		Pruned	Percentage	Pruned	Percentage
Conv1–1	87.6M	43.8M	50%	37.668M	86%
Conv1–2	1850M	925M	50%	767.75M	83%
Conv2–1	920M	460M	50%	349.6M	76%
Conv2–2	1850M	925M	50%	573.5M	62%
Conv3–1	920M	230M	50%	142.6M	62%
Conv3–2	1850M	462.5M	50%	272.875M	59%
Conv3–3	1850M	462.5M	25%	319.125M	69%
Conv4–1	920M	230M	25%	167.9M	73%
Conv4–2	1850M	462.5M	25%	291.375M	63%
Conv4–3	1850M	462.5M	25%	328.375M	71%
Conv5–1	462.42M	231.21M	25%	157.2228M	68%
Conv5–2	462.42M	462.42M	100%	365.3118M	79%
Conv5–3	462.42M	462.42M	100%	309.8214M	67%
FC6	102.76M	0.512M	0.40%	0.512M	100%
FC7	16.78M				
FC8	4.1M				
Total	15458.5M	5820.362M	37.70%	4083.636M	26.42%

5 Conclusion

In this paper, we proposed a multi-grained deep convolutional neural network pruning approach. Different from the popular approaches only pruning at single scale, a weight-level pruning method is also taken into account based on the employment of filter-level pruning in our network pruning framework. The experimental results show that our approach compresses the parameters in AlexNet and VGG16 by 19.75× and 22.53×, which are 2.05 and 5.89 higher than the previous pruning approaches, respectively. In addition, we have greatly improved the speed of the network approaches compared with the weight-level pruning. And we will carry out the future work in two aspects: dynamic pruning and improving pruning efficiency.

References

1. Cheng, Y., Wang, D., Zhou, P., Zhang, T.: A survey of model compression and acceleration for deep neural networks. Front. Inf. Technol. Electron. Eng. **19**, 64–77 (2017)
2. Simonyan, K., Zisserman, A.: Very deep convolutional networks for large-scale image recognition. In: ICLR, pp. 1–14 (2014)
3. Cheng, J., Wang, P.-S., Li, G., Hu, Q.-H., Lu, H.-Q.: Recent advances in efficient computation of deep convolutional neural networks. Front. Inf. Technol. Electron. Eng. **19**, 64–77 (2018)
4. Denton, E.L., Zaremba, W., Bruna, J., LeCun, Y., Fergus, R.: Exploiting linear structure within convolutional networks for efficient evaluation. In: Advances in Neural Information Processing Systems, pp. 1269–1277 (2014)
5. Jaderberg, M., Vedaldi, A., Zisserman, A.: Speeding up convolutional neural networks with low rank expansions (2014)
6. Kim, Y.-D., Park, E., Yoo, S., Choi, T., Yang, L., Shin, D.: Compression of deep convolutional neural networks for fast and low power mobile applications. In: ICLR (2015)
7. Han, S., Pool, J., Tran, J., Dally, W.: Learning both weights and connections for efficient neural network. In: Advances in Neural Information Processing Systems, pp. 1135–1143 (2015)
8. Hwang, K., Sung, W.: Fixed-point feedforward deep neural network design using weights +1, 0, and −1. In: 2014 IEEE Workshop on Signal Processing Systems (SiPS), pp. 1–6 (2014)
9. Anwar, S., Hwang, K., Sung, W.: Fixed point optimization of deep convolutional neural networks for object recognition. In: 2015 IEEE International Conference on Acoustics, Speech and Signal Processing (ICASSP), pp. 1131–1135 (2015)
10. Yim, J, Joo, D., Bae, J., Kim, J.: A gift from knowledge distillation: fast optimization, network minimization and transfer learning. In: Proceedings of the IEEE Conference on Computer Vision and Pattern Recognition, pp. 4133–4141 (2017)
11. Sun, G., Liang, L., Chen, T., et al.: Network traffic classification based on transfer learning [J]. Computers and Electrical Engineering, pp. 1–8, (2018)
12. Howard, A.G., et al.: Mobilenets: efficient convolutional neural networks for mobile vision applications (2017)
13. Sandler, M., Howard, A., Zhu, M., Zhmoginov, A., Chen, L.-C.: Mobilenetv2: inverted residuals and linear bottlenecks. In: Proceedings of the IEEE Conference on Computer Vision and Pattern Recognition, pp. 4510–4520 (2018)

14. Zhang, X., Zhou, X., Lin, M., Sun, J.: Shufflenet: an extremely efficient convolutional neural network for mobile devices. In: Proceedings of the IEEE Conference on Computer Vision and Pattern Recognition, pp. 6848–6856 (2018)
15. Guo, Y., Yao, A., Chen, Y.: Dynamic network surgery for efficient dnns. In: Advances in Neural Information Processing Systems, pp. 1379–1387 (2016)
16. Molchanov, P., Tyree, S., Karras, T., Aila, T., Kautz, J.: Pruning convolutional neural networks for resource efficient inference (2017)
17. He, Y., Zhang, X., Sun J.: Channel pruning for accelerating very deep neural networks. In: Proceedings of the IEEE International Conference on Computer Vision, pp. 1389–1397 (2017)
18. Liu, Z., Li, J., Shen, Z., Huang, G., Yan, S., Zhang, C.: Learning efficient convolutional networks through network slimming. In: Proceedings of the IEEE International Conference on Computer Vision, pp. 2736–2744 (2017)
19. Han, S., Mao, H., Dally, W.: Deep compression: compressing deep neural networks with pruning, trained quantization and huffman coding. In: ICLR (2015)
20. Lebedev, V., Lempitsky, V.: Fast convnets using group-wise brain damage. In: Proceedings of the IEEE Conference on Computer Vision and Pattern Recognition, pp. 2554–2564 (2016)
21. Yang, T.-J., Chen, Y.-H., Sze, V.: Designing energy-efficient convolutional neural networks using energy-aware pruning. In: Proceedings of the IEEE Conference on Computer Vision and Pattern Recognition, pp. 5687–5695 (2017)
22. LeCun, Y., Denker, J.S., Solla, S.A.: Optimal brain damage. In: Advances in Neural Information Processing Systems, pp. 598–605 (1990)
23. Hassibi, B., Stork, D.G.: Second order derivatives for network pruning: optimal brain surgeon. In: Advances in Neural Information Processing Systems, pp. 164–171 (1993)
24. Wen, W., Wu, C., Wang, Y., Chen, Y., Li, H.: Learning structured sparsity in deep neural networks. In: Advances in Neural Information Processing Systems, pp. 2074–2082 (2016)
25. Li, H., Kadav, A., Durdanovic, I., Samet, H., Graf, H.P.: Pruning filters for efficient convnets (2016)
26. Hu, H., Peng, R., Tai, Y.-W., Tang, C.-K.: Network trimming: a data-driven neuron pruning approach towards efficient deep architectures, pp. 1–9 (2016)
27. Luo, J.-H., Wu, J.: An entropy-based pruning method for cnn compression (2017)
28. Luo, J.-H., Wu, J., Lin, W.: Thinet: a filter level pruning method for deep neural network compression. In: Proceedings of the IEEE International Conference on Computer Vision, pp. 5058–5066 (2017)
29. Zou, J., Rui, T., Zhou, Y., Yang, C., Zhang, S.: Convolutional neural network simplification via feature map pruning. Comput. Electr. Eng. **70**, 950–958 (2018)
30. Lin, M., Chen, Q., Yan, S.: Network in network. In: ICLR (2014)
31. Jia, Y., et al.: Caffe: convolutional architecture for fast feature embedding. In: Proceedings of the 22nd ACM International Conference on Multimedia, pp. 675–678 (2014)
32. Krizhevsky, A., Sutskever, I., Hinton, G.E.: Imagenet classification with deep convolutional neural networks. In: Advances in Neural Information Processing Systems, pp. 1097–1105 (2012)

Donggan Speech Recognition Based on Convolution Neural Networks

Haiyan Xu, Yuren You, and Hongwu Yang[✉]

College of Physics and Electronic Engineering, Northwest Normal University,
Lanzhou 730070, China
yanghw@nwnu.edu.cn

Abstract. Donggan language, which is a special variant of Mandarin, is used by Donggan people in Central Asia. Donggan language includes Gansu dialect and Shaanxi dialect. This paper proposes a convolutional neural network (CNN) based Donggan language speech recognition method for the Donggan Shaanxi dialect. A text corpus and a pronunciation dictionary were designed for of Donggan Shannxi dialect and the corresponding speech corpus was recorded. Then the acoustic models of Donggan Shaanxi dialect was trained by CNN. Experimental results demonstrate that the recognition rate of proposed CNN-based method achieves lower word error rate than that of the monophonic hidden Markov model (HMM) based method, triphone HMM-based method and DNN- based method.

Keywords: Donggan language · Donggan speech recognition · Convolutional neural network · Acoustic model

1 Introduction

With the proposal of "One Belt and One Road", the relations between central Asian countries and China are gradually strengthened. As one of the nationalities along the route, Donggan nationality also determines the unique status and role of Donggan language in "One Belt and One Road" language plan. Therefore, it is extremely important for us to study Donggan language. Donggan language belongs to the Sino-Tibetan language branch, it is a variant of Shanxi-Gansu dialect. Donggan language originally had the difference between Dongyu dialect (now Shaanxi Guanzhong dialect), Lanzhou dialect, Hezhou dialect (now Gansu Linxia dialect), Lotus City dialect (now Gansu Qin'an dialect), and Xining dialect [1]. After a long time, Fusion, now Donggan language is divided into two major dialects: Gansu dialect and Shaanxi dialect. After a long time of integration, Donggan language is now mainly divided into two dialects including Gansu dialect and Shaanxi dialect. It is mainly distributed from the junction of the roads on both sides of the Chuhe River in Kyrgyzstan and Kazakhstan [2] to the west and east of Bishkek in the capital of Kyrgyzstan. Donggan language is the first successful case in which phonemes are used to spell Chinese. It is composed of 38 letters spelled with Russian letters, and the written language is based on Gansu dialect [2].

© Springer Nature Singapore Pte Ltd. 2019
X. Cheng et al. (Eds.): ICPCSEE 2019, CCIS 1058, pp. 577–584, 2019.
https://doi.org/10.1007/978-981-15-0118-0_44

Speech recognition is the process of converting speech signal into corresponding text. Early speech recognition is mainly for digital and isolated word [3]. Later, with the development of speech recognition, Linear Predictive Coding (LPC) [4] technology and dynamic programming planning [5] algorithm was applied to speech recognition, which effectively solved the problem of the generation of the speech signal model and the unequal length of the speech signal [6]. At the same time, we realized the feature extraction of the speech signal. After that, the technology based on Dynamic Time Warping (DTW) [7] became mature, Vector Quantization (VQ) and Hidden Markov model theory [8] were proposed, and some breakthroughs were made in speech recognition. At present, with the proposal of deep learning method, the application of deep learning method for speech recognition has become a research hotspot [9–11]. However, the problem of non-specific human differences and noise effects in speech signals has not been well solved. The proposed convolutional neural network solves the above problems. Compared with deep neural network, convolutional neural network extracts local information of speech signal by introducing convolution, and then improves the robustness of the model to features through aggregation. Therefore, based on the previous work, this paper proposes the research of Donggan speech recognition based on CNN. The results show that the recognition rate of Donggan speech based on CNN is better than that of the other three models.

2   Donggan Speech Recognition Framework

2.1   DNN-Based Acoustic Model Training

The speech recognition framework proposed in this paper is shown in Fig. 1, which Mainly includes the training stage and the stage of recognition; The training stage includes acoustic model training and language model training. Recognition is the process of decoding and outputting text by combining acoustic model, language model and dictionary.

The acoustic model is to find the most likely word sequence W after a speech sequence O is given, so that it has the highest matching degree with the speech signal [12]. DNN acoustic model training is to use DNN in the acoustic modeling of speech recognition and replace the traditional Gaussian Mixture Modeling (GMM) with DNN to estimate the posterior probability of the acoustic state of Donggan language [12]. DNN is generated by continuously stacking the hidden layers from the bottom up. In the pre-training of Donggan language acoustic model, each layer of DNN was trained using a layer-by-layer unsupervised training method. Only one layer was trained at a time, while the other layer maintained the parameters at the time of initialization, which can ensure that the parameters of each layer were optimal. Then, the trained data are taken as the input of the next layer, which can reduce the error of inputting the data directly after the multi-layer neural network to the next layer. In the process of DNN acoustic model training, the input is the feature extraction module to extract the acoustic features. The output of hidden layer is activated by sigmoid function. Softmax function is used to classify the final output layer. Assuming that the training sample is marked z, then

$$soft\max(net_i) = \frac{e^{net_i}}{\sum_{j-1}^{K} e^{net_i}} \tag{1}$$

where net_k is given by:

$$net_k = \sum_{j=1}^{nH} w_{kj}b_j \tag{2}$$

$$z_k = soft\max(net_k) = \frac{e^{net_k}}{\sum_{j=1}^{K} e^{net_j}} \tag{3}$$

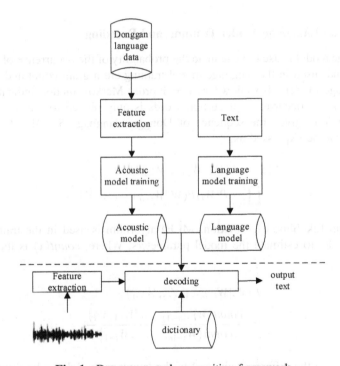

Fig. 1. Donggan speech recognition framework

2.2 CNN-Based Acoustic Model Training

CNN adopts pooling processing technology, which has better robustness against slight changes of signals caused by different speakers or different styles of speakers. At the same time, the convolutional layer adopts sparse local connection [13], which reduces the training parameters and avoids overfitting. CNN mainly includes input layer, convolution layer, pooling layer, nonlinear layer and output layer. The CNN input is the frame sequence after extracting the features of the original voice signal. The output is the probability of word sequence i under the condition of given frame W sequence x. The convolutional layer input is a vector of fixed size and can be thought of as a sequence of vectors or frames. Different from the fully connected network, the input of each node in the convolutional layer is only the local area of the upper neural network.

The pooling layer is obtained by the convolution layer of the previous layer through down-sampling. Common pooling methods include Ave-pooling and Max-pooling [14]. The down-sampling method can reduce the complexity of the algorithm and has better robustness. In CNN model training, log-likelihood function is used, and formula (4) is the training process as a parameter θ: x is the input, t is the mark.

$$L(\theta) = \sum_{n}^{N} \log(p(i_n|x_n, \theta)) \tag{4}$$

2.3   N-Gram Language Model Training and Decoding

The language model is used to calculate the probability of the occurrence of a sentence. The commands used in the language model training are n-gram count and n-gram. N-gram language model, also known as n − 1 order Markov model, indicates that the probability of the occurrence of current words is only related to the previous n − 1 words. Therefore, given the sequence of Donggan language $S = W_1, W_2, \ldots W_i$, the probability can be expressed as:

$$P(S) = P(W_1, W_2, \ldots W_k)$$
$$= \prod_{i=1}^{k} P(W_i) P(W_i|W_{i-n+1}, \ldots, W_{i-1}) \tag{5}$$

Maximum Likelihood Estimation (MLE) algorithm is used in the training of the language model to estimate the model parameters. Where, $count(x)$ is the word frequency of x

$$P(W_i|W_{i-(n+1)} \cdots, W_{i-1})$$
$$= \frac{count(W_{i-(n-1)} \cdots, W_{i-1}, W_i)}{count(W_{i-(n-1)} \cdots, W_{i-1})} \tag{6}$$

According to the different values of n, the language model can be divided into 1-gram, 2-gram, 3-gram and so on. With the increase of n value, the more accurate the model is, the larger the computation is. In this paper, 3-gram language model is trained for decoding.

Decoding is the process of converting input speech feature vectors into character sequences based on trained acoustic models, language models and dictionary [15].

Viterbi algorithm is used to find the optimal path during decoding. Given a sequence of observations $Y = \{y_1, y_2, \ldots, y_n\}$, Viterbi algorithm can be represented by Eq. (7)

$$x_1, x_2, \ldots, x_N = \arg \max_{x \in X} P(x_1, x_2, \ldots, x_N | y_1, y_2 \ldots, y_N)$$

$$= \arg \max_{x \in X} \prod_{i=1}^{N} P(y_i | x_i) P(x_i | x_{i-1}) \tag{7}$$

where the $x_1, x_2, \ldots, x_N$ is state sequence.

3 Corpus Construction

3.1 Corpus Design

In this paper, we select 5 male college students as corpus recorders to record donggan language corpus. Each recorder records 400 sentences of corpus, totaling 2,000 sentences. The recording of corpus is carried out in a professional recording studio. At last, the recorded voice data is saved in the format of mono channel with the speaker serial number plus the voice serial number in the format of 16 kHz sampling rate and 16-bit sampling accuracy.

3.2 Corpus Labelling

Donggan language has 24 initials and 32 finals, the number of initials and finals are moderate, so the vowels are selected as the recognition unit. Since Donggan language is a variant of Chinese, the annotation method of Donggan language is designed with reference to the text annotation of Chinese speech recognition when labeling Donggan language. The labeling of Donggan language text corpus is completed by manual labeling. Table 1 is a reference table for some Donggan language initials and Chinese Pinyin. Finally, the marked text is saved for each voice file.

Table 1. Donggan language and Chinese pinyin.

Donggan language	Б/б	П/п	М/м	Ф/ф	В/в	З/з
Pinyin	b	p	m	f	v	z
Donggan language	Ц/ц	С/с	Д/д	Т/т	Н/н	Л/л
Pinyin	c	s	d	t	n	l

3.3 Dictionary Design

The mapping relationship between the modeling unit of acoustic model and the language model is obtained through the pronunciation dictionary. The acoustic model and the language model are connected to form a searching state space for decoder to decode. Since there is no uniform word segmentation rule in Donggan language, the pronunciation dictionary of Donggan language is designed with syllables as the unit rather than words. Donggan pronunciation dictionary contains all the words in the training corpus and the test corpus.

4 Experimental Results and Analysis

4.1 Experimental Data

The corpus selects the corpus of 5 male speakers, each speaker contains 400 sentences and the corpus contains a total of 2000 sentences. 1600 sentences are used as training corpus, 100 sentences are used as cross-training, and the remaining 300 sentences are used as test corpus. The classification of training corpus, test corpus and verification corpus is divided into groups according to the phonetic number. The prepared corpus including voice files and labeled text files are saved as the corresponding directory. Then we modify the path information.

4.2 Experiment

In order to compare the speech recognition results under different acoustic models, this paper trains three acoustic models, including monophonic HMM, triphone HMM and DNN model. In the training of monophonic, the acoustic characteristics of 13-dimensional MFCC are firstly extracted, and then the Cepstral Mean Variance Normalization (CMVN) technology is used to preprocess the acoustic characteristics of 13-dimensional MFCC. Then the 39-dimensional MFCC characteristics are obtained through difference as the input of the model. During the training process, 40 iterations are carried out. For the first 10 iterations of the iteration, forced alignment is carried out once every iteration. From the 10th iteration to the 20th iteration, forced alignment was carried out once every iteration, and from the 20th iteration to the 38th iteration, forced alignment was carried out once every iteration.

In the process of training the triphone HMM, firstly we extracted the 13-dimensional MFCC characteristics of Donggan language Training corpus, and then Linear Discriminant Analysis (LDA) and maximum likelihood linear transformation (MLLT) are added in the training process. Which are used to optimize the acoustic model to obtain the MLLT+LDA triphone model. Finally, the adaptive training (SAT) is used to obtain the triphone acoustic model of LDA+MLLT+SAT. The experiment was iterated 40 times, and a forced alignment was performed when iterating to the 10th, 20th, and 30th. In addition, alignment was performed once before DNN model training.

In the process of training the DNN model, because the number of hidden layer nodes and the choice of hidden layer number will affect the recognition result [16], the number of nodes is too small, the network learning solves the problem with insufficient information, and it is difficult to cover all the rules in the sample. If there are too many nodes, the training time will increase and decrease the generalization ability of the network. Therefore, this paper selects 4 hidden layers DNN model for research. The number of hidden layer nodes is1024, and the initial learning rate is 0.008. With the increase of iteration times, the learning rate will decrease according to the degree of relative improvement of the objective function. DNN training adopts FBank feature with better effect. The extracted 40-dimensional FBank is processed by CMVN, then the feature is spliced to obtain the vector of 11 frames as the input of DNN. The DNN acoustic model is trained using a back propagation (BP) algorithm.

The training process of CNN model is similar to that of DNN, including input layer, 2 convolution layer, 3 maximum pooling layer, 4 full connection layer and output layer. The input is the Fbank feature of 11 frames and the first-order and second-order difference, and the number of nodes in the output layer is 2432. There are two filters in the convolutional layer, and the output of each filter is a vector of 19 dimensions, and then a maximum value is found in every three Numbers, and finally a matrix of 6*128 is obtained as the input of the second convolutional layer. In this paper, Sigmoid activation function is used in the full connection layer, and the initial learning rate is 0.008.

In order to verify the reliability of the experiment, this paper divided 2000 sentences of Donggan language corpus into five parts on average, namely T1–T5. Then, five groups of data were cross-verified. The results of five experiments corresponded to t1–t5 according to the test data.

Table 2. Word error rate

Acoustic model	WER (%)				
	T1	T2	T3	T4	T5
Monophonic	17.13	20.98	18.69	17.06	18.55
Triphone	15.60	17.19	15.59	16.89	16.89
DNN	15.52	16.45	15.13	16.60	15.85
CNN	15.06	15.69	14.96	16.23	15.59

Table 2 contains the word error rate of donggan language speech recognition under different acoustic models. It can be seen from the table that the recognition accuracy of DNN model has been improved to a certain extent compared with the Monophonic model and the Triphone model, especially compared with the Monophonic model, the recognition rate has been significantly improved. CNN adds convolutional layer and pooling layer in the network structure, it has a higher recognition rate compared with DNN. However, due to the shortage of Donggan language corpus, it can be seen from Table 2 that although DNN has improved the speech recognition rate relative to Monophonic and Triphone models, the improvement is not significant. At the same time, the cross-experimental results show that no matter how the test set is transformed, the acoustic model trained by CNN achieves the best recognition result.

5 Conclusion

In this paper, Donggan language is taken as the research object. Firstly, Donggan language corpus and dictionary are designed. Then, Donggan language acoustic model and language model are trained. From the experimental results, based on the deep learning Donggan speech recognition results is best, however, by the limitation of Donggan language corpora, Donggan language based on CNN relative speech recognition based on HMM speech recognition results only a modest increase, so, at the back of the experiment we can expand the corpus or migration study ways to improve the Donggan speech recognition rate.

References

1. Lin, T.: Research on Donggan Language. China Social Science Press (2012)
2. Wang, S.: Survey and Research on the Chinese and Asian Donggan Dialect. Commercia (2015)
3. Furui, S.: History and development of speech recognition. In: Chen, F., Huggins (eds.) Speech Technology, pp. 1–18. Springer, Boston (2010). https://doi.org/10.1007/978-0-387-73819-2_1
4. Hai, J., Joo, E.M.: Improved linear predictive coding method for speech recognition. In: Conference on Joint Conference of the Fourth International Conference on Information, Communications & Signal Processing (2003)
5. Sakoe, H., Chiba, S.: Dynamic programming algorithm optimization for spoken word recognition. IEEE Trans. Acoust. Speech Signal Process. **26**(1), 43–49 (2003)
6. Jing, Z., Qin, B.: DTW speech recognition algorithm of optimization template matching. In: World Automation Congress (2012)
7. Muda, L., Begam, M., Elamvazuthi, I.: Voice recognition algorithms using Mel Frequency Cepstral Coefficient (MFCC) and Dynamic Time Warping (DTW) techniques. Ttps **2** (2010)
8. Omer, A.E.: Joint MFCC-and-vector quantization based text-independent speaker recognition system. In: International Conference on Communication (2017)
9. Hinton, G., Deng, L., Yu, D., et al.: Deep neural networks for acoustic modeling in speech recognition: the shared views of four research groups. IEEE Signal Process. Mag. **29**(6), 82–97 (2012)
10. Nguyen, Q.B., Vu, T.T., Chi, M.L.: Improving acoustic model for English ASR System using deep neural network. In: IEEE Rivf International Conference on Computing & Communication Technologies-research (2015)
11. Hu, W., Fu, M., Pan, W.: Primi speech recognition based on deep neural network. In: IEEE International Conference on Intelligent Systems. IEEE (2016)
12. Karáfidt, M., Baskar, M.K., Veselý, K., et al.: Analysis of multilingual BLSTM acoustic model on low and high resource languages. In: 2018 IEEE International Conference on Acoustics, Speech and Signal Processing (ICASSP), pp. 5789–5793. IEEE (2018)
13. Abdel-Hamid, O., Mohamed, A., Jiang, H., et al.: Convolutional neural networks for speech recognition. IEEE/ACM Trans. Audio Speech Lang. Process. **22**(10), 1533–1545 (2014)
14. Zeiler, M.D., Fergus, R.: Stochastic pooling for regularization of deep convolutional neural networks. arXiv preprint arXiv:1301.3557 (2013)
15. Rashmi, S., Hanumanthappa, M., Reddy, M.V.: Hidden Markov Model for speech recognition system—a pilot study and a naive approach for speech-to-text model. In: Agrawal, S.S., Devi, A., Wason, R., Bansal, P. (eds.) Speech and Language Processing for Human-Machine Communications. AISC, vol. 664, pp. 77–90. Springer, Singapore (2018). https://doi.org/10.1007/978-981-10-6626-9_9
16. Dighe, P., Luyet, G., Asaei, A., et al.: Exploiting low-dimensional structures to enhance DNN based acoustic modeling in speech recognition. In: IEEE International Conference on Acoustics (2016)

Design of Polynomial Fuzzy Neural Network Classifiers Based on Density Fuzzy C-Means and L2-Norm Regularization

Shaocong Xue, Wei Huang[✉], Chuanyin Yang, and Jinsong Wang

Tianjin University of Technology, Tianjin 300384, China
huangwabc@163.com

Abstract. In this paper, polynomial fuzzy neural network classifiers (PFNNCs) is proposed by means of density fuzzy c-means and L2-norm regularization. The overall design of PFNNCs was realized by means of fuzzy rules that come in form of three parts, namely premise part, consequence part and aggregation part. The premise part was developed by density fuzzy c-means that helps determine the apex parameters of membership functions, while the consequence part was realized by means of two types of polynomials including linear and quadratic. L2-norm regularization that can alleviate the overfitting problem was exploited to estimate the parameters of polynomials, which constructed the aggregation part. Experimental results of several data sets demonstrate that the proposed classifiers show higher classification accuracy in comparison with some other classifiers reported in the literature.

Keywords: Polynomial fuzzy neural network classifiers · Density fuzzy clustering · L2-norm regularization · Fuzzy rules

1 Introduction

The past decades have witnessed the rapid development of classifier that is an essential part in the field of pattern recognition. Diverse classifiers have been constructed to improve the classification accuracy such as Support Vector Machine (SVM), Relevance Vector Machine (RVM), K-Nearest Neighbor (KNN), Possibilities Fuzzy C-means Clustering (PFC), and so on. SVM avoids over-fitting by seeking out segregated hyperplanes that maximize the edge width, which has generated good results in many pattern classifications. However, SVM is not free from limitations with a large number of support vectors and Mercer condition. As for RVM, only a few parameters are needed to optimize in the process of training, but the matrix inversion still brings a huge amount of computation. Worse more, insufficient capacity to handle imbalanced training data also influences the adaptation in some occasion [1]. KNN classifier is commonly exhibited in the supervised learning classification of machine learning. Nevertheless, the traditional Euclidean distance used in the classifier is not always applicable to the actual situation, which affects the classification accuracy [2]. PFC algorithm initialized by Pal et al. [3] solves the problem of overlapping clustering centers to some extent, yet its performance is sensitive to initialization.

© Springer Nature Singapore Pte Ltd. 2019
X. Cheng et al. (Eds.): ICPCSEE 2019, CCIS 1058, pp. 585–596, 2019.
https://doi.org/10.1007/978-981-15-0118-0_45

The emergence of neural networks brings new possibilities to pattern recognition. Research about artificial neural networks such as Multi-Layer Perceptron (MLP), have been fully developed in the early age. Nandi et al. [4] applied Back Propagation (BP) based MLP for signal recognition classifier. Huang Chunlin et al. [5] improved BP by using fusion inversion algorithm and then trained three-layer MLP network. Whatever, slow in learning and easy to fall into local optimum, the disadvantage of MLP neural network can't be ignored. Later, hidden element and BP neural network appeared one after another. Buhmann introduced radial basis function into the design of neural networks to simulate the local response characteristics of biological neurons. Radial Basis Function Neural Network (RBFNN) has been widely used in various pattern recognition and classification problems because of its simple structure, strong non-linear approximation ability, fast convergence and global convergence. In the classical RBFNN model, Gauss function, linear polynomial and BP algorithm attend their duties in the premise part, conclusion part and aggregation part. Gauss function has strong local approximation ability, but it can't deal with high-dimensional problems very well. BP algorithm converges slowly and easily falls into local optimum, resulting in a large amount of calculation [9]. In the classical RBFNN model, Gauss function, linear polynomial and BP algorithm attend their duties in the premise part, conclusion part and aggregation part. Gauss function has strong local approximation ability, but it can't deal with high-dimensional problems very well. BP algorithm converges slowly and easily falls into local optimum, resulting in a large amount of calculation [9].

In this paper, we propose Polynomial fuzzy neural network classifiers (PFNNCs) by means of density fuzzy c-means and L2-norm regularization to overcome the aforementioned limitations. The outstanding features of the proposed PFNNCs are as follows:

First, the proposed PFNNCs can support an identification of relative high-order nonlinear relations between input and output. In the design of PFNNCs, two types of polynomials are used to construct the consequence part of PFNNCs, while most of conventional models only use linear polynomial such as RBFNNs.

Second, the proposed PFNNCs can alleviate the overfitting problem. The coefficients of polynomials of PFNNCs are estimated by using L2-norm regularization. In the conventional classifiers, the coefficients of linear polynomial are generally estimated by means of least square method.

Finally, the proposed PFNNCs are developed based on "good" information granulation. Instead of conventional fuzzy clustering method, we use density fuzzy c-means (DFCM) algorithm to realize information granulation. In comparison with fuzzy clustering method, the DFCM can deal with high-dimensional problems well.

This paper is arranged as follows. Section 2 recalls the classic algorithm of FCM and DCM. Section 3 provides an overall description of a detailed design methodology of PFNNCs. Section 4 presents the experimental results. Finally, concluding remarks are drawn in Sect. 5.

2 Related Work

2.1 Fuzzy C-Means

Fuzzy c-means (FCM) algorithm is a partition-based clustering algorithm. The FCM algorithm is an improvement of the ordinary C-means algorithm. It divides the data into some kind of fuzzification. The classification result is judged by calculating the membership degree of the data, which effectively increases the classification accuracy [11].

Traditional FCM algorithm is easy to understand and easy to operate. However, the FCM algorithm is prone to fall into local optimum because of the inappropriate selection of initialization centers, and the inadequate interference processing of the noise in the FCM algorithm makes the classification results deviate easily. In addition, if the location of the initialization centers are not selected properly, the computational complexity of the algorithm will be greatly increased, thus reducing the efficiency of the algorithm.

2.2 Density-Based Clustering

Density-based Clustering (DENCLUE) is a clustering algorithm based on a set of density distribution functions. The core idea of DENCLUE is to model the density of every data in data space by using kernel density estimation.

On the basis of generalizing other clustering methods, DENCLUE algorithm takes full account of the need for global optimization. The algorithm determines the clustering center based on the density distribution of the input data points, so that it can find clusters of arbitrary shape, and also can deal with noise better.

Therefore, it is possible to use DENCLUE algorithm to improve FCM algorithm. The optimized algorithm can give full play to the global optimum of DENCLUE algorithm, and can obtain more accurate initialization of clustering centers, reduce the number of searching clustering centers, which is helpful to improve the efficiency and accuracy of the algorithm.

2.3 Classification Algorithm

In the construction of PFNNCs, classification algorithm is deeply needed to realize the ultimate destination. Traditional classification algorithms can be divided into two categories: direct method and indirect method. With high computational complexity, direct classification is often adopted to solve problems of small scale. Here we use indirect classification, which is realized by combining multiple binary classifiers. Specifically, for the problem of k categories, k classifiers are trained in the training set. Each classifier divides the training set into $i(i = 1, 2, \ldots, k)$ class and non-i classes. Classes mark their classes as 1 and non-classes mark their classes as -1.

3   PFNNCs

The PFNNCs constructed in this paper, using the idea of neural network, include three functional modules, which are the premise, conclusion and aggregation stages, as shown in Fig. 1. The premise part calculates a reasonable membership matrix by using the proposed DFCM algorithm, and transmits it to the aggregation part as the weight of hidden layer in the PFNNCs. The conclusion part uses polynomials. In the aggregation part, the L2 norm regularization is used to estimate the parameters. Consequently combined with the binary classifier and the discriminant function to calculate the classification results.

3.1   DFCM Algorithm of the Premise Part

In order to overcome the shortcomings of traditional FCM algorithm, such as easily falling into local optimum due to improper initialization and excessive computation, a DFCM algorithm is proposed. The algorithm uses DENCLUE algorithm to optimize the traditional FCM algorithm, which makes the initialized clustering centers more accurate, greatly reduces the computational load caused by searching the clustering centers, and gives full play to the advantages of DENCLUE algorithm in dealing with noise and considering global optimum.

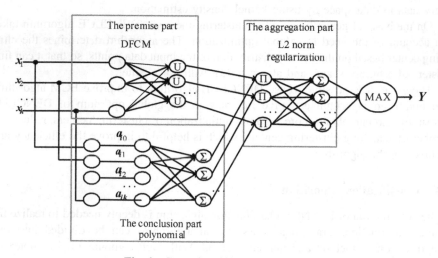

Fig. 1.   General architecture of PFNNC

The specific steps of DFCM algorithm are shown in Table 1.

$$U_{ik} = \frac{1}{\sum\limits_{j=1}^{c} \left(\frac{d_{ik}}{d_{jk}}\right)^{\frac{2}{m-1}}} \tag{1}$$

Where U_{ik} represents the probability that the k-th data point belongs to the class i, and forms the membership matrix U.

$$U_{ik} = \frac{1}{\sum\limits_{j=1}^{c} \left(\frac{d_{ik}}{d_{jk}}\right)^{\frac{2}{m-1}}} \tag{2}$$

Where d_{ik} denotes the spatial distance between the i-th central spot and the k-th data, m represents the weight index usually with a value 2.

$$V_{ij} = \frac{\sum\limits_{k=1}^{n} (U_{ik})^m x_{kj}}{\sum\limits_{k=1}^{n} (U_{ik})^m} \tag{3}$$

Where V_{ij} represents the size of the first central point in the j dimension and forms a clustering central matrix V. x_{kj} denotes the i-th dimension of the k-th input data.

Table 1. DFCM algorithm

Input: data set X
Output: membership matrix U
1. Initialize cluster center set V by using the algorithm in Table 2.
2. Initialize the parameters, in which the membership matrix U satisfies the formula (1).
3. According to cluster center set V, the membership matrix U is improved by formula (2).
4. According to matrix U, the clustering center set V is improved by formula (3).
5. According to matrix U, the objective function E is calculated by formula (4).
6. Repeat 3, 4, 5 steps until the objective function converges to the minimum or reaches the upper limit of iteration times.

$$E = \left\| U^{(r+1)} - U^r \right\| = max_{i,k} \left| U_{ik}^{(r+1)} - U_{ik}^r \right| \tag{4}$$

DFCM initializes clustering centers with kernel function estimation method, and regards each data point as a high density indicator for the surrounding area. DENCLUE is used to propose a set of points according to the density, which is used as the initial clustering center of DFCM clustering. The specific steps are shown in Table 2.

$$D_i = \sum_{k=1}^{n} \exp\left[-\frac{\|x_i - x_k\|^2}{(0.5r_1)^2} \right] \tag{5}$$

$$r_1 = \frac{1}{2} min_k \{ max_i \|x_i - x_k\| \} \tag{6}$$

Where D_i represents the density of each input data, and x_i represents the i-th input data. r_i symbols the distance parameter used in the calculation.

Table 2. Algorithm for initializing clustering center of DFCM

Input: data set X
Output: Processed cluster center set V
1. According to data set X, the distribution of data around each data is calculated by using density index formulas (5) and (6).
2. Choose the data with the highest density index as the first clustering center and record its density index as si.
3. According to the current density index and the selected clustering centers, the density index is modified by formula (7) and (8) to remove the interference of the selected clustering centers.
4. Repeat steps 2 and 3 until all clustering centers are found and the initialized clustering centers are output.

$$D_i = D_i - D_{si} \exp\left[-\frac{\|x_i - x_{si}\|}{(0.5r_2)^2}\right], i \neq si \tag{7}$$

Where si represents the subscription of the last determined clustering center.

$$r_2 = \frac{1}{2}r_1 \tag{8}$$

3.2 Polynomials of the Conclusion Part

The conclusion part is constructed by polynomials. Here, two types of polynomials are considered:

(1) Linear polynomial:

$$y_i = a_{i0} + a_{i1}x_1 + \ldots + a_{ik}x_k \tag{9}$$

(2) Quadratic polynomial:

$$y_j = a_{j0} + a_{j1}x_1 + \ldots + a_{jk}x_k + a_{j(k+1)}x_1^2 + \ldots$$
$$+ a_{j(2k)}x_k^2 + a_{j(2k+1)}x_1x_2 + \ldots + a_{j((k+2)(k+1)}x_{k-1}x_k \tag{10}$$

Where y represents the output data, $x_1, x_2, \cdots, x_k$ denotes the input variables respectively. And $a_{im}(m = 0, 1, \ldots, k)$ denotes the coefficient of polynomial, respectively.

3.3   L2-Norm Regularization of the Aggregation Part

L2-norm regularization is applied to the construction of aggregation part instead of BP algorithm with slow convergence speed, local optimum and large amount of calculation. At the same time, the use of regularization can decrease the possibility of overfitting occurring frequently in classification. Figure 1 depicts a general architecture of PFNNCs. As shown in Fig. 1, the nodes of aggregation part are the product of the fuzzy set (i.e. membership matrix U) obtained by DFCM algorithm and the linear expression function transferred by the conclusion part. For convenience, we consider a linear format of polynomial in the following way:

$$\hat{y}_j = \sum_{i=1}^{n} (a_{i0} + a_{i1}x_1 + \ldots + a_{ik}x_k)U_{ij} \tag{11}$$

Simply, the expression can be described as follows:

$$Y = AW \tag{12}$$

Where

$$Y = [y_1 \ y_2 \cdots y_m]^T \tag{13}$$

$$A = [a_{10} \ldots a_{n0} \quad a_{11} \ldots a_{n1} \ldots a_{1k} \ldots a_{nk}]^T \tag{14}$$

$$W = \begin{bmatrix} \hat{U}_{11} \ldots \hat{U}_{n1} & x_{11}\hat{U}_{11} \ldots x_{11}\hat{U}_{n1} \ldots & x_{k1}\hat{U}_{11} \ldots x_{k1}\hat{U}_{n1} \\ \vdots & & \\ \hat{U}_{1m} \ldots \hat{U}_{nm} & x_{1m}\hat{U}_{rm} \ldots x_{1m}\hat{U}_{rmn} \ldots & x_{kn}\hat{U}_{1m} \ldots x_{kn}\hat{U}_{nm} \end{bmatrix} \tag{15}$$

According to the LSE method, the coefficient $a_{im}(m = 0, 1, \ldots, k)$ can be calculated by the following expression:

$$A = (W^T W)^{-1} W^T Y \tag{16}$$

To alleviate the overfitting problem, we use L2-norm regularization as follows:

$$A = (W^T W + \lambda E)^{-1} W^T Y \tag{17}$$

Where E is the unit matrix, λ represents a predetermined minimal number.

In the aggregation part of PFNNCs, two classifiers are used to classify, and the input train values are 1 and -1. As shown in Fig. 1, an output matrix of m * k is formed after aggregating one time with L2-norm regularization, and after aggregating K times. Among them, M represents the number of samples and K represents the type of samples. Take the number of columns where the maximum value of each row of the matrix is located, that is, the final output value of the PFNNCs.

4 Experiments

The experiment is divided into two stages: in the first stage, we apply PFNNCs to ORL international face database to verify its classification accuracy in face recognition; in the second stage, we apply PFNNCs to more international standard databases for classification.

4.1 Experiment of ORL International Face Database

The ORL database contains 400 face images from 40 people in different states. They differ in position, rotation, scale and expression. Each image is digitized and the whole gray level is displayed between 0 and 255 by 112 * 92 pixel array.

Table 3. The classification results of ORL dataset

Fuzzy rules	PCA threshold	Results of the train set	Results of the test set
2	90	100.0 ± 0.0	97.5 ± 1.8
	95	100.0 ± 0.0	97.0 ± 2.1
3	90	100.0 ± 0.0	98.0 ± 2.1
	95	100.0 ± 0.0	97.0 ± 3.3
4	90	100.0 ± 0.0	98.0 ± 2.1
	95	100.0 ± 0.0	97.0 ± 3.3
5	90	100.0 ± 0.0	$\mathbf{98.5 \pm 2.2}$
	95	100.0 ± 0.0	97.5 ± 3.5
6	90	100.0 ± 0.0	97.5 ± 3.5
	95	100.0 ± 0.0	97.5 ± 3.5

The experiment consists of three steps. In the first step, we divide the data into 60% and 40% parts, 60% of the data is used as training set, and 40% of the data is used for testing. Therefore, we use 240 face images of 40 people as training set and 160 face images as test set. All images are gray level, which is converted into data and used in experiments. In the second step, PCA is used to preprocess the data and extract face features. The percentage threshold is set at 90% and 95%. In the third step, PFNNCs is designed and trained. The number of classification rules in the second classifier is controlled between 2 and 6, and face matching and recognition are carried out.

Table 3 shows the classification results of ORL dataset when PFNNCs are exhibited. The Classification accuracy curves are vividly depicted in Figs. 2 and 3 with different PCA threshold. It illustrates that when the percentage threshold of PCA is determined, the average classification accuracy of test set increases first and then decreases with the increase of the fuzzy parameters, and the critical point is fuzzy rule 5. When the percentage threshold of PCA is 95% and the classification fuzzy rules are set to 5 and 6, the classification accuracy reaches 97.5% and the variance fluctuation is 3.5%. When the percentage of PCA is 90% and the fuzzy rule is 5, the accuracy reaches 98.5% and the variance fluctuation value is 2.2%. Therefore, when the percentage of

PCA is 90% and the fuzzy rule is 5, the classification accuracy is the highest, which is 98.5%.

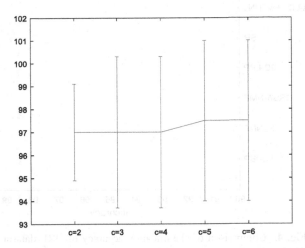

Fig. 2. Classification accuracy curve of ORL dataset (PCA threshold = 95%)

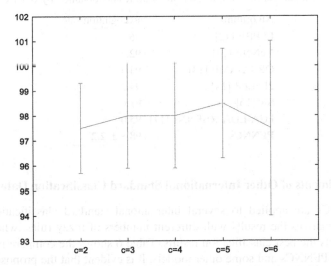

Fig. 3. Classification accuracy curve of ORL dataset (PCA threshold = 90%)

Table 4 summarizes the comparative results between the proposed PFNNCs and some other previous models. The corresponding graph is shown in Fig. 4. It is evident that the proposed PFNNCs leads to better performance.

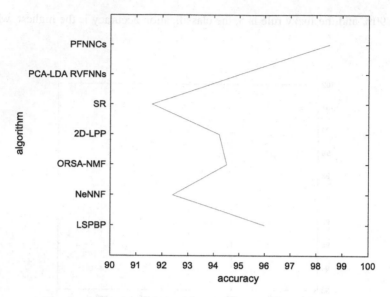

Fig. 4. Comparison of classification accuracy of ORL dataset

Table 4. Comparison of classification rate with results obtained by other classifiers

Algorithm	Recognition rate
LSPBP [12]	96
NeNNF [13]	92.4
ORSA-NMF [14]	94.5
2D-LPP [14]	94.2
SR [16]	91.6
PCA-LDA RBFNNs [17]	95.0
PFNNCs	**98.5 ± 2.2**

4.2   Experiments of Other International Standard Classification Data Sets

Here, PFNNCs are applied to several international standard classification datasets. Table 5 summarizes the results with different numbers of fuzzy rules, where the bold type represents the best classification results. Table 6 shows the comparative results of the proposed PFNNCs and some other models. It is evident that the proposed PFNNCs lead to better classification accuracy in comparison with some "common" models reported in the literatures.

Table 5. Classification results of Liver, Iris, Balance dataset

Fuzzy rules	Liver	Balance	Pima	Iris
2	70.3 ± 2.9	89.2 ± 1.3	77.4 ± 2.3	**96.7 ± 0.0**
3	69.6 ± 3.1	89.8 ± 1.3	**77.5 ± 2.0**	96.0 ± 0.9
4	66.8 ± 2.4	90.4 ± 0.9	76.3 ± 1.9	95.7 ± 0.9
5	66.7 ± 3.7	90.8 ± 0.9	75.1 ± 1.4	95.0 ± 0.0
6	66.8 ± 4.0	**91.8 ± 0.8**	75.2 ± 2.0	94.3 ± 1.9

Table 6. Comparison of classification rate with results obtained by other classifiers

Datasets	MLP [18]	RVM [18]	SVM [19]	KNN [20]	PFC [20]	**PFNNCs**
Liver	67.7	69.1	69.6	62.9	–	**70.3 ± 2.9**
Balance	–	–	–	84.4	–	**91.8 ± 0.8**
Pima	73.1	74.8	74.7	70.3	–	**77.5 ± 2.0**
Iris	66.4	–	–	94.6	93.3	**96.7 ± 0.0**

5   Conclusions

In this study, polynomial fuzzy neural network classifiers (PFNNCs) by means of density fuzzy c-means and L2-norm regularization are proposed. Combining the advantages of FCM algorithm and DENCLUE algorithm, the precondition part of the model proposes DFCM algorithm to adapt to high-dimensional classification problems; the conclusion part uses polynomials; the aggregation part uses L2-norm regularization algorithm to estimate parameters innovatively, which reduces the amount of calculation and alleviates the occurrence of over-fitting. The proposed model can be applied not only to face recognition, but also to other pattern recognition and classification problems.

Acknowledgement. This work was supported in part by the National Natural Science Foundation of China under Grant 61673295, by the Natural Science Foundation of Tianjin under Grant 18JCYBJC85200, and by the National College Students' innovation and entrepreneurship project under Grant 201710060041.

References

1. Sun, G., Chen, T., Su, Y., et al.: Internet traffic classification based on incremental support vector machines. Mob. Netw. Appl. **23**(4), 789–796 (2018)
2. Mohamed, T.M.: Pulsar selection using fuzzy KNN classifier. Future Comput. Inf. J. **3**(1), 1–6 (2018). https://www.sciencedirect.com/science/article/pii/S2314728817300776
3. Mei, J.P., Chen, L.: Fuzzy clustering with weighted medoids for relational data. Pattern Recogn. **43**(5), 1964–1974 (2010)
4. Nandi, A.K., Azzouz, E.E.: Automatic analogue modulation recognition. Sig. Process. **46**(2), 211–222 (1995)

5. Yang, G., Wagn, L., Dai, L.Z., Yang, H., Lu, R.X.: Self-organizing learning of RBF neural network based on AQPSO. Control Decis. **33**(9), 1631–1636 (2018)
6. Buhmann, M.D.: Radial basis functions. Acta Numerica **9**(5), 1–38 (2000)
7. Zhang, Z.Z., Qiao, J.F., Yu, W.: An on-line adaptive RBF network structure optimization algorithm based on LM algorithm. Control Decis. Making **32**(7), 1247–1252 (2017)
8. Yan, X.D., Wu, X.S.: Collaborative representation of adaptive gabor features face recognition algorithm. Sens. Micro Syst. **37**(3), 118–122 (2018)
9. Sharma, P., Arya, K.V., Yadav, R.N.: Efficient face recognition using wavelet-based generalized neural network. Signal Process. **93**, 1557–1565 (2013)
10. Li, C., Chiang, T.W.: Complex fuzzy model with PSO-RLSE hybrid learning approach to function approximation. Int. J. Intell. Inf. Database Syst. **5**(4), 409 (2011)
11. Hung, W.-L., Chang, Y.-C.: A modified fuzzy C-Means algorithm for differentiation in MRI of ophthalmology. In: Torra, V., Narukawa, Y., Valls, A., Domingo-Ferrer, J. (eds.) MDAI 2006. LNCS (LNAI), vol. 3885, pp. 340–350. Springer, Heidelberg (2006). https://doi.org/10.1007/11681960_33
12. Aroussi, M.E., Hassouni, M.E., Ghouzali, S., et al.: Local steerable pyramid binary pattern sequence LSPBPS for face recognition method. Signal Process. **5**(4), 281–284 (2009)
13. Guan, N.Y., et al.: NeNMF: an optimal gradient method for non-negative matrix factorization. IEEE Trans. Signal Process. **60**(6), 2882–2898 (2012)
14. Guan, N.Y., et al.: Online nonnegative matrix factorization with robust stochastic approximation. IEEE Trans. Neural Netw. Learn. Syst. **23**(7), 1087–1099 (2012)
15. Yang, J., Zhang, D., Frangi, A.F., et al.: Two-dimensional PCA: a new approach to appearance-based face representation and recognition. IEEE Trans. Pattern Anal. Mach. Intell. **26**(1), 131–137 (2004)
16. Hu, D., Feng, G., Zhou, Z.: Two-dimensional locality preserving projections (2DLPP) with its application to palmprint recognition. Pattern Recogn. **40**(1), 339–342 (2007)
17. Oh, S.K., Yoo, S.H., Pedrycz, W.: Design of face recognition algorithm using PCA-LDA combined for hybrid data pre-processing and polynomial-based RBF neural networks: design and its application. Expert Syst. Appl. **40**, 1451–1466 (2013)
18. Tipping, M.E.: The relevance vector machine. Adv. Neural. Inf. Process. Syst. **12**, 652–658 (2000)
19. Vapnik, V.: The Nature of Statistical Learning Theory. Springer, New York (1995)
20. Tahir, M.A., Bouridane, A., Kurugollu, F.: Simultaneous feature selection and feature weighting using Hybrid Tabu Search/K-nearest neighbor classifier. Pattern Recogn. Lett. **28**(4), 438–446 (2007)
21. Mei, J.-P., Chen, L.: Fuzzy clustering with weighted medoids for relational data. Pattern Recogn. **43**(5), 1964–1974 (2010)

Pedestrian Detection Method Based on SSD Model

Xin Li$^{(\boxtimes)}$, Xiangao Luo, and Haijiang Hao

Guangxi Key Laboratory of Embedded Technology and Intelligent System,
Guilin University of Technology, Guilin 541004, China
1996019@glut.edu.cn

Abstract. Pedestrian detection has a wide range of applications in daily life, and many fields require pedestrians to conduct detection with high precision and speed, which is an urgent problem to be solved. The traditional pedestrian detection method improves the detection performance by improving the classification algorithm and extracting more effective features. In this paper, a pedestrian detection method is proposed based on single shot multibox detector (SSD) model, which replaces the basic network part of SSD model with inception network structure with smaller parameters, faster running speed and stronger nonlinear expression ability. A high-performance network model for pedestrian detection was based on improved SSD. The experimental results show that the proposed method is faster than the original model, and the average precision of pedestrian recognition and location is 89.6%, which is 2.6% higher than the original model.

Keywords: Pedestrian detection · Single shot multibox detector model · Inception network

1 Introduction

Science and technology have changed people's way of life and provided great convenience for our lives. Intelligent has become a trend, and computer vision is an important research field of Artificial Intelligence (AI). Due to its very wide range of applications and its enormous commercial value, it has become a favorite in industry and academia. After years of development, pedestrian detection has achieved significant improvements in detection accuracy and speed, but there is still a high room for improvement in the accuracy of pedestrian detection when solving practical problems. The reason why the accuracy of pedestrian detection differs greatly from the recognition accuracy of the human eye is that the appearance of the pedestrian changes greatly due to factors such as wearing, lighting, background, body posture, and occlusion [1], which makes the recognition difficult. Pedestrian detection is used in security, unmanned [2], surveillance and robotics.

On the other hand, due to the rise of Deep-learning [3], Deep Neural Networks (DNN) are widely used in many Artificial Intelligence applications, including speech recognition, image classification, text understanding and many other fields. Although DNN achieves higher precision on many AI tasks, its cost is computationally

© Springer Nature Singapore Pte Ltd. 2019
X. Cheng et al. (Eds.): ICPCSEE 2019, CCIS 1058, pp. 597–607, 2019.
https://doi.org/10.1007/978-981-15-0118-0_46

complex. How to design a lightweight network model [4–7] has become a hot research topic, such as two-stage detector RCNN, fast RCNN, faster RCNN [8], etc., single-stage detector SSD [9], YOLO [10], etc., to help target detection reach an unprecedented peak.

In the 2005 CVPR, the Histogram Oriented of Gradient (HOG) gradient direction histogram proposed by Dalal [11] is the most widely used pedestrian feature descriptor. Its perfect detection performance on the MIT standard pedestrian database has led to the use of it to detect pedestrians in many fields of application and technology. Xie [12] and others pointed out that the current pedestrian detection method is computationally intensive, the pedestrian feature extraction is complex, and the test results are susceptible to complex background. They added a selective attention layer to the traditional CNN to simulate the selective attention function of the human eye, filter complex backgrounds, and highlight pedestrian characteristics. Gao [13] and others pointed out the problem of poor robustness of pedestrian detectors based on artificial extraction features. Based on the YOLO network structure, combined with the characteristics of pedestrians showing a small aspect ratio in the image, clustering selects the appropriate number and specifications of candidate frames, improves the YOLO network structure, and forms a network structure suitable for pedestrian detection.

The rest of the paper is organized as follows: Sect. 2 introduces pedestrian detection based on SSD, Sect. 3 presents experiments and results analysis, and the last part is conclusions.

2   SSD-Based Pedestrian Detection

2.1   SSD Model

Since the SSD target detection model does not have time-consuming region generation and feature re-sampling steps, it directly performs convolution operations on the entire image, and predicts the category and corresponding coordinates of the objects contained in the image, which greatly improves the detection speed. At the same time, the accuracy of target detection is greatly improved by using a small-sized convolution kernel, multi-scale prediction, and the like.

The SSD network consists of two network structures, a Base network and an Auxiliary network. The Basic network part is mainly an improvement of the VGG16 network structure with high classification accuracy in the field of image classification. The network of the classification layer is removed to retain the first five convolution layers, and then the convolution layer is used instead of the fully connected layers Fc6 and Fc7. The Auxiliary network is a convolution network structure for target detection added on the basis of the basic network, namely Conv6, Conv7, Conv8 and Conv9. The size of these layers is gradually reduced so that multi-scale prediction can be performed. As shown in Fig. 1 below:

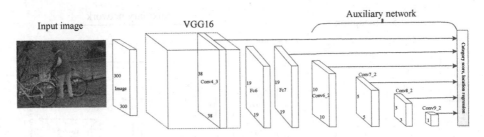

Fig. 1. SSD network structure

Each additional secondary network layer produces a fixed set of predictions through a series of convolution kernels. For a feature layer of m × n × p (p is the number of channels, m, n are sizes), each auxiliary layer predicts it with a 3 × 3 × p convolution kernel and produces a score for a certain category. Or, the position offset of the object relative to the default bounding box, and the corresponding values are predicted respectively at m × n positions.

The SSD model predicts k bounding boxes at each location of the feature map and simultaneously predicts the score at which the object category appears at this location and the offset of the object position relative to the bounding box. Thus, c × k (c is the number of categories) and 4 × k position offsets are predicted at the position of each feature map. This results in a total of (c + 4) k filters that are applied around each location in the feature map, yielding (c + 4) kmn outputs for a m × n feature map. Finally, the output result is subjected to Non-Maximum Suppression (NMS) to obtain the final predicted value of the object type and position information in the image.

2.2 Improved SSD Model

The SSD target detection model uses the VGG16 network as the Basic network, but the VGG16 network model has many parameters and takes up most of the running time in the feature extraction process. And in the forward propagation process, the loss of information in the transformation process due to the existence of nonlinear transformation.

The improved SSD method replaces the VGG base network portion with an Inception network. After a large number of target feature maps are obtained, the feature maps of different levels are combined by the feature fusion part to obtain the fusion feature map for use as the basis for prediction, and finally the classification and location information of the objects in the image are obtained. The overall structure of the model for improving the SSD method is shown in Fig. 2:

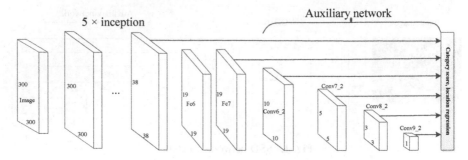

Fig. 2. The overall structure of the improved SSD model

The Inception model modifies the traditional convolution layer in the network, which can effectively alleviate the problem of deep neural network parameter space, easy over-fitting, deep network structure gradient dispersion, and model performance degradation. The Inception module is based on the consideration of multiple different sizes of convolution kernels to enhance the resilience of the network, using 1×1, 3×3, 5×5 convolution kernels and 3×3 Average Pooling respectively. More nonlinear transformations are established while reducing the parameters, making CNN more capable of learning features. At the same time, Batch Normalization (BN) is added to each convolution layer, so that the input of each layer of neural network maintains the same distribution, accelerates the training speed of the convolution network, and improves the classification accuracy after convergence. The structure of the Inception model is shown in Fig. 3 below:

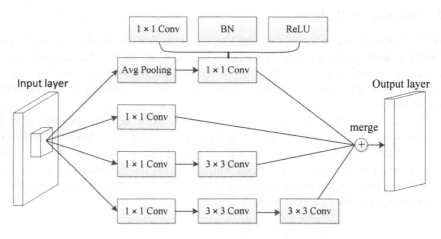

Fig. 3. Inception model structure

2.3   Loss Function

The loss function of the improved model is basically similar to the loss function of the original SSD. The location coordinates are predicted by the Smooth L1 loss function between the prediction box and the real box parameters, and the Softmax function predicts the classification confidence. The total loss function is the weighted sum of the loss of location and the loss of confidence:

$$L(x, c, l, g) = \frac{1}{N}(L_{conf}(x, c) + \alpha L_{loc}(x, l, g)) \tag{1}$$

$$L_{conf}(x, c) = -\sum_{i \in Pos}^{N} x_{ij}^{p} \log(\hat{c}_{i}^{p}) - \sum_{i \in Neg} \log(\hat{c}_{i}^{0}) \tag{2}$$

$$L_{loc}(x, l, g) = \sum_{i=Pos}^{N} \sum_{m \in \{cx, cy, w, h\}} x_{ij}^{k} smooth_{L1}(l_{i}^{m} - \hat{g}_{i}^{m}) \tag{3}$$

$$\hat{c}_{i}^{p} = \frac{\exp(c_{i}^{p})}{\sum_{p} \exp(c_{i}^{p})} \tag{4}$$

$$\hat{g}_{j}^{cx} = (g_{j}^{cx} - d_{i}^{cx})/d_{i}^{\omega}, \quad \hat{g}_{j}^{cy} = (g_{j}^{cy} - d_{i}^{cy})/d_{i}^{h} \tag{5}$$

$$\hat{g}_{j}^{\omega} = \log\left(\frac{g_{j}^{\omega}}{d_{i}^{\omega}}\right), \quad \hat{g}_{j}^{h} = \log\left(\frac{g_{j}^{h}}{d_{i}^{h}}\right) \tag{6}$$

Where x_{ij}^{p} is the indicator that the i-th default bounding box matches the j-th ground truth bounding box of category p, c indicates the real category, l indicates the boundary coordinates of the prediction frame, g indicates the boundary coordinates of the real frame, and N indicates the number of matching default bounding boxes. The parameter α is used to adjust the ratio between the confidence loss and the positioning loss, which is generally 1 by default. *Pos* represents a positive example in the sample, *Neg* represents a counterexample in the sample, x_{ij}^{k} indicates whether the i-th default bounding box matches the j-th ground truth bounding box with respect to the category k, and $\hat{c}_{i}^{p}$ indicates the probability that the target in the i-th prediction frame is the p-th category, $\hat{g}_{j}^{m}$ represents the real box, and l_{i}^{m} represents the prediction box.

The positioning loss is the loss of Smooth L1, and the formula is as follows:

$$smooth_{L1} = \begin{cases} 0.5x^2 & if \ |x| < 1 \\ |x| - 0.5 & otherwise \end{cases} \tag{7}$$

2.4 Network Model Evaluation Index

This paper mainly uses three evaluation indicators to measure the quality of the network model: Intersection Over Union (IOU), Average Precision (AP), Frames Per Second (FPS).

The IOU is the intersection of the detection frame and the real candidate area of the object and the union of the two. The formula is as follows:

$$IOU = \frac{area\left(B_i \cap B_{gt}\right)}{area\left(B_i \cup B_{gt}\right)} \tag{8}$$

In the above formula, B_i represents the i-th detection area of the prediction result, B_{gt} represents the real candidate area based on the training sample, and *area* represents the area of the designated image area. When the intersection ratio IOU is greater than a certain threshold A, the detection frame is considered to be a positive inspection, otherwise it is considered to be a false detection. For any real candidate area, it can only be checked once.

For the M detection frames and the N real candidate frames, the number of positive inspections TP, the number of missed inspections FN, and the number of false detection FP can be obtained. As shown in Table 1 below:

Table 1. Classification result confusion matrix

Condition	Test outcome	
	Positive	Negative
Positive	True positive (TP)	False negative (FN)
Negative	False positive (FP)	True negative (TN)

The relationship is as follows:

$$\begin{aligned} TP + FN &= N \\ TP + FP &= M \end{aligned} \tag{9}$$

In the test set, the detection accuracy, detection speed and parameter amount before and after the model improvement are analyzed to compare the performance of the model. The test set image is input into the trained network, the pedestrian position in the map is detected and the detection result is recorded. When the IOU of the bounding box in the labeled data corresponding to the target bounding box of the model and the test set is greater than or equal to the set threshold, the detection result is considered correct, otherwise it is regarded as a detection error. The accuracy of the evaluation index of the detection precision rate (P), recall rate (R), Average Precision (AP) are defined as:

$$P = \frac{TP}{TP + FP} \tag{10}$$

$$R = \frac{TP}{TP + FN} \tag{11}$$

$$AP = \int_0^1 P(R)dR \tag{12}$$

where P(R) is the cumulative recall rate on the abscissa and precision rate on the ordinate.

The precision rate and the recall rate are the basic two evaluation indicators. The precision rate reflects the proportion of the correctly detected samples in the detection frame. The recall rate reflects the proportion of the correctly determined samples in the real box. The average precision is a single point of limitation for precision and recall, reflecting the global performance of the network model.

3   Experiment and Result Analysis

3.1   Lab Environment

This article uses 8 GB memory, NVIDIA GTX1050TI GPU as the hardware platform, operating system is Windows 7, parallel computing framework version is CUDA 9.0, deep neural network acceleration library is CUDNN v7.0. This paper uses the Python programming language to implement the SSD target detection model on the Keras deep learning framework and complete the training and verification of the model.

3.2   Data Set

This article uses the INRIA pedestrian database, which is currently the most used static pedestrian detection database, providing the original image and the corresponding annotation file. The training set has 614 positive samples (including 2416 pedestrians) and 1218 negative samples; the test set has 288 positive samples (including 1126 pedestrians) and 453 negative samples. Most of the human body in the picture is in a standing position with a height of more than 100 pixels, which occupies a large area of the image, various pedestrian attitudes, sufficient lighting conditions, and high definition, which is suitable for pedestrian detection. By default, the IOU threshold is set to 0.5 to determine the positive check in these databases.

3.3   Model Training

In order to effectively train the network and better predict the results, the training data and the test data are equally scaled to 300 × 300. The weight training method is Adaptive moment estimation (Adam) with a minimum batch size of 8. The initial learning rate is 0.0003, and 80 iterations of the entire training sample are trained. The parameter attenuation value is 0.0005, the momentum factor is 0.9, and the Non-Maximum Suppression is 0.45.

3.4   Pedestrian Detection Model Performance Analysis

3.4.1   Comparative Experiment of Different Basic Network Models

In this section, the network model structure is optimized according to the above data set and model training parameters, and the size of the training data set is increased and the model parameters are adjusted to optimize the system. Five network models different from the SSD basic network are designed respectively. The number of Inception network layers is 1, 2, 3, 4 and 5 respectively. The detailed network structure is shown in Table 2:

Table 2. Basic network models for different Inception layers

Basic network structure	Convolution layer and Inception layer distribution	Parameter amount
VGG-16	2Conv + 2Conv + 3Conv + 3Conv + 3Conv	3.1×10^9
4 Conv_block + 1 Inception	2Conv + 2Conv + 3Conv + 3Conv + Inception	3.0×10^9
3 Conv_block + 2 Inception	2Conv + 2Conv + 3Conv + 2 Inception	2.5×10^9
2 Conv_block + 3 Inception	2Conv + 2Conv + 3 Inception	1.9×10^9
1 Conv_block + 4 Inception	2Conv + 4 Inception	1.8×10^9
5 Inception	5 Inception	1.5×10^9

The basic network models of different Inception layers are tested in the test set. The loss function of the training is shown in Fig. 4 below. The five different improvements to the underlying network VGG16 converge faster, and the parameter size of the network model is greatly reduced by improvements to different convolution network blocks. Among them, the convergence speed is different between different improvement methods, but the network structure of 5 Inception has the least amount of network structure parameters, so the network structure of 5 Inception is selected to improve the optimal network model for SSD pedestrian detection.

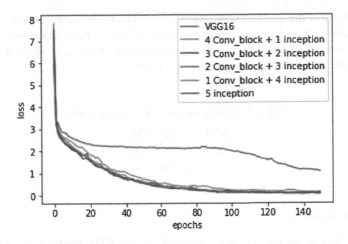

Fig. 4. Comparison of loss functions for different network models

3.4.2 Pedestrian Detection Model Comparison Experiment

This paper mainly uses the improved SSD pedestrian detection method to conduct experiments. The basic network of the improved SSD model is the 5 Inception network structure described in Sect. 3.4.1. The threshold is set to 0.5. When the IOU is greater than 0.5, it is positive, otherwise it is negative. Check. Figure 5 is a P-R graph of the original SSD network model and the improved model test. It can be seen from the figure that the accuracy of the improved model is higher than that of the original SSD model, and the maximum recall rate is less than 1, indicating that both have missed samples. However, the recall rate of the improved model is larger than the original model, indicating that the missed detection rate is small.

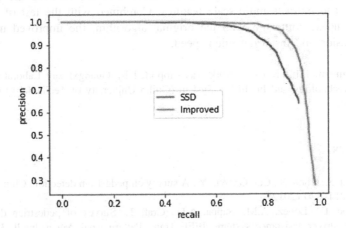

Fig. 5. P-R curve of the original SSD and the improved model

Table 3 shows the analysis of experimental results under different network models. The improved method has an average precision of 89.6% and a detection speed of 31 frames per second, which is higher than the SSD300 network, and meets the precision and real-time requirements of pedestrian detection.

Table 3. Comparison of experimental results

Model	AP%	Fps
SDD	86.8	23
Improved	89.6	31

3.5   Result Analysis

This paper first improves the basic network of the SSD model and compares the different improved models. Through the convergence speed and parameter size of different network models, it is confirmed that the 5 Inception network structure is the optimal network, which improves the training speed of the network and the recognition ability of pedestrians. Then compared with the original SSD network based on the improved SSD network model, the experimental results show that the average precision of the improved network for pedestrian detection is 2.6% higher than SSD, and the detection speed is 34.7% faster.

4   Conclusion

In order to improve the accuracy and detection speed of pedestrian detection, this paper proposes a pedestrian detection method based on improved SSD. This method replaces the VGG16 network structure by the Inception module, reduces the amount of parameters, and extracts multi-scale features. Combined with the test of objectivity evaluation index, compared with the original algorithm, the improved method has higher precision and faster detection speed.

Acknowledgment. This research work was supported by Guangxi key Laboratory Fund of Embedded Technology and Intelligent System (Guilin University of Technology) under Grant No. 2017-2-5.

References

1. Su, S., Li, S., Chen, S., Cai, G., Wu, Y.: A survey on pedestrian detection. Chin. J. Electron. **40**(04), 814–820 (2012)
2. Geronimo, D., Lopez, A.M., Sappa, A.D., Graf, T.: Survey of pedestrian detection for advanced driver assistance systems. IEEE Trans. Pattern Anal. Mach. Intell. **32**(7), 1239–1258 (2010)

3. Jia, Y.: Research on Object Detection and Tracking Algorithm Based on Deep Feature Learning. Xi'an University of Science and Technology, Xi'an (2017)
4. Szegedy, C., et al.: Going deeper with convolutions (2015)
5. Ioffe, S., Szegedy, C.: Batch normalization: accelerating deep network training by reducing internal covariate shift (2015)
6. Szegedy, C., Vanhoucke, V., Ioffe, S., Shlens, J., Wojna, Z.: Rethinking the inception architecture for computer vision. In: Computer Vision and Pattern Recognition (2016)
7. Szegedy, C., Ioffe, S., Vanhoucke, V., Alemi, A.: Inception-v4, inception-ResNet and the impact of residual connections on learning (2016)
8. Ren, S., He, K., Girshick, R., Sun, J.: Faster R-CNN: towards real-time object detection with region proposal networks. IEEE Trans. Pattern Anal. Mach. Intell. **39**(6), 1137–1149 (2017)
9. Liu, W., et al.: SSD: single shot multibox detector. In: European Conference on Computer Vision (2016)
10. Redmon, J., Divvala, S., Girshick, R., Farhadi, A.: You only look once: unified, real-time object detection (2015)
11. Dalal, N., Triggs, B.: Histograms of oriented gradients for human detection. In: 2005 IEEE Computer Society Conference on Computer Vision and Pattern Recognition, CVPR (2005)
12. Xie, L., Ji, G., Peng Q., Luo, E.: Application of preprocessing convolutional neural network in pedestrian detection. J. Front. Comput. Sci. Technol. **12**(116(05)), 32–42 (2018)
13. Gao, Z., Chen, J., Li, Z.: Pedestrian detection method based on YOLO network. Comput. Eng. **44**(5), 215–219 (2018)

Optimizing Breast Mass Segmentation Algorithms with Generative Adversarial Nets

Qi Yin, Haiwei Pan[✉], Bin Yang, Xiaofei Bian, and Chunling Chen

Harbin Engineering University, Harbin, People's Republic of China
panhaiwei@hrbeu.edu.cn

Abstract. Breast cancer is the most ordinary malignant tumor in women worldwide. Early breast cancer screening is the key to reduce mortality. Clinical trials have shown that Computer Aided Design improves the accuracy of breast cancer detection. Segmentation of mammography is a critical step in Computer Aided Design. In recent years, FCN has been applied in the field of image segmentation. Generative Adversarial Networks is also popularized for its ability on generate images which is difficult to distinguish from real images, and have been applied in the image semantic segmentation domain. We apply the Dilated Convolutions to the partial convolutional layer of the Multi-FCN and use the ideas of Generative Adversarial Networks to train and correct our segmentation network. Experiments show that the Dice index of the model D-Multi-FCN-CRF-Adversarial Training on the datasets InBreast and DDSM-BCRP can be increased to 91.15% and 91.8%.

Keywords: Breast mass segmentation · GAN · Dilated convolutions · Adversarial training

1 Introduction

The global incidence of breast cancer has risen since the late 1970s. One woman has breast cancer in eight women in the United States. China is not a high-risk country for breast cancer, but it should not be optimistic. Breast cancer is the first malignant tumor in China. According to the National Cancer Center and the Ministry of Health's Disease Prevention and Control Bureau released in 2012, the data of breast cancer in 2009 showed that the incidence of breast cancer in the national cancer registration area ranked first in female malignant tumors, and the incidence of breast cancer in women (rough rate) nationwide the total is 42.55/100,000, the city is 51.91/100,000, and the rural area is 23.12/100,000. Since the 1990s, global breast cancer mortality has declined, which is because the development of breast cancer screening work, the proportion of early cases increase; and the development of comprehensive treatment of breast cancer, improve the efficacy.

Mammography, which is used for screening and diagnosis of breast cancer, is the most basic and important breast disease imaging examination method, can be detected early breast cancer. This tool is primarily used to detect early signs of breast cancer. In general, mammogram images contain areas with low contrast and complex structural backgrounds that may be difficult to detect at early stages due to indistinguishable from

© Springer Nature Singapore Pte Ltd. 2019
X. Cheng et al. (Eds.): ICPCSEE 2019, CCIS 1058, pp. 608–620, 2019.
https://doi.org/10.1007/978-981-15-0118-0_47

surrounding normal tissue. Therefore, automated lumps detection is a challenge in Computer Aided Design (CAD) systems. Mammography has been proven to be a valid method for breast cancer early detection and diagnosis, which can conspicuous reduces mortality [1].

Dalmiya [2] proposed a research method for segmentation used wavelet transform and K-means clustering. Cordeiro et al. [3] proposed a GrowCut algorithm and achieved relatively good results. Traditional mass segmentation rely on manual selection of features, build classifiers based on models and learning features from tumors [4, 5]. Convolutional neural network (CNN) is a common one medical image segmentation method. But less research on breast mass's segmentation with deep learning [6–8]. Neeraj Dhungel used CRF with re-weighted tree to improve the segmentation network [9]. They also used multiple deep belief networks (DBNs), GMM, and use improved SVM for segmentation [10]. A work used CNN's output as a potential function, make the most progressive performance [11]. Zhu et al. [12] proposed the end-to-end training of FCN and CRF as RNN and added an adversarial loss prevention model over-fitting to increase the Dice index.

Generative adversarial networks (GAN) provides one method to generate images which is difficult to distinguish from real images [13–15]. GAN uses an optimized discriminator network to discriminate between real images and generated images, which prompts the segmentation network to generate images that look more real. In the field of domain adaptation, the trained discriminator network is used to distinguish images from different domains [16, 17] and to improve image segmentation. A trained discriminator network is used to distinguish between manual segmentation and generated segmentation [18]. This segmentation method has also been used in medical field to segment prostate cancer in MRI images [19] and organs in chest X-ray images [20]. Goodfellow et al. [13] proposed a Generative Adversarial Nets (GAN). Luc et al. [18], for the first time, use GAN in the field of image semantic segmentation, with the training ideas of generative adversarial networks to correct high-order inconsistency in image segmentation and improve image segmentation accuracy. Yang et al. [21] used the idea of adversarial networks on the medical image segmentation problem of liver, and obtained Dice 95%. Moeskops et al. [22] used the idea of adversarial network on Brain's medical image segmentation problem and achieved good results.

A newer method of CNN to segmentation is to use dilated convolutions, it makes the weight of the convolutional layer be distributed sparsely on the larger receptive field [23, 24]. Dilated CNN is an effective way to get large receptive field with a limited number of training weights and convolutional layers, without using subsampled layers. The dilated convolutions proposed by Yu and Koltun [23] is the first to apply the dilated convolutions in the field of image semantic segmentation. They improved the segmentation performance of the algorithm by using dilated convolutions to change VGG-16. Chen et al. [25–28] also proposed a series of algorithms DeepLab to improve the application of dilated convolutions.

Inspired by [12, 18, 23, 29, 30], we propose an improved model. First, apply dilated convolutions to the convolution layer of Multi-FCN. Second, add adversarial network for adversarial training to correct the high order inconsistency between Ground Truth segmentation maps and the images generated by the segmentation network. Third, we validate our model D-Multi-FCN-CRF-Adversarial Training, achieves state-of-the-art results on the datasets InBreast and DDSM-BCRP.

2   Breast Mass Segmentation Algorithms

With the development of artificial neural networks, artificial intelligence technology has entered people's lives, and deep learning has become the research direction of many researchers, and has been continuously applied in various fields. Among them, in recent years, the development is more enthusiastic in the field of computer vision. Image semantic segmentation and image recognition in the field of computer vision are hot topics. We have transplanted the method of image semantic segmentation into our field of medical image processing. Practice can prove that this method can solve the problem of medical image segmentation in computer-aided diagnosis today. Among the many algorithms, the full convolutional network and the conditional random field are the most frequently used algorithms in the field of image segmentation, and various improvements are proposed based on this.

2.1   FCN and CRF as RNN

Full Convolution Network (FCN) [31] was a successful image segmentation model, which preserved the predicted spatial structure. FCN consists of convolution, deconvolution [32] or pooling of each layer. From the segmentation results, we can clearly know what objects are segmented. The last three layers in the CNN network are one-dimensional vectors, are no longer use convolution calculation way, so the two-dimensional information is lost. While the Full Convolution Network, all the three layers can be converted to 1×1 convolution kernels which corresponding to the multi-channel convolutional layer of the equivalent vector length, make after three layer also all adopt the convolution computation. In the whole model, all of them are convolution layers, and there is no vector. The recognition of FCN is pixel-level recognition, and each pixel of the input image has a corresponding judgment mark on the output, indicating what object/category this pixel is most likely to belong to.

Conditional Random Field (CRF) is a common structure learning method. FCN is the pixel-to-pixel mapping, so every pixel in the final output image is marked with classification, and these classification are simply regarded as different variables. Each pixel establishes a connection with other pixels. Then you will get a "Complete Graph". Through the energy function optimization solution, the recognition and judgment that obviously do not conform to the facts are removed and replaced with reasonable explanations. Finally, the optimization of the semantic prediction results of FCN images can be obtained and the final semantic segmentation results can be generated.

Embedding the CRF process originally isolated from the deep model training into the neural network, that is, integrating the FCN and CRF process into an end-to-end system. And what that does is that the energy function of the CRF final prediction result can be directly used to guide the training of FCN model parameters to achieve better image semantic segmentation results. The greatest significance of CRF as RNN

[33] is to regard the CRF solution inference process as RNN correlation operation, which is regarded as the combination of convolution layer, Softmax layer and other neural network layers, embedded in CNN model, and achieves the real algorithm fusion of the two.

FCN convolution layer, we will lose pixel information by use the pooling layer. For the pixel-level image segmentation problem, each pixel information is very important for the segmentation result, so how to reduce the information loss is especially important.

2.2 Dilated Convolutions

Recent studies on convolution networks with semantic segmentation [31] analyze filter dilated, but choose not to use it. Reference [34] use dilated to simplify the architecture [31]. In contrast, [23] a new convolutional network is proposed, which can systematically aggregate multi-scale context information.

DCNNs it has proven to be simple and successful for semantic segmentation [31, 35]. However, since there is a pooling layer in the network whose step size is not 1, the spatial resolution of the feature map will gradually decrease. The remedial measure taken is to add deconvolution layer [31], which requires extra computational time and memory. Dilated convolutions is a convolutional idea proposed to solve the problem that shrink image (subsampled) will reduce image resolution and information loss in image semantic segmentation. The benefit of dilated convolutions is that it can increases the receptive field without information lose which due to the pooling layer, so that each convolution output contains a wide range of information. The dilated convolution can fully aggregate the context information without losing the receptive field and resolution [23].

The receptive field is used to indicate the size of the range of perception of the original image by different neurons inside the network, or the size of the mapped area on the original image of the pixel on the feature maps output by each layer of convolution. For the calculation of the receptive field, we use the following Eq. (1):

$$F_b = (F_f - 1) \times s + k \tag{1}$$

Where F_b means backward receptive field, F_b means forward receptive field, s is the stride, and k is the kernel size.

In the image segmentation problem based on the FCN idea, the output image is the same size as the input. However, since there are several pooling layers of stride > 1 in FCN, therefore, the higher the number of layers in the network is the more original image information will be contained in a pixel of the feature map, in other words, the larger receptive field will be. But, the cost of this is that the image resolution will be reduced. Because of the pooling layer, the size of the next layer feature map will small, but we should make the final output image the same size as the input image, so the network layer in the back segment of the FCN, the feature map must magnify (upsampling) to restore the reduced feature map to the original size. In this process, it is impossible to completely restore the information lost during the pooling process, thus result in the loss of information, the accuracy is reduced.

Without the pooling layer, in the subsequent network layer, the receptive field of the smaller size convolution kernel will be small; Since the receptive field of the convolution kernel of small size is small, then increase the size of the convolution kernel; However, the larger the convolution kernel size, the more the calculation will increase; We can make the original compact convolution kernel 'fluffy', but the convolution kernel needs to calculate the same point, that is, the fluffy position is fully filled in 0, and according to the original calculation method of the convolution kernel. Since the effective calculation point is unchanged, so the computation doesn't change. In addition, the size of the feature map of each layer is unchanged, so the image information is saved. Through the use of dilated convolutions, our k × k traditional convolution kernel becomes into the size of the k_d convolution kernel, with rate r and introduce r − 1 zeros between the values of the convolution kernel, only nonzero values are considered in the calculation, which does not increase the amount of calculation and expands the receptive field. The calculation formula for the size of the convolution kernel k_d with fluffy rate r is Eq. (2):

$$k_d = k + (k - 1)(r - 1) \tag{2}$$

The receptive field of the dilated convolutions is exponential growth

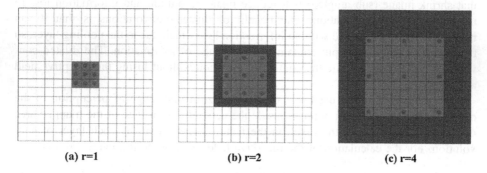

(a) r=1   (b) r=2   (c) r=4

Fig. 1. Dilated convolutions. (Color figure online)

Figure 1(a) corresponds to 3 × 3, r = 1 convolution kernel, and the same as the traditional convolution operation; Fig. 1(b) corresponds to a convolution kernel of 3 × 3, r = 2 the actual effective convolution kernel size is 3 × 3, but rate = 2, which is equivalent to a 5 × 5 convolution kernel, 9 points in the figure the weight are not 0, the rest are 0, only 9 red points are convoluted, and the rest are skipped. It can be seen that although the kernel size is only 3 × 3, the receptive field of this convolution has increased to 7 × 7; Fig. 1(c) corresponds to 3 × 3, r = 4 kernel, r = 4 which is equivalent to 9 × 9 kernel, the weight of 9 points are not 0, the rest are 0, only 9 red points are convoluted. As you can see, although the kernel size is only 3 × 3, the receptive field has increased to 15 × 15.

In the field of computer vision, the optimization effect of dilated convolutions has been verified in the image segmentation technology. For color image dataset ImageNet, segmentation networks using dilated convolutions can improve the performance of image segmentation. But in the research I have learned, no one has applied this improvement to mammography, so is this method suitable for our mammography grayscale images? The specific analysis of the difference between grayscale image and color image in the field of computer vision is that the color channel of the matrix information of the image is RGB channel or single channel. For grayscale images we will be relatively simpler in data processing, and we can perceive more sensitive data information by applying dilated convolutions.

We will verify our ideas in specific experiments. We apply the dilated convolutions to the convolution layers of multi-FCN. See the experiments section for details.

2.3 Adversarial Training for Segmentation Network

In the article, Goodfellow et al. [13] describe the generative confrontation network in this way: a gang that manufactures counterfeit money and a policeman who seizes the case, respectively, a generation model and a discriminant model in the generated confrontation network; The generation model constantly improves its ability to falsify the truth, and the discriminant model continuously improves its ability to distinguish between authenticity. In the game of generating models and discriminating models, the criminal gang and the police constantly improve their abilities until the counterfeit and real coins manufactured cannot be distinguish. This kind of confrontational training is also a manifestation of synergy. We use the idea of Generative Confrontation Network (GAN) to construct our confrontation training network to optimize our segmentation network, aiming to correct the high order inconsistency between segmentation graph and Ground Truth (GT) segmentation maps, and improve our segmentation model.

In the left side of the overall network architecture is the Segmentation network (Splitter S), and on the right is the Adversarial network (Discriminator D), which is equivalent to the generator and discriminator in the traditional GAN. The adversarial network captures high order exterior information and the Ground Truth (GT) segmentation maps was distinguished from the output of the segmentation network. To better guide the splitter prediction, the adversarial network to provide additional loss function, during the training process, the parameters of the segmentation network are updated. The overall network architecture see Fig. 2.

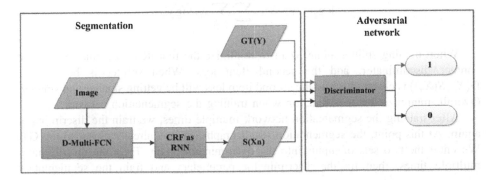

Fig. 2. Overall network architecture.

The output of the segmentation network is the input of the discriminator network. The discriminator network is used to distinguish between generating segmentation map or manually segmenting map. X_n represents the original image, Y_n is the X_n's Ground Truth label map, $S(X_n)$ is the class probability map. After segmenting in the segmentation network, we have two combinations input to the adversarial network. The first is X_n and Y_n, and the second is X_n and $S(X_n)$. Adversarial network outputs 1 when the input is the first combination, and 0 is output when the input is the second combination. The Discriminator and Splitter alternately train separately.

First train Discriminator, we define the loss function as Eq. (3):

$$\sum_{n=1}^{N} \log_1 [D(X_n, Y_n), 1] + \log_1 \{[1 - D(X_n, S(X_n))], 1\} \tag{3}$$

$D(x, y)$ Is the result of Discriminator, is a probability value of 0 to 1, indicate the probability that the second item y belongs to the GT. Where $log_1(x, y)$ is a known binary cross-entropy loss as Eq. (4):

$$\log_1(\hat{y}, y) = -[y \log \hat{y} + (1 - y) \log (1 - \hat{y})] \tag{4}$$

When training the discriminator alone, in order to get the best discriminator, we need to reduce the loss of the adversarial network, which is a good distinction between $S(X_n)$ and Y_n. When training discriminator, the first item in the loss function $D(X_n, Y_n)$ is getting closer and closer to 1, and the second item $1 - D(X_n, S(X_n))$ gets closer and closer to 1, we minimize the loss function, and when it converges, fix the discriminator network parameters and train the segmentation network.

Then train the segmentation network, we define the loss function as Eq. (5):

$$\sum_{n=1}^{N} \log_2 [S(X_n), Y_n] + \lambda \log_1 \{[1 - D(X_n, S(X_n))], 0\} \tag{5}$$

Where $log_2(x, y)$ is a known multi-class cross-entropy, calculating the loss for each pixel as Eq. (6):

$$\log_2(\hat{y}, y) = -\sum_{i=1}^{H \times W} \sum_{c=1}^{C} y_{ic} \log \hat{y}_{ic} \tag{6}$$

When training splitter alone, want to minimize the first item loss, and in order to contact discriminator, add the second item loss. When splitter is better, $1 - D(X_n, S(X_n))$ is closer to 0, and the second item loss will be getting smaller and smaller. Overall, minimize this loss function when training the segmentation network.

After training the segmentation network multiple times, we train the discriminator again. At this point, the segmentation result graph $S(X_n)$ is already closer to the GT. We enter the two sets of inputs into the discriminator again. Train the discriminator multiple times, then fix the discriminator parameters and train the segmentation network again. At this point we still have to make the first item of the loss function of the segmentation network smaller and smaller, and the second item loss function is also because of the new discriminator, this loss has become larger, we still have to minimize this loss. Train the segmentation network several times, and then we will train the discriminator to better distinguish between GT and $S(X_n)$ under the new and more accurate segmentation. The splitter and discriminator are trained alternately for a few times shown in Generative Adversarial Networks (GAN) algorithm, until the discriminator cannot easily distinguish between GT and the result of the Segmentation network. With proper training, reasoning no longer requires an adversarial network. Segmentation network performance is improved, and we can get a higher performance segmentation network.

Previous experiments have proved that GAN has been well applied to the semantic segmentation of images, and can better optimize image segmentation networks by using higher-order information. Recently, this segmentation method is used to segmentation prostate cancer in MRI in the field of medical images [19] and organs in chest the X-ray [20]. Yang et al. [21] used the idea of GAN on the medical image segmentation problem of liver, and Moeskops et al. [22] used the idea of GAN on the medical image segmentation problem of Brain. Have achieved good results.

So, can we apply GAN on mammography to optimize our segmentation network? By comparing the datasets of other applications with the characteristics of our datasets, in the face of the complex background of the chest, liver and brain complement images, we can make good use of GAN to improve our segmentation performance, then we can also optimize the segmentation results for breast X-ray images with relatively clear background contours. We will continue to verify in subsequent experiments.

3 Experiments

We use the public dataset InBreast and DDSM-BCRP to validate the proposed model. From DDSM-BCRP [40] dataset selects 156 images in the training set and 160 images in the test set. The dataset InBreast [36] is a mammography dataset that provides accurate contours. The dataset contains 116 mass regions, the training set contains 58 images, and the testing set selects the remaining 58 images, which is the same as the method mentioned in [9–11]. We enhanced the training set by flipping, which get a training set of 232 images [12]. For the preprocessing of the original image, we use the image enhancement technique for the extracted ROI images [12, 37].

Our experimental equipment on Python 3.6, TensorFlow-GPU 1.7. The size of each picture is fixed as $H \times W = 40 \times 40$, the number of pixels on each image is $i = 1600$, the number of categories is $C = 2$, epoch $= 5000$ and the Dice indicator is defined as $\frac{2 \times TP}{2 \times TP + FP + FN}$, with learning rate 0.003 [38]. The ϵ used for weights of CRF as RNN is 0.5 in the dataset. For mean field approximation or the CRF as RNN, we use 5 iterations in training and 10 iterations in testing [12]. See Table 1 for the setting of FCN. The number of parameters for each layer in the parameter setting of the FCN basically similar to the parameter of CNN used in the work [11]. The dilated convolution of rate $= 2$ to improve four FCNs models with different convolution kernels. Use the activation function tanh.

Table 1. Network parameters of Multi-FCN.

Net layers	Net number			
	FCN1	FCN2	FCN4	FCN3
First layer	K_s = 2 × 2 rate = 2	K_s = 3 × 3 rate = 2	K_s = 4 × 4 rate = 2	K_s = 5 × 5 rate = 2
Second layer	K_s = 2 × 2 Max pooling 2 × 2	K_s = 3 × 3 Max pooling 2 × 2	K_s = 4 × 4 Max pooling 2 × 2	K_s = 5 × 5 Max pooling 2 × 2
Third layer	K_s = 2 × 2 Max pooling 2 × 2	K_s = 3 × 3 Max pooing 2 × 2	K_s = 4 × 4 Max pooling 2 × 2	K_s = 5 × 5 Max pooling 2 × 2
Fourth layer	K_s = 9 × 9 rate = 2	K_s = 8 × 8 rate = 2	K_s = 7 × 7 rate = 2	K_s = 7 × 7 rate = 2
Fifth layer	40 × 40 de-conv filter + Softmax	40 × 40 de-conv filter + Softmax	40 × 40 de-conv filter + Softmax	40 × 40 de-conv filter + Softmax

About the setting of the adversarial network is shown in Fig. 3. Our discriminator is set up as a 4-layer neural network, and the last layer is blended through Linear. The activation function is Relu = 0.2. According to the training model of [18], our network update performs an alternate training of the splitter and the discriminator every iteration, which is slightly different from the idea of [18].

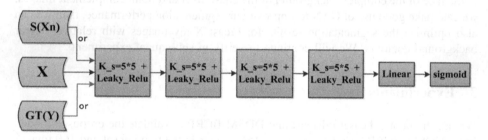

Fig. 3. Adversarial network.

As can be seen from the Fig. 4, is the result of InBreast. The red line is the result of the Dice change in the training set, while the blue line is the result of Dice change in the test set. Our test results can reach our expected results after enough iterations, and can effectively prevent over-fitting. But at the same time, it can be seen that in the process of training, due to the existence of adversarial training, our experiment will be unstable.

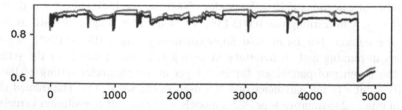

Fig. 4. Changes of Dice in the training set and test set. (Color figure online)

We verify the algorithm models of multiple combinations, respectively is D-Multi-FCN-CRF; Multi-FCN-CRF-Adversarial Training; Multi-FCN-CRF-Adversarial Training, and validate them on the dataset. It is proved that the D-Multi-FCN-CRF-Adversarial Training model with dilated convolutions and Adversarial Training is the best experiment. And we conduct experiments on the model without Adversarial Training, during the training, the Train Dice index once reached 98%, there is a serious over-fitting phenomenon. Compared with the application of FCN, FCN also has over-fitting problems during the experiment. Experiments show that through the adversarial training, can effectively reduce over-fitting in the datasets InBreast and DDSM-BCRP. From the results show in Table 2, the use of CNN in medical image sets is superior to traditional methods based on manual feature extraction [5]. The Dice index of experiments [9–11] is best to get 90% of the results. Reference [12] added adversarial-loss [39], our model D-Multi-FCN-CRF-Adversarial Training result is superior to theirs.

Table 2. Comparison of experimental results.

Method	DDSM-BCRP Dice (%)	InBreast Dice (%)
TRW deep learning [9]	89	89
Deep learning [10]	87	88
FCN	90.21	89.48
Deep learning + CNN [11]	90	90
Multi-FCN-CRF with Adversarial-loss [12]	91.30	90.97
Our model (D-Multi-FCN-CRF)	91.35	91
Our model (Multi-FCN-CRF-Adversarial Training)	91.5	91.08
Our model (D-Multi-FCN-CRF-Adversarial Training)	**91.8**	**91.15**

To further demonstrate our experimental results, we visualize the segmentation results of DDSM-BCRP's 1000th and 4800th iterations, as shown in Figs. 5 and 6. Three columns of images, the first column is the $S(X_n)$ of the segmentation network, the second column is the original image X_n, and the third column is the Ground Truth (GT) segmentation maps. You can see that the segmentation of the tumor margins was more effective when epoch = 4800 than when epoch = 1000.

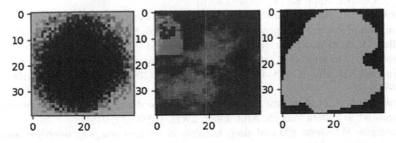

Fig. 5. Segmentation result of the epoch = 1000.

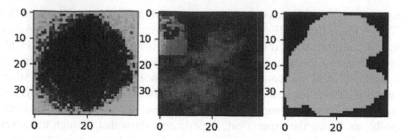

Fig. 6. Segmentation result of the epoch = 4800.

4 Conclusion

We use dilated convolutions in the convolutional layer of Multi-FCN to eliminate the loss of pixel information caused by the pooling layer, and expand the receptive field does not increase the computation; Improved model with Generative Adversarial Nets (GAN) training. Experiments have shown that our model D-Multi-FCN-CRF-Adversarial Training could improve the accuracy of mammography segmentation on the public datasets InBreast and DDSM-BCRP. The best Dice value are 91.15% and 91.8%.

At the same time, there are some problems to be solved in the experiment. Because the public dataset InBreast and DDSM-BCRP are small, the training results are not ideal, and the phenomenon of over-fitting is still easy to occur in the process. The setup of the adversarial network may be too simple, and the instability of GAN during training can also have an impact on the outcome.

Acknowledgements. This work was supported by the National Natural Science Foundation of China under Grant No. 61672181, No. 51679058, Natural Science Foundation of Heilongjiang Province under Grant No. F2016005.

References

1. Oeffinger, K.C.: Breast cancer screening for women at average risk: 2015 guideline update from the American Cancer Society. JAMA **314**(15), 1599–1614 (2015)
2. Dalmiya, S.: Application of wavelet based k-means algorithm in mammogram segmentation. Int. J. Comput. Appl. **52**(15), 15–19 (2016)
3. Cordeiro, F.R.: An adaptive semi-supervised Fuzzy GrowCut algorithm to segment masses of regions of interest of mammographic images. Appl. Soft Comput. **46**, 613–628 (2016)
4. Beller, M., Stotzka, R.: An example-based system to support the segmentation of stellate lesions. In: Meinzer, H.P., Handels, H. (eds.) Bildverarbeitung fur die Medizin 2005, Informatik aktuell, pp. 475–479. Springer, Berlin (2005). https://doi.org/10.1007/3-540-26431-0_97
5. Cardoso, J.S.: Closed shortest path in the original coordinates with an application to breast cancer. Int. J. Pattern Recogn. Artif. Intell. **29**(1), 1555002 (2015)
6. Greenspan, H.: Guest editorial deep learning in medical imaging: overview and future promise of an exciting new technique. IEEE Trans. Med. Imaging **35**(5), 1153–1159 (2016)

7. Kallenberg, M.: Unsupervised deep learning applied to breast density segmentation and mammographic risk scoring. IEEE Trans. Med. Imaging 35(5), 1322–1331 (2016)
8. Zhu, W., Lou, Q., Vang, Y.S., Xie, X.: Deep multi-instance networks with sparse label assignment for whole mammogram classification. In: Descoteaux, M., Maier-Hein, L., Franz, A., Jannin, P., Collins, D.L., Duchesne, S. (eds.) MICCAI 2017. LNCS, vol. 10435, pp. 603–611. Springer, Cham (2017). https://doi.org/10.1007/978-3-319-66179-7_69
9. Dhungel, N., Carneiro, G.: Tree RE-weighted belief propagation using deep learning potentials for mass segmentation from mammograms. In: IEEE International Symposium on Biomedical Imaging, pp. 760–763. IEEE (2015)
10. Dhungel, N., Carneiro, G.: Deep structured learning for mass segmentation from mammograms. In: IEEE International Conference on Image Processing, pp. 2950–2954. IEEE, Quebec City (2015)
11. Dhungel, N., Carneiro, G., Bradley, A.P.: Deep learning and structured prediction for the segmentation of mass in mammograms. In: Navab, N., Hornegger, J., Wells, W.M., Frangi, A.F. (eds.) MICCAI 2015. LNCS, vol. 9349, pp. 605–612. Springer, Cham (2015). https://doi.org/10.1007/978-3-319-24553-9_74
12. Zhu, W.: Adversarial deep structural networks for mammographic mass segmentation. CoRR (2016)
13. Goodfellow, I.J., Pouget-Abadie, J.: Generative adversarial nets. In: International Conference on Neural Information Processing Systems, pp. 2672–2680. MIT Press (2014)
14. Radford, A.: Unsupervised representation learning with deep convolutional generative adversarial networks. Computer Science (2015)
15. Wolterink, J.M.: Generative adversarial networks for noise reduction in low-dose CT. IEEE Trans. Med. Imaging 36(12), 2536–2545 (2017)
16. Ganin, Y.: Domain-adversarial training of neural networks. Mach. Learn. Res. 17(1), 2096–2030 (2017)
17. Kamnitsas, K., et al.: Unsupervised domain adaptation in brain lesion segmentation with adversarial networks. In: Niethammer, M., et al. (eds.) IPMI 2017. LNCS, vol. 10265, pp. 597–609. Springer, Cham (2017). https://doi.org/10.1007/978-3-319-59050-9_47
18. Luc, P.: Semantic segmentation using adversarial networks. CoRR (2016)
19. Kohl, S.: Adversarial networks for the detection of aggressive prostate cancer. CoRR (2017)
20. Dai, W.: Scan: structure correcting adversarial network for chest x-rays organ segmentation. CoRR (2017)
21. Yang, D., et al.: Automatic liver segmentation using an adversarial image-to-image network. In: Descoteaux, M., Maier-Hein, L., Franz, A., Jannin, P., Collins, D.L., Duchesne, S. (eds.) MICCAI 2017. LNCS, vol. 10435, pp. 507–515. Springer, Cham (2017). https://doi.org/10.1007/978-3-319-66179-7_58
22. Moeskops, P., Veta, M., Lafarge, M.W., Eppenhof, K.A.J., Pluim, J.P.W.: Adversarial training and dilated convolutions for brain MRI segmentation. In: Cardoso, M.J., et al. (eds.) DLMIA/ML-CDS -2017. LNCS, vol. 10553, pp. 56–64. Springer, Cham (2017). https://doi.org/10.1007/978-3-319-67558-9_7
23. Yu, F.: Multi-scale context aggregation by dilated convolutions (2015)
24. Wolterink, J.M., Leiner, T., Viergever, M.A., Išgum, I.: Dilated convolutional neural networks for cardiovascular MR segmentation in congenital heart disease. In: Zuluaga, M. A., Bhatia, K., Kainz, B., Moghari, M.H., Pace, D.F. (eds.) RAMBO/HVSMR-2016. LNCS, vol. 10129, pp. 95–102. Springer, Cham (2017). https://doi.org/10.1007/978-3-319-52280-7_9
25. Chen, L.C.: Semantic image segmentation with deep convolutional nets and fully connected CRFs. Comput. Sci. 4, 357–361 (2014)

26. Chen, L.C.: DeepLab: semantic image segmentation with deep convolutional nets, atrous convolution, and fully connected CRFs. IEEE Trans. Pattern Anal. Mach. Intell. **40**(4), 834–848 (2016)
27. Chen, L.C.: Rethinking atrous convolution for semantic image segmentation. CoRR (2017)
28. Chen, L.C.: Encoder-decoder with atrous separable convolution for semantic image segmentation. CoRR (2018)
29. Zhu, W.: Co-occurrence feature learning for skeleton based action recognition using regularized deep LSTM networks. In: AAAI, vol. 2 (2016)
30. Zhu, W., Miao, J.: Hierarchical extreme learning machine for unsupervised representation learning. In: International Joint Conference on Neural Networks, pp. 1–8. IEEE (2015)
31. Long, J.: Fully convolutional networks for semantic segmentation. IEEE Trans. Pattern Anal. Mach. Intell. **39**(4), 640–651 (2014)
32. Zeiler, M.D., Krishnan, D.: Deconvolutional networks. In: 2010 IEEE Computer Society Conference on Computer Vision and Pattern Recognition, pp. 2528–2535. IEEE (2010)
33. Zheng, S.: Conditional Random Fields as Recurrent Neural Networks. CoRR, pp. 1529–1537 (2015)
34. Chen, L C.: Attention to scale: scale-aware semantic image segmentation. CoRR, pp. 3640–3649 (2015)
35. Sermanet, P.: OverFeat: integrated recognition, localization and detection using convolutional networks. Eprint Arxiv (2013)
36. Ines, C.: INbreast: toward a full-field digital mammographic database. Acad. Radiol. **19**(2), 236–248 (2012)
37. Ball, J.E., Bruce, LM.: Digital mammographic computer aided diagnosis (CAD) using adaptive level set segmentation. In: International Conference of the IEEE Engineering in Medicine Biology Society. IEEE (2007)
38. Kingma, D.: Adam: a method for stochastic optimization. Computer Science (2014)
39. Szegedy, C.: Intriguing properties of neural networks. Computer Science (2013)
40. Heath, M., Bowyer, K.: Current status of the digital database for screening mammography. In: Digital mammography, pp. 457–460. Springer, Dordrecht (1998). https://doi.org/10.1007/978-94-011-5318-8_75

Multiple Music Sentiment Classification Model Based on Convolutional Neural Network

Jing Yang[1], Fanfu Zeng[1], Yong Wang[1(✉)], Hairui Yu[2],
and Le Zhang[3]

[1] Harbin Engineering University, Harbin 150001, China
wangyongcs@hrbeu.edu.cn
[2] China Shipbuilding Industry System Engineering Research Institute,
Beijing 100094, China
[3] Dalian Shipbuilding Industry Co., Ltd., Dalian 116000, China

Abstract. The network community is a platform for people to communicate. In order to accurately analyze the emotions displayed in music community, this paper proposes a convolutional neural network classification model based on multi-dimensional emotions. Firstly, to solve the problem of feature extraction of emotion words under similar sentence patterns, it proposed a multi-emotion classification method and emotion vector splicing method that conform to music community emotion characteristics. Secondly, aiming at the coexistence of multiple categories of emotions in music comment text, it applied an emotional value measurement method based on music characteristics. Finally, the classification model was constructed with combining methods of emotion vector splicing and emotion value measurement. Through experimental analysis, this model is proved to have good performance in accuracy.

Keywords: Network community · Music community · TextCNN ·
Multiple sentiment · Sentimental value

1 Introduction

With network communities exploding to change the relationship between users and the web. People create and share their content through comments and other means, and interact quickly with others in different ways. Tang [1] uses emojis in comments as tag values to train the model and classify texts with sentiments. Qu [2] has attempted to use the comment data containing user rating information as weak labeling information to train the model and improve the accuracy of classification. Experiments show that the sentiments contained in comments can effectively reflect the polarity of users' sentiments, and how to accurately extract the sentimental features of texts has become one of the hot issues at present.

For the music community, the existing sentimental analysis methods mainly focus on the analysis of lyrics [3], which fail to reflect the sentiments of the audience and are not applicable to the classification of pure music. Analyzing user reviews can subjectively reflect people's feelings about songs. Because music comment information is mixed with sentiment, the existing sentiment analysis model cannot accurately describe

© Springer Nature Singapore Pte Ltd. 2019
X. Cheng et al. (Eds.): ICPCSEE 2019, CCIS 1058, pp. 621–632, 2019.
https://doi.org/10.1007/978-981-15-0118-0_48

this phenomenon. From the above problems, it can be seen that the existing sentiment analysis methods have no understanding of music field and no sentiment classification model applicable to music community. In order to accurately analyze music community comment text sentiment, this paper proposes a convolutional neural network sentiment classification model based on multiple music sentiment. At the same time of reasonably dividing the emotional categories of songs, it provides users with more accurate recommendation services.

2 Related Work

2.1 Domain-Oriented Sentiment Classification

Ghiassi et al. [4] conduct a related study on Twitter-specific dictionaries, reveal meaningful double phrases and three phrases phrases can improve the accuracy of the categoryifier. Fan et al. [5] constructe a film and television domain dictionary which combined with weak marked information of the score, which improves the accuracy of film commentary sentiment analysis. Levorashka et al. [6] create the topic product product to determine the polarity of the topics discussed in the tweet and tried to use hybrid techniques by combining vocabulary and machine learning. The results of the study show that the method can achieve scalability in a distributed environment. Leal et al. [7] classify the mail polarity by using machine learning methods. Kouloumpis [8] define and interpret three sentimental analyses in Twitter positiveness, negativene and neutrality, by adding semantics as an additional feature to the sentimental analysis training set to enhance the accuracy of the classification, to provide more choices for sentiment analysis and the direction of research.

Through the introduction, we can find that the sentimental domain method has strong pertinence and high accuracy. However, the multiple sentiments of music commentary information make the sentimental analysis method in the music field missed, and the analysis method is not fully adapted to the field.

2.2 Sentiment Analysis Based on Deep Learning

Bengio et al. [9] construct a binary language model by neural networks to map the words into vectors and propose an n-gram model. Kim [10] propose the use of convolutional neural networks for sentence modeling to solve the problem of sentiment classification, and verify the use of Word2Vec as the input of CNN-non-static training model can perform better in classification. Chen et al. [11] constructe a feature dictionary which contains the vector representation of part of speech and positive and negative sentiments combine the sentiment dictionary and convolutional neural network to improve the classification accuracy of convolutional neural networks in the microblogging field. Lv et al. [12] constructe convolutional neural network models by word vectors carry the features of adjacent words, and achieved considerable results in the classification of short texts.

Through the introduction, we can find that the deep learning has a good effect in text sentiment analysis. It does not require a large number of tag training data sets, which saves the cost of manual manual extraction of features, and can better utilize features such as word order, but also can not better distinguish feature words with different similarities with high similarity.

We consider the actual song category and the feature extraction of music-transferred information under similar sentence patterns, and propose a multi-sentiment division method based on music field and sentiment vector splicing method. Considering the phenomenon of song commentary and multi-sentiment coexistence, we propose a multi-sentimental value measurement method based on music characteristics. Finally, base on the commentary features of the music community, we combine the advantages of the domain-based sentiment analysis and the sentiment analysis method based on the deep learning framework, and integrate them into the convolutional neural network to propose a convolutional neural system based on multi-faceted sentiments to improve the model classification performance.

3   Multiple Music Sentiment Classification Model Based on CNN

3.1   Sentimental Analysis in the Music Field

The sentimental expression in the comment is determined by the type of music the user is listening to. Traditional music classifications, such as pop music, rock music, and melody, do not reflect the sentimental characteristics of music revealed in lyrics and tunes, nor can it be used to analyze the sentimental characteristics of comments. The sentiments in music are sometimes more diverse than the human expressions in reality. For example, the song - El Dorado the first half is quiet and the second half is passionate. Simply using the sentimental word of calm cannot summarize all sentimental content. At the same time, the user reviews triggered by this song are also colorful. For example, the three comments under the song - El Dorado are shown in Fig. 1.

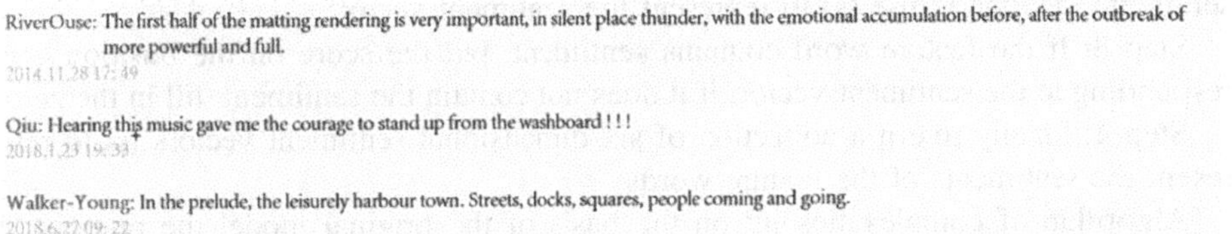

RiverOuse: The first half of the matting rendering is very important, in silent place thunder, with the emotional accumulation before, after the outbreak of more powerful and full.
2014.11.28 17:49

Qiu: Hearing this music gave me the courage to stand up from the washboard ! ! !
2018.1.23 19:33

Walker-Young: In the prelude, the leisurely harbour town. Streets, docks, squares, people coming and going.
2018.6.27 09:22

Fig. 1. Three comments under the song - El Dorado.

It can be found that with simple sentimental classification, "silent", "powerful", "outbreak", "courage", and "leisurely" all contain distinctly different sentiments, and the expression is not clear.

This section will divide the multiple sentiments in the music community according to the multiple sentimental characteristics of the music field. Using this approach, we can analyze the multiple sentiments in the comments and get a vector that can represent the multiple sentiments of the comment. Finally, the splicing of sentiment vectors is completed by combining the vectors of multiple sentiments to improve the accuracy of sentiment feature extraction.

Multiple Sentiments Partation

Markham budd's music ee theory [13] proposed that music can reflect sentiments, proving the feasibility of music sentiment division. Currently, The most widely recognized musical sentimental division is the Hevner sentimental ring [14], which divides sentiment into eight categories.

It can be found that music can reflect people's sentiments. The existing ways of classifying sentiments have the influence of cultural differences between China and the west and the classification methods are too brief to cover the categories of multiple sentiments. Therefore, based on Hevner's sentimental circle, for the text Context, its set of basic language values (seed words) are: senti = {1: "happy", 2: "warm", 3: "quiet", 4: "excited", 5: "sentimental", 6: "Sorrow"}.

For the seed word set Senti = {$(S_1, v_1), (S_2, v_2)...(S_6, v_6)$}, $S_i (i = 1,2...6)$ represents a seed word, The emotional weight represented by v_i (i = 1, 2...6). For the music review corpus segmentation, preprocessing, calculate the TF-IDF value of each feature word. Considering that TF-IDF indicates the frequency at which the current feature word appears in the text, words with high TF-IDF are more valuable. In this paper, according to the size of TF-IDF, the feature word Top 10000 is taken as the emotional word. For the sentiment word set s, the cosine similarity of the i-th sentiment word Si and the seed word S_j (j = 1, 2...6) is calculated separately, which is represented by $Sim_{i,j}$. By sim_{ij} multiplying the emotional weight of the seed word S_j, the emotional weight of the emotional word for the seed word is obtained, and then the emotional dictionary Sentiment in the music community field is constructed. The specific process is as follows.

Step 1: Data preprocessing, using jieba participles and stopping words;

Step 2: Match each word in the text with the sentiment in each category (senti) in the sentiment dictionary; use (s) to represent the sentiment vector;

Step 3: If the feature word contains sentiment, put the score on the position corresponding to the sentiment vector, if it does not contain the sentiment, fill in the zero;

Step 4: Finally return a collection of six-dimensional sentiment vectors (s) to represent the sentiments of the feature words.

Algorithm of complex lies in: on the basis of the original model the process of adding a matching and splicing emotion vector, when there is no emotional factors in the text, the same time complexity and the original model, with the increase of emotional factors contained in the text, the algorithm's time complexity O (n) is the highest, considering the music community in this essay, so the training model of time will not very big.

Word Vector Combined with Multiple Sentimental Factors

Text vectorization can be expressed in two forms: one-hot representation and distributed representation. Considering that one-hot feature words were independent from each other before and the connection between words could not be taken into account, this paper users Word2Vec to realize the process of word vectorization.

The probability function of the language model can be written as follow.

$$p(Context(w)|w) = \prod_{u \in Context(w)} p(u|w) \tag{1}$$

(u) represents a word in the context of (w). For the vector (w) of each central word of a song comment (S), train (w) with the surrounding words (u) within the distance of the central word. Therefore, the model relies heavily on corpus and only takes semantic information into account, ignoring words containing different sentiments under the same sentence pattern.

In this paper, characteristic words are manually labeled with sentiment to solve the problem. First of all, we train the data under six sentiment labels separately. We splice the sentiment vectors of this category at the end of each vector, and then conducted vectorization processing on the training data. The word vectors without sentiment were complemented by zero, so that we could solve the classification problem of different sentiment words under the same sentence pattern.

n = (0, 0, 0, 0, 0, 0)

$$w = \begin{cases} (t \oplus s) & s \neq \emptyset \\ (t \oplus n) & s = \emptyset \end{cases} \tag{2}$$

$\oplus$ indicates vector stitching, (t) is the semantic vector to be categoryfield, and (s) is the sentiment vector. When (s) does not exist, the vector is guaranteed to be equal by zero.

3.2 Sentiment Classification Model Based on Multi-faceted CNN

In this paper, multi-sentiment classification is combined in the input layer and the connected layer to solve the problem of sentiment classification under similar sentence patterns and the coexistence of multi-sentiment classification, so as to improve the accuracy of the existing community multi-sentiment classification. (see Fig. 2).

$$\alpha_i = \arg\min_\alpha L(\alpha, D) = ||x_i - D\alpha||_2^2 + \lambda ||\alpha||_1,$$

$$h_{m,j} = \max_{i \in N_m} \alpha_{i,j}, \, for \, j = 1, \cdots, K \tag{3}$$

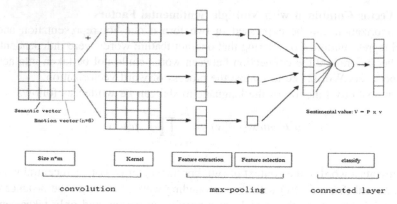

Fig. 2. CNN model of this paper. The input layer is the word vector w in Sect. 3.1, combined into matrix as the network input. Convolutional layer uses convolution kernel of different height Windows to extract information of feature vector combination of different number of fields in parallel, and obtain multiple feature maps for feature selection of pooling layer. Pooling layer reduces the size of feature map obtained by convolution layer, which is also called sampling layer: maximum sampling is adopted in this paper (see formula 3). Connected layer calculation analyzes the sentiment of information by calculating the sentiment value, and the connection categoryfiler outputs the sentiment classification results.

Connection Layer Design. The connected layer categories the characteristic results selected by pooling layer. Considering the complexity of music comment information and the diversity of sentiments, this paper uses multinomial logistic regression to classify the input according to the multi-sentiment classification in Sect. 3.1. Calculate the probability of sentimental distribution of each song.

Multinomial logistic regression represents the sentimental tendency of songs [15]. The sentimental distribution of each song can be expressed by the probability distribution of the song on music sentimental category.

$$P_i = \frac{e^{y_i}}{\sum_{k-1}^{J} e^{y_k}} = \frac{e^{w_i x}}{\sum_{k-1}^{J} e^{w_k x}} \quad y \in \{1, 2, 3, 4, 5, 6\} \tag{4}$$

Multinomial logistic regression only categoryifies songs under the most probable sentiment label by comparing the probability of each sentiment. In fact, the same song does not contain a single sentiment, but a more complex and diversified sentiment. For example, in a song, the "happiness" label accounts for 40%, and the "sadness" label accounts for 38%. At this time, the sentiment expressed in this song actually includes both happiness and sadness. It is obviously inaccurate to directly classify the song as "happiness". In view of this situation, this paper puts forward the concept of sentimental value.

Definition 5 (sentimental value) sentimental value is used to classify the final sentimental categories of songs. (v) represents the weight of the ith sentiment, (P$_i$) is the probability of such sentiment in formula (4), and (V) represents the final weight of the song.

$$V = \sum_{i=1}^{6} V_i P_i \tag{5}$$

Based on the weight v of six sentiments and the classification of convolutional neural network, this paper calculates the final sentimental value of songs with multiple sentiments, and finally divides the final sentimental category of songs according to the range of sentimental value.

Step 1: Put the vector returned by the connected layer into softmax function to calculate the probability distribution of the six categories;

Step 2: for the probability P under each category, get the label of the category;

Step 3: If the characteristic word contains sentiment, assign the score under sentiment to the corresponding position of sentiment vector; otherwise, if it does not contain sentiment, then zero is added.

Step 4: obtain the final sentimental value of the text by multiplying v under the category label with probability P obtained from softmax and adding them together.

4 Experiments

4.1 Experimental Environment

Experimental platform: system type: Windows 10 64-bit operating system
Processor: Inter (R) Core (TM) i7-7700hq
Memory: 8G
Compilation environment: Pyhon 3.6
Development environment: JetBrains PyCharm

4.2 Data Set and Data Preprocessing

The data set adopted in this paper is to crawl the netease cloud music song list hot comments by Python crawler. including training data and test data. According to the six sentimental labels mentioned in this paper, the data are divided into six categories, which correspond to the song evaluation of the corresponding sentimental labels in music community. A total of 10517 songs' hot comments are crawled, among which the hot comments of 3000 songs are taken as test data, and the rest of the data are used to train the model (Table 1).

Table 1. Data set.

Label	Training data	Test data
Excited	1376 songs	500 songs
Happy	1225 songs	500 songs
Warm	1154 songs	500 songs
Quiet	1316 songs	500 songs
Sentimental	1273 songs	500 songs
Sorrow	1173 songs	500 songs

This paper uses the exact mode of jieba for segmentation. The data format after word segmentation is as follows (see Fig. 3).

```
A sunny day
Higher; Listen; Child; mother; Happiness;
Want; wave. The first; old; 90; Lucky; Age; Mandarin; music; Jay Chou. Peak; Jay Chou.
Accompany; Mature; Look at; old; Witness; generation; Wonderful;
Junior high school; Listen; Jay Chou. Students. Laugh; A few years; Listen; Tears; All
night; Lost; Time; Defeat; Red lips and white teeth;
Listen; Primary school. fourth grade; Printed; brain; Scene; Forever; Sweep the floor.
classroom; Cleanliness; Paranoid; teacher; Times; changed; Opportunities; warped; Class;
Courage; rainy;
```

Fig. 3. It shows the data format after jieba participle.

4.3   Evaluation Criteria

In this paper, we use sentiment classification. For the calculation of the accuracy of multi-classification, one calculation method is to take all the categories into account at one time to calculate the accuracy of category prediction. Another method is to calculate the accuracy of each category, and then carry out arithmetic average to get the accuracy of the test set. The first method is micro-average, and the second method is macro-average [16].

4.4   The Experimental Results

In the input layer, this paper uses the word2vec model in Gensim, an open source third-party Python toolkit, to model word segmentation results, vectorize words, and train the model. The dimension of feature vector is 200. First, we obtain the song comments under the song list of the above six sentimental categories, and divide the data into two parts, one for the training model and one for the test.

Cosine Similarity

Use "happy" as the seed word to enumerate the cosine similarity between the contrast method and the words under the method.

Table 2. Cosine similarity.

Sentimental word	Word2vec feature Top5/Cosine similarity	Sentiment splicing feature Top5/Cosine similarity
	lively/0.71653456	lively/0.71653456
	sad/0.70253175	smile/0.60165411
happy	kindergarten/0.57608277	delighted/0.57814431
	Children's song/0.54781321	kindergarten/0.57608277
	La la la/0.49176182	Children's song/0.54781321

Table 2 shows that the cosine similarity of the characteristic word "sad" calculated using the Word2vec model is 0.70253175 for the emotional word "happy" before emotional splicing. The traditional Word2vec speculates on the surrounding words of the central word, just considering the meaning of the word, "happy" and "sadness" exist in the situation shown in Fig. 3, and the two characteristic words are considered to be relatively consistent. Therefore, the traditional Word2vec model does not distinguish well between feature words that are easy to have the same sentence pattern. After using the method of emotional stitching, "smile" is spliced with emotion vector, the cosine similarity grows to 0.60165411, and the feature word "sad" is excluded from the feature word Top5. It shows that using the method of emotional splicing to splicing emotional labels on feature words containing emotions, the model can take into account the meaning of the words and better consider the emotional factors, which improves the accuracy of emotional division.

Comparison of Classification Results

Firstly, considering that the text data to be analyzed are comments from music community, most of the comments are short texts, which are characterized by multiple features, repeatability and incomplete sentences. This paper uses three commonly used deep learning models to train and test the comment text of music community, and compares the classification effect of different models on the short text data of music community. The models involved in the comparison are TextCNN model, TextRNN model and LSTM model. The same data set was used for training and testing under the three models respectively, and the same word vector was input as the input layer of the model. The classification effect of the three models was evaluated by calculating macro average and micro average (Fig. 4).

In this experiment, three sets of data sets (Netease cloud music, QQ music and Kugou music) were tested respectively. When the processed word vector is used as the input layer of CNN, the vector matrix USES multi-category logistic regression to calculate the probability of six sentiment categories, and calculates the final value of songs through the formula of sentimental value mentioned in the paper, so as to complete the multi-sentiment division (Fig. 5).

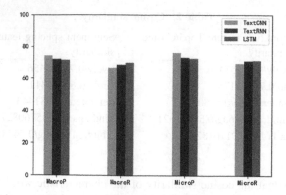

Fig. 4. TextCNN has the highest accuracy and LSTM has the highest recall rate. As the comment text more for short text, so the LSTM cannot reflect its advantage of memory ability in addition, review the text in the repeatability is very strong and incomplete words would happen, so in terms of classification accuracy TextCNN the characteristics of the local optimal value of the feature extracting of effect is higher than that of the latter two model classification effect. Considering that this paper is an emotion analysis oriented to music field, the importance of accuracy is higher than recall rate. Therefore, TextCNN model with the best performance in classification accuracy is selected.

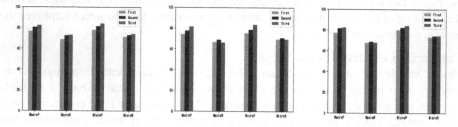

Fig. 5. First, the traditional Word2vec + CNN method does without any other changes to the words. Second, Based on the traditional method, in the input layer of convolutional neural network, we use the sentiment vector splicing mentioned in this paper to conduct experiments. Third, in the connected layer of convolutional neural network, we processed multiple sentiments through the formula of affective value for experiments.

In the experiment, after the sentiment vector was spliced, the accuracy was significantly improved compared with that before. The macro average and micro average were improved by about 4%, which solved the problem of feature extraction of different sentiments in similar sentence patterns. After adding affective value classification, the accuracy of the model increased slightly. It shows that when questions like "happy" and "sad" are included with a fair probability in a comment, sentimental value can be more carefully dealt with under the problem of song classification.

5 Conclusion

Considering that the existing sentiment analysis methods are not fully suitable for the classification criteria of sentiment in music field and the phenomenon of multi-sentiment coexistence in comments, this paper proposes a multi-sentiment classification model of music based on convolutional neural network. Firstly, we proposed a method of multi-sentiment partition to adapt to the music community sentiment features and the method of sentiment vector splicing. It can scientifically summarize the multiple sentiments contained in music and improve the accuracy of sentiment feature extraction. Secondly, combining with the structural characteristics of convolutional neural network, we proposed an sentimental value measurement method based on music characteristics. It scientifically analyzes the sentiments in the comments and classifies them accurately. Finally, we combined the multi-sentiment classification method, the sentiment vector splicing method and the sentiment value measurement method into the CNN to construct the classification model. Experimental results show that this method can accurately and reasonably classify the sentimental categories of songs, and thus provide users with more accurate recommendation services.

Acknowledgement. The research work was supported by the National Natural Science Foundation of China under Grant No. 61672179, 61370083 and 61402126, The Youth Foundation of Heilongjiang Province of China under Grant No. QC2016083, the Fundamental Research Funds for the Central Universities under Grant No. HEUCF180606, and the Innovative Talents Research Special Funds of Harbin Science and Technology Bureau under Grant No. 2016RQQXJ128.

References

1. Tang, D.Y., Qin, B., Liu, T.: Deep learning for sentiment analysis: successful approaches and future challenges. Wiley Interdiscip. Rev. Data Min. Knowl. Discovery **5**(6), 292–303 (2015)
2. Qu, L.Z., Gemulla, R., Weikum, G.: A weakly supervised model for sentence-level semantic orientation analysis with multiple experts. In: Proceedings of the 2012 Joint Conference on Empirical Methods in Natural Language Processing and Computational Natural Language Learning, pp. 149–159 (2012)
3. Shao, X.: Music emotion classification research based on music content and lyrics. Comput. Technol. Dev. **8**(3), 1720–1730 (2015)
4. Ghiassi, M., Skinner, J., Zimbra, D.: Twitter brand sentiment analysis: a hybrid system using n-gram analysis and dynamic artificial neural network. Expert Syst. Appl. **40**(16), 6266–6282 (2013)
5. Fan, Z.: Emotional analysis of film reviews based on dictionary and weak labeling information. Comput. Appl. **35**(11), 38–42 (2018)
6. Levorashka, A., Utz, S., Ambros, R.: What's in a like? Motivations for Pressing the like button. In: Proceedings of the Tenth International AAAI Conference on Web and Social Media, vol. 19, no. 12, pp. 2149–2158 (2016)
7. Jorge, A.M., Leal, J.P., Anand, S.S., Dias, H.: A study of machine learning methods for detecting user interest during web sessions. ACM **89**(4), 149–157 (2014)

8. Kouloumpis, E., Wilson, T., Moore, J.: Twitter sentiment analysis: the good the bad and the OMG! In: Fifth International Conference on Weblogs and Social Media, pp. 538–541 (2011)
9. Bengio, Y., Ducharme, R., Vincent, P., et al.: A neural probabilistic language model. J. Mach. Learn. Res. **3**(2003), 1137–1155 (2003)
10. Kim, Y.: Convolutional neural networks for sentence categoryification. In: 2014 Conference on EMNLP, pp. 1746–1751 (2014)
11. Chen, Z.: Chinese emotion analysis based on convolution neural network and word emotion sequence features. Chin. J. Inf. Technol. **29**(6), 131–137 (2015)
12. Lv, C.: Emotional classification model based on CNN and word proximity features. Comput. Eng. **5**(44), 182–187 (2018)
13. Budd, M.: Values of art: pictures, poetry and music. J. Aesthetics Art Criticism **57**(1), 76–78 (1999)
14. Hevner, K.: Expression in music: a discussion of experimental studies and theories. Psychol. Rev. **42**(2), 186–204 (1935)
15. Yang, X.: Automatic construction and optimization of emotion dictionary based on Word2Vec. Comput. Sci. **44**(1), 42–47 (2017)
16. Pandarachalil, R., Sendhilkumar, S., Mahalakshmi, G.S.: Twitter sentiment analysis for large-scale data: an unsupervised approach. Cogn. Comput. **7**(2), 254–262 (2015)
17. Sun, G., Song, Z., Liu, J., et al.: Feature selection method based on maximum information coefficient and approximate Markov blanket. Zidonghua Xuebao/Acta Automatica Sinica **43**(5), 795–805 (2009)
18. Sun, G., Lang, F., Xue, Y.: Chinese chunking method based on conditional random fields and semantic classes. J. Harbin Inst. Technol. **43**(7), 135–139 (2011)

Integrated Navigation Filtering Method Based on Wavelet Neural Network Optimized by MEA Model

Zhu Tao$^{(\boxtimes)}$, Saisai Gao, and Ying Huang

PAP Engineering University, Xi'an, Shaanxi, China
375144587@qq.com

Abstract. In the experiment of combined navigation filtering using wavelet neural network, the initial parameters of the network have the influence of randomness on network convergence and navigation accuracy. A combined navigation filtering method based on wavelet neural network optimized by mind evolution algorithm is proposed. Firstly, the efficient global search ability of the mind evolution algorithm was used to quickly and accurately obtain the initial parameters of the appropriate wavelet neural network, and then the optimized wavelet neural network was applied to directly predict the position and velocity error data. This method is different from the traditional filtering method, while avoiding the drawbacks of the neural network. The simulation experiments with wavelet neural network and GA-wavelet network were carried out. The results show that the proposed method can effectively improve the accuracy of the integrated navigation system and provide a feasible path for combined navigation filtering.

Keywords: Integrated navigation · Data fusion · Wavelet neural network · Mind evolution algorithm

1 Introduction

Wavelet neural network is a research hotspot in recent years. It is a combination of wavelet theory and artificial neural network. It has powerful data analysis and nonlinear mapping functions. In this regard, we can use wavelet neural network to integrate information on navigation parameters. Different from traditional filtering algorithm [1, 2], by using the nonlinear prediction function of the wavelet neural network, the navigation parameter error of the next stage can be predicted and corrected, and without building mathematical model from traditional filtering algorithm. With strong nonlinear prediction and fault tolerance, it can be well applied in integrated navigation.

Through experimental verification and theoretical analysis [3, 4], the main problem of wavelet neural network is the randomness of the initial parameters of the network. The weights of the input layer to the hidden layer of the wavelet neural network and the initial parameters of the threshold are determined by the wavelet scale parameters, if the wavelet network Inappropriate initialization of scale and displacement parameters may result in non-convergence of results. The weight of the hidden layer to the output layer

© Springer Nature Singapore Pte Ltd. 2019
X. Cheng et al. (Eds.): ICPCSEE 2019, CCIS 1058, pp. 633–644, 2019.
https://doi.org/10.1007/978-981-15-0118-0_49

and the threshold initial parameter itself are uncertain, and each time the result is random, it is difficult to grasp the degree of good or bad. Moreover, the wavelet's scale and the initial value of the displacement parameter not only determine the weight and threshold of the network, but also affect the prediction ability of the wavelet neural network.

In this regard, the Mind Evolutionary Algorithm is proposed to optimize the wavelet neural network. The thought evolution algorithm is a further improvement of the genetic algorithm [5]. For the problems of premature, random walk and slow convergence of the evolutionary algorithm, new convergence and XOR operations are added on the basis of the genetic algorithm, which improves the overall efficiency of the algorithm. The literature [6–8] uses the thought evolution algorithm to optimize BP neural network and apply it in their respective fields. The literature [9] uses genetic algorithm as the learning algorithm of wavelet neural network and is applied in integrated navigation.

In this paper, considering the shortcomings of BP neural network and the problems of genetic algorithm, the mind evolution algorithm is proposed to optimize the initial parameters of wavelet neural network and apply it to integrated navigation. Firstly, the method uses the global search ability of the thought evolution algorithm to quickly and accurately obtain the initial parameters of the appropriate wavelet neural network, and then uses the optimized wavelet neural network to directly predict the position velocity error information after the data is solved. This method is different from the traditional filtering method while avoiding the drawbacks of the neural network.

2   Wavelet Neural Network

Wavelet neural network is a combination of wavelet theory and artificial neural network [10]. It replaces the excitation function of the hidden layer of artificial neural network with the wavelet basis function of wavelet theory. Therefore, it has the multi-resolution layering feature of wavelet analysis in time-frequency space. The powerful nonlinear fitting of the neural network and other functions. Compared with BP neural network, the wavelet neural network has a simple and clear network structure design, and has a clear theoretical basis. The network training can avoid local convergence or non-convergence, which makes the wavelet neural network have a wide range of applications. Its network structure is shown in Fig. 1.

Assume that the number of input signal samples is $q(1, 2, \ldots, n)$, the number of input layer nodes is I, the number of hidden layer nodes is H, and the number of output layer nodes is M.

The number of input layer and output layer nodes is determined by the actual input and output. The number of hidden layer nodes of the wavelet neural network can be adaptively determined by experience or formulation.

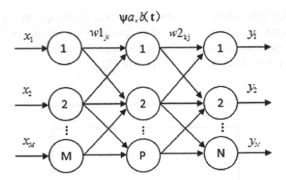

Fig. 1. Wavelet neural network model

Establish a wavelet neural network activation function and get the network output as

$$y_k = \sum_{j=1}^{J} w2_{kj}^* \phi \left(\frac{\sum\limits_{i=1}^{I} w1_{ji}x_i - b_j}{a_j} \right)$$

Wavelet neural network learning methods generally use an improved gradient descent method. During training, the momentum term is added to the correction algorithm of weight and threshold, and the correction value of the previous step is used to smooth the learning path, avoiding falling into local minimum values and speeding up the learning.

However, wavelet neural networks also have the disadvantage of BP neural networks. The weights and threshold initial parameters of the input layer to the hidden layer of the wavelet neural network are determined by the wavelet scale parameters. If the scale and displacement parameters of the wavelet network are not properly initialized, the result may not converge. The weight of the hidden layer to the output layer and the threshold initial parameter itself are uncertain, and each time the result is random, it is difficult to grasp the degree of good or bad. Moreover, the wavelet scale and the initial value of the displacement parameter not only affect the weight and threshold, but also affect the prediction ability of the wavelet neural network.

3   MEA Model Optimization Wavelet Neural Network

3.1   Mind Evolutionary Algorithm

Evolutionary algorithm is a kind of heuristic random search algorithm developed by combining computer science with biological evolution, such as genetic algorithm and evolutionary strategy. The thought evolution algorithm is a further improvement of the genetic algorithm. For the problems of premature, random walk and slow convergence of the evolutionary algorithm, new convergence and XOR operations are added on the basis of the genetic algorithm, so that the algorithm can maintain the structure. The parallelism and operational independence increase the overall efficiency of the algorithm. Its main system framework is shown in Fig. 2.

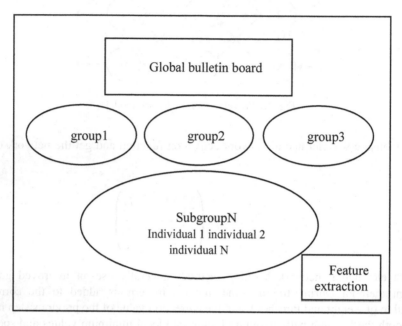

Fig. 2. Structure of thought evolution algorithm

The main idea of the thought evolution algorithm is: firstly, a certain number of individuals are randomly obtained from all the solutions, and the individuals with the highest scores and some temporary individuals are searched according to the individual scores; then each of the individuals with high scores is centered on each Some new individuals are generated around the individual, thereby obtaining a number of superior sub-populations and temporary sub-populations; then converging operations within each sub-population until maturity, and the highest score of the sub-population is used as the score of the sub-population; The scores of each sub-group are posted on the global bulletin board, and the alienation operation is performed between the sub-groups to complete the process of replacement, abandonment, and individual release in the sub-population between the superior sub-population and the sub-population, thereby obtaining the most of all individuals. Excellent individuals and their scores.

3.2 MEA Model Optimization Wavelet Neural Network

The wavelet evolution and wavelet parameters and the displacement parameters, initial weights and thresholds of the wavelet neural network are processed by the thought evolution algorithm. First, according to the topology of the wavelet neural network, the space is mapped to the coding space, and each code corresponds to a solution to the problem. Then the reciprocal of the mean square error of the training set is selected as the score function of each individual and the population. Using the thought evolution algorithm, after continuous iteration, the optimal individual is output, and the wavelet scale and displacement parameters, initial weight and threshold are used as training. Wavelet neural network. The specific steps are shown in Fig. 3.

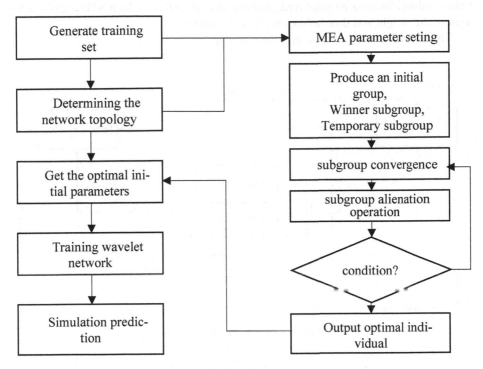

Fig. 3. Step flow chart

4 Filter Method Design

In the design of the combined structure, the position and speed combination in the loose coupling mode is adopted, and the SINS is the main navigation system, the GPS is the auxiliary navigation system. The GPS navigation data is used to correct the SINS navigation parameters in real time.

When selecting the input of the network, the indirect filtering method is adopted, that is, the error amount of the navigation parameters of each subsystem of the integrated navigation system is used as the prediction object.

When using the output of the network, the feedback correction method is adopted, that is, the prediction of the network output is fed into the inertial navigation system, and directly input into the inertial mechanical programming equation to correct the navigation parameters.

In the structural design of the network, according to the required navigation parameters and their dimensions, two parallel wavelet neural networks are used to train the
position and velocity error information respectively, and the number of input and output
nodes of each network is 3, corresponding to the three dimensions of the data. The hidden
layer of the network can satisfy the error condition by adaptively changing the number of
cells. The number of hidden layer nodes in the network is 6. The wavelet basis function
selects the function morlet, and the expression is $\psi(x) = \cos(1.75x)\exp(-x^2/2)$. The
training algorithm uses an improved gradient descent method, which adds a momentum
term to the weight and threshold correction algorithm.

The overall filtering design is shown in Fig. 4.

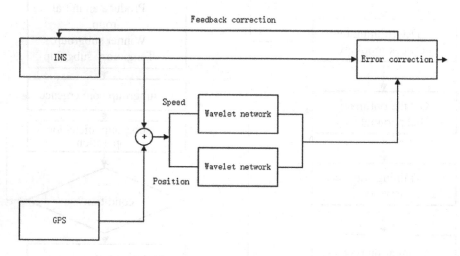

Fig. 4. Scheme structure design

In the training phase of the network, the SINS and GPS measured navigation data
are first solved, and the position and velocity information of the SINS and GPS are
obtained as the training sample set. Taking the difference between SINS and GPS
navigation parameters as input, the actual error of SINS output position velocity
information is the expected output, and the real output of the network is used as the
prediction error of SINS. The mean square error of the output is minimized by a certain
number of trainings.

In the prediction stage of the network, the difference between the SINS and the GPS
navigation parameters is input to the trained wavelet neural network, and the output of
the network is used as the error prediction value of the SINS at that time, and fed back
to the mechanical equation of the inertial navigation system to correct the system
navigation parameter.

In the actual navigation, the maximum effective time of every other network prediction, using the predicted value of the network in the previous period as the real data for offline training, during the training time, GPS can be used as the main navigation system, and the position and speed information of the GPS output is used as Navigate the output track to ensure that the completed navigation information is obtained.

5 Simulation Analysis

The simulation experiment uses a nine-axis IMU module and a GPS receiver module. The accuracy of the gyro of the IMU module is 0.05 °/s, the accelerometer accuracy is 0.01 g, the positioning accuracy of the GPS module is 2.5 m, and the speed measurement accuracy is 0.1 m/s. Fix the two on the same platform and make a uniform linear motion with a speed of 1 m/s in the north direction. The simulation time is 300 s. Analyze the data obtained by the experiment, using the data of the first 200 s as the training sample, and the data of the last 100 s as the test sample, and processing the data according to the design scheme, and obtaining the INS error corrected by the optimized wavelet neural network of the MEA model, and the real error is performed. Compared. In order to verify the superiority of the method, a reference experiment using wavelet neural network and genetic algorithm to optimize the wavelet neural network as the combined navigation filtering method was carried out. The results are as follows (Figs. 5, 6, 7, 8, 9 and 10).

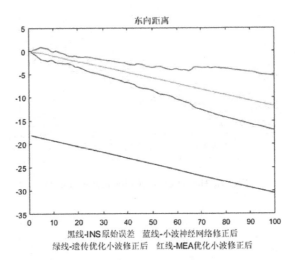

Fig. 5. Eastward distance error (Color figure online)

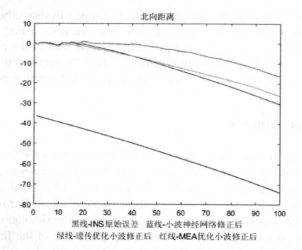

北向距离

黑线-INS原始误差   蓝线-小波神经网络修正后
绿线-遗传优化小波修正后   红线-MEA优化小波修正后

Fig. 6. Northward distance error (Color figure online)

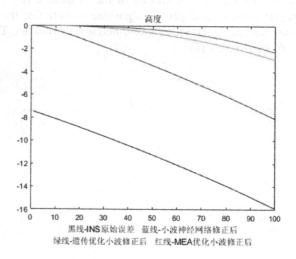

高度

黑线-INS原始误差   蓝线-小波神经网络修正后
绿线-遗传优化小波修正后   红线-MEA优化小波修正后

Fig. 7. Height error (Color figure online)

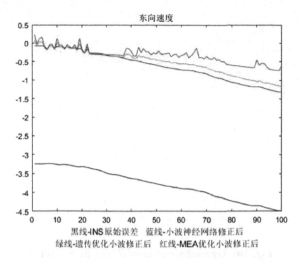

Fig. 8. Eastward speed error (Color figure online)

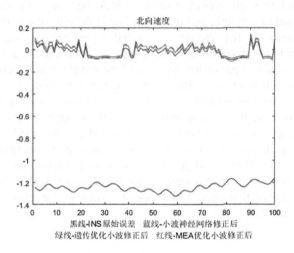

Fig. 9. Northward speed error (Color figure online)

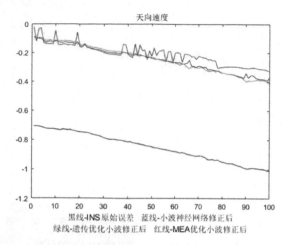

天向速度

黑线-INS原始误差　蓝线-小波神经网络修正后
绿线-遗传优化小波修正后　红线-MEA优化小波修正后

Fig. 10. Upward speed error (Color figure online)

The above figure shows the result of the position and velocity error of the last 100 s. The black line is the original error of the INS, the blue line is the error corrected by the wavelet neural network, and the green line is the error corrected by the genetic wavelet neural network. It is the error corrected by the MEA model to optimize the wavelet neural network. According to the simulation results in the above figure, it can be seen that the nonlinear prediction function of the wavelet neural network can greatly reduce the position and velocity error of the integrated navigation. On this basis, the wavelet neural network optimized by the genetic algorithm has a greater improvement in accuracy. However, the method proposed by the thought evolution algorithm to optimize the wavelet neural network is the best, and the corrected error value is the smallest. And when the number of iterations is the same, the simulation time of wavelet neural network is 11 s, the time of genetic wavelet neural network is 22 s, and the time of optimization of wavelet neural network by MEA model is 14 s. Subjective evaluation often has quantization error, and its specific mean square error value is shown in the following Table 1.

Table 1. Mean square error results of position and velocity

	Eastward speed m/s	Northward speed m/s	Up speed m/s	Eastward distance m	Northward distance m	Height m	Simulation time s
SINS raw data	4.0248	0.9478	1.0462	25.3390	50.2643	13.2498	–
Wavelet network	0.7854	0.0222	0.2424	8.9187	11.6813	3.6654	11
GA wavelet network	0.5167	0.0045	0.2437	5.9248	10.5487	0.9274	22
MEA wavelet network	0.2179	0.0016	0.1056	2.7973	4.4251	0.6176	14

Based on the above simulation results, the following conclusions can be drawn:

Firstly, theoretically, wavelet neural networks avoid the introduction of new errors in the establishment of mathematical models, so they have certain advantages. Genetic algorithms and thought evolution algorithms can effectively improve the problems of wavelet neural networks. Experimental results show that wavelet neural networks are used. The prediction of the position velocity information can greatly reduce the navigation error. The wavelet neural network optimized by the genetic algorithm is better. The method of optimizing the wavelet neural network by the thought evolution algorithm is the best and can be applied to it. Improve the accuracy of integrated navigation.

Secondly, for the problems of wavelet neural network, genetic algorithm and thought evolution algorithm under the same iteration times, the thought evolution algorithm is much lower than the genetic algorithm because of the characteristics of sub-population parallel processing, which can effectively improve the timeliness of combined navigation. In the simulation experiment, the maximum effective time of prediction within the allowable range of error is 3 min. In real navigation, the latest data must be used for offline training every other maximum effective time. The simulation results show that the data of 5 min only requires 14 s of training time, so During the training time, GPS can be used as the main navigation system, and the position and speed information of the GPS output is used as the navigation output trajectory to ensure the full navigation.

6 Conclusion

In the experiment of combined navigation filtering using wavelet neural network, the initial parameters of the network have the influence of randomness on network convergence and navigation accuracy. This paper proposes a combined navigation filtering method based on thought evolution algorithm to optimize wavelet neural network. This method is different from the traditional filtering method, avoids the difficulty of establishing the mathematical model, and uses the thought evolution algorithm to optimize, avoiding the problems of wavelet neural network. The simulation experiments are carried out with reference to wavelet neural network and genetic wavelet network. The results show that the method can effectively improve the accuracy of the integrated navigation system. At the same time, due to the advantages of parallel processing, the simulation time is shorter, ensuring navigation timeliness and filtering for combined navigation. Provide a new viable path.

References

1. Lin, X., Li, R., Gao, Q.: Integrated Navigation and Information Fusion Method. National Defence Industry Press, Beijing (2017)
2. Qin, H., Cong, L., Sun, X.: Accuracy improvement of GPS/MEMS-INS integrated navigation system during GPS signal outage for land vehicle navigation. J. Syst. Eng. Electron. 23(02), 256–264 (2012)

3. Efitorov, A., Shiroky, V., Dolenko, S.: A Neural Network of Multireotion Wavelet Analysis. Springer, Heidelberg (2018)
4. Hu, Z., Liu, L.: Applications of wavelet analysis in differential propagation phase shift data de-noising. Adv. Atmos. Sci. 31(04), 825–835 (2014)
5. Yue, P.: Exploration of TSP based on genetic algorithm. Mod. Inf. Technol. 3(04), 10–12 (2019)
6. Wang, S., Wang, J., Wang, Y., Ma, W.: Short-term load forecasting of BP neural network based on improved genetic algorithm. Foreign Electron. Meas. Technol. 38(01), 15–18 (2019)
7. Guo, Q., Zheng, Y., Zhu, W., Jin, B.: Research on high strength steel forming based on BP neural network genetic algorithm [J/OL]. Mater. Sci. Technol. 1–9 (2019)
8. Sun, P., Cai, R., Xie, C., Yi, Z.: Evaluation of slope stability based on genetically optimized neural network [J/OL]. Modern Electron. Technol. 2019(05), 75–78 (2019)
9. Lin, X.: Algorithm of GPS/SINS integrated navigation system based on genetic wavelet neural network. Ordnance Ind. Autom. 30(04), 42–45 (2011)
10. Kui, D.: Neural Network Design. Mechanical Industry Press, Beijing (2012)

Survey of Methods for Time Series Symbolic Aggregate Approximation

Lin Wang[1], Faming Lu[1(✉)], Minghao Cui[1], and Yunxia Bao[2]

[1] College of Computer Science and Engineering,
Shandong University of Science and Technology, Qingdao 266590, China
fm_lu@163.com
[2] College of Mathematics and System Science,
Shandong University of Science and Technology, Qingdao 266590, China

Abstract. Time series analysis is widely used in the fields of finance, medical, and climate monitoring. However, the high dimension characteristic of time series brings a lot of inconvenience to its application. In order to solve the high dimensionality problem of time series, symbolic representation, a method of time series feature representation is proposed, which plays an important role in time series classification and clustering, pattern matching, anomaly detection and others. In this paper, existing symbolization representation methods of time series were reviewed and compared. Firstly, the classical symbolic aggregate approximation (SAX) principle and its deficiencies were analyzed. Then, several SAX improvement methods, including aSAX, SMSAX, ESAX and some others, were introduced and classified; Meanwhile, an experiment evaluation of the existing SAX methods was given. Finally, some unresolved issues of existing SAX methods were summed up for future work.

Keywords: Time series · SAX · Symbolic representation · Data mining

1 Introduction

Time series is a sequence of numerical values obtained by performing equal time interval observations on a certain physical quantity. It is widely used in our daily life. For example, in the medical field, continuous monitoring of patients' electrocardiogram (ECG) and electroencephalogram (EEG) activities can promptly detect abnormalities in patients' conditions [1, 2]. In the financial field, observing stock price movements, interest rates, etc., can predict stock returns to a certain extent, or detect abnormalities in stock manipulation [3, 4]. In the climatic environment, monitoring rainfall, river water levels or temperature changes can provide a basis for climate change analysis [5, 6]. In the field of industrial manufacturing, collecting and analyzing the production equipment data in the actual production process can timely detect and alert equipment abnormalities, reduce equipment failure rate, and improve yield and production efficiency [7, 8].

The dimension of the time series is determined by its time span and time interval, so the data volume of the time series is usually large and the dimension is high. The analysis and mining of the original data directly will greatly increase the computational

X. Cheng et al. (Eds.): ICPCSEE 2019, CCIS 1058, pp. 645–657, 2019.
https://doi.org/10.1007/978-981-15-0118-0_50

complexity. The usual way to solve this problem is to reduce dimensionality of time series data. There are some mature dimensionality reduction methods, such as Discrete Fourier Transform (DFT) [9], Discrete Wavelet Transform (DWT) [10], piecewise linear approximation (PLA) [11], piecewise aggregate approximation (PAA) [12], symbolic aggregate approximation (SAX) [13], etc. This paper focuses on the most mature and most mature symbolic aggregate approximation methods and its improved methods.

2 Analysis of the SAX Method

The SAX method is a PAA-based symbolic representation method proposed by Lin and Keogh [13]. The SAX method requires time series data to approximate a normal distribution, which can perform fast and efficient data dimensionality reduction on time series, and the method can ensure that the distance between similar modes in the symbol space satisfies the lower bounds [14] requirement of the true distance and could prevent false dismissals from happening. The SAX method has the following four steps.

Step1: Normalize the original time series and convert it to a sequence with a mean of 0 and a standard deviation of 1.

Step2: Dimensionality reduction by PAA method. Then, we need to use Eq. (1) to convert the original time series $C = \{c_1, c_2, \ldots, c_n\}$ of length n into a vector $\bar{C} = \{\bar{c}_1, \bar{c}_2, \ldots, \bar{c}_w\}$ represented in the w-dimensional space, as shown in Fig. 1. In Eq. (1), $\bar{c}_i$ is the i-th element in time series $\bar{C}$, and the c_j is the j-th element in time series C. Simply stated, the n-dimensional original time series C is first divided into w equal-sized segments, and then the average value of each segment is calculated, and the original segment is replaced by the mean to form a new w-dimensional sequence $\bar{C}$.

$$\bar{c}_i = \frac{w}{n} \sum_{j=\frac{n}{w}(i-1)+1}^{\frac{n}{w}i} c_j \tag{1}$$

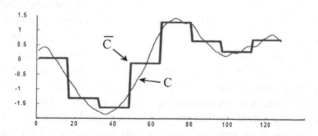

Fig. 1. PAA method dimension reduction representation

Step3: Symbolic representation. Select the alphabet size of symbols α, and then find the "breakpoints" β_i of the interval in Table 1. The "breakpoints" that will produce α equal-sized areas under Gaussian curve. Then, the average value obtained by the PAA method in Step 2 are mapped to the corresponding letter, and finally the sequence $\bar{C}$ is discretized into $\hat{C} = \{\hat{c}_1, \hat{c}_2 \ldots, \hat{c}_w\}$, as shown in Fig. 2.

Table 1. The breakpoints for values of α from 3 to 10.

$\beta_i \backslash \alpha$	3	4	5	6	7	8	9	10
β_1	−0.43	−0.67	−0.84	−0.97	−1.07	−1.15	−1.22	−1.28
β_2	0.43	0	−0.25	−0.43	−0.57	−0.67	−0.76	−0.84
β_3	−	0.67	0.25	0	−0.18	−0.32	−0.43	−0.53
β_4	−	−	0.84	0.43	0.18	0	−0.14	−0.25
β_5	−	−	−	0.97	0.57	0.32	0.14	0
β_6	−	−	−	−	1.07	0.67	0.43	0.25
β_7	−	−	−	−	−	1.15	0.76	0.52
β_8	−	−	−	−	−	−	1.22	0.84
β_9	−	−	−	−	−	−	−	1.28

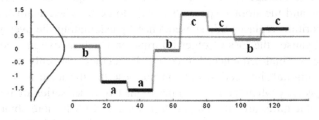

Fig. 2. Symbolic representation of a time series, the original time series is discretized into symbol sequences: baabccbc

Step4: Distance Measures. Given two time series Q and C of length n, after the above three steps, the original time series transformed into a letters sequence $\hat{Q}$ and $\hat{C}$ of length w. They redefine the distance calculation equation. Equation (2) defines the distance between two letters sequences, and Eq. (3) defines the distance between two letters. $\hat{q}_i$ and $\hat{c}_i$ are the i-th elements in $\hat{Q}$ and $\hat{C}$, respectively, and they are letters such as a, b or c. The value of β_i could be found in Table 1. For example, when $\alpha = 4$, dist(a, b) = 0, dist(a, c) = 0.67.

$$MINDIST\left(\hat{Q}, \hat{C}\right) \equiv \sqrt{\frac{n}{w}} \sqrt{\sum_{i=1}^{w} \left(dist(\hat{q}_i, \hat{c}_i)\right)^2} \tag{2}$$

$$dist(\hat{q}_i, \hat{c}_i) = \begin{cases} 0, & |\hat{q}_i - \hat{c}_i| \leq 1; \\ \beta_{max(\hat{q}_i, \hat{c}_i)-1} - \beta_{min(\hat{q}_i, \hat{c}_i)}, & otherwise \end{cases} \tag{3}$$

The MINDIST function in Eq. (2) returns the minimum distance of the original time series represented by the two words, which has been proved that lower bounds the true Euclidean distance [14]. The lower bound distance is usually measured by tightness of lower bound (TLB). The definition of TLB is Eq. (4). The value of TLB is between 0 and 1. The closer the value is to 1, the closer the lower bound distance is to the true distance, and the smaller the error.

$$TLB = \frac{lower\ bound\ distance}{true\ euclidean\ distance} \tag{4}$$

Although the SAX method can quickly reduce dimension, high efficiency calculations, etc., it also has several obvious shortcomings. First, the SAX method requires that the data approximate a Gaussian distribution, and the SAX method is required to first select the number of segments w and the alphabet size of symbols. The larger the value of w, the finer the feature representation and the more feature information is included. Conversely, the smaller the value of w, the coarser the feature representation, the more information will be lost. Therefore, there is a trade-off between the quality of the approximate representation and the magnitude of the dimensional reduction. Similarly, the smaller the α, the weaker the compactness of the lower boundary, but as α becomes larger, the lower bound compactness becomes stronger and the required storage spaces increases. So the compactness of the approximate representation and the storage space also needs to be weighed against. Second, since the PAA algorithm replaces this subsequence by calculating the average of the subsequences, it causes the loss of other information (such as extreme points and variances) in the original sequence. Moreover, the method can only reflect the overall trend of the original time series, but can not describe the local information of each segment. Especially when the mean values of the two times series are the same but the morphological features are different, the limitation of SAX is more obvious. Third, the distance between the same symbol and the adjacent symbol is 0. Even if the time series represented by the same and adjacent symbols have different morphological features, the distance between them is also zero. These shortcomings may lead to time series analysis and the generation of errors in mining tasks.

In response to the above SAX deficiency, scholars have proposed a number of improvements. These methods can be mainly divided into two categories. The first category is an improved method for adaptive segmentation proposed for SAX segmentation parameter selection and distance calculation between symbols. The second category is an improved approach to the enhanced sequence representation proposed for the loss of information from the original sequence using the PAA algorithm. Below we will elaborate on the improved methods of SAX.

3   Improved Methods of SAX

3.1   Improvement of Adaptive Segmentation and Similarity Calculation Method

In view of the need to select the segmentation parameters first when using the SAX method, many scholars have proposed improved methods. Literature [15] propose a new generic framework to compute adaptive Segmentation Based Symbolic Representations (SBSR) of time series. Instead of using PAA, the method uses adaptive piecewise constant approximation (APCA) [16] to perform discretized symbolic representations. The APCA method segments a time series into a series of variable length subsequences, each of which is represented by a two-tuple consisting of the mean of the data in the subsection and the rightmost time scale value. Although the method can be adaptively segmented, the method requires twice as much storage as the PAA method and has a higher time complexity.

Literature [17, 18] introduced a novel adaptive symbolic approach based on the combination of SAX and k-means algorithm which called adaptive SAX (aSAX). They use the alphabet size of symbols as parameter kin k-means algorithm and the "breakpoints" under Gaussian curve as the clever initialization of cluster intervals. Then according to the k-means algorithm to obtained the "adaptive breakpoints". This method performs well on high Gaussian distribution and the lack of Gaussian distribution datasets.

It is also to solve the problem that the time series must be equal-length segmentation in the SAX algorithm. Literature [19] propose a vector symbolic algorithm based on segmentation algorithm for time series (SMSAX). First, they use the triangular threshold method to extract the feature of time series of random samples which is sampled randomly. Then calculate the maximum compression ratio of time series, extract the segmentation point as the time window width, and then find the segmentation mode of the time series. Finally, using vector of mean and volatility of subsequences to symbolic them. But they did not give a specific distance calculation formula.

In order to solve the problem that the distance between adjacent symbols in SAX is zero, the literature [20] proposes an improved similarity calculation method. This method takes into account the minimum distance between two adjacent symbols. As shown in Fig. 3, dist(a, b) = value2 + value4, dist(b, c) = value1 + value3, dist(a, c) = value4 + value3 + value0. The method satisfies the distance lower bound requirement such that the distance between adjacent symbols is no longer zero and the lower bound is tighter. However, this method is because the principle of triangular inequality is violated, so only the similarity can be calculated and the distance cannot be calculated.

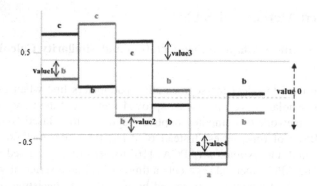

Fig. 3. Neglected distance between two characters

Although these methods have achieved some results in some data, they still use the mean of the sequence segments as the basis for the approximate representation. The calculation of the symbol approximation distance still does not consider the difference inside each sequence segment.

3.2   Enhanced Sequence Representation

The PAA method used in SAX only considers the loss of the original time series information caused by the mean can be divided into two categories, one is the lack of statistical features, and the other is the lack of morphological features. In response to the lack of these two kinds of information, scholars added the two types of information lost on the basis of SAX to enhance the representation of the sequence. So we divided these improved methods into two categories, one is SAX improvement based on statistical features, another is an improved method based on morphological features.

Improved Methods of SAX Based on Statistical Features
For the loss of statistical feature information, the literature [21] proposed an extended symbol aggregation approximation method (ESAX) based on sequence maxima, minima and mean, which maps the maximum and minimum values of each subsequence. The symbols are also added to the symbol sequence, causing one SAX character of each subsequence to become three, as shown in Fig. 4. This can find some similar patterns more efficiently and accurately. However, this method not only expands the importance of the extreme point, artificially causes the distortion of the characteristic information, but also increases the computational cost, and whether the calculation of the distance satisfies the lower bound requirement remains to be proved.

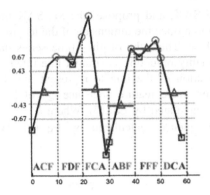

Fig. 4. Symbolic sequence after adding the maximum and minimum values is ACF-FDFFCAABFFFFDCA

The same is true for the SAX method to describe the incompleteness of time series information. The literature [22] proposed the statistics feature vector symbolic (SFVS), which adds variance to describe the divergence of the sequence based on SAX. The method considers the overall characteristics of the time-series symbols as vectors, and the mean and variance of each time segment are respectively used as components describing the mean and divergence. The overall feature of the timing symbol is the vector sum of the two components, as shown in Eq. (5), where $\hat{X}_i$ represents the symbol vector of the i-th sub-sequence, $\hat{x}_{i1}$ and $\hat{x}_{i2}$ represent the mean and variance of the i-th sub-sequence, respectively.

$$\hat{X}_i = \hat{x}_{i1} \cdot i + \hat{x}_{i2} \cdot j \tag{5}$$

The distance calculation function $dist()$ between such two symbol vectors is calculated by the following Eq. (6). In Eq. (6), $\hat{X}_i$, $\hat{X}_j$ are two symbol vectors, and d represents the dimension of the symbol vector. Because the symbol vector has only two dimensions: mean and variance, the value of d is 2.

$$dist(\hat{X}_i, \hat{X}_j) = \sqrt{\sum_{d=1}^{2} \left(\hat{x}_{id} - \hat{x}_{jd}\right)^2} = \sqrt{\left(\hat{x}_{i1} - \hat{x}_{j1}\right)^2 + \left(\hat{x}_{i2} - \hat{x}_{j2}\right)^2} \tag{6}$$

The distance calculation function of this method is MINDIST(·) similar to Eq. (2). This method is beneficial for more accurate analysis in applications of time series pattern recognition. However, if the two sub-sequences have the same mean and variance, but the morphological characteristics are different, the method does not distinguish them well. Moreover, it is not reasonable to use the dividing points of the mean to divide the variance.

Improved Methods of SAX Based on Shape Features
In view of the loss of shape features information of the original time series, literature [23] comprehensively considers the mean and slope to enhance the sequence

representation based on SAX, and proposes the SF_SAX method. The method first uses the PAA algorithm to reduce the dimension of the length n time series Q to obtain a sequence Q' of length w. Then, the original time series data in each subsequence obtained after dimensionality reduction is straight-line fitted by least squares method. Use the slope approximation of the line to represent the shape feature of the time segment, so that another shape feature sequence Q'' of length w is obtained. The element value q_i'' in Q'' calculated by Eq. (7), Where $k = \frac{n}{w}$. $\bar{q}_i$ is the fitted line slope value of the i-th subsequence, calculated by Eq. (8), Where $j_0 = k(i-1)+1$, $i = 1, 2, \ldots, w$.

$$q_i'' = \frac{k-1}{2}\bar{q}_i \qquad (7)$$

$$\bar{q}_i = \frac{k * \sum_{j=j_0}^{k*i} j * q_j - \left(\sum_{j=j_0}^{k*i} j\right)\left(\sum_{j=j_0}^{k*i} q_j\right)}{k * \sum_{j=j_0}^{k*i} j^2 - \left(\sum_{j=j_0}^{k*i} j\right)^2} \qquad (8)$$

Then according to the numerical distribution of Q'', the data space is equally divided, and then converted into a character sequence according to the SAX method. The symbol components represented by the mean and slope are then converted into sequence symbol vectors according to the method in [22]. Given two time series Q and C of the same length n, their dimensionality reduction and symbolization are represented as $\hat{Q}, \hat{C}$ by the SF_SAX method. For any l, $1 < l < w$, there are $\hat{q}_l = \widehat{A_l^q} \cdot i + \widehat{S_l^q} \cdot j$, $\hat{c}_l = \widehat{A_l^c} \cdot i + \widehat{S_l^c} \cdot j$, $\widehat{A_l^q}$ is a symbol obtained by the SAX method, $\widehat{S_l^q}$ can be understood as the symbol corresponding to the slope. Then their distance calculation method for satisfying the lower bounding is defined in Eqs. (9) (10), where function $dist()$ is calculated in the same way as the Eq. (3).

$$MINDIST(\hat{Q}, \hat{C}) = \sqrt{\frac{n}{w}}\sqrt{\sum_{l=1}^{w}\left(\left(dist\left(\widehat{A_l^q}, \widehat{A_l^c}\right)\right)^2 + \mu\left(dist\left(\widehat{S_l^q}, \widehat{S_l^c}\right)\right)^2\right)} \qquad (9)$$

$$\mu = \frac{\left(1+\frac{w}{n}\right)^2}{9\left(1-\frac{w}{n}\right)} \qquad (10)$$

The method uses the mean and the slope to jointly describe the trend of the data, and the slope is converted into a domain to achieve symbolization, which can better identify sequences with morphological differences. However, this method is to fit the segment by the least squares method. When the compression is large and the sequence fluctuation is large, the fitting effect may be poor.

The same is to solve the problem that SAX can't identify similar mean values but different trends. In [24], a time series symbolization method based on trend distance (SAX_TD) is proposed. Several typical trends are defined in the paper: (a) slight increase, (b) slight decrease, (c) significant increase, (d) significant decrease, (e) overall increase, (f) overall decline, (g) first decline and then rise, (h) rise first After falling.

The trend factor is defined in the paper: the difference between the start and end values of each subsequence and its mean. The trend distance is used to indicate the trend change. The new symbol is not defined in the literature, but the trend is quantitatively measured by calculating the distance of the trend, called the trend distance.

The function $td()$ for calculating the trend distance is Eq. (11).

The symbols q and c represent two equal-length time series segments, the trend distance $td(q, c)$ between them is defined as Eq. (11), where t_s, t_e are the starting and ending time points of q and c, respectively. $\Delta q(t)$ and $\Delta c(t)$ are the difference between q and c at time t and their mean, which can be calculated by Eqs. (12) (13), $\bar{q}$ and $\bar{c}$ are the average values of the time series segments q and c, respectively.

$$td(q, c) = \sqrt{(\Delta q(t_s) - \Delta c(t_s))^2 + (\Delta q(t_e) - \Delta c(t_e))^2} \tag{11}$$

$$\Delta q(t) = q(t) - \bar{q} \tag{12}$$

$$\Delta c(t) = c(t) - \bar{c} \tag{13}$$

This method adds a trend factor to the representation of the time series: $Q :_{\Delta q(1)} \widehat{q_1}_{\Delta q(2)} \widehat{q_2}_{\Delta q(3)} \cdots_{\Delta q(w)} \widehat{q_w}_{\Delta q(w+1)}$, and $\Delta q(w+1)$ represents the last trend factor. Specific representation method example: $_{0.2}f_{1.2}e_{-0.1}a_{-1.2}c_1d_{-0.2}b_{-0.3}$. The distance calculation Equation for this method is TDIST, as shown in Eq. (14), where the function $dist()$ is also the same as Eq. (3).

$$TDIST(Q, C) = \sqrt{\frac{n}{w}}\sqrt{\sum_{i=1}^{w}\left((dist(\hat{q}_i, \hat{c}_i))^2 + \frac{w}{n}(td(q_i, c_i))^2\right)} \tag{14}$$

Although the algorithm constructs a trend distance, it can detect time series with the same mean but different trends. However, the symbol sequence obtained by this method has a dimensional approximation twice that of the classical SAX method. This approach also expands the importance of the trend and is not suitable for more volatile data.

For the SAX algorithm, the similarity between time series cannot be distinguished when the symbols are consistent in each sequence segment. The time series symbol aggregation approximation method based on the beginning and end distance (SAX_SM) is proposed in [25]. Similar to the literature [24], the starting point of the subsequence is used to construct the trend. However, the time series approximation method in [25] is different from the literature [24]. In [25], when the subsequence mean is calculated, the starting point of the subsequence is removed, and then the corresponding characters of each subsequence are obtained according to the SAX method. The start and end points are added to the character sequence as shown in Eq. (15), where s_i and e_i represent the starting index value and ending index value of the i-th subsequence, respectively

$$\tilde{X} = \left(\left(s_1, \widehat{X_1}, e_1 \right), \left(s_2, \widehat{X_2}, e_2 \right), \ldots, \left(s_w, \widehat{X_w}, e_w \right) \right) \tag{15}$$

The starting distance defined in the literature is as shown in the Eq. (16), where p and q are sub-sequences of equal length, p_s, q_s is the starting value of p and q, p_e, q_e is the end value of p and q. Given two time series Q and C, the distance calculation method between them that the party satisfies the lower bounding requirement is as shown in Eq. (17), where the function $dist()$ is also the same as Eq. (3).

$$smd(p, q) = \sqrt{(p_s - q_s)^2 + (p_e - q_e)^2} \tag{16}$$

$$Dist(Q, C) = \sqrt{\frac{n}{w}} \sqrt{\sum_{i=1}^{w} \left(dist(\hat{q}_i, \hat{c}_i)^2 \right) + \frac{w}{n} \left(\sum_{i=1}^{w} smd(q_i, c_i)^2 \right)} \tag{17}$$

Different from the literature [24] and [25], the literature [26] proposes to use numerical derivatives to describe the morphological characteristics of sub-sequences (NSM). Numerical derivatives is $s_i' = s_i - s_{i-1}$. If the numerical derivative is a positive number, it means a rise, a negative number means a decrease, and if it is 0, it means no change. Symbolization of time series is still using the SAX method. Then use DTW to measure the similarity between the numerical derivatives. Given two time series Q and C, the total similarity calculation method is Eq. (18), where $\hat{q}_i$ and $\hat{c}_i$ respectively represent the i-th element of the sequence obtained by symbolizing the time series Q and C using the SAX method, Q_i' and C_i' represent the numerical derivative sequence of the time series Q and C, respectively, the function $dist()$ is also the same as Eq. (3).

$$Dist(Q, C) = \sqrt{\frac{n}{w}} \sqrt{\sum_{i=1}^{w} \left(dist(\hat{q}_i, \hat{c}_i)^2 + \frac{w}{n} DTW(Q_i', C_i') \right)} \tag{18}$$

The symbolic representation of time series is not given in this paper, and because the similarity calculation combined with the DTW method, it will lead to the limitation of long calculation time.

4   Experimental Comparison and Analysis

In this section, we will present the result of our experimental analysis and compare. We performed the experiments on several diverse time series datasets, which are provided by the UCR Time Series repository [27]. So far, The UCR Time Series Archive contains a total of 128 time series datasets, including ECG datasets, signal datasets acquired by sensors, image contour datasets and some motion datasets, and other types of datasets. A brief description of some datasets is in Table 2.

Table 2. A brief description of some data sets in UCR time series Archive

Type	Name	Train	Test	Class	Length	Data donor/editor
Image	DistalPhalanxTW	400	139	6	80	L. Davis & A. Bagnall
Senor	Earthquakes	322	139	2	512	A. Bagnall
ECG	ECG5000	500	4500	5	140	Y. Chen & E. Keogh
EOG	EOGHorizontalSignal	362	362	12	1250	E. Keogh & H. A. Dau
Motion	CricketX	390	390	12	300	A. Mueen & E. Keogh
Device	ElectricDevices	8926	7711	7	96	A. Bagnall & J. Lines
Spector	Strawberry	613	370	2	235	K. Kemsley & A. Bagnall
Spectrum	SemgHandSubjectCh2	450	450	5	1500	C.-C. M. Yeh
Trajectory	GestureMidAirD1	208	130	26	vary	H. A. Dau
Simulated	SyntheticControl	300	300	6	60	R. Alcock & Y. Manolopoulos

The experimental analysis shows that the dimension of the ESAX method is three times that of the SAX dimension after dimension reduction. The classification error of the ESAX method is similar to the SAX method. The ESAX method works better than the SAX method in pattern matching. Use the one-Nearest Neighbor (1NN) method to compare the classification accuracy of these symbolic methods. The SAX_TD method has a lower classification error than the ESAX method. The SAX_TD method is better than the ESAX and SAX methods in terms of parameter expansion (robustness). And in terms of dimensionality reduction, the SAX_TD and SAX methods are indistinguishable, but they are better than the ESAX method. The average classification accuracy of the SAX_SM algorithm is slightly higher than the SAX_TD algorithm, and the computational cost is slightly lower than the SAX_TD algorithm. SF_SAX and SFVS are better than ESAX in terms of anomaly detection and similarity queries. But because of the extra calculation of the symbolic representation of another feature, their computational cost is higher than the SAX algorithm. Because the NSM method uses DTW to calculate the distance, it consumes more time than SAX_TD, SAX, etc., but its classification error rate is lower than the SAX_TD and so on.

5   Conclusions and Future Work

In this review, we introduce time-series symbolization methods and analysis their respective strengths and weaknesses. Although the time series symbolic approximation method has made great progress in the past, most scholars are improving methods for one or two defects of classical SAX. The improved method in adaptive segmentation does not consider the enhancement sequence. It is indicated that the improved method in the enhanced sequence representation also does not consider the adaptive segmentation problem. If we want to do adaptive segmentation or add characters to the representation of the enhancement sequence, it will increase the computational complexity and storage space. There is a trade-off problem here. Symbolizing time series is a means of processing data in mining and analysis. Our ultimate goal is to make time series classification, pattern matching and anomaly detection problems simple and efficient.

Acknowledgement. This work was supported by the National Natural Science Foundation of China [grant numbers 61602279, 61472229]; Shandong Province Postdoctoral Innovation Project [grant number 201603056]; the Sci. & Tech. Development Fund of Shandong Province of China [grant number 2016ZDJS02A11 and Grant ZR2017MF027]; the SDUST Research Fund [grant number 2015TDJH102]; and the Fund of Oceanic telemetry Engineering and Technology Research Center, State Oceanic Administration (grant number 2018002).

References

1. Annam, J.R., Surampudi, B.R.: Inter-patient heart-beat classification using complete ECG beat time series by alignment of R-peaks using SVM and decision rule. In: International Conference on Signal & Information Processing. (2017)
2. Yang, P., Dumont, G., Ansermino, J.M.: Adaptive change detection in heart rate trend monitoring in anesthetized children. IEEE Trans. Biomed. Eng. **53**(11), 2211–2219 (2006)
3. Rathnayaka, R.M.K.T., Seneviratne, D.M.K.N., Wei, J., Arumawadu, H.I.: A hybrid statistical approach for stock market forecasting based on artificial neural network and ARIMA time series models. In: International Conference on Behavioral (2015)
4. Golmohammadi, K., Zaiane, O.R.: Time series contextual anomaly detection for detecting market manipulation in stock market. In: IEEE International Conference on Data Science & Advanced Analytics (2015)
5. Papagiannopoulou, C., Decubber, S., Miralles, D.G., Demuzere, M., Verhoest, Niko E.C., Waegeman, W.: Analyzing granger causality in climate data with time series classification methods. In: Altun, Y., et al. (eds.) ECML PKDD 2017. LNCS (LNAI), vol. 10536, pp. 15–26. Springer, Cham (2017). https://doi.org/10.1007/978-3-319-71273-4_2
6. Itoh, N., Kurths, J.: Change-point detection of climate time series by nonparametric method. In: Proceedings of the World Congress on Engineering and Computer Science, vol. 1, pp. 445–448 (2010)
7. Xing, W., Lin, J., Patel, N., Braun, M.: A self-learning and online algorithm for time series anomaly detection, with application in CPU manufacturing. In: ACM International on Conference on Information & Knowledge Management (2016)
8. Feng, C., Li, T., Chana, D.: Multi-level anomaly detection in industrial control systems via package signatures and LSTM networks. In: IEEE/IFIP International Conference on Dependable Systems & Networks (2017)
9. Agrawal, R., Faloutsos, C., Swami, A.: Efficient similarity search in sequence databases. In: Lomet, D.B. (ed.) FODO 1993. LNCS, vol. 730, pp. 69–84. Springer, Heidelberg (1993). https://doi.org/10.1007/3-540-57301-1_5

10. Chan, K.P., Fu, W.C.: Efficient time series matching by wavelets. In: International Conference on Data Engineering (1999)
11. Keogh, E., Chu, S., Hart, D., Pazzani, M.: An online algorithm for segmenting time series. In: Proceedings 2001 IEEE International Conference on Data Mining, pp. 289–296. IEEE (2001)
12. Keogh, E., Chakrabarti, K., Pazzani, M., Mehrotra, S.: Dimensionality reduction for fast similarity search in large time series databases. Knowl. Inf. Syst. **3**(3), 263–286 (2001)
13. Lin, J., Keogh, E., Lonardi, S., Chiu, B.: A symbolic representation of time series, with implications for streaming algorithms. In: ACM Sigmod Workshop on Research Issues in Data Mining & Knowledge Discovery (2003)
14. Faloutsos, C., Ranganathan, M., Manolopoulos, Y.: Fast subsequence matching in time-series databases. ACM (1994)
15. Hugueney, B.: Adaptive segmentation-based symbolic representations of time series for better modeling and lower bounding distance measures. In: Fürnkranz, J., Scheffer, T., Spiliopoulou, M. (eds.) PKDD 2006. LNCS (LNAI), vol. 4213, pp. 545–552. Springer, Heidelberg (2006). https://doi.org/10.1007/11871637_54
16. Keogh, E., Chakrabarti, K., Pazzani, M., Mehrotra, S.: Locally adaptive dimensionality reduction for indexing large time series databases. ACM SIGMOD Rec. **30**(2), 151–162 (2001)
17. Pham, N.D., Le, Q.L., Dang, T.K.: Two novel adaptive symbolic representations for similarity search in time series databases. In: 2010 12th International Asia-Pacific Web Conference, pp. 181–187. IEEE (2010)
18. Pham, N.D., Le, Q.L., Dang, T.K.: HOT *a*SAX: a novel adaptive symbolic representation for time series discords discovery. In: Nguyen, N.T., Le, M.T., Świątek, J. (eds.) ACIIDS 2010. LNCS (LNAI), vol. 5990, pp. 113–121. Springer, Heidelberg (2010). https://doi.org/10.1007/978-3-642-12145-6_12
19. 陈湘涛, 李明亮, 陈玉娟: 基于分割模式的时间序列矢量符号化算法. 计算机工程 **37**(4), 55–57 (2011)
20. Muhammad Fuad, M.M., Marteau, P.-F.: Enhancing the symbolic aggregate approximation method using updated lookup tables. In: Setchi, R., Jordanov, I., Howlett, R.J., Jain, L.C. (eds.) KES 2010. LNCS (LNAI), vol. 6276, pp. 420–431. Springer, Heidelberg (2010). https://doi.org/10.1007/978-3-642-15387-7_46
21. Lkhagva, B., Suzuki, Y., Kawagoe, K.: Extended SAX: extension of symbolic aggregate approximation for financial time series data representation. DEWS2006 4A-i8 7 (2006)
22. Zhong, Q., Cai, Z.: Symbolic algorithm for time series data based on statistic feature. Chin. J. Comput. **31**(10), 1857–1864 (2008)
23. 李海林, 郭崇慧: 基于形态特征的时间序列符号聚合近似方法. 模式识别与人工智能 **24**(5), 665–672 (2011)
24. Sun, Y., Li, J., Liu, J., Sun, B., Chow, C.: An improvement of symbolic aggregate approximation distance measure for time series. Neurocomputing **138**, 189–198 (2014)
25. 季海娟, 周从华, 刘志锋: 一种基于始末距离的时间序列符号聚合近似表示方法. 计算机科学 **45**(06), 222–227 (2018)
26. 李海林, 梁叶: 基于数值符号和形态特征的时间序列相似性度量方法. 控制与决策 **32**(3), 451–458 (2017)
27. Dau, H., et al.: The UCR Time Series Classification Archive (2018). https://www.cs.ucr.edu/~eamonn/time_series_data_2018/

Multimodal 3D Convolutional Neural Networks for Classification of Brain Disease Using Structural MR and FDG-PET Images

Kun Han, Haiwei Pan[✉], Ruiqi Gao, Jieyao Yu, and Bin Yang

Harbin Engineering University, Harbin, People's Republic of China
panhaiwei@hrbeu.edu.cn

Abstract. The classification and identification of brain diseases with multimodal information have attracted increasing attention in the domain of computer-aided. Compared with traditional method which use single modal feature information, multiple modal information fusion can classify and diagnose brain diseases more comprehensively and accurately in patient subjects. Existing multimodal methods require manual extraction of features or additional personal information, which consumes a lot of manual work. Furthermore, the difference between different modal images along with different manual feature extraction make it difficult for models to learn the optimal solution. In this paper, we propose a multimodal 3D convolutional neural networks framework for classification of brain disease diagnosis using MR images data and PET images data of subjects. We demonstrate the performance of the proposed approach for classification of Alzheimer's disease (AD) versus mild cognitive impairment (MCI) and normal controls (NC) on the Alzheimer's Disease National Initiative (ADNI) data set of 3D structural MRI brain scans and FDG-PET images. Experimental results show that the performance of the proposed method for AD vs. NC, MCI vs. NC are 93.55% and 78.92% accuracy respectively. And the accuracy of the results of AD, MCI and NC 3-classification experiments is 68.86%.

Keywords: Alzheimer's disease · MRI · FDG-PET ·
Convolutional neural networks · Residual networks · Deep learning ·
Image classification

1 Introduction

Alzheimer's disease, also known as senile dementia, is a central nervous system degenerative disease with insidious onset and chronic progression. Clinical manifestations of amnesia, loss of language ability are the most common type of dementia in the elderly. It is estimated that there are around 90 million AD patients in the world, with the number of AD patients expected to reach 300 million by 2050 [1, 2]. The ADNI report shows that there are as many as 50 million AD patients worldwide in 2018 [3]. The current research progress on AD is slow and the cause cannot be confirmed. It is generally found that AD is advanced, and even treatment will not have much effect [4]. Therefore, early diagnosis of AD is a better way to inhibit the rapid

© Springer Nature Singapore Pte Ltd. 2019
X. Cheng et al. (Eds.): ICPCSEE 2019, CCIS 1058, pp. 658–668, 2019.
https://doi.org/10.1007/978-981-15-0118-0_51

development of the disease or even avoid the disease. Mild Cognitive Impairment may be intermediate between AD and health. The annual conversion rate from MCI to AD is 10%–15% [5]. There are also studies shows that patients with MCI are more likely to develop AD than those who have not previously had MCI [6]. In the diagnosis of brain diseases, brain medical images have become one of the most important tools for doctors to diagnose and make decisions, such as computed tomography (CT), magnetic resonance imaging (MRI) and positrons emission tomography (PET) [7]. These brain medical imaging methods have been widely used in the diagnosis of AD [8]. Machine learning algorithms have good effects in analyzing Alzheimer's pathological medical images. Among numerous machine learning methods, deep learning has been showing the state-of-the-art performance in the recent years [9]. Recent studies also have shown that diagnosing AD is more accurate than experienced clinicians by using deep learning methods [10].

The previous studies shows that most methods require the extraction of pathological features of AD from brain medical images except neural networks, such as: Hippocampus, temporal lobe, amygdala and other regions of interest (ROI). Gray et al. [11] proposed an ROI-based method for extracting sagittal, coronal, and cross-sectional regions of interest in a three-dimensional brain image for multi-region PET image information combination for classification prediction. Gray et al. [12] proposed spatial segmentation of 3D brain images into 83 anatomical regions based on MRI images of subjects, and then extracted the average of the regions of interest in the FDG-PET images of the same subject using this region of interest as a template. The signal strength is finally predicted by the Support Vector Machine (SVM) based on the extracted intensity characteristics. Similarly, Garali et al. [13] used the ROI method in the study to map the entire image to 116 anatomical regions of interest, and extracted 21 regions of interest into the SVM and random forest (Random Forest). Perform classification prediction to evaluate performance. Silveira et al. [14] used the traditional machine learning method Boosting classification algorithm to mix simple classifiers and classify the whole brain PET images into AD and MCI. Liu et al. [15] achieved Convolutional Neural Networks (CNNs) [16] and Recurrent Neural Networks (RNNs) two deep learning methods to classify AD based on brain FDG-PET images. In addition, compared to traditional 2D images, Hosseini-Asl et al. [17] showed that 3D CNNs can capture the characteristics of brain structure better than 2D CNNs for 3D brain medical images. Compared with the traditional 2D CNNs, some studies have improved and extended it to 3D CNNs. Jianxu et al. [18] used a 3D convolution method to segment 3D biological images in ISBI (International Symposium on Biomedical Imaging) competition, which achieved good results. Similarly, Cicek et al. [19] used the 3D convolution operation instead of the normal 2D convolution operation to segment the Xenopus kidney image. The experimental results show that better performance can be obtained with 3D convolutional operation. In the study of the direction of Alzheimer's disease, Korolev et al. [20] based on the Residual Networks (ResNet) framework model [21] and the VoxResNet framework model [22], 3D convolution operations on three-dimensional brain images. This attempt is instructive, which shows that 3D CNNs can achieve good classification results without the need for extensive and complex preprocessing steps on the original image.

Compared with existing methods, there are some potential problem: (1) Complex preprocessing of the original image of the subject does not necessarily extract features better, and will inevitably bring additional overhead. (2) Learning features with machines can be more efficient and perform better than manual functions. (3) Compared with 2D CNNs, 3D CNNs can capture the spatial features of 3D brain structure more comprehensively and accurately. (4) Learning with multimodal data can learn more interesting knowledge than using a single modal data. In this paper, we propose a 3D convolutional neural network architectures combined with MR images and PET images to identify AD. It consists of two parts: the first part is two residual networks based on ResNet and VoxResNet with similar structure, performing 3D convolution operations on MR images and FDG-PET images to extract the respective modal images respectively. The second part is to fuse the above two models for extracting features. The information obtained by combining the modal features is finally classified by deep neural networks. We demonstrate that the proposed model does not require the raw handcrafted feature and complex image preprocessing steps. The proposed method uses two modal images to achieve better classification. And our resulting classification performance for AD vs. NC, MCI vs. NC is 93.55% and 78.92% accuracy. And the accuracy of the results of AD, MCI and NC tri-classification experiments is 68.86%.

We examine the proposed network based on the data from the Alzheimer's Disease Neuroimaging Initiative project. It provides a large number of MR images and PET images of patients with AD. We used it to test the performance of one-versus-one classification and 3-classification of our proposed network architecture experiments. In our study, we test performance of the proposed models by using different modal image feature information to be more comprehensive and accurate to achieve relatively good classification results without complex preprocessing. In the future we will also be in data augmentation, oversampling and the MCI data set is further divided into sMCI (stable MCI) and pMCI (progressive MCI) classifications for optimization and improvement.

2   Data and Materials

2.1   Data Description

Data set for the experiment is derived from Alzheimer's Disease Neuroimaging Initiative database (adni.loni.usc.edu). The ADNI is a public interest website for the Alzheimer's Neuroimaging Initiative, which was founded in 2003 as a public-private partnership. It contains MR images, PET images and other biological markers. The structural MR image has a strong ability to express the anatomy of the brain, and can perform better features in the biological anatomy and texture of the brain tissue. The PET image has a strong ability to express metabolic activity in the brain, and can better detect the characteristics of metabolic abnormalities in the pathological region. Therefore, in this paper, the corresponding structural MR image and PET image two modal images are selected as input data for all subjects. All subjects are searched and downloaded for AD, MCI and NC, strictly ensuring that all subjects downloaded must contain data from two modal images: MR images and PET images. In order to ensure the robustness and uniformity of the data, the MR image selected in this paper is the MRI of the T1 weight structure including post-processing. And the PET image is the image of FDG-PET (18F isotope mark) containing post-processing. Based on the experience of previous research experiments, the experimental data set in this paper is divided into three classes (AD, MCI, and NC). The data set contains a total of 379 subjects, and different modal images of each subject are must be conformed to the above rules. In order to prevent potential "leaks" of data set, we choose the image of the most recent time of each subject among the MRI scans. In FDG-PET images, we select the first image folder in the process of "_Coreg_Avg_Standardized_image_and_Voxel_size_br_" to be the input image data corresponding to the subject. The total number of image data in the data set is 758, and each image we select is three-dimensional spatial data in NIFTI format.

These subjects of data set are divided into three classes specifically: 114 of AD patients, 132 of MCI and 133 of NC. It is worth noting that each subject has 1 MRI scan, 1 PET image and a classification label. This means that there are a total of 379 MRI scans and 379 PET images. For the three classes of data set, we run 10-fold cross-validations with 10 different fold splits each time to get better approximation of prediction performance. We perform both one-versus-one classification and 3-classification tasks in the following proportions: there are about 103 AD, 119 MCI and 120 NC subjects are randomly selected as train set, 11 AD, 13 MCI and 13 NC subjects as a test set in each cross-validation. To normalize the brain image data, we process all MR images by FSL tool to skull stripping, in addition, normalize the voxel values to 0–255 and resize the input image size to 128*128*128 of MRI scans and PET images respectively. Item acquisition protocols, image post-acquisition preprocessing procedures and other detailed descriptions of ADNI subject cohorts can be found at http://www.adni-info.org. Researchers on this site participated in the contribution of ADNI, but did not participate in the analysis and writing of this paper.

2.2 Method

There are two major steps in the proposed framework: (1) Multimodal information extraction: two residual networks with similar structures, performing 3D convolution operations on structural MR images and FDG-PET images to extract their own modal information respectively. (2) Multimodal information fusion: The above two 3D ResNet used to extract features are model-fused, and the information obtained by combining the two modal features is classified by deep neural networks finally.

For the feature extraction of two image modes of structural MR image and FDG-PET image, the proposed method will use the improved 3D ResNet model based on 2D ResNet network. Since both the structural MR image and the FDG-PET image are three-dimensional spatial image data, such data has three-dimensional spatial characteristics, so the 3D convolution operation can be used to better extract spatially correlated three-dimensional image data features. In order to ensure that the model parameter training can converge correctly and avoid gradient explosion and gradient dispersion, the proposed network model is based on the VoxResNet architecture prototype.

Based on the above advantages, we propose a 3D ResNet model based on the VoxResNet architecture to extract features from different modal images. The proposed 3D ResNet feature extraction network not only inherits the advantages of the ResNet architecture, but also comprehensively and accurately captures the 3D image feature information. The 3D ResNet architecture for feature extraction is shown in Fig. 1. The proposed feature extraction model has 19 layers containing six VoxRes blocks. The first four VoxRes blocks each contain 64 filters for convolution. The last two VoxRes blocks each contain 128 filters for convolution. Then 3D max-pooling operations are performed on 128 feature maps, and the result vector is flattened. A single mode 3D image with a size of 128*128*128*1 is input in the proposed network. After the feature extraction of this model, a 1*1024 single modal feature vector is generated as an output, and it is the feature vector of this subject's modality.

It is also important to note that because of the model size and the limitations of GPU memory, we modify the batchsize is 3 in both one-versus-one architecture and 3-classification. The 3D residual convolution operation used by the model used to extract the features, so that the storage and calculation of the hyperparameters of the proposed network have brought the hardware to the limit state. For this reason we have not used the batchnorm layer, although it could improve the accuracy of the model classification by using batchnorm layer.

After extracting the structural MR image and PET image features separately, we merged the two 3D ResNet models. As shown in Fig. 2, the two feature vectors are concatenated into a fused vector of size 1*2048, which is the vector of the final multimodal feature. The purpose of feature fusion is to classify this feature vector as the input to a deep dense neural network. A two-layer deep neural network is then used for the final classification test. Because the proposed framework is a multimodel fusion into a single network model architecture, we take a unified training after building a good network. The first hidden layer contains 256 neurons in the fully connected network. The number of neurons in the output layer is determined according to the number of classifications (two or three). Finally we use the softmax function to convert the categorized output values into relative probabilities. The fusion model we propose uses a back propagation algorithm.

It is worth noting that we use single-modal images (MR images only or PET images only) for comparison experiments in order to compare the performance of the proposed model. When classifying with single-mode images, we use the proposed 3D ResNet model for feature extraction. It is then classified using a two-layer fully connected network. The first hidden layer contains 128 neurons. The number of neurons in the output layer is determined based on the classification result. Similarly, in the final classification we use the softmax layer and the whole networks uses the backpropagation algorithm.

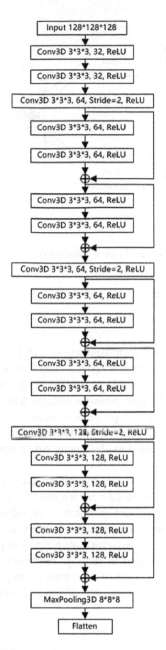

Fig. 1. 3D ResNet architecture

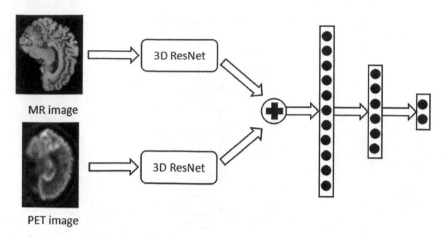

Fig. 2. Feature information fusion.

3   Experimental Results

3.1   Experimental Setup

The proposed networks is built with Tensorflow and Keras in the hardware environment of an i7-8700 k CPU and two NVIDA 1080ti GPUs. The data set that we use in the experiments containing 114 of AD patients, 132 of MCI and 133 of NC that have been preprocessed. We run 10-fold cross-validations with 10 different fold splits each time to get better approximation of prediction performance. We train all binary classification models using Adam optimizer with learning rate of 1e-4 and batchsize of 3 for 70 epochs. And 3-classification models using Adam optimizer with rate of 1e-4 but batchsize of 2 for 70 epochs.

Because of the limitation of model hyperparameters calculation, we train the 3D ResNet model and the fully connected network model on GPU 1, and the other model learning and parameter training are performed on GPU 0. The global step, train accuracy, train loss, train time and other parameters are saved in the log for each training step. A test evaluation is performed for every 5 steps of the training and the test accuracy and test loss are recorded in the log. In the training of binary-classification using the multimodal image, the time of each training step is approximately 1.1 s. And the time of each training step for three-classification is approximately 1.8 s.

3.2   Result

We perform three binary-classification tasks (AD vs. MCI, AD vs. NC and MCI vs. NC). In order to prevent the "ambiguity area" problem caused by the multi-classification using multiple binary-classifications, we also tried to perform multi-classification (Three classifications) experiments, AD vs. MCI vs. NC. The above four experiments are performed by three ways: (1) Classification using only MR modal images. (2) Classification using only PET modal images. (3) Classification using multimodal images. The experimental results are shown in Table 1. By comparing the experimental results, we know that using the 3D ResNet model proposed by our model to extract the features of the image, the performance of classification with MR images is better than that of PET images. It is obvious that the accuracy of classification in combination with the extraction of two image modal features is significantly higher than that of single-modal images alone. In the experimental results, AD vs. NC classification is the best for the two-classification, and the accuracy of the MCI vs. NC classification is relatively low. As we suspect, the accuracy of multi-classification task is the worst. So the multi-classification task is still a challenge.

Table 1. Experimental results.

ACC	AD vs. MCI	AD vs. NC	MCI vs. NC	AD vs. MCI vs. NC
MRI	74.83%	87.10%	72.52%	65.01%
PET	69.11%	80.65%	69.73%	61.49%
MRI + PET	79.17%	93.55%	78.92%	68.86%

It is worth noting that we can identify whether the training is over-fitting based on the accuracy of the training. As shown in Fig. 3 AD vs. NC, when the training rate is stable to 1.00 and then no longer changes and the model is trained to around 50 epochs, we can firmly believe that the training has reached overfitting in Fig. 3(a). At the same time, in Fig. 3(b), the accuracy of the test will not increase but will slightly within a certain range.

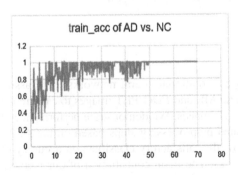

(a) The accuracy of training.

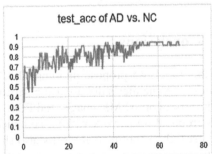

(b) The accuracy of testing.

Fig. 3. The accuracy of AD vs. NC.

From the results in Table 2, We compare the performance of the proposed method with other published methods. It is worth noting that the reason for the different experimental results is not only because of the different model structure and feature extraction methods, but also because different ADNI subjects are selected. In addition, the randomized data set of cross-validation verification is also a potential impact. The proposed method has the following three advantages: First, we use 3D convolution operation to extract features from 3D image data set, which enables the model to more accurately discover the spatial characteristics of 3D images. Second, we use ResNet to build a model to extract features, because the performance of ResNet is better than the ordinary CNN model. Third, we use two modal features to fuse, so that we can more fully explore more interesting features to achieve better performance. Liu et al. [15] combined with CNNs and RNNs methods for PET classification using PET images. Silveira et al. [14] used boosting AD diagnosis using PET images. Gray et al. [12] use multi-region of FDG-PET data for classification. Korolev et al. [20] used the heuristic 3D convolution method to simulate vgg-16 and ResNet models separately using MR images. Daoqiang et al. [23] use multi-modal multi-task learning for joint prediction of multiple regression and classification.

Table 2. Comparison of proposed method with existing methods.

	AD vs. NC		MCI vs. NC	
	AUC	ACC	AUC	ACC
Liu et al. [15]	95.3%	91.2%	83.9%	78.9%
Silveira et al. [14]	–	90.9%	–	79.6%
Gray et al. [12]	90.0%	81.6%	73.0%	70.2%
Korolev et al. [20]	88.0%	79.0%	67.0%	63.0%
Korolev et al. [20]	87.0%	80.0%	65.0%	61.0%
Daoqiang et al. [23]	92.30%	86.87%	71.90%	66.62%
Our method	95.14%	93.55%	78.37%	78.92%

4   Conclusions

In this paper, we propose multimodal 3D convolutional neural networks for Classification of Alzheimer's disease using structural MR and FDG-PET images to capture the rich features of 3D MR images and FDG-PET images. We demonstrate performance of the proposed framework without complicated preprocessing steps based on the ADNI data set. Our experimental results also show the satisfactory classification accuracy of multimodal 3D convolutional neural networks. The main contribution of our method: (1) The input image data does not require complex preprocessing steps. (2) the 3D ResNet model can extract the spatial information of 3D images better than other methods. (3) Using multiple modal data to detect classification more comprehensively and accurately.

In our future work, we hope to get better classification results by adding information from other different modalities and upgrading existing model structures. At the same time, we will also try to extract more interesting regions in the image by a powerful mask which achieved by machine-training, according to this mask to increase the attention of the region of interest, which means that we will focus on those more important regions of interest to achieve more excellent classification effect.

Acknowledgements. This work was supported by the National Natural Science Foundation of China under Grant No. 61672181, No. 51679058, Natural Science Foundation of Heilongjiang Province under Grant No. F2016005. We would like to thank our teacher for guiding this paper. We would also like to thank classmates for their encouragement and help.

References

1. Thung, K.H., Wee, C., Yap, P.T., Shen, D.: Neurodegenerative disease diagnosis using incomplete multi-modality data via matrix shrinkage and completion. NeuroImage **91**, 386–400 (2014)
2. Zhan, L., et al.: Comparison of 9 tractography algorithms for detecting abnormal structural brain networks in Alzheimers disease. Front. Aging Neurosci. **48**(7), 401–408 (2015)
3. The State of The Art of Dementia Research: New Frontiers. https://www.alz.co.uk/research/world-report-2018/. Accessed Sept 2018
4. Manivannan, W., Li, S., Akbar, S., Zhang, J., Trucco, E., McKenna, S.J.: Gland segmentation in colon histology images using hand-crafted features and convolutional neural networks. In: IEEE International Symposium on Biomedical Imaging, vol. 4, pp. 1405–1408 (2016)
5. Jia, W., Li, F., Hu, Q.: Automatic segmentation of liver tumor in CT images with deep convolutional neural networks. J. Comput. Commun. **11**(3), 146–151 (2015)
6. Litjens, G., et al.: Deep learning as a tool for increased accuracy and efficiency of histopathological diagnosis. Nat. Sci. Rep. **6**, 26286 (2016)
7. Haiwei, P., Pengyuan, L., Qing, L., Qilong, H., Xiaoning, F., Linlin, G.: Brain CT image similarity retrieval method based on uncertain location graph. IEEE J. Biomed. Health Inform. **18**(2), 574–584 (2014)
8. Linlin, G., Haiwei, P., Xiaoqin, X., Zhiqiang. Z., Qing L., Qilong, H.: Graph modeling and mining methods for brain images. Multimedia Tools Appl. **75**(15), 9333–9369 (2016)
9. Linlin, G., et al.: Brain medical image diagnosis based on corners with importance-values. BMC Bioinform. **18**(1), 505 (2017)
10. Shen, D., Wu, G., Suk, H.I.: Deep learning in medical image analysis. Ann. Rev. Biomed. Engl. **19**(1), 221–248 (2017)
11. Gray, K.R., Wolz, R., Heckemann, R.A., Aljabar, P., Hammmers, A., Rucckert, D.: Multi-region analysis of longitudinal FDG-PET for the classification of Alzheimer's disease. NeuroImage **60**(1), 221–229 (2012)
12. Gray, K.R., Wolz, R., Keihaninejad, S.: Regional analysis of FDG-PET for using in the classification of Alzheimer's disease. In: IEEE International Symposium on Biomedical Imaging, vol. 3, pp. 1082–1085 (2011)
13. Garali, I., Adel, M., Bourennane, S., Guedj, E.: Region-based brain selection and classification on pet images for Alzheimer's disease computer aided diagnosis. In: IEEE International Conference on Image Processing, vol. 10, no. 9, pp. 27–30 (2015)
14. Silveira, M., Marques, J.: Boosting Alzheimer disease diagnosis using PET images. In: IEEE International Conference on Pattern Recognition (ICPR), vol. 8, pp. 2556–2559 (2015)
15. Liu, M., Cheng, D., Yan, W.: Classification of Alzheimer's disease by combination of convolutional and recurrent neural networks using FDG-PET images. Front. Neuroinf. **12**(6), 1–12 (2018)

16. Zeiler, M.D., Fergus, R.: Visualizing and understanding convolutional networks. In: Fleet, D., Pajdla, T., Schiele, B., Tuytelaars, T. (eds.) ECCV 2014. LNCS, vol. 8689, pp. 818–833. Springer, Cham (2014). https://doi.org/10.1007/978-3-319-10590-1_53

17. Hosseini-Asl, E., Keynton, R., El-Baz, A.: Alzheimer's disease diagnostics by adaptation of 3D fonvolutional network. In: IEEE International Conference on Image Processing, vol. 7, pp. 126–130 (2016)

18. Jianxu, C., Liu, Y., Yizhe, Z., Alber, M.: Combining fully convolutional and recurrent neural networks for 3D biomedical image segmentation. Neural Inf. Process. Syst. 6(9), 1–9 (2016)

19. Çiçek, Ö., Abdulkadir, A., Lienkamp, S.S., Brox, T., Ronneberger, O.: 3D U-Net: learning dense volumetric segmentation from sparse annotation. Neural Inf. Process. Syst. 21(6), 1–8 (2016)

20. Korolev, S., Safiullin, A., Belyaev, M., Dodonova, Y.: Residual and plain convolutional neural networks for 3D brain MRI classification. In: IEEE International Symposium on Biomedical Imaging, vol. 6 (2017)

21. Kaiming, H., Xiangyu, Z., Shaoqing, R., Jian, S.: Deep residual learning for image recognition. In: The IEEE Conference on Computer Vision and Pattern Recognition (CVPR), vol. 6, pp. 770–778 (2016

22. Voxresnet: deep voxelwise residual networks for volumetric brain segmentation. https://arxiv.org/abs/1608.05895/. Accessed Aug 2016

23. Daoqiang, Z., Dinggang, S.: Multi-modal multi-task learning for joint prediction of multiple regression and classification variables in Alzheimer's disease. Neuroimage 18(9), 895–907 (2012)

Comparative Study of Combined Fault Diagnosis Schemes Based on Convolutional Neural Network

Mei Li[1], Zhiqiang Huo[2], Fabien CAUS[3], and Yu Zhang[2(✉)]

[1] China University of Geosciences, Beijing, China
[2] University of Lincoln, Lincoln, UK
YZhang@lincoln.ac.uk
[3] IMT Mines, Albi, France

Abstract. In this paper, comparative combined fault diagnosis schemes are studied including vibration analysis, acoustic signal analysis and thermal image analysis based on the Convolutional Neural Network (CNN). The advantage of the CNN structure is that it does not need manual feature extraction or selection, which requires prior knowledge of specific machinery dynamics. The vibration and acoustic signals were transformed into spectrograms, which are effective for the diagnostic analysis by using CNN. Comparatively, the thermal images were directly analyzed using CNN. The effectiveness of the CNN-based diagnosis methods was investigated through the analysis of different experimental data, i.e., vibration, acoustic signals and thermal images, which were collected from a test rig where different types of faults are induced on the roller bearing and shaft. The results show that the thermal image analysis and acoustic signal analysis could achieve relatively higher accuracy rate compared to vibration analysis. Moreover, the advantage is easy-deployment because of the non-contact way during signal acquisition. With the CNN-based fault diagnosis method for the three different signals collected, the accuracy of different signal predictions for combined faults can be compared, and the effective method can be applied to fault diagnosis of other industrial rotating machinery.

Keywords: Fault diagnosis · Rotating machinery ·
Convolutional neural networks

1 Introduction

Bearing and shaft are among the most fundamental and important components in rotating machinery. In industry, a minor defection or degradation of rolling bearings may lead to the unexpected breakdown of the entire system resulting in downtimes and financial losses. Therefore, fault diagnostics has become a very practical solution to minimize these issues. According to McKinsey Global Institute report, industrial companies will benefit from fault diagnostics by saving at least $ 630 billion per year in 2025 [1]. Techniques of machine health monitoring facilitate early fault detection and isolation, and further greatly improve the reliability of the diagnosis system.

© Springer Nature Singapore Pte Ltd. 2019
X. Cheng et al. (Eds.): ICPCSEE 2019, CCIS 1058, pp. 669–681, 2019.
https://doi.org/10.1007/978-981-15-0118-0_52

Plentiful condition monitoring approaches are available to characterize malfunctions in the machinery by continuously observing its running states, such as vibration, acoustic, thermal, electrical current signals, etc. Among those, the vibration signal is one type of measurement that has been widely used in the field of condition monitoring. It is directly associated with mechanical phenomenon whereby oscillations occur about a place of interest by mounting an accelerometer [2]; nevertheless, it is an intrusive technique and this technique cannot be employed after the installment of the complex equipment. Comparatively, acoustic monitoring and thermal image monitoring are easy implement and non-intrusive methods.

To accurately characterize fault symptoms associated with performance degradation hidden in measured non-stationary signals, advanced signals processing and feature extraction techniques are normally needed [3–5], which particularly need fine-tuning methods and prior knowledge. Convolutional Neural Network (CNN), as an increasingly growing deep learning technique, has been successfully used in fault diagnosis by automatically learning and representing inherent knowledge from data sets [6]. CNN has proven to be more suitable for image analysis and classification. In this paper, the performance of CNN-based combined diagnosis methods are compared, where vibration, acoustic signals, and thermal image signals are individually applied for identifying health conditions of rotating machinery. For evaluation, vibration and acoustic signals are transformed into spectrograms, which provide salient time-frequency domain information, and make the most use of the advantages of CNN. The results of comparative study using different types of data sets (vibration, acoustic and thermal signals) are also compared and discussed.

The rest of this paper is organized as follows: Sect. 2 presents the related works on data-driven fault diagnosis. Section 3 introduces the methodology for analyzing and diagnosing different conditions of rotating machinery. Section 4 shows the experimental results through the analysis of different signatures using CNN. Finally, conclusions are drawn in Sect. 5.

2 Related Literature

Fault detecting and diagnosis normally include three steps: signal acquisition, feature extraction and fault classification. Starting from the data acquisition part, there are many techniques for measuring physical conditions of rotating machinery [7], such as vibration monitoring, acoustic monitoring, thermal image monitoring. Among them, vibration signal analysis, current, and acoustic signal analysis are most popular. The vibration signal analysis is well-established compared to many signal processing techniques for condition monitoring of rotating machines. The acoustic signal analysis is gradually being accepted and applied in fault detection as a non-contact method. It is very effective for the early detection of faults and has the potential to be a powerful fault diagnostics tool to identify various types of progressing faults [8]. Thermal image analysis is a very convenient and non-contact method which has been widely used for diagnosing key components in rotating machinery by monitoring the evolution of temperature change, especially useful with the emergence of incipient failures. Thermal image based fault diagnosis has also attracted increasing attention over the recent decades [9].

After data acquisition, in the step of feature extraction, raw signals are usually transformed into desired forms that contain the time or frequency-domain information of interest. Feature extraction techniques are indispensable for this purpose [10, 11]. For discriminating fault symptoms and distinguishing different conditions of the rolling bearing, the commonly used features include time domain features (kurtosis, root mean square, skewness, etc.), frequency domain features [12], and time-frequency domain features [13, 14]. [15] and [16] combined time domain features with artificial neural network (ANN). The basis of frequency domain analysis is the analysis of the vibration signals' frequency spectrum. Frequency domain features are applied in [17]. In [18], time domain features and frequency domain features were combined based on information fusion and an ANN model was trained for fault diagnosis. [19] trained a SVM model with feature extraction using PCA method. Nevertheless, achieving accurate fault detection always needs a lot of prior knowledge about specific machinery running conditions, fault types and signal processing techniques [20, 21]. Meanwhile, on the basis of extracted features, fault recognition and classification methods are further needed to facilitate the decision-making about the severity and type of early fault in the machinery.

As one of artificial intelligence (AI) techniques, machine learning is an effective solution for fault recognition by learning from historic data, especially from extracted features, without being explicitly programmed [22]. There are many typical machine learning techniques such as K-nearest neighbors, support vector machine [23], and neural networks [24]. Among them, neural networks are more effective in fault classification and identification [25] with less manual feature extraction. Particularly, CNN is a one of popular machine learning methods that attempts to model hierarchical representations behind data and classify patterns via multiple layers. CNN has been widely applied in a wide range of applications in computer vision, such as image classification, object detection, segmentation and face recognition.

3 Methodology

3.1 The Principle of CNN

In 1998, the first Convolutional Neural-Network named LeNet-5 was proposed by Yann LeCun applied to document Recognition. But not until 2012, CNN began popular all over the world; most of them focus on computer vision, automatic speech recognition, and natural language processing.

The major advantage of CNN is that it does not need manually extract features from the image. CNNs use a variety of multilayer perceptrons (MLP) designed to require minimal preprocessing. They have multi hidden layers besides input and output layer.

There are two phases of the classification, training phase and test phase. The network learns to extract features by training. While training, we just feed the images to the network (pixel values). The input layer takes the image; the parameters of the hidden layers are automatically adjusted by the error feedback of the output and the real classification. This is a process of model self-training. After self-training, it can recognize the type of input image and return its classification result.

The biggest distinction of CNN is the convolution layer which can identify displacement, scaling and other forms of distortion invariant images. Different filters are used to detect different objects. Compared with fully connected MLP, CNN saves a lot of parameters because it is a fixed set of parameters for one filter.

Figure 1 is an example of convolution with a 3×3 filter and stride 1. The image is 5×5. Then the convolved feature map is 3×3. Suppose we want to learn 9 features, which mean we have 9 filters in one convolution layer. There are $3 \times 3 \times 9 = 81$ parameters. For a fully connected neural network with 9 neurons, there will be $5 \times 5 \times 9 = 225$ parameters.

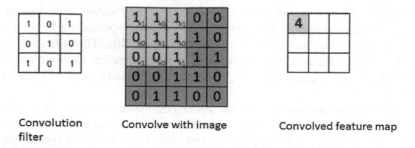

Convolution filter   Convolve with image   Convolved feature map

Fig. 1. Example of convolution

The neurons of convolution layer always use ReLU (rectify linear unit) activation function which is much faster than with tanh or sigmoid activation function. The ReLU activation function is shown in Eq. (1). The structure of a neuron is shown is Fig. 2.

$$f(x) = \max(0, x) \tag{1}$$

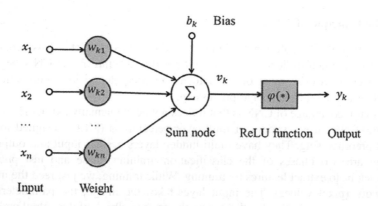

Fig. 2. The structure of a neuron

The subsampling is to reduce its dimensions and overfitting. One of the techniques of subsampling is max pooling which means selecting the highest pixel value from a region. For example, a max-pooling layer of size 2 × 2 will select the maximum pixel intensity value from 2 × 2 regions. Figure 3 shows an example of max-pooling of size 2 × 2 and stride 2.

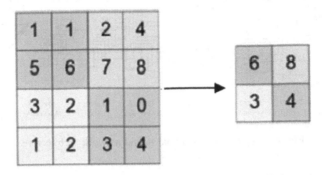

Fig. 3. Example of max-pooling of size 2 × 2 and stride 2

When convolution or subsampling, zero padding is often used which preserves as much information about the input. The formula for calculating zero padding for any given convolution layer is shown in (2).

$$P = ((O - 1)S + K - W)/2 \qquad (2)$$

where P is the padding, O is the output height/length, S is the stride, K is the filter size and W is the input height/length.

The fully connected layer also referred to dense layer, is to flatten the high-level features that are learned by convolutional layers and combining all the features. It predicts the input class label to the output layer by using a softmax classifier.

3.2 The Applied CNN Model

The method proposed in this paper is based on the CNN fault diagnosis method. In most bearing fault diagnosis methods, the originally collected signals cannot be directly processed. Therefore, it is necessary to pre-process the data. The function of data pre-processing is to extract the characteristics of the signal from a large amount of data. In this paper, the thermal image signals are directly analyzed using CNN while the vibration and acoustic signals are transformed into spectrograms.

Data preprocessing can facilitate CNN identification and training. Once all the original signals are converted into the form of signals, CNN can be trained to classify these images. CNN is effective and promoted in image recognition. This study designed a CNN model to solve the image classification task of fault diagnosis.

In the proposed model, the model for image classification consists of their alternating convolutional and pooling layers with a fully connected layer. Figure 4 shows a CNN architecture where an image is fed to the network as an input then goes through multiple convolution, subsampling, and fully connected layers and finally outputs the classification or identification of the image [26].

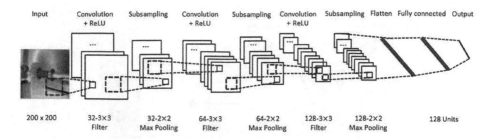

Fig. 4. Typical convolutional neural network architecture

4 Experimental Study

The experimental data used in this paper are collected from an experimental test rig, named GUNT PT500, as shown in Fig. 5. In this study, we simulated compound faults on bearings and shafts. Specifically, four conditions (A, B, C, and D) of roller bearings are used, which are healthy condition, single-point fault on inner race, outer race, and roller respectively. We also consider two conditions of shaft, namely balanced and unbalanced shafts. The hybrid conditions of roller bearing and shaft are simulated, and the three sensors are used to synchronously collect data under three different rotating speeds. Herein, three types of sensors are separately applied to collect vibration signals, acoustic signals, and thermal images.

Fig. 5. GUNT PT500 test rig

4.1   Test Set-up

The equipment includes an experimental simulator, GUNT PT500, as shown in Fig. 5. It allows simulating different conditions of rolling bearings and unbalanced shaft. This enables generation of known experimental parameters to the key component in the structure and collecting vibration signals using accelerometer sensors. Also utilized on the console comprise of an infrared camera FLIR Ax5 with its software ResearchIR and sound recorder Olympus WS-853. The camera is fixed 25 cm from the bearing, and the recorder is maintained on a support 45 cm from the shaft mounted between the roller bearing and the motor.

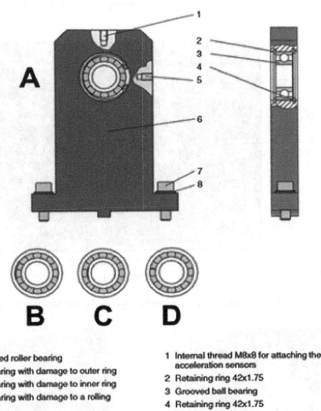

A Undamaged roller bearing

B Roller bearing with damage to outer ring

C Roller bearing with damage to inner ring

D Roller bearing with damage to a rolling element

The roller bearing identification is located on the outer ring

1 Internal thread M8x8 for attaching the acceleration sensors

2 Retaining ring 42x1.75

3 Grooved ball bearing

4 Retaining ring 42x1.75

5 Internal thread M8x8 for attaching the acceleration sensors

6 Bearing block

7 Fixing bolt (M8x25)

8 Plain washer

Fig. 6. Bearing block and different conditions of bearings and setup parameters in the test rig [27].

4.2   Data Set

There are 24 sets of data. There are different bearings A, B, C, D as shown in Fig. 6 with the balanced and unbalanced shaft, under 3 different speeds, 1000 r.p.m., 2000 r.p. m., and 3000 r.p.m. respectively.

ResearchIR recorded the infrared video of each fault first, and then each video was divided into 500 thermal images by Photoshop. The acoustic signals were recorded Olympus WS-853. Adobe Audition software converts each faulty MP3 audio signal into a wave file. Then python split each wave file into 500 pieces and transforms each into a spectrum; GUNT PT500 embedded accelerator sensor and its software PT500.04 recorded vibration signals, then we use Excel software to change them to the work-sheet. Spectrograms are acquired by Python program. Figure 7 is an example of spectrograms of acoustic signals for the bearing with inner race fault (Fig. 8).

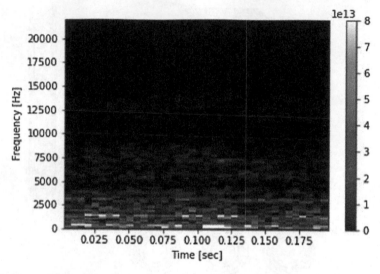

Fig. 7. Example of spectrogram of acoustic signals for the bearing with inner race fault.

Fig. 8. Examples of thermal images for the bearing B and C with a speed of 1000 r.p.m.

For each kind of signals, there are 12000 samples, 7680 of them for training, 1920 of them for validation, 2400 of them for testing. Due to the different characteristics of the three signals, 7680 training samples were divided into different batches and batch sizes for better diagnostic results. Table 1 shows the epochs and batch size parameter settings of different signals after experimental optimization.

Table 1. Different parameter settings of three signals.

Signal types	Batch size	Epochs
Acoustic signal	32	20
Thermal image	32	15
Vibration signal	64	30

4.3   Results

The results have shown that these infrared image analysis has the best classification accuracy which is 100%, acoustic signals 92.5%, and then vibration signals 80.5%. As an example, Fig. 9 shows the training and validation accuracy and loss of vibration signals. Figure 10 presents the training and validation loss of vibration signals. The CNN model constructed in this paper takes the accuracy of mechanical fault classification as the evaluation standard of the model. When the pictures of test or verification set are sent to the system for classification, the system will recognize the images and give the predicted category probability; thus adjusting the model to an optimized one by improving fault classification accuracy rate.

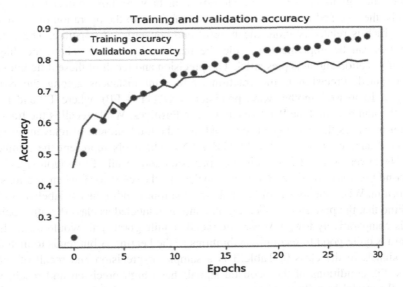

Fig. 9. Training and validation accuracy of vibration signals

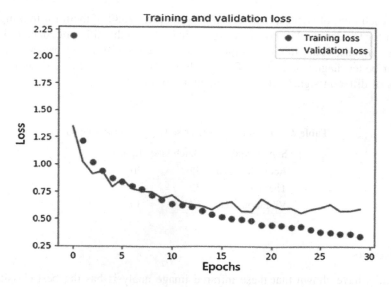

Fig. 10. Training and validation loss of vibration signals

Apart from thermal images, vibration signals and acoustic signals are simultaneously collected based on the same experimental set up, with three kinds of signal acquisition method. The CNN model has different feature extraction capabilities for different signals. Therefore, when different signals from the same device are to achieve the best model, the parameter settings of the model are different, and the obtained classification accuracy is also different.

Through the experimental validation, it proved that the accuracy for infrared images and acoustic signals are very high. Moreover, as they are non-contact fault diagnosis methods, the received noise is less and no interference to the operation of the machine itself, which is relatively more meaningful and convenient to operate. The detection techniques can be implemented while the machine is running. Then, the diagnosis result of vibration signal is presented. The precision and recall of these different signals are computed. Precision is the fraction of relevant instances among the retrieved instances. Expressed another way, precision = TP/(TP + FP), where TP and FP represent the numbers of True Positives and False Positives. While recall is the fraction of relevant instances that have been retrieved over the total amount of relevant instances. Expressed another way, recall = TP/(TP + FN), where FN represent the numbers of False Negatives. Table 2 is a sample of precision and recall of vibration signals. The four conditions (A, B, C, and D) are labeled as classes 0 to 3 when the axis is in equilibrium. When the axis is out of balance, the four conditions are labeled as class 4 to 7. Find that the precision for bearing A condition, labeled as class 0, with a balanced shaft is comparatively low, however the recall is quite good. This would mean that the program is quite good to recognize vibrations of the bearing A but tends to include too many signals in the class 0. Table 3 is a sample of precision and recall of acoustic signals. Six conditions of the acoustic signals have high precision and recall, which shows the model is quite good to recognize acoustic signals.

Table 2. Precision and recall of vibration signals.

Class	Precision	Recall
Class 0	0.54	0.86
Class 1	0.76	0.74
Class 2	0.88	0.86
Class 3	0.87	0.63
Class 4	0.85	0.68
Class 5	0.90	0.76
Class 6	0.90	0.86
Class 7	0.77	0.82
Average	0.80	0.78

Table 3. Precision and recall of acoustic signals.

Class	Precision	Recall
Class 0	0.98	1.00
Class 1	0.97	0.96
Class 2	1.00	0.99
Class 3	0.74	0.75
Class 4	0.99	0.99
Class 5	0.96	0.98
Class 6	0.73	0.73
Class 7	1.00	0.99
Average	0.93	0.92

CNN has also shown high efficiency because it detects significant features without manual feature extraction. This method can be adapted to the real industry in the future to reduce breakdowns and maintenance costs.

5   Conclusions

This paper has compared three monitoring techniques. Sound signals, vibration signals and thermal images are collected by different devices. After corresponding preprocessing of the signals, CNN was used to classify the signals. The comparison results can provide useful insights leading to practical solutions for fault diagnostic of rotating machinery. It aims to improve machinery runtime and reduce maintenance costs. Collecting infrared images is an effective method in combined fault diagnosis. In this paper, from the input of 12000 images, the CNN gives different accuracy values for the infrared images, vibration signals, and acoustic signals. The non-contact methods, including infrared image and acoustic signal analysis, have shown high accuracy. For future work, real-time fault diagnosis will be carried out. Further study on multi-sensor fusion technology will be investigated to improve the accuracy of fault diagnosis.

References

1. McKinsey and company: The Internet of Things mapping the value beyond the hype. https://www.mckinsey.com/business-functions/digital-mckinsey/our-insights/the-internet-of-things-the-value-of-digitizing-the-physical-world,last. Accessed 1 Sept 2018
2. Janssens, O., et al.: Convolutional neural network based fault detection for rotating machinery. J. Sound Vibr. **377**, 331–345 (2016)
3. Chahal, B., Ahmad, S., Rana, A.S., Verma, A., Goyat, N.S.: Fault diagnosis of bearing by the application of acoustic signal. Invertis J. Sci. Technol. **5**, 40–44 (2012)
4. Huo, Z., Zhang, Y., Shu, L., Gallimore, M.: A new bearing fault diagnosis method based on fine-to-coarse multiscale permutation entropy, laplacian score and SVM. IEEE Access **7**, 17050–17066 (2019)
5. Ashish, V.: Review on thermal image processing tecniques for machine condition monitoring. Int. J. Wireless Commun. Netw. Technol. **3**, 49–53 (2014)
6. Xia, M., Li, T., Xu, L., Liu, L., de Silva, C.W.: Fault diagnosis for rotating machinery using multiple sensors and convolutional neural networks. IEEE/ASME Trans. Mechatron. **23**(1), 101–110 (2018)
7. Touret, T., Changenet, C., Ville, F., Lalmi, M., Becquerelle, S.: On the use of temperature for online condition monitoring of geared systems–a review. Mech. Syst. Signal Process. **101**, 197–210 (2018)
8. Adam, G.: Fault diagnosis of single-phase induction motor based on acoustic signals. Mech. Syst. Signal Process. **117**, 65–80 (2019)
9. Janssens, O., Van de Walle, R., Loccufier, M., Van Hoecke, S.: Deep learning for infrared thermal image based machine health monitoring. IEEE/ASME Trans. Mechatron. **23**(1), 151–159 (2018)
10. Dai, X., Gao, Z.: From model, signal to knowledge: a data-driven perspective of fault detection and diagnosis. IEEE Trans. Ind. Inf. **23**(1), 2226–2238 (2013)
11. Zhang, Y., Bingham, C., Yang, Z., Ling, B.W.K., Gallimore, M.: Machine fault detection by signal denoising—with application to industrial gas turbines. Measurement **58**, 230–240 (2014)
12. Yuan, L., He, Y., Huang, J., Sun, Y.: A new neural-network-based fault diagnosis approach for analog circuits by using kurtosis and entropy as a preprocessor. IEEE Trans. Instrum. Meas. **59**(3), 586–595 (2010)
13. Chen, J., et al.: Wavelet transform based on inner product in fault diagnosis of rotating machinery: a review. Mech. Syst. Signal Process. **70**, 1–35 (2016)
14. Lei, Y., Lin, J., He, Z., Zuo, M.J.: A review on empirical mode decomposition in fault diagnosis of rotating machinery. Mech. Syst. Signal Process. **35**(1), 108–126 (2013)
15. Hu, Q., Zhang, S., Yang, S.: Variable condition bearing fault diagnosis based on time-domain and artificial intelligence. In: Applied Mechanics and Materials, vol. 203, pp. 329–333 (2012)
16. Sreejith, B., Verma, A.K., Srividya, A.: Fault diagnosis of rolling element bearing using time-domain features and neural networks. In: IEEE Region 10 and the Third International Conference on Industrial and Information Systems, pp. 1–6 (2016)
17. Mao, K., Wu, Y.: Fault diagnosis of rolling element bearing based on vibration frequency analysis. In: 2011 Third International Conference on Measuring Technology and Mechatronics Automation, pp. 198–201 (2011)
18. Jiang, Z., Jiao, W., Meng, S.: Fault diagnosis method of time domain and time-frequency domain based on information fusion. In: Applied Mechanics and Materials, vol. 300, pp. 635–639 (2013)

19. Cao, M., Pan, H., Chang, X.: Research on automatic fault diagnosis based on time-frequency characteristics and PCASVM. In: International Conference on Ubiquitous Robots and Ambient Intelligence, pp. 593–598 (2016)
20. Jardine, A.K., Lin, D., Banjevic, D.: A review on machinery diagnostics and prognostics implementing condition-based maintenance. Mech. Syst. Signal Process. **108**, 1483–1510 (2006)
21. Younus, A.M., Yang, B.S.: Intelligent fault diagnosis of rotating machinery using infrared thermal image. Expert Syst. Appl. **39**(2), 2082–2091 (2012)
22. Liu, R., Yang, B., Zio, E., Chen, X.: Artificial intelligence for fault diagnosis of rotating machinery: a review. Mech. Syst. Signal Process. **108**, 33–47 (2018)
23. Konar, P., Chattopadhyay, P.: Bearing fault detection of induction motor using wavelet and Support Vector Machines (SVMs). Appl. Soft Comput. **11**(6), 4203–4211 (2011)
24. Babu, T.R., Sekhar, A.S.: Shaft crack identification using artificial neural networks and wavelet transform data of a transient rotor. Adv. Vib. Eng **9**, 207–214 (2010)
25. Xie, Y., Zhang, T.: Fault diagnosis for rotating machinery based on convolutional neural network and empirical mode decomposition. Shock Vibr. (2017)
26. Sharma DataCamp/aditya: Convolutional Neural Networks in Python with Keras. https://www.datacamp.com/community/tutorials/convolutional-neural-networks-python. Accessed 1 Sept 2018
27. PT 500 machinery diagnostic system. https://www.gunt.de/index.php?option=com_gunt&task=gunt.list.category&lang=en&category_id=77. Accessed 25 Mar 2015
28. Huo, Z., Zhang, Y., Francq, P., Shu, L., Huang, J.: Incipient fault diagnosis of roller bearing using optimized wavelet transform based multi-speed vibration signatures. IEEE Access **5**, 19442–19456 (2017)

Author Index

Printed in the United States
By Bookmasters